Applications of Modeling and Machine Learning in Additive Manufacturing

Applications of Modeling and Machine Learning in Additive Manufacturing

Guest Editors

Tuhin Mukherjee

QianruWu

Basel • Beijing • Wuhan • Barcelona • Belgrade • Novi Sad • Cluj • Manchester

Guest Editors

Tuhin Mukherjee
Mechanical Engineering
Iowa State University
Ames
United States

QianruWu
Electrical & Automation
Engineering
Nanjing Normal University
Nanjing
China

Editorial Office
MDPI AG
Grosspeteranlage 5
4052 Basel, Switzerland

This is a reprint of the Special Issue, published open access by the journal *Materials* (ISSN 1996-1944), freely accessible at: https://www.mdpi.com/journal/materials/special_issues/S48FKEHPEK.

For citation purposes, cite each article independently as indicated on the article page online and as indicated below:

Lastname, A.A.; Lastname, B.B. Article Title. *Journal Name* **Year**, *Volume Number*, Page Range.

ISBN 978-3-7258-3746-5 (Hbk)
ISBN 978-3-7258-3745-8 (PDF)
https://doi.org/10.3390/books978-3-7258-3745-8

Cover image courtesy of Tuhin Mukherjee

Contents

About the Editors

Tuhin Mukherjee

Tuhin Mukherjee is an Assistant Professor of Mechanical Engineering at Iowa State University, USA. Previously, he was a Postdoctoral Researcher in the Department of Materials Science and Engineering at Pennsylvania State University, USA from where he also got his Ph.D. He is the author of many papers in leading journals including Nature Reviews Materials, Nature Materials, and Progress in Materials Science. He authored an edited book entitled "The Science and Technology of 3D Printing" (MDPI, 2021) and a textbook on "Theory and Practice of Additive Manufacturing" (Wiley, 2023). He served as a Guest Editor for the journals "Computational Materials Science", "Materials", and "Science and Technology ofWelding and Joining". He is an Editorial Board Member of the journals "Science and Technology ofWelding and Joining" and "Welding Journal".

Qianru Wu

Dr. QianruWu is an Assistant Professor at the Jiangsu Key Laboratory of 3D Printing Equipment and Manufacturing, School of Electrical & Automation Engineering, Nanjing Normal University, Nanjing, China. Her research focuses on additive manufacturing, with expertise in mechanistic modeling, machine learning, residual stresses and distortion, and finite element analysis. She specializes in directed energy deposition-arc processes and leverages neural networks for predictive modeling and process optimization.

Editorial

Recent Progress in Computational and Data Sciences for Additive Manufacturing

Tuhin Mukherjee [1,*] and Qianru Wu [2]

[1] Department of Mechanical Engineering, Iowa State University, Ames, IA 50011, USA
[2] Jiangsu Key Laboratory of 3D Printing Equipment and Manufacturing, School of Electrical & Automation Engineering, Nanjing Normal University, Nanjing 210023, China; qrwu@nnu.edu.cn
* Correspondence: tuhinm@iastate.edu

Received: 25 January 2025
Revised: 25 February 2025
Accepted: 4 March 2025
Published: 6 March 2025

Citation: Mukherjee, T.; Wu, Q. Recent Progress in Computational and Data Sciences for Additive Manufacturing. *Materials* **2025**, *18*, 1177. https://doi.org/10.3390/ma18051177

Additive manufacturing (AM), often referred to as 3D printing, is a preferred technique for producing components that are challenging to manufacture through conventional methods. This approach facilitates the direct fabrication of complex parts in a single step from a 3D digital model. Today, AM components are widely utilized across various sectors, including healthcare, aerospace, the automotive industry, energy, the marine sector, and consumer products [1]. Examples of such components include custom medical implants tailored to individual patients, aero-engine parts, intricate geometries with internal channels, lattice structures, and materials designed with location-specific chemical compositions, microstructures, and properties [2]. The materials used include metals, polymers, ceramics, and composites. Among these, metal and alloy printing is advancing most rapidly due to its growing applications, high demand, and ability to produce specialized, functional parts. Various AM techniques are employed based on the material, geometry, and complexity of the desired component [3]. For instance, powder bed fusion (PBF) and directed energy deposition (DED) are commonly utilized for metallic parts. These processes involve melting fine layers of powder or wire feedstocks using high-energy sources like lasers, electron beams, or electric arcs, followed by solidification to form the final part. Similarly, different processes are used in the industry for printing polymers, ceramics, and composite materials.

In AM, a reduction in defects, maintaining geometric consistency, and control of microstructure and mechanical properties cannot be achieved by time-consuming and expensive experimental trials because of the involvement of many variables with a large parameter window. Physics-based computational models are often used as an alternative. Figure 1 provides a few examples of using such models in AM. However, the evolution of microstructures, properties, and defects depends on many complex physical processes, and the mechanistic understanding of many of these processes is not fully developed. The use of data science techniques such as machine learning can automate several steps, including process monitoring, defect detection, sensing, and process control, and can help in the selection of appropriate processing conditions to improve structure and properties. This would minimize the need for human intervention and significantly improve the process efficiency, productivity, and part quality, and reduce materials and energy waste and cost.

Topics in this Special Issue include applications of modeling and machine learning for the novel design of additively manufactured products, additive manufacturing processes, alloy design, tailoring microstructures, customized mechanical and chemical properties, improved creep resistance, fatigue life, and serviceability, reducing defects and residual stresses and distortion. The scope of this Special Issue also includes all AM processes for alloys, ceramics, and polymers.

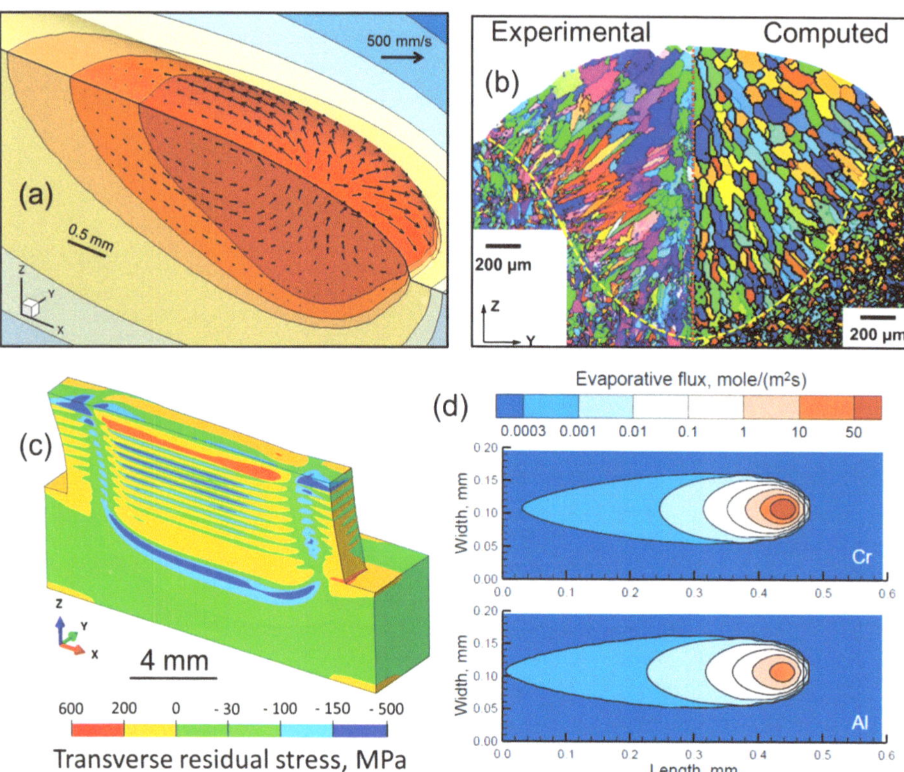

Figure 1. Results from mechanistic models of metal additive manufacturing showing (**a**) 3D temperature and velocity fields during directed energy deposition, (**b**) grain structure during directed energy deposition that matches with experimental data, (**c**) 3D residual stress distribution during directed energy deposition, and (**d**) evaporative flux of alloying elements such as Cr and Al during powder bed fusion. These figures are owned by the authors.

Topics in this Special Issue include applications of modeling and machine learning for the novel design of additively manufactured products, additive manufacturing processes, alloy design, tailoring microstructures, customized mechanical and chemical properties, improved creep resistance, fatigue life, and serviceability, reducing defects and residual stresses and distortion. The scope of this Special Issue also includes all AM processes for alloys, ceramics, and polymers.

This Special Issue contains a total of 12 articles including 11 research articles [4–14] and a review paper [15] on the applications of process monitoring, modeling, and statistical analysis in metal additive manufacturing. The 11 research articles uniquely contribute to six distinct areas: (i) the prediction of temperature fields, (ii) keyhole and molten pool geometry calculations, (iii) the estimation of part geometry, (iv) the determination of part surface characteristics, (v) defects and anomaly detection, and (vi) the prediction of mechanical properties, as discussed below.

Sajadi et al. [7] and Fagersand et al. [9] explore computational approaches for predicting temperature fields in metal AM. Sajadi et al. [7] introduce a physics-informed online learning framework for real-time temperature prediction in metal AM using physics-informed neural networks (PINNs). By integrating physics-based inputs and loss functions, the model adapts dynamically to unseen process conditions, demonstrating superior performance in critical regions like the heat-affected zone and melt pool. The approach highlights the role of hyperparameters, such as learning rate and batch size, in optimizing performance for diverse conditions. Fagersand et al. [9] use deep learning, specifically multilayer perceptrons (MLPs), to predict temperature history in wire arc additive manufacturing of aluminum bars. Training on finite element simulations, their models achieve low error rates under baseline conditions but show reduced accuracy for new process parameters, particularly varying scanning speeds. Together, these studies emphasize data- and physics-driven frameworks for the prediction of temperature fields during metal AM.

Wu et al. [5] and Dong et al. [10] use deep learning techniques for predicting molten pool and keyhole geometries. Wu et al. [5] focus on DED, developing surrogate models based on recurrent neural networks (RNNs) like LSTM, Bi-LSTM, and GRU to predict melt pool characteristics. Their models achieve high accuracy, with an R^2 of 0.98 for peak temperature prediction and over 0.88 for melt pool geometry. Dong et al. [10] investigate laser PBF, using a computer vision tool leveraging the BASNet deep learning model to segment keyhole morphologies from X-ray images. Achieving average accuracies of 91.24% for the keyhole area and 92.81% for the keyhole boundary shape, this tool automates the labeling process, enabling faster and more reliable analysis of keyhole dynamics. Together, these studies identify the potential of deep learning to reliably predict molten pool and keyhole geometries during metal AM.

The article by Hermann et al. [11] introduces a novel decision-making workflow to optimize process parameters for laser DED. Acknowledging the limitations of current analytical, numerical, and machine learning methods in predicting optimal parameters, the authors propose a Gaussian Process Regression (GPR) model. This model predicts the geometry of single DED tracks based on input parameters while incorporating uncertainty quantification (UQ). By leveraging UQ and expert user knowledge, the workflow facilitates the inverse task of identifying parameter sets that minimize deviations between desired and actual track geometries. The GPR model, trained and validated using 379 experimental track cross-sections, demonstrates its efficacy through two illustrative test cases. This approach reduces reliance on trial-and-error experimentation, enabling a more systematic and user-centric method to achieve precise track geometries in laser DED process.

Zhou et al. [6] investigate the use of laser DED for repairing structural aluminum alloys. The study addresses challenges such as uneven material flow and defects caused by unequal powder particle size. A multiscale, multiphysics model integrating discrete element and finite volume methods is developed to analyze the fluid dynamics and thermal behavior of the molten pool during the repair process. Additionally, a macroscale thermomechanical model evaluates stress evolution and verifies the structural integrity of deposited layers. Mengesha et al. [14] explore the surface hardness and scratch resistance of electroless nickel plating on additively manufactured composite components. Using K-means clustering and Taguchi's design of experiments, the study quantifies scratch widths and evaluates their relationship to hardness levels. Enhanced characterization through SEM imaging improves analysis accuracy, highlighting the potential of Ni-plating for improving surface properties in industrial applications.

Sinha et al. [8] address the challenge of gas porosity in laser PBF. By combining mechanistic modeling and experimental data, the authors propose a dimensionless gas porosity index to predict and mitigate pore formation. Tested against independent data, the index achieves 92% accuracy for alloys like stainless steel 316, Ti-6Al-4V, Inconel 718, and AlSi10Mg, with AlSi10Mg being the most prone to porosity. A gas porosity map is developed for practical process optimization. In contrast, Khan et al. [13] focus on defect detection in metal AM through machine learning and optical tomography (OT). Using layer-wise OT imaging, a Random Forest Classifier identifies anomalies, validated against CT data. The model achieves a 99.98% detection accuracy, correlating with 79.40% of

CT-detected defects. Together, these studies advance defect mitigation in AM through mechanistic modeling and machine learning.

The studies by Pazireh et al. [4] and Scime et al. [12] explore data-driven methodologies to predict the mechanical properties of AM parts. Pazireh et al. [4] investigated the effects of toolpath patterns, geometries, and layering on the mechanical properties of DED parts. Using finite element simulations, a linear mixed-effects model, principal component analysis, and self-organizing map clustering, they identify important relationships between process parameters and residual stresses, strains, and mechanical properties. Scime et al. [12] focus on qualifying laser PBF parts using in situ sensor data, integrating powder bed imaging, machine health metrics, and laser paths. Machine learning models trained on over 6000 tensile specimens predict tensile properties with a 61% error reduction compared to non-data-driven approaches. Both studies underscore the importance of leveraging machine learning and data analysis to predict mechanical properties in AM parts.

This Special Issue of "Materials" attracted numerous submissions, and the final publication consists of 12 high-quality peer-reviewed articles. In addition to highlighting exciting advancements, the articles in this Special Issue identified several contemporary scientific, technological, and economic challenges that require immediate attention. They emphasized the need for future research and development to enable the production of high-quality parts in a cost-effective way. It is evident that work in these critical areas of AM, especially focusing on modeling and machine learning, is still in its early stages, and needs further research and development.

Author Contributions: The authors had equal contributions. All authors have read and agreed to the published version of the manuscript.

Conflicts of Interest: The authors declare no conflicts of interest.

References

1. Mukherjee, T.; DebRoy, T. *Theory and Practice of Additive Manufacturing*, 1st ed.; John Wiley & Sons: Hoboken, NJ, USA, 2023; ISBN 978-1-394-20227-0.
2. Mukherjee, T. Recent progress in process, structure, properties, and performance in additive manufacturing. *Sci. Technol. Weld. Join.* **2023**, *28*, 941–945. [CrossRef]
3. Mukherjee, T.; Elmer, J.; Wei, H.; Lienert, T.; Zhang, W.; Kou, S.; DebRoy, T. Control of grain structure, phases, and defects in additive manufacturing of high performance metallic components. *Prog. Mater. Sci.* **2023**, *138*, 101153. [CrossRef]
4. Pazireh, S.; Mirazimzadeh, S.E.; Urbanic, J. Application of Linear Mixed-Effects Model, Principal Component Analysis, and Clustering to Direct Energy Deposition Fabricated Parts Using FEM Simulation Data. *Materials* **2024**, *17*, 5127. [CrossRef] [PubMed]
5. Wu, S.-H.; Tariq, U.; Joy, R.; Mahmood, M.A.; Malik, A.W.; Liou, F. A Robust Recurrent Neural Networks-Based Surrogate Model for Thermal History and Melt Pool Characteristics in Directed Energy Deposition. *Materials* **2024**, *17*, 4363. [CrossRef] [PubMed]
6. Zhou, X.; Pei, Z.; Liu, Z.; Yang, L.; Yin, Y.; He, Y.; Wu, Q.; Nie, Y. Multiscale Simulation of Laser-Based Direct Energy Deposition (DED-LB/M) Using Powder Feedstock for Surface Repair of Aluminum Alloy. *Materials* **2024**, *17*, 3559. [CrossRef] [PubMed]
7. Sajadi, P.; Dehaghani, M.R.; Tang, Y.; Wang, G.G. Physics-Informed Online Learning for Temperature Prediction in Metal AM. *Materials* **2024**, *17*, 3306. [CrossRef] [PubMed]
8. Sinha, S.; Mukherjee, T. Mitigation of Gas Porosity in Additive Manufacturing Using Experimental Data Analysis and Mechanistic Modeling. *Materials* **2024**, *17*, 1569. [CrossRef] [PubMed]
9. Fagersand, H.M.; Morin, D.; Mathisen, K.M.; He, J.; Zhang, Z. Transferability of Temperature Evolution of Dissimilar Wire-Arc Additively Manufactured Components by Machine Learning. *Materials* **2024**, *17*, 742. [CrossRef]
10. Dong, W.; Lian, J.; Yan, C.; Zhong, Y.; Karnati, S.; Guo, Q.; Chen, L.; Morgan, D. Deep-Learning-Based Segmentation of Keyhole in In-Situ X-ray Imaging of Laser Powder Bed Fusion. *Materials* **2024**, *17*, 510. [CrossRef] [PubMed]
11. Hermann, F.; Michalowski, A.; Brünnette, T.; Reimann, P.; Vogt, S.; Graf, T. Data-Driven Prediction and Uncertainty Quantification of Process Parameters for Directed Energy Deposition. *Materials* **2023**, *16*, 7308. [CrossRef] [PubMed]

12. Scime, L.; Joslin, C.; Collins, D.A.; Sprayberry, M.; Singh, A.; Halsey, W.; Duncan, R.; Snow, Z.; Dehoff, R.; Paquit, V. A Data-Driven Framework for Direct Local Tensile Property Prediction of Laser Powder Bed Fusion Parts. *Materials* **2023**, *16*, 7293. [CrossRef] [PubMed]
13. Khan, I.A.; Birkhofer, H.; Kunz, D.; Lukas, D.; Ploshikhin, V. A Random Forest Classifier for Anomaly Detection in Laser-Powder Bed Fusion Using Optical Monitoring. *Materials* **2023**, *16*, 6470. [CrossRef] [PubMed]
14. Mengesha, B.N.; Grizzle, A.C.; Demisse, W.; Klein, K.L.; Elliott, A.; Tyagi, P. Machine Learning-Enabled Quantitative Analysis of Optically Obscure Scratches on Nickel-Plated Additively Manufactured (AM) Samples. *Materials* **2023**, *16*, 6301. [CrossRef] [PubMed]
15. Johnson, G.A.; Dolde, M.M.; Zaugg, J.T.; Quintana, M.J.; Collins, P.C. Monitoring, Modeling, and Statistical Analysis in Metal Additive Manufacturing: A Review. *Materials* **2024**, *17*, 5872. [CrossRef] [PubMed]

 *materials*

Review

Monitoring, Modeling, and Statistical Analysis in Metal Additive Manufacturing: A Review

Grant A. Johnson [1,2], Matthew M. Dolde [1,2], Jonathan T. Zaugg [1,2], Maria J. Quintana [1,2,3] and Peter C. Collins [1,2,3,4,*]

[1] Department of Materials Science and Engineering, Iowa State University, Ames, IA 50011, USA; grantajo@iastate.edu (G.A.J.); mdolde@iastate.edu (M.M.D.); jzaugg@iastate.edu (J.T.Z.); mariaqh@iastate.edu (M.J.Q.)
[2] Ames National Laboratory, Ames, IA 50011, USA
[3] Center for Advanced Non-Ferrous Structural Alloys (CANFSA), USA
[4] Center for Smart Design and Manufacturing, Iowa State University, Ames, IA 50011, USA
* Correspondence: pcollins@iastate.edu

Abstract: Despite the significant advances made involving the additive manufacturing (AM) of metals, including those related to both materials and processes, challenges remain in regard to the rapid qualification and insertion of such materials into applications. In general, understanding the process–microstructure–property interrelationships is essential. To successfully understand these interrelationships on a process-by-process basis and exploit such knowledge in practice, leveraging monitoring, modeling, and statistical analysis is necessary. Monitoring allows for the identification and measurement of parameters and features associated with important physical processes that may vary spatially and temporally during the AM processes that will influence part properties, including spatial variations within a single part and part-to-part variability, and, ultimately, quality. Modeling allows for the prediction of physical processes, material states, and properties of future builds by creating material state abstractions that can then be tested or evolved virtually. Statistical analysis permits the data from monitoring to inform modeling, and vice versa, under the added consideration that physical measurements and mathematical abstractions contain uncertainties. Throughout this review, the feedstock, energy source, melt pool, defects, compositional distribution, microstructure, texture, residual stresses, and mechanical properties are examined from the points of view of monitoring, modeling, and statistical analysis. As with most active research subjects, there remain both possibilities and limitations, and these will be considered and discussed as appropriate.

Keywords: additive manufacturing; monitoring; modeling; statistics

Citation: Johnson, G.A.; Dolde, M.M.; Zaugg, J.T.; Quintana, M.J.; Collins, P.C. Monitoring, Modeling, and Statistical Analysis in Metal Additive Manufacturing: A Review. *Materials* **2024**, *17*, 5872. https://doi.org/10.3390/ma17235872

Academic Editor: Federico Mazzucato

Received: 23 September 2024
Revised: 18 November 2024
Accepted: 21 November 2024
Published: 29 November 2024

1. Introduction

In the last couple of decades, the additive manufacturing of metallic materials has been the subject of significant research and investment, including considerable growth in the number of manufacturing systems available commercially and the development and growth of AM methods. There have been numerous research initiatives and activities adding to the available literature that have enabled the wider community to develop new understandings regarding the differences in processing and material state [1–3] between AM materials and traditionally manufactured ones (e.g., casting, welding, forging). Increasingly, new monitoring and computational modeling methods for AM processes, which permit the capture of qualitative and quantitative data as part of a "digital thread" for advanced manufacturing, are being developed and disseminated [4]. Tables 1 and 2 show a summary of the state of the art of monitoring and modeling different variables and characteristics of AM parts.

Table 1. Summary of state of the art on monitoring different variables and build characteristics of additive manufacturing methods.

AM Method	Variable/Characteristic Monitored							
	Feedstock	Processing Parameters	Melt Pool	Chemistry	Defects	Residual Stress and Distortion	Microstructure and Texture	Mechanical Properties
Laser PBF	A	B	A	B	A	A	B	B
Electron Beam PBF	A	B	A	B	A	A	X	X
Laser Powder DED	B	B	B	B	A	A	B	X
Laser Wire DED	X	B	B	X	X	B	X	X
Electron Beam Wire DED	B	B	B	X	A	B	X	X
WAAM	B	A	A	X	A	A	B	B

Note: A = Commonly studied, B = Limited research, X = No publications.

Table 2. Summary of state of the art on modeling different variables and build characteristics of additive manufacturing methods.

AM Method	Variable/Characteristic Modeled							
	Feedstock	Processing Parameters	Melt Pool	Chemistry	Defects	Residual Stress and Distortion	Microstructure and Texture	Mechanical Properties
Laser PBF	A	A	A	B	A	A	A	A
Electron Beam PBF	A	A	A	B	A	A	A	A
Laser Powder DED	B	A	A	B	B	B	B	A
Laser Wire DED	B	A	B	B	B	A	B	B
Electron Beam Wire DED	B	B	B	B	B	A	A	B
WAAM	B	A	A	B	B	A	A	A

Note: A = Commonly studied, B = Limited research.

Several alloys have dominated research activities and the industrial use of AM parts (e.g., titanium alloys, superalloys, aluminum alloys, refractory-based alloys, steels, and stainless steels) [5–9]. Given the governing thermophysical properties of the metals (e.g., melting point, boiling point, coefficient of thermal expansion, thermal conductivity, density, reflectivity), and their significant variability (e.g., thermal conductivity can range from <0.1 to ~4.0 W/cm·K), along with the details of the energy sources and deposition systems (e.g., wavelengths of lasers, beam profiling, scanning strategy), the various combinations of materials and deposition systems can either result in depositions that are "feasible" (i.e., high-quality builds) or "infeasible" [10,11]. If the alloying elements in a material system are incompatible with the processing physics, the parts can have high levels of defects, including defects that either appear during deposition or evolve/appear during post-deposition thermal treatments. In an extreme case, the build may not be possible to produce, experiencing material failures, such as delamination or cracking at length scales corresponding to the line-by-line or layer-by-layer processing length scales. Research continues to mitigate deleterious defects in materials with prints that are generally thought to be "infeasible" [12], thus enabling the AM of new and existing alloy systems; e.g., for some high-strength aluminum alloys that are susceptible to solidification cracking, the addition of inoculants has been successful in promoting homogeneous nucleation to limit solidification cracking [13]. Recognizing that deposition feasibility is an important concept, some researchers are directly studying feasibility over multidimensional compositional spaces, rapidly identifying interesting alloy pathways, and providing a key concept to enable future gradient materials [14].

For this review paper, the monitoring, modeling, and statistical analysis methods for AM of metallic materials are classified according to two distinct perspectives, namely (i) the processing methods and (ii) the material states that are produced. Overviews of these perspectives are shown in Figure 1. The processing of AM (Figure 1a) encompasses a—the feedstock (i.e., powder, wire), b—the energy source (i.e., laser, electron beam, plasma arc), and c—the melt pool (including spattering and vapor plume). By changing any of these processing aspects, the final material state of the part (Figure 1b) will change accordingly. The material state includes d—defects, e—compositional distribution, f—microstructure and texture, and g—residual stress and distortion and their influence on mechanical properties.

For monitoring, modeling, and statistical analysis, Figure 1 will be evaluated in depth individually. These evaluations will look like "heat maps" to elucidate perceived strengths and weaknesses on the same color scale—light to dark, respectively. For example, Figure 2 for monitoring considers the relative number of papers related to monitoring each aspect of processing and material state found in the literature. Figure 3 is similar to Figure 2, such that the color is relative to what has been modeled most and least frequently, while also taking into account the accuracy of the state-of-the-art models and simulations. Figure 4 evaluates the availability of data for each aspect of processing and material state, which can be captured through monitoring techniques and used to validate models and simulations.

In the context of this review paper, monitoring (Figure 2) is meant to imply real-time monitoring during the process, which can be difficult to define in a way that is either accepted broadly or applied easily. Of importance, we consider monitoring based upon the physics permitted and not any existing constraint of engineering systems; i.e., we consider what may be possible, rather than being limited to what exists. For the purposes of this review, the term "monitoring" is defined as (i) the use of measurement devices of varying modalities to (ii) obtain quantitative measurements that are spatially and temporally registered and that record (iii) interpretable signals of meaningful physical phenomena associated with the process and/or material state that (iv) significantly impact the quality of the material subsequent to the deposition and which, when aggregated, (v) have the potential to form a critical detail recorded in a digital thread (the term digital thread is one of two terms (the other term is a digital twin) that arises from the richness of the digital transformation associated with Industry 4.0. Digital twin refers to the concept of obtaining an exact replica of a part where, for each voxel (volume element), the manufacturing and material state attributes would be known and tied to a specific part—1:1. Digital thread refers to not only the lifetime of a part (i.e., the physical object is likely to deviate from the digital twin during service) but opens the door for statistical treatments of multiple parts). Such a digital thread could be used for quality control prior to placing a part into service, lifetime management for parts during service, or as information to support next-generation materials, processing, and design. Monitoring AM processes is important to ensure the quality of manufactured parts by rejecting those with an excessive number or type of defects (e.g., porosity, distortion, lack-of-fusion defects) before they are inserted into service and subjected to operational conditions (e.g., forces, stresses, temperatures) [10,15,16]. With the number of interdependent variables in AM, monitoring different features of the build process is a logical, though largely aspirational, goal.

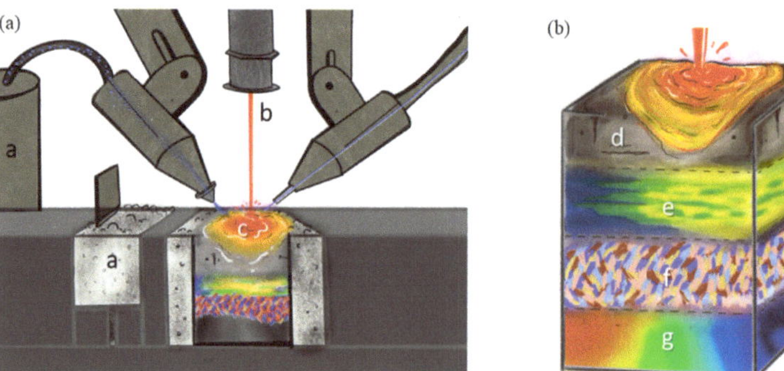

Figure 1. Schematic of an additive manufacturing process, separated into (**a**) processing and (**b**) material state, showing a—feedstock (powder, wire), b—energy source (laser or electron beam), c—melt pool, spattering and vapor plume, d—defects (spherical porosity, lack of fusion defects), e—compositional distribution, f—microstructure and texture, and g—residual stresses and distortion.

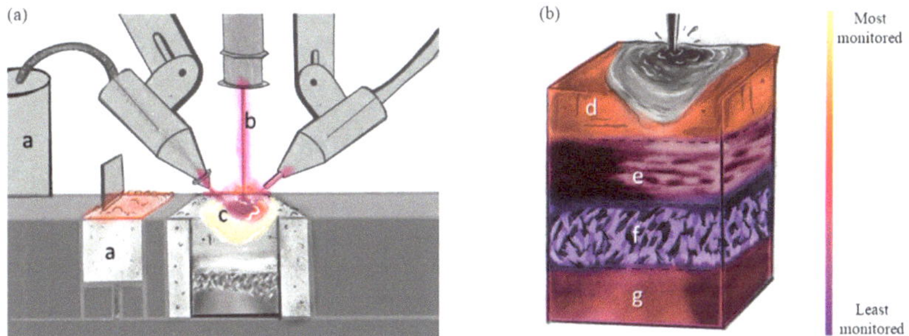

Figure 2. Schematic of an additive manufacturing process, separated into (**a**) processing and (**b**) material state, with an emphasis on monitoring, showing a—feedstock (powder, wire), b—energy source (laser or electron beam), c—melt pool, spattering and vapor plume, d—defects (spherical porosity, lack of fusion defects), e—compositional distribution, f—microstructure and texture, and g—residual stresses and distortion.

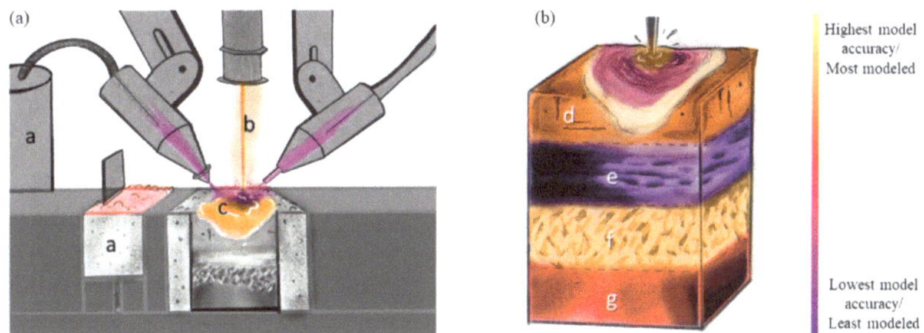

Figure 3. Schematic of an additive manufacturing process, separated into (**a**) processing and (**b**) material state, with an emphasis on modeling, showing a—feedstock (powder, wire), b—energy source (laser or electron beam), c—melt pool, spattering and vapor plume, d—defects (spherical porosity, lack of fusion defects), e—compositional distribution, f—microstructure and texture, and g—residual stresses and distortion.

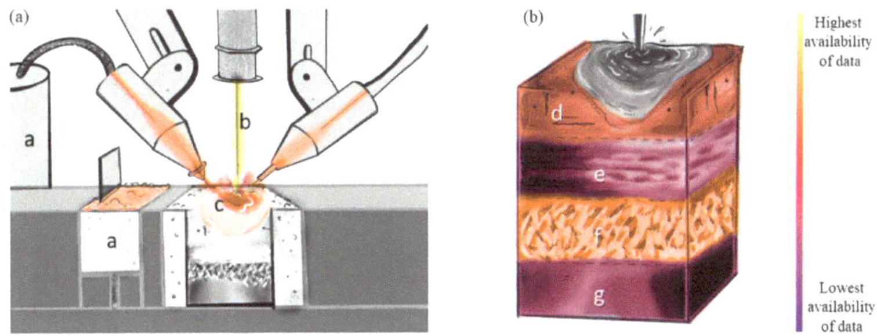

Figure 4. Schematic of an additive manufacturing process, separated into (**a**) processing and (**b**) material state, with an emphasis on statistics (or availability of data for statistical analysis), showing a—feedstock (powder, wire), b—energy source (laser or electron beam), c—melt pool, spattering and vapor plume, d—defects (spherical porosity, lack of fusion defects), e—compositional distribution, f—microstructure and texture, and g—residual stresses and distortion.

Modeling and integrated computational materials engineering (ICME) is considered an increasingly important aspect of AM and is currently being used to predict the microstructure and properties of a part before building one [17–23]. Modeling (Figure 3) is a form of abstraction, and in the materials science discipline, it normally consists of mathematical relationships that are used to define how the material state reacts given specific processing conditions. Modeling how processing parameters influence the resulting material state and properties helps to ensure a proper understanding of the physics that exists during AM processing and that the manufactured parts have the desired properties.

Statistics have been used in the analysis of monitoring and modeling results to aid in interpreting correlations between processing and material state (Figure 4). By statistically analyzing these additively manufactured parts, manufacturers can have increased confidence that the parts they make are meeting or exceeding the specifications set by design or standards. In a survey, up to 47% of manufacturers reported that having a lack of confidence in the quality of additively manufactured parts was enough reason to not pursue this form of manufacturing [24]. The use of statistical methods and uncertainty quantification tools can help identify critical manufacturing signatures that are important to monitor and model. For processing, leveraging statistics helps ensure confidence that monitored data are both reasonable and representative. By statistically refining monitoring data, improving models to further understand the material state and resulting materials properties should be possible and result in increased confidence that the final parts are made as expected [25,26].

Important to all three components of this review paper (monitoring, modeling, and statistical analysis) are the concepts of length and time scales. The operating physical processes in the variably sized process zones span across length scales (10^{-9} to 10^{-1} m) and operate both during processing as well as following deposition (i.e., continued crack propagation in parts with high degrees of residual stress), and thus, time scales are also spanned (10^{-8} to 10^{+7} s).

This paper is organized under two broad primary sections: Section 2 is on processing, and Section 3 is on material state. Relevant subsections to each primary section, such as "Feedstock" (Section 2.1), are then described. Embedded in each subsection are details associated with the state-of-the-art in monitoring, modeling, and statistical analysis.

2. Processing in Additive Manufacturing

Additive manufacturing processes use computers to control the delivery of both the material feedstock and a quantity of energy into a position that is co-registered in time and space. Under the right conditions (e.g., sufficient energy density, thermophysical properties), the energy and feedstock combine to create a molten pool, whose interaction with the environment and realization of other complex physics results in the Material State (Section 3). Thus, in this section, we consider (Section 2.1) Feedstock and Environment; (Section 2.2) Energy Source and Thermal Distributions; and (Section 2.3) Melt Pool, which includes a discussion of the vapor plume as it is integrally coupled with the dynamics of the molten pool.

2.1. Feedstock and Environment

In fusion-based AM processes, such as powder bed fusion (PBF), powder-blown directed energy deposition (DED), and wire-fed based additive manufacturing approaches, including wire arc additive manufacturing (WAAM) and electron beam additive manufacturing (EBM), the feedstock is either powder or wire. This feedstock and base material then undergo highly localized fusion events (melting and solidification) that add material, volume by volume, to form the finished part (Figures 1a and 2a). The PBF process starts by spreading a layer of powder onto a build plate (or a previous layer) to be selectively melted using either a laser or an electron beam as the energy source. Another AM process, DED, uses wire or powder that is fed into the focal point of a laser or electron beam.

While the objective regarding monitoring associated with this review paper is to focus on digital means of in situ monitoring, we will briefly note the classical methods of ex situ characterization and quantitative measurements of the starting material that will be used in the process. Classically, powder is characterized using a variety of techniques to assess its size, shape, density, and flowability. Typically, these techniques involve samples for optical or electron imaging, weight distributions using sieves, time-based techniques to determine mass flowrates, and geometrical measurements such as the angle of repose [27–32]. In both powder and wire, bulk chemical analysis and surface chemical analysis can be performed.

2.1.1. Feedstock and Environmental Monitoring

The in situ monitoring of feedstock (Figure 2a) can be conducted but is not currently widely implemented. In principle, powder-based feedstock can be monitored using various methods. For example, one in situ monitoring technique uses optical techniques to image the powder during or following powder spreading [33]. These images can provide information regarding potential powder oxidation from differences in the color of the particles [34], cross-contamination also using color differences [35], uneven layer thickness using image intensity as a surrogate of a semi-insulating layer of variable thickness [36], and any machine issues or defects caused by the spreader [37] or anomalously large particles or other large debris. Recent work by Tan et al. [38] suggests that it may be possible to use principal component analysis to infer the quantitative signatures of conventional particle attributes (e.g., density, friction, interparticle forces) through measurements of the "avalanche" angle of powder against the powder spreader, which could be used as a continuous convolutional term of some of these fundamental metrics, typically measured externally in batches. Beyond these optical techniques, limited research has explored the use of eddy current measurements to monitor discontinuities in the feedstock layer [39].

Contrary to powder bed-based AM techniques, which provide spatially resolved information that could be used to correlate the powder to be processed with the process zone/material state, the monitoring of feedstock presents greater challenges for DED processes, as any correlation must be made through a time-resolved measurement of incoming material that is captured by the molten pool at some incremental time step later. Further complicating powder-blown DED techniques is the fact that only a fraction of the incoming powder is captured by the molten pool. Despite these challenges, researchers have demonstrated that the flow of powder can be monitored as it is blown into the melt pool using high-speed optical imaging systems [40]. Powder-based DED processes have the potential to record optical snapshots of powder within the powder feed system, including angles of repose or when the powder is entering the powder feed mechanism. Using such images, it would be possible to obtain information related to size, morphology, roughness, oxidation, or cross-contamination from limited analysis areas, i.e., the surfaces of volumes of powder [41]. Powder mass flow has also been studied using in-line acoustic measurements [42]. In certain materials systems and processing conditions where there is sufficient contrast for the modality of information in the images (e.g., light) and depth of information in the images, high-speed imaging of the powder flow approaching and interacting with the melt pool could have a sufficient resolution such that properties of the powder can be extracted (e.g., morphology, surface quality, particle size) [43]. The limitations include difficulties when the melt pool temperatures and emissivity prevent quantitative interpretable information from being collected.

While wire-based AM techniques do not enjoy as wide a variety of in situ monitoring techniques as powder-based systems, some techniques do exist. Optical monitoring of the wire can be conducted at the point where the wire feedstock enters the melt pool. Of particular interest to some researchers is the degree of wire deflection [44], as deflection is reported to be directly coupled with the tendency to form defects or achieve dimensionally accurate parts. However, beyond these techniques, there is limited literature available that describes the monitoring of wire feedstock for wire-based processes. The lack of research does not negate the importance of the subject. For example, a recent thesis by Ng Chi-Ho [45] demonstrated that the presence of surface contaminates (e.g., soaps used in wire drawing) could result in defects in additively manufactured builds, indicating that the chemical analysis of incoming wire could be an important parameter to monitor. Similarly, other researchers have indicated that internal defects within wires might lead to defects in the deposition. Looking forward, in this review, we suggest that an opportunity space exists to conduct research and development activities to monitor the wire using, for example, ultrasonic testing, optical imaging, instruments to monitor shape, and/or techniques capable of chemical analysis such as X-ray fluorescence (XRF) to detect composition anomalies.

In structural metals, the composition of the depositions is known to have a significant effect on the resulting microstructure, properties, and performance [46–48]. When AM approaches are adopted, the deposition composition will invariably differ from the "certificate of composition" that is associated with the material feedstock. The chemistry will change due to capture/alloying from contaminates or impurities carried by the feedstock (e.g., the soaps in the wire drawing mentioned previously), preferential evaporation [47], or ingestion of gases present in the chamber atmosphere (e.g., moisture, oxygen, nitrogen). Further, the atmospheric conditions may have an effect on material feed and build quality in other ways, such as altering powder flowability in powder bed systems, which can lead to predicting or inferring powder or part quality [49]. Thus, it is plausible that an important aspect would be to monitor the atmosphere within the build chamber and, if possible, the vapor plume (described in Section 2.3). Table 3 presents common monitoring techniques for each variable or characteristic discussed in this review, along with a rating for the amount of data collected for future use in modeling and statistical analysis.

Table 3. List of common monitoring techniques and the available data coming from these techniques.

Label	Description	Monitoring Technique *	References	Data Availability **	References
a	Feedstock	PBF_1, DED_2, and $WAAM_2$: optical monitoring, compositional measurements	[34–37,42,44]	XX	[50–52]
b	Energy source	$Laser_6$; electron $beam_5$; wire $arc_{1,3}$: multi-sensor electrical monitoring systems Heat $flow_1$: NIR, thermal, and thermocouple	[53–60]	XXX	[61–65]
c	Melt pool	Melt $pool_{1,3}$: optical, thermal, X-ray Vapor $plume_{1,4}$: optical and chemical $Spatter_{1,4}$: optical and thermal	[66–91]	XXX	[92–95]
d	Defects	$Optical_{1,3}$, $Thermal_{1,3}$, X-$ray_{1,3}$, $Acoustic_{1,3}$	[90,96–105]	XX	[63,106,107]
e	Compositional distribution	LIBS-based systems on vapor $plume_2$, ultrasonic/$acoustic_6$	[82,84,108–110]	X	[111,112]
f	Microstructure	IR_2 and $ultrasonic_2$	[113–118]	XX	[111,115]
g	Residual Stress and Distortion	Residual $stress_2$: ultrasonic $Distortion_{1,3}$: optical/DIC and displacement sensor	[119–128]	X	[125,129,130]
h	Mechanical Properties	Machining $forces_5$, $ultrasonic_{5,6}$, or $acoustic_6$	[131–137]	X	[25,138]

* The availability of monitoring a parameter or characteristic is represented by 1—common in laboratory scale, 2—limited in laboratory scale, 3—common in commercial scale, 4—limited in commercial scale, 5—limited research, 6—potential applications but not commonly researched. **: The X-XX-XXX denotes a ranking of the amount of data that can be gathered from in situ monitoring techniques: X—limited data gathered, XX—medium amount of data gathered, XXX—large amounts of data gathered.

2.1.2. Feedstock Modeling

As with monitoring, the two most common AM feedstocks that are modeled are powder and wire (Figures 1a and 3a). Approaches associated with wire modeling are often inherited partially from welding practices. Welding and AM share similar physics, so the overlap in concepts and models is understandable.

Arguably the most extensive effort, and thus most complete software that is available for certain aspects of powder modeling, is an extension of LAMMPS (Large-scale Atomic/Molecular Massively Parallel Simulator), which was created at Sandia National Laboratory [139]. The extension is known as LIGGGHTS (LAMMPS Improved for General Granular and Granular Heat Transfer Simulations) [140]. LIGGGHTS uses the discrete element method (DEM) to model the motion of particles [141] and is combined with computational fluid dynamics (CFD) to include the flow of surrounding liquid. This modeling technique gives the ability to model powder particle movement during raking and is a start to modeling complex interactions in the feedstock (Figure 5). DEM models will often use one of the following functions to describe contact mechanics: Hertzian, Johnson–Kendall–Roberts, or Derjaguin–Muller–Toporov [142,143]. The main difference between these models is how the cohesion is modeled. LIGGGHTS has been used to determine the powder flowability in DED [144]. LIGGGHTS has also been used to model how irregular powder can influence its flowability properties during AM [141], as well as modifying the surface finish during powder spreading in PBF [145]. In addition to LIGGGHTS, there are other software that have been developed for the purpose of modeling and simulating parts of the AM process and material state. These are shown in Table 4 in addition to common model formulations.

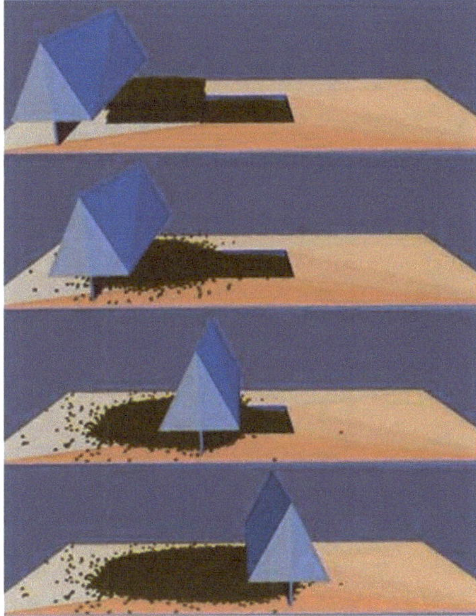

Figure 5. Discrete element model (DEM) simulation of additive manufacturing powder rake over time. Reprinted with permission from [141].

Wire AM techniques have also been modeled. However, this modeling has been primarily performed by the welding community. Adapting and modifying these models to AM may be possible for the techniques that use wire as feedstock. Many of the models related to wire techniques include more information related to the energy source; therefore, most of the models will be discussed later. However, as these models have matured over the years, more complicated and realistic models have been created. One example is a model of a twin-wire gas metal arc welding (GMAW) process [146]. For this process, two wires are used in the welding process. In welding, and therefore in AM, using two or more feedstocks is of some interest, especially in alloys that have significant changes in microstructure and properties with changes in feedstock to control local properties in AM (e.g., duplex stainless steels [147]), and thus, these additional model developments are of benefit to the wider AM community.

Table 4. List of modeling formulations and some software for AM.

Label	Description	Formulation	Software	References
a	Feedstock: powder feed, raking, and wire feeding	Discrete element method (DEM) with computational fluid dynamics (CFD); numerical models	LIGGGHTS	[140]
b	Energy source	Numerical models; finite element method (FEM) for deformation AM processes	N/A	[148–150]
c	Melt pool	FEM or finite volume method (FVM) CFD; material point method (MPM); numerical models; lattice Boltzmann method	AdditiveFOAM, ExaMPM, TruchasPBF	[151–154]
d	Defects	CFD; numerical models	N/A	[11,155]
e	Composition variation	Numerical models	N/A	[156–158]
f	Microstructure	Kinetic Monte Carlo (KMC); cellular automata (CA); phase field (PF); Potts model; CALPHAD Scheil-Gulliver model; numerical models	AMCAFE, AMPE, ExaCA, MEUMAPPS-SS	[14,159–167]
g	Residual stress	FEM	See Table 5	[22]

2.1.3. Feedstock Statistics

Collecting, storing, and using the spatially- and temporally-rich data from process monitoring is computationally expensive. In an unconstrained processing environment, researchers might desire to store very large quantities of data and, as noted for the case of monitoring feedstock, could record high-speed optical images of all the powder entering the system (Figures 1a and 4a). One can conceive that a sophisticated real-time image analysis program might then automatically detect size, morphology, and inhomogeneities. While this is lacking in the literature, and thus represents an aspirational possibility, once collected, it is necessary to consider the probabilities and statistics of such measurement [52]. For certain measurements, it is possible to leverage statistical descriptors that exist, such as particle size distribution (PSD) values of D10, D50, and D90 for the powder feedstock, where the designation (e.g., D10) signifies that 10% of the total powder is finer than the corresponding size (e.g., D10 of 15 μm indicates that 10% of the powder is finer than 15 μm). However, a single measurement of PSD for a well-mixed powder sample will invariably differ from the PSD for powder following a period of time where free-settling can occur [168] and thus will likely differ from the PSD of in situ measurements, where the integrated PSD over time (i.e., a PSD(t)) should approach measurements for a lot of powder. Similar time variabilities are expected for other property metrics, such as sphericity. However, in principle, it should be possible to develop a time-dependent model based upon powder flow, size, density, shape, free settling, etc., apply assumptions about the distributions of the model parameters (e.g., Gaussian or log-normal for powder size), and develop statistics associated with the dynamics of the feedstock [50,51]. To the authors' awareness, such work has not been conducted for the dynamics of powder flow.

Another critical component to characterizing and analyzing the feedstock is absorptivity. Since factors such as the powder size distribution have an impact on absorptivity, direct measurement is difficult [169]. In a study concerning the measurements of absorptivity of metallic powder, two different PSD models were created, and changing the distribution from a Gaussian to a bimodal distribution made a significant difference in the measured absorptivity [24]. Further complicating the issue, the absorptivity of a single layer of spherical particles is higher than a flat surface, as incoming light can be scattered by the spheres followed by various interactions with other particles, effectively creating multiple scattering events. Therefore, if a piece of the as-built material or a small batch of the powder is taken aside for absorptivity testing, two different results could be obtained. This issue highlights just how important statistics are and how, if neglected or used inappropriately, the information gathered from monitoring or used in modeling can be markedly off from the ground truth.

2.2. Energy Source and Thermal Distributions

Common fusion-based AM processes use a laser beam, electron beam, or an electrical arc as energy sources (Figure 1a). The energy source is not only the main contributor to the overall heat input into the system but is also the one that is easiest to monitor and dynamically control as part of the process (the other heat source would be associated with enthalpies of reactions and phase transformations). The primary energy source is responsible for fusing the feedstock and enabling a stable melt pool. The accurate control of the incident power and shape of the energy source are important, as their control enables the process to be optimized, thereby increasing the repeatability of the process and the attending properties of the additively manufactured components.

Similar to feedstock materials, considering the classical approaches to measuring attributes of the energy source is useful. Regarding lasers, their power at various points along a beam path can be measured using a thermopile. Such data can be used to develop calibration curves, confirm incident power prior to/following depositions, or diagnose the "health" of AM systems. Similarly, certain operators may wish to record the profile of their beam prior to and/or after depositions. In such instances, beam profilers can be used. Regarding electron beams, their power can be either calculated using the settings or measured more accurately using, for example, a Faraday cup. The latter is preferred when aspects of the electron guns, such as apertures, are subject to evolution/drift/degradation over time.

2.2.1. Energy Source Monitoring

By using certain optical components, including particular configurations of beam splitters and/or partially reflective mirrors, collecting laser power dynamically during a deposition is possible (Figure 2a). Under certain manufacturing conditions, operators may wish to monitor the shape of a beam and could adopt beam splitters or partially reflective mirrors to monitor the shape of the incident beam. Published research on direct energy source power monitoring is limited or non-existent for laser and electron beams. Measurements in electron beams mostly fall in the category of voltage differences between the electron source and the grounded build plate [170] or currents in lenses that can be correlated with the positions of the beam. Research on monitoring the arc in wire arc processes is far more mature since the technology is derived from welding. Multi-sensor setups to monitor multiple parameters of the arc are currently used in laboratory systems [54] as well as in commercial systems [55]. These setups involve monitoring the voltage, current, sound, radio frequency, and temperature of the arc [54]. Acoustic monitoring during AM builds has shown to be effective for defect detection in multiple systems, but experimentation to detect other processing parameters such as laser power has been explored only in laboratory settings [53].

While monitoring the energy source can provide quantitative information regarding the energy entering the process zone, only a fraction of the incident energy is available to the material to undergo the necessary thermophysical processes (e.g., fusion for liquid-based AM) to achieve the desired material state. The general categories of energy loss are known and include (i) energies that are lost through inefficiencies of "coupling" between the energy source and the material, such as reflection and scattering from the incoming material and/or scattering/absorption in a plume above the substrate; and (ii) energies that are lost due to traditional heat transfer balances, including conduction, convection, radiation, and the thermodynamics of phase transformations. The heat flow during an AM build can influence the final microstructure and properties [171,172]; thus, from the perspective of the material state, it is arguably more important to conduct in situ monitoring of the temperature distributions (e.g., gradients, time-dependent heat flows) than the energy of the system. In situ monitoring approaches for heat flow include the use of thermocouples [56,172] to monitor the temperature at specific locations over a period of time. Thermocouple-based temperature monitoring can take place at the build plate [57], close-to-the-line depositions [58], or even within a deposition by inserting (i.e., "harpooning" [173]) the thermocouple into the melt pool or directly depositing on top of the thermocouple [174]. Infrared (IR) imaging and pyrometers can measure the temperature of the current layer as well as previous sections of the build [59,60]. Monitoring the build temperature for PBF can be accomplished after the rake of a layer of new powder and can help estimate the layer quality while also monitoring the temperature at specific areas [36]. Table 3 presents common monitoring techniques for each variable or characteristic discussed in this review, along with a rating for the amount of data collected for future use in modeling and statistical analysis.

2.2.2. Thermal Modeling

The complex thermal gyrations associated with AM that exist as a result of the repetitive motion of the energy source relative to previous built material represents a unique challenge when seeking to understand and describe a thermal history, which has an effect on the material state and therefore the properties. There are several methods that are used to model the spatially and temporally dependent thermal histories, ranging from simple heat source models (Figure 3a) to more complex finite element method (FEM) models (Table 4) [175].

The first models for the heat flow of a moving heat source in welding were presented by Rosenthal [148]. These equations provided a starting point for refinement by several researchers in the welding community over many years [149]. Other numerical models have been developed for many types of arc welding methods, including gas tungsten arc welding (GTAW) [176] and plasma arc welding [177]. However, many of these models can be modified and used for new applications, such as different AM processes.

Heat source models are often used in conjunction with mechanical models to optimize welding parameters for different alloys and predict distortion [178,179]. For example, Chen et al. [150] used CFD and finite volume method (FVM) models to predict the weld pool and thermal dynamics during welding, which further enabled the prediction of optimized weld parameters such as arc current and speed for different thicknesses and materials.

In these multi-physics models, one of the key parameters is the functional forms of the energy distributions, which will vary by the nature of the energy source. Several common energy distributions exist, such as Gaussian, top hat, and inverse Gaussian. The shape of the laser can influence the solidification structure of the alloy by changing the thermal gradients, interfacial velocities, and thermal gyrations [180].

2.2.3. Energy Source Statistics

Monitoring aspects of the energy source is critical to understanding the conditions the material experiences, but as energy sources vary from system to system, the actual data record may seem not to be transferrable from machine to machine, a problem that can be overcome using a combination of dimensional and statistical analysis [61]. To describe the energy source more accurately between machines, the energy density is used [63–65], which is a convolution of the power of the energy source and the volume of the melt pool (Figure 4). Figure 6 shows a general relationship between the characteristic material temperature and the energy density required for manufacturing each material. This figure was developed by collecting experimental details from more than 100 papers across all types of additive manufacturing processes and materials classes. As the reader might surmise, few of the papers from which these data have been extracted present their experimental settings as energy densities. However, each contained sufficient details regarding exposure time (e.g., speed or time), energy input, and volume processed to extract a nominal energy density. The authors recognize the exceptional variability that such an analysis of research papers incorporates. However, despite the uncertainty, it is noteworthy that this correlation is not only valid for metal-based AM but seems to be general for all variants of additive manufacturing. This general trend provides a good first step for what processing variables should be chosen to additively manufacture different materials, regardless of the system used. However, it is possible to extend beyond this simple relationship by, for example, performing a dimensional analysis on the processing inputs that can further abstract the AM process. An example of this approach as applied to SLM has been demonstrated in [61].

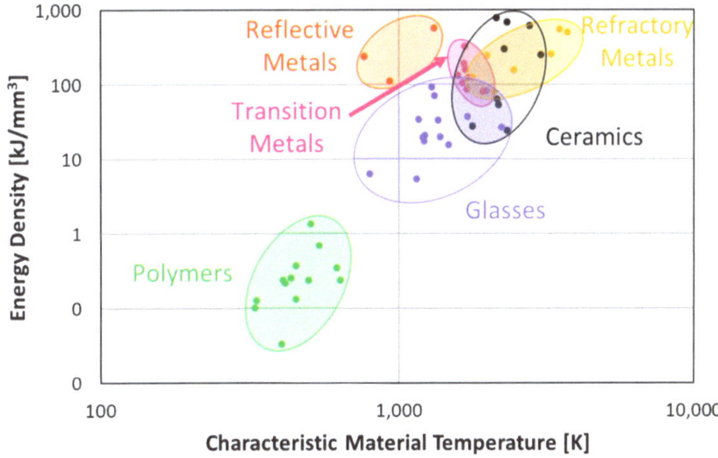

Figure 6. Ashby-like diagram of volumetric energy density versus characteristic material temperature for various material types categorized by color. Recreated with permission from [181].

Convoluting energy source variables such as power, shape, and speed can result in higher-order dimensionless variables. By listing all the process variables in a system and systematically combining them into new dimensionless input variables, the total input variables needed will decrease (by the total amount of fundamental dimensions in the system) [61]. By representing the inputs in this new way, the number of supposed process variables needing to be tested to provide a whole picture can be decreased. Without dimensional analysis, two research groups with different source power limits would not be able to replicate experiments between them effectively. However, by manipulating other dimensional process variables, the two groups demonstrated that they could effectively circumvent this power difference by achieving similar dimensionless variables by now having the same system [62]. This would allow for a research lab with an EBPBF and another lab with an LPBF system; while having completely different energy sources, they could still maintain similar build conditions by targeting similar higher-order variables such as energy density, even though their process variables are quite different.

Another method for optimizing process parameters involves the creation of code surrogates [182]. As opposed to running full simulations, these code surrogates are both faster and less complex than simulations and thus are ideal for rapid and broad testing to identify regions of viability and not necessarily completely replace simulations. A case study evaluated using code surrogates in AM [182]. From a small data set, the most promising surrogate code, a Gaussian process (GP) code surrogate, was able to predict the region of viable energy densities that most closely matched the region of viable energy densities made from a much larger data set [182]. Using this GP code surrogate could be a beneficial approach to finding a region of viable process parameters, cutting down computation time, and allowing researchers to obtain samples faster.

2.3. Melt Pool

Fusion-based AM depends on a heat source melting the feedstock and forming a molten pool of material (Figure 1a) that solidifies upon cooling. However, prior to solidification, there are physical attributes and dynamics associated with the melt pool that are of interest to those who seek to control the process and material state. Among the most important physical parameters and dynamics are (i) the physical shape of the molten pool, including both what lies above the deposition plane that is governed by wetting and what lies below the deposition plane that is governed by heat transfer, thermal gradients, and interface motion; (ii) dynamics associated with a keyhole (if present), including keyhole collapse; and (iii) dynamics associated with convection and the capture/retention of defects within the liquid.

Notably, not all of these attributes and/or dynamics are quantifiable using existing techniques. Even classical measurement techniques, such as the measurement of wetting angle or velocity of solid–liquid interfaces, are often limited to analogs [183–185] rather than the direct measurement of the materials of interest. Some limited ex situ techniques can be used for the direct measurement of the melt pool. Optical or electron imaging can detect the melt pool size and shape depending on the alloy.

2.3.1. Melt Pool and Vapor Plume Monitoring

Whereas for feedstock and energy sources, brief discussions of existing measurement techniques are merited, there is little "classical" work on melt pool size. Thus, there exists a robust body of modern and emerging peer-reviewed work on melt pool monitoring (Figures 2a and 7 and Table 3) in the literature, as it represents an area of active research by many groups [66,71,81,98,186–188]. Not only do many commercial AM systems have melt pool detection and monitoring systems [189], but research groups have developed and integrated their own monitoring techniques into a wide variety of systems. Optical monitoring is the most common approach for both laboratory and commercial systems, as it can be deployed using low-cost complementary metal oxide semiconductor (CMOS) cameras mounted coaxially or off-axis [66,67]. An analysis of optically monitoring the melt pool can include detection methods to separate the melt pool from solidified material, spatter (or small amounts of liquid material ejected from the melt pool), and other sources of optical emissions [66,76]. As the data sets are large, new approaches for automated image processes and quantitative analysis range from simple pixel intensity segmentation [190] to machine learning or neural networks for the automatic detection of melt pool boundaries [68]. In addition to the low-cost CMOS camera detectors, high-speed imaging is also used for melt pool monitoring in multiple laboratory systems [191], although there is a concurrent increase in the size of the data sets and a decrease in ease/speed of data processing.

Thermal and IR-based sensing techniques are also widely explored to conduct melt pool monitoring. Thermal imaging is used in both on- and off-axis monitoring configurations, identical to optical imaging [69,70]. Melt pool size determination is based on the liquidus-solidus transition point and determining if the temperature of a particular pixel or cluster is above or below this transition point [71]. Thermal monitoring data can be used to train machine learning models that are subsequently used to simulate melt pool properties and automatic liquid–solid boundary detection models, as will be discussed later [72]. A limitation associated with thermal and IR sensors is the accuracy of the calibrations to relate the intensity of a pixel cluster with known transition temperatures (e.g., solidus, liquidus). There is also the potential complication of metal vapor plume gases coating optics or other monitoring components [113]. Along with traditional thermal imaging that measures surface temperature during the AM process, emerging work on the use of X-ray radiography to monitor the sub-surface temperatures and melt pool dimensions during deposition has been conducted [192].

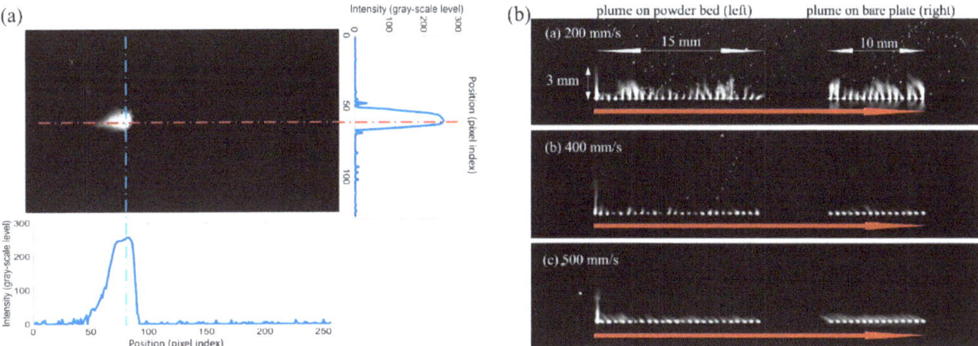

Figure 7. Examples of monitoring melt pool and vapor plume. (**a**) Vision-based in situ monitoring results of melt pool detection in LBPF, (**b**) effect of scan speed on vapor plume in LBPF, the scan direction is indicated by the red arrow. Figure 7a is reprinted from [66] under Creative Commons Attribution License (CC BY); Figure 7b is reprinted from [188] under Creative Commons Attribution License (CC BY).

X-ray-based monitoring techniques are emerging in the AM space. These techniques allow for the real-time viewing of phenomena such as powder mixing within the melt pool [74], the liquid–solid boundary [193], and the evolution of spatter [73]. X-ray backscatter detection can also be used for monitoring the melt pool as well as the plume and spatter during a deposition [75]. However, these methods are limited by the depth of the X-ray, the resolution desired, the richness (size) of the data, and the geometry of the parts being deposited. Thus, these X-ray-based techniques more widely provide in situ data to better understand the governing physics and develop accurate models rather than necessarily providing real-time in situ process monitoring.

Methods other than optical, thermal, and X-ray-based imaging for monitoring the melt pool of AM processes are not as well studied but can still be useful. Ultrasonic measurements can detect melt pool dimensions due to the sharp decrease in the shear modulus and density in the liquid phase [77]. Eddy currents and radio frequency emissions have also been studied for melt pool monitoring [75].

During fusion-based AM processes, after melting the metallic feedstock (both powder and wire), the material continues to heat, and if the evaporation temperature is reached or exceeded, evaporation occurs, forming a vapor plume [194]. Continued heating can cause the metal vapor to become plasma and be added to the plume [194]. The infrared imaging of the plume in PBF processes has been used to support the development of statistical studies to detect melt pool instabilities [78]. Optical imaging is also a popular technique for monitoring the vapor plume during a build and has been studied in multiple systems [79–81]. The plume, consisting of vaporized metal from the melt pool (Figure 7b) [188], provides a physically relevant foundation upon which other advanced monitoring techniques for the chemical composition of the build have been developed, including, notably, laser-induced breakdown spectroscopy techniques (LIBS). This monitoring method has been used for detecting specific elements during laser-based AM builds [82–85]. Schlieren imaging is an optical imaging technique that can monitor density gradients and flow in gases. Schlieren imaging can be used to monitor the shielding gas flow as well as processing by-products and evaporation during the AM process [195]. Optical emission spectroscopy is a similar technique for monitoring the vapor above a deposition that can be used for monitoring the melt pool, surface, and subsurface conditions [86].

Spatter in AM is the ejection of material from the fusion zone. Spatter can consist of liquid droplets ejected from the melt pool or unmelted feedstock that is blown away from the melt pool zone [87,196]. Optical monitoring with image thresholding to classify spatter during a build has been carried out in multiple systems [81,87,88]. Optical imaging often uses image processing to determine the location of the spatter. The optical imaging of spatter patterns can be used to calibrate models to predict spatter locations and distributions in future builds [91]. X-ray imaging can give higher resolution monitoring of the spatter, as the optical emissions of the melt pool and plume do not need to be segmented out from the image for the analysis [197]. X-ray-based spatter monitoring has been conducted in multiple laboratory AM setups [89,90]. Optical imaging is also used to monitor spatter patterns to validate models [91].

Many of these monitoring techniques, including optical and thermal imaging, can be integrated and optimized using techniques such as machine learning, genetic algorithms, artificial intelligence, and others to aid in the analysis of the large amounts of data generated during monitoring, reduce the time required for such analysis, and provide predicted information for future decisions. One example is a study using a deep learning-based approach to create a model to analyze the melt pool in an LPBF system [198]. Other authors have reportedly used integrated imaging combined with a neural network to monitor the melt pool either using optical or thermal methods in DED systems [199,200].

2.3.2. Melt Pool Modeling

As stated earlier, there have been significant advancements in the understanding of melt pools in AM via in situ studies, including X-ray imaging of thin sheets [201,202], which have provided essential information for melt pool modeling (Figures 3a and 8) [18]. Several methods and software packages exist that can model melt pools (Table 4), and they often use FEM or FVM for CFD, e.g., Flow-3D [203]. Flow-3D, a commercially available software package, can model and simulate melt pools to elucidate the influence of processing parameters on defects [155]. Other models for melt pools include smoothed particle hydrodynamics (SPH) [204], which can also be used to help with defect modeling [155]. FEM and similar modeling methods are particularly well suited for heat flow because of energy source–material interactions and are commonly used to simulate and study the complex thermal histories associated with AM processes [205–209].

ExaAM, an effort of the U.S. Department of Energy's Exascale Computing Project (ECP), has created several software packages for AM modeling [210]. The models and software packages that have been created are intended to be integrated together, as seen in Figure 8, which is an integrated thermal and solidification simulation. A few software packages from ExaAM are especially well-suited to simulate and study the melt pool, including AdditiveFOAM, Truchas/TruchasPBF, and ExaMPM [151–153,211]. AdditiveFOAM is built on OpenFOAM [154] and uses CFD to calculate fluid flow and heat transfer in the melt pool and to simulate the solidification but does not include microstructural evolution and is limited to the liquid-to-solid phase transformation. Similarly, Truchas, made for casting solidification simulations, and its powder bed fusion counterpart TruchasPBF use FVM CFD to model thermal histories in AM. The thermal histories simulated in AdditiveFOAM and TruchasPBF can then be used to simulate grain structures through an ICME framework when connected to a microstructure model. This will be discussed in Section 3.3. Another interesting bit of code that has been developed by the ECP project is ExaMPM. ExaMPM uses the material point method (MPM), a variant of the particle-in-cell (PIC) method [153]. ExaMPM is able to resolve the physics and dynamics associated with the complex interactions found in melt pools, including those between and among solid, liquid, vapor, and powder. One exciting possibility is the inclusion of interactions between the laser and matter. This software is intended to be able to model many of the complex interfaces found in AM melt pools (Table 4).

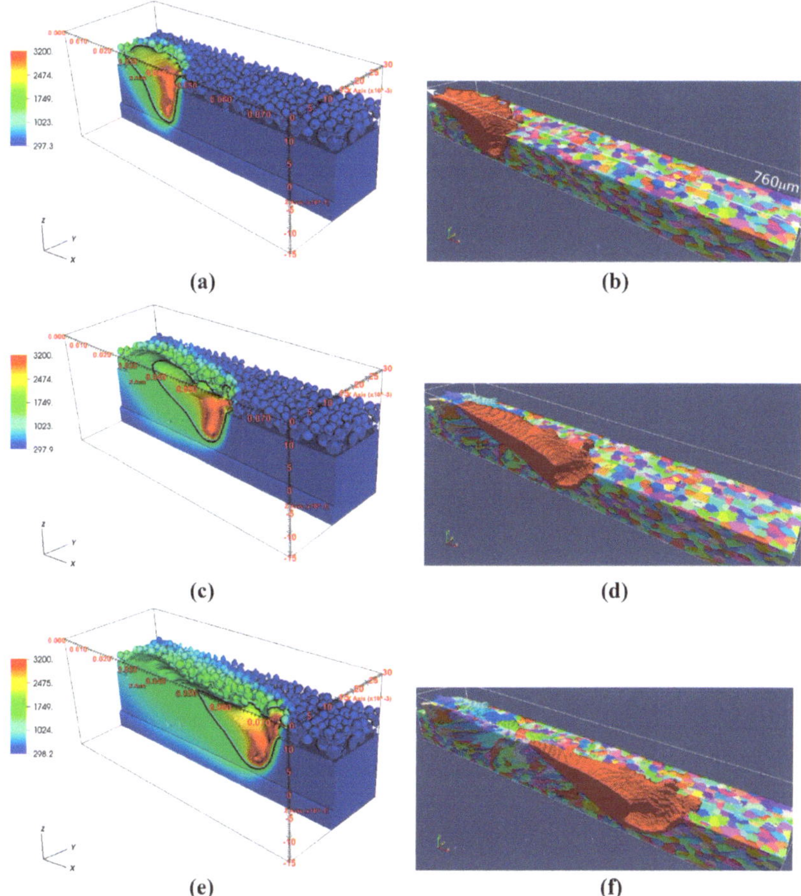

Figure 8. Integrated thermal profile (**left column**) and solidification (**right column**) model at 327 µs (**a,b**), 674 µs (**c,d**), and 967 µs (**e,f**). Reprinted with permission from [18].

As mentioned previously, the interaction of the energy source with the feedstock and substrate material can often lead to dynamic and competitive phenomena, such as vaporization in a plume or ejection of molten material, known as spatter. Since these phenomena are difficult to model and simulate directly, creating simplified models that ignore certain details of the physics to ascertain the result rather than the process is useful [157,212]. One example of a simplification of complicated physics in the energy source–feedstock interaction is the use of the Langmuir equation [156] to describe vaporization and gettering (solute loss and pickup) [19,213,214]. Models of metal constituent vaporization can be tied to thermal models to estimate material and solute loss in AM builds [158]. Other events that result from the dynamic nature of the melt pool in AM include the formation of a keyhole, as well as material ejection [215–217]. These events can lead to chemical differences in the build as a result of process variations [218,219].

2.3.3. Melt Pool Statistics

The use of statistics to analyze aspects of AM processes, such as the melt pool, can involve both the quantification of the uncertainties associated with monitoring data, as well as the validation of models (Figure 4a) [92–94]. Sensitivity analyses consider the various sources of uncertainty that accumulate from a model and how these various sources contribute to the overall uncertainty. Sensitivity analyses of differing AM processes have shown that small changes to machine settings have a considerable effect on the overall result [220]. With the use of thermal cameras to monitor the size of melt pools, sensitivity analyses can inform on which parameters vary the size of the melt pool, which in turn can help validate models [95].

Understanding which elements may preferentially vaporize during an AM process, as well as their relative rates of loss, enables the prediction of the composition of the final build. One study [214] created a model of how much aluminum would evaporate during the Ti-6Al-4V build. Sensitivity analysis was completed on this model as a first step to ensure the total error in model predictions was not greater than the expected results. This was a novel and better approach than was standard at the time, but the model had to be validated with the previous evaporation of aluminum in titanium data. Using a monitoring technique such as LIBS to measure vaporized elements in the plume of the sample could better inform these models, though few researchers have attempted to integrate LIBS directly into an AM system. Further sensitivity analyses could be conducted to understand how varying the process variables (and by extension the melt pool) can influence preferential evaporation.

Gathering statistics from spatter has been accomplished through monitoring the plume [221]. Using a support vector machine or SVM (a form of supervised learning algorithm), variations of the plume frame to frame from an inline infrared camera are tracked. When large variations are present, the material is likely in an unstable state (e.g., spattering, keyholing). By training this SVM on the monitored data from earlier in the build on whether the spatter has occurred, the model can predict not only when the spatter occurs but also the potential creation of defects if the spatter lands back on a susceptible surface. One interesting example of the interplay between monitoring, modeling, and statistics is exemplified by new research into the use of electromagnetic techniques to monitor the plume and spatter. By aggregating spatter data from cameras and considering particle statistics, a new model has been developed, supporting the theory that new instrumentation can be deployed.

One emerging technology that leverages statistical methods is the detection of spatter in real time. A maximum-entropy double-threshold image processing algorithm that is based on a genetic algorithm (MEDTIA-GA) has been used to recognize spatter from monitoring data, and its results were compared to three other traditional threshold segmentation methods (Otsu's method, the triangle threshold segmentation algorithm, and K-means clustering algorithm) [222]. The processing time (and thus the computational overhead) of this novel GA approach was, at worst, an order of magnitude lower than these other methods. The MEDTIA-GA method has also demonstrated an ability to not succumb to segmentation errors such as noise sensitivity, spatter conglutination, and spatter omission [222]. Emerging statistical methods like these are instrumental to both improving confidence in obtained data and making statistical analysis more approachable by reducing computational costs.

3. Material State of Additively Manufactured Materials

The term "material state" of a component is related to the concept of material state awareness (MSA), which is defined as the "digitally enabled reliable nondestructive quantitative materials/damage characterization regardless of scale" [1–3]. From a traditional materials science perspective, this includes but is not limited to the following attributes: composition, solute distributions, microstructure (phases, their size, distribution, and correlations), crystallographic texture, and the presence of defect structures (e.g., dislocations, porosity, interfaces, cracks) across all length scales and couples to more macroscopic non-traditional materials science attributes, such as surface roughness and the shape/topology of the part or component. Each of these attributes (both microscopic and macroscopic) is informed in part or completely by the processing and can have large effects on the properties and performance of the material.

It is worth briefly considering how these aspects of the material state influence the properties of the material. Here (Figure 9), we introduce properties in a manner somewhat analogous to Maslow's hierarchy of needs [223], where, as we proceed from the most foundational design properties (e.g., elasticity and plasticity) to the properties that become critical when materials are put into service (e.g., fracture, fatigue), we correlate them with the most critical aspects of the microstructural state.

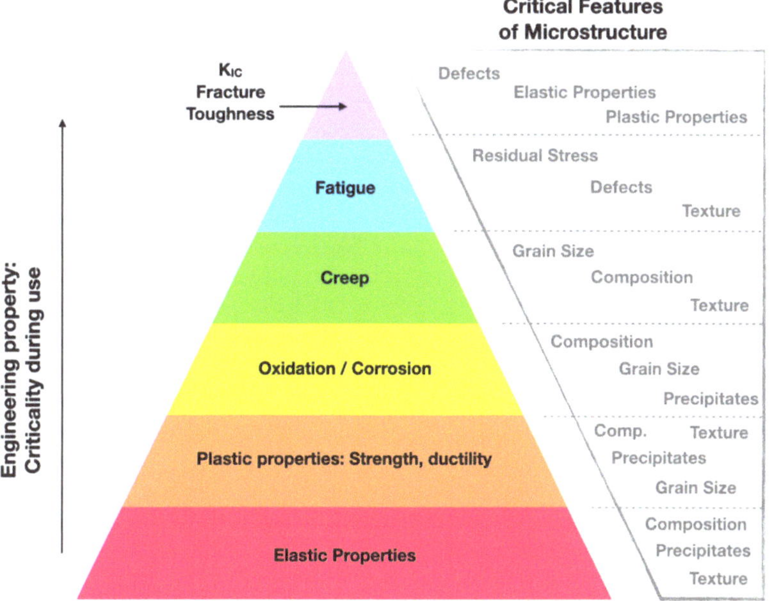

Figure 9. Hierarchy of mechanical properties with associated critical microstructural features (based on Maslow's hierarchy of needs [223]).

From this type of hierarchical map, we can begin to understand the process-structure-property correlations. For example, additive manufacturing can produce a wide variety of defects depending on the processing parameters. Defects can result from extremes in the energy source (e.g., laser power that is either too low or too high), which cause abnormalities in the melt pool, such as keyholing and balling [11,155,224]. Excessive residual stresses originate from the complex thermal gyrations and can cause cracking [206,225–227] either during the build or during relaxation in periods of time (e.g., weeks) following deposition and can have a deleterious effect on many critical mechanical properties, including ductility and fatigue [15,228], even though strength may be increased due to Taylor hardening [138]. Further, residual stresses can lead to distortion and can be very spatially dependent due to complex thermal histories [225,227,229]. While it is tempting to assert that the composition of the part is set by (or equivalent to) the certified composition of the feedstock, through previous discussions in this paper, it is emphatically noted that since AM is a dynamic process, extreme processing conditions can change the composition locally or globally. These compositional changes can further influence the local and overall properties of the material [219]. The microstructure can also have a marked effect on the properties of a material. Given that the microstructure of any arbitrary material is strongly influenced by its thermal history, it follows that the energy source and melt pool represent the most important corresponding processing parameters in AM [20,210,230]. Collectively, these interrelated material state attributes govern the properties (discrete measurements) and performance (statistical distributions of properties) of the material. Therefore, while the mechanical properties are not normally included in the material state, they have been added to this section in this review. Therefore, monitoring and modeling AM builds is important while considering and analyzing the data created statistically to understand how the material state develops during AM and what the resulting material state will be.

3.1. Defects

As with most engineered materials (except, for example, materials with engineered porosity levels), the presence of defects is generally considered to be undesirable, attributing to a general reduction in many properties including mechanical, electrical, and thermal. The AM process can result in a variety of defects, some of which are not found in materials produced using other processes. The defects commonly found in AM (Figure 1b) include porosity, lack-of-fusion defects, delamination, cracking, and balling [231].

Defects within AM can be detected and measured post-deposition through both destructive and non-destructive evaluation (NDE) techniques. Common measurement methods include X-ray imaging, ultrasonic testing, and microscopy [232]. Standards exist for some types of flaw characterizations and techniques for detecting them using NDE, including X-ray computed tomography, eddy currents, acoustic emission, and X-ray backscatter [233]. Classic metallographic characterization is useful for evaluating some types of defects such as lack-of-fusion (LOF) defects, spherical porosity, and hot cracking. Depending on the length scale, these defects are detectable and may be quantitatively measured using either optical or electron imaging techniques.

3.1.1. Defect Monitoring

Monitoring defects within AM is, quite appropriately, an active area of research, as their presence can greatly diminish the properties of the final part (Figure 2b). Monitoring and detecting aspects of these defects (e.g., location, size, type) during a build and developing strategies for the mitigation/elimination/repair of defects are important aspects of ensuring part quality. Setups that are common for in situ process monitoring techniques of a build are similar to those for defect detection within the material state. The coaxial imaging of the melt pool can map porosity volume and location using a trained convolutional neural network, which can then be confirmed using X-ray computed tomography (CT) [96]. High-speed optical signatures during a build can be correlated with ex situ characterization of defects [97]. Optical imaging has been used to monitor balling [98] and detect spatter and holes in a PBF build [99]. Optical emission spectroscopy has been used in some studies for porosity monitoring [100].

Defects including porosity, LOF defects, delamination, and cracking all include either a separation of the layers in a build or the build not being fully dense. As the thermal conductivity of these voids, whether internally under vacuum or entrapped gas, is significantly lower than the surrounding material, measurable differences in the thermal signatures are observed. Thermal imaging has been successful in detecting defects by showing discontinuity in thermal signatures [234,235]. Multiple sensor setups have been developed using a combination of optical and thermal imaging techniques for the purpose of monitoring builds for defect detection [236].

The use of X-ray imaging for defect monitoring is an emerging detection method that typically uses X-rays produced from a synchrotron. While using a synchrotron is not feasible for most commercial applications and systems, as the samples need to be very thin, research has been conducted using this technique to show the physics associated with the formation of different types of defects, such as keyholing or entrapment in low-temperature regions of the molten pool (Figure 10) [237]. The high-speed X-ray imaging of the keyhole threshold and morphology has been studied [101]. Research into keyhole porosity monitoring using a synchrotron has shown ground truth observations with high resolution [102].

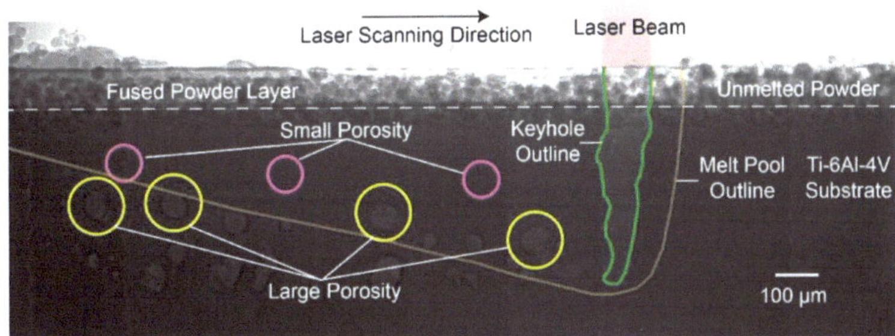

Figure 10. In situ X-ray imaging showing defect formation and melt pool outline detection in LPBF. Figure is reprinted from [237] under Creative Commons Attribution License (CC BY).

Spattering can cause defects in a build such as not fully reintegrating to the melt pool, creating uneven thickness for subsequent layers, and changing the composition of the part, as the spatter can have a high oxygen content [90]. Spatter monitoring for the purpose of defect detection has been successful using high-speed optical imaging during a PBF process [103].

As mentioned in Section 2.3, techniques such as machine learning, genetic algorithms, and artificial intelligence have aided the analysis of melt pools during AM processing. In addition, these techniques have aided in the detection of defects and certain characteristics such as origin, size, and morphology. One study captured and processed high-speed optical imaging, including the signatures of the spatter. Another example is the MEDTIA-GA algorithm, mentioned in Section 2.3.3, which was used to automatically detect spatter signatures [222]. Other examples include machine learning combined with artificial neural networks [238], convolutional neural networks [239], support vector machines [240], and tree algorithms [241] to detect defects during printing. Multi-sensor setups have been shown in LPBF systems to identify defect formation using a deep convolution neural network (CNN) [242]. Monitoring the bead dimensions of a deposition can be paired with neural network models such as a dimensionless artificial neural network (DI-ANN) to be used for defect monitoring in a DED system [243].

Acoustic monitoring, a common non-destructive evaluation technique, uses a sensor to pick up generated elastic waves in a material. Acoustic monitoring setups have been shown in multiple laboratory settings. A study using acoustic monitoring has shown success in detecting LOF defects within a build along with high-speed imaging and photodiode data [244]. Acoustic monitoring of defect events during a build was performed and then correlated to the type of defect [104]. For example, acoustic signatures from cracks have been monitored successfully during a build [105]. Table 3 presents common monitoring technique defects.

3.1.2. Modeling Defect Formation

The most common defects within metal AM builds are LOF defects, keyholing, and spherical pores. Keyholing can be avoided by ensuring the combination of the power source and speed (and thermophysical properties of the material) are such that the energy density does not lead to excessive vaporization [224]. However, energy densities that are too low can also result in the generation of other defects [107]. Within a single AM process such as LPBF, even the laser scanning strategy can change the amount of spherical defects present [218]. Due to the thermal physics and fluid dynamics of the melt pool, as well as the thermal gradients and viscosity of the liquid, pores can become trapped. Spot scan strategies have differing melt pool morphologies, which result in the retention of fewer spherical defects. Thus, modeling the physics of the melt pool is inherently important to the prediction of porosity (Figures 3b and 11) [155], as the thermal gradient present (driving force for pore movement by convective flows) is outweighed by the drag, resulting in pores that are not able to rise to the surface and be eliminated. Such modeling has been conducted with a moderate degree of success [245] by abstracting the shape of the melt pool to have one "depth" parameter, but due to the three-dimensional morphology of melt pools, using more complex shape parameters, as well as including diffusion of heat from previous passes, can produce results that are closer to reality [224].

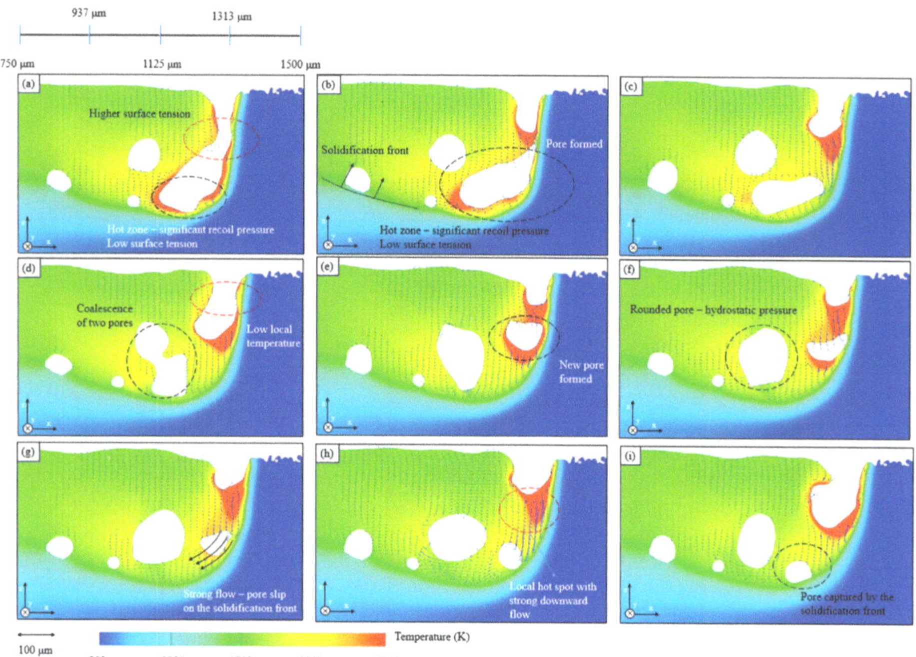

Figure 11. Finite volume method (FVM) model of melt pool with keyholing and pore formation over time (**a**) t = 2.395 ms, (**b**) t = 2.4 ms, (**c**) t = 2.415 ms, (**d**) t = 2.43 ms, (**e**) t = 2.445 ms, (**f**) t = 2.45 ms, (**g**) t = 2.455, (**h**) t = 2.47 ms, and (**i**) t = 2.495 ms. Figure is reprinted from [155] under Creative Commons Attribution License (CC BY).

3.1.3. Defect Statistics

The statistics associated with the defects (Figure 4b), including the conditions under which they form, their formation frequency, and their geometric metrics (i.e., size, proximity, and location), are necessary to predict the performance of any given part [63]. Considering the potential quantity of parts that could be manufactured for industrial application, proper statistical treatments enable the systematic study and implementation of strategies to reduce/eliminate defects, design topologies/shapes to meet the expected service demands (e.g., mechanical loads) [106], or even classify parts as "acceptable" or "reject" based upon monitoring signatures, especially once paired with machine learning algorithms [239]. These methods are necessary due to the variability that metal additive manufacturing is subjected to. For instance, one case study looked at the repeatability of an SLM printing process and found it to be acceptable under the guidelines of engineering standards for dimensional and geometrical analysis [246]. However, a large-scale industrial study [247] of manufactured tensile samples found that nearly two percent of samples, though nominally geometrically similar, failed catastrophically. These failures were correlated with parts in which there were clusters of LOF defects whose configuration significantly degraded the structural integrity of the test coupons. The Beese Research Group [248–251] has published pioneering research that incorporates full stress triaxiality and the Lode angle parameter to develop models to understand how defects (and defect clusters) interact with the design/topology of the part to understand deformation and failure. Regarding in situ quality control, the probability of some attribute of defect formation (e.g., size, number, location) can be correlated with whether an anomalously poor property might be expected, and thus, the part should be categorized as a failed part prior to use [245,252,253]. After studying the concentration of defects within a build process, a probabilistic simulation, such as a Monte Carlo simulation, could be run to determine the percentage of parts that match differing thresholds for the probability of success/failure [224].

Notably, such statistical approaches can incorporate data from modeling, measurements, or a hybrid of both data types. For example, in situ monitoring in combination with a machine learning algorithm could be used to set thresholds for differing variables such as defect size, location, and the density of the defects being detected. This has been demonstrated by optimizing the energy density of prints to minimize the formation of defects such as pores, lack of fusion defects, and keyholes [254–256]. This combination could also statistically determine when enough defects are in proximity to one another for catastrophic failure to occur during printing or during operation.

3.2. Compositional Distribution

In metals and alloys, the chemical composition has a very significant effect on the mechanical, electrical, thermal, and physical properties. Thermal processing and solidification can lead to chemical segregation within a part [253]. Further, in additive manufacturing, different elements have different propensities to evaporation (or gettering), leading to selective loss (or gain) of elemental species, resulting in compositions (Figure 1b) in the build that differ from the feedstock input [257].

Classic ex situ approaches to measuring the composition of AM samples include energy dispersive spectroscopy (EDS), X-ray fluorescence (XRF), inductive coupled plasma (ICP), or X-ray diffraction (XRD). These techniques are often limited to some degree. For example, EDS, XRD, and ICP are often destructive. Similarly, EDS and XRD require particular details to be met for each specimen, whether it is a polished surface (EDS) or size/flatness (XRD).

3.2.1. Elemental Monitoring

The elevated temperatures in the melt pool can cause substantial metal vaporization [217]. Monitoring chemistry then becomes important for understanding and predicting the compositional distribution in the build during printing (Figure 2b, Table 3). As mentioned previously, LIBS is a characterization method based on detecting a light spectrum of metallic vapor (Figure 12) [82]. A study using a custom in situ LIBS setup monitored varying compositions of WC in a NiFeBSi matrix [82]. This study showed direct LIBS measurements at the melt pool as well as post-deposition trailing for the cladding head on the hot but solidified metal. Optical emission spectroscopy (OES) is similar to LIBS where the vapor is monitored and different light wavelengths are calibrated to different elements, but in traditional LIBS, the solid is directly ablated to vapor, while OES can be calibrated to monitor the metal going from the solid to liquid phase (melting) and liquid to solid phase (evaporation) [110]. For example, the vaporized metal composition has been monitored using laser-induced plasma emission spectroscopy during a Ti-Al binary AM build, showing the Al content at different locations [84]. Using OES, the preliminary results showed that differences in composition in a Ni-based alloy could be measured [110]. Other research involving OES monitoring was conducted in a Cr and tool steel system [258].

Although the most promising in situ monitoring techniques for compositional analysis are based on LIBS or OES, alternative technologies have been explored. For example, researchers have demonstrated that a mass spectroscopy-based technique can measure the ejecta and atmosphere near the melt pool [108]. Using custom setups, energy-dispersive XRF has been shown to monitor Cr and Ni in the vapor produced during an LPBF process [109]. In the future, AM processes using an electron beam could use existing electron microscopic techniques, including either an in situ EDS detector or a backscatter detector to monitor the composition of the deposition with either a wide area averaging or spatially resolving compositional fluctuations within the build. Some other characterization techniques such as acoustic or ultrasonic testing could be applied to in situ monitoring setups if the composition varied enough to influence the elastic properties of the alloy.

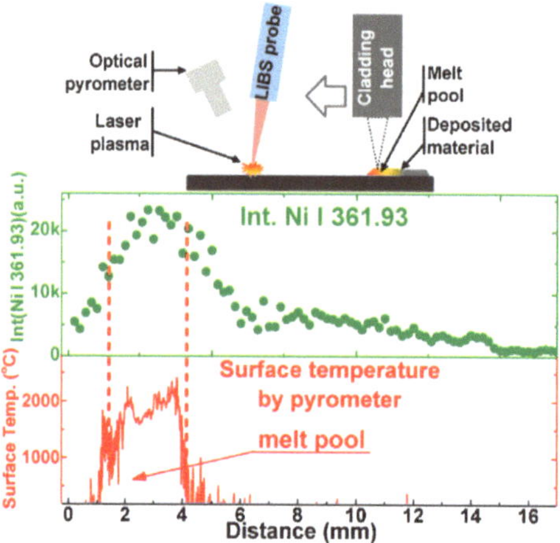

Figure 12. Schematic of laser-induced breakdown spectroscopy (LIBS) to optimize data collection of composition and temperature. Figure is reprinted from [82] under Creative Commons Attribution License (CC BY).

3.2.2. Modeling Composition

As O'Donnell et al. [219] showed, processing parameters can lead to changes in the local or overall composition of an AM build. Therefore, understanding how the energy source and processing parameters influence the composition is important. As stated in Section 2.3, there have been efforts to understand vaporization and gettering using the Langmuir equation [19,156,213,214]. While this equation has been used to identify and understand trends in compositional changes based on thermal histories [158], it has not been applied to understand how processing parameters can influence the local composition of a build (Figure 3b). The thermal [259,260], solidification [219,261], and mechanical properties [219,262] of a material can change with changing composition. Understanding and modeling compositional changes as a result of the processing parameters may improve the accuracy of future models of AM parts that seek to understand local microstructure and properties.

3.2.3. Statistics of LIBS Data

Accurately identifying the composition of local areas within a part is inherently useful to ensure the part is built to specification [111,112]. As mentioned, the primary in situ monitoring method used to obtain this data is LIBS. To analyze LIBS data in real time, however, spectra from the samples need to be generated, and characteristic peaks for differing elements need to be found and deconvolved, which can be nontrivial. Similar confidence intervals could be generated, as was mentioned for feedstock analysis, but another method could be implemented to reduce the computation time of deconvolving peaks. Using prior data of spectra from both the pure elements, as well as binary or higher-component alloy standards, independent regions of each element can be flagged [263]. The intensities associated with the spectra of the pure elements can be compared to the intensities of the standards, permitting the creation of a function that maps spectral signatures with the composition of the measured specimens. Further, a machine learning algorithm might be developed to ingest real-time LIBS data, then use this heuristic to not only analyze spectra faster but also provide information regarding statistics of specimens over builds. With even minimal updates to this methodology or computational power, this would be close to achieving real-time analytical speeds that would allow the full compositional distribution of data throughout the build process (Figure 4b).

3.3. Microstructure and Texture

The microstructure of metal parts made using fusion-based AM processes can be vastly different from traditionally manufactured materials and can vary significantly between different processes. The microstructure is often set by the solidification rate and the thermal gradient, which are governed by the equation $R = (1/G)(\partial T/\partial t)$, where R is the solidification velocity, G is the thermal gradient, and $\partial T/\partial t$ is the cooling rate. The microstructure can also evolve over time after solidification due to phase transformations and heat accumulation.

Measuring and characterizing the microstructure of AM samples is commonplace. Optical and electron microscopy, XRD, and EBSD are all methods used to image and measure the microstructure of a sample post-deposition. Sample preparation techniques are important for all of these methods in order to ensure accurate results. Microstructural measurements can include the phases present, fraction of the phases present, grain or feature size, grain or feature geometry, and texture (Figure 1b).

3.3.1. Microstructural Monitoring

The microstructure of an AM build governs the material properties [138,264–267]. Owing to various reasons (e.g., practicality, time, chemical handling, technological limitations), most monitoring techniques cannot directly observe the microstructure of a build (Figure 2b, Table 3). In addition to the most obvious limitations associated with implementing "traditional characterization" in situ—for example, the difficulties of polishing, chemically etching, and imaging a microstructure to mimic optical microscopy—any microstructure that could be revealed from a surface characterization technique may not necessarily correspond to the microstructure at that precise location following the completion of an arbitrary build. Indeed, microstructures are known to evolve due to remelting and subsequent thermomechanical gyrations, phase transformations, and microstructural evolutions. Despite these difficulties, there exists a strong base of knowledge from the casting and welding communities, as well as arising from computational efforts, to provide the materials scientist guidelines to set their process parameters to affect particular material states. For example, for the thermal gradient ($G = |\nabla T|$) and solidification rate ($R = (1/G)(\partial T/\partial t)$), interfacial velocity can be used to predictably set processing conditions to either promote columnar grains exhibiting texture or equiaxed grains without texture [268–271]. Even with this knowledge, the aforementioned issues lead to most monitoring techniques targeted at observing the microstructure using indirect measurements to classify the microstructure of an in situ build. Thermal imaging has been used in a few laboratory setups to observe the build and calibrate thermal data to microstructural formation and evolution. For example, in an electron beam PBF system, IR thermal imaging was used successfully to observe the build and the thermal gradients and match the IR intensity data to the microstructure at different locations in the final part [113]. A system using a combination of sensors including an IR camera was able to monitor melt pool temperature, real-time cooling rates, and thermal maps and then, using this data, was able to create a closed-loop system to control the microstructure by controlling the cooling rate of the deposition [114]. Similar approaches have been used for general microstructural details such as grain size based on melt pool size and other thermal characteristics of an AM build [115,118]. Ultrasonic inspection techniques enable the inference of details of microstructures, such as an assessment of the phases present, their fraction, grain sizes, and whether the crystallographic texture is present, as differences in each result in attending variations in the measured signals, and, when anisotropy is present, result in directionally dependent responses, including in the wave attenuations [116,117]; however, acoustic waves are generally interpreted in a more general or bulk manner rather than a site-specific manner. Perhaps the most direct method is an emerging approach based upon spatially resolved acoustic spectroscopy (SRAS) [131–134], which has been applied to the surfaces of additively manufactured specimens, demonstrating that it is possible to spatially map texture from as-deposited surfaces [272]. Similarly, for hybrid variants of AM, there is a possibility to use force feedback machining to map grains [135–137], though there are several engineering challenges to deploying this at the time of this paper.

3.3.2. Modeling the Material State

The execution of complex models of different AM processes requires multiple iterations of calculations at a wide range of time or length scales, from predicting local thermal history to simulating the effects of residual stress. However, errors and uncertainties can transfer and compound rapidly between each of a series of integrated computations. Therefore, a methodology to quantify the error in models and ensure the models' validity is necessary. A group has created such a process as a part of the Exascale Additive Manufacturing project (ExaAM) [273] (Table 4). Upon modeling the deposition of an AM sample using their Frontier Exascale Computer, select regions were chosen to predict their properties, and experimental samples were made to compare to the simulation results. After running their uncertainty quantification algorithm, they predicted that their compounded error should not be statistically significant, which, if correct, would help validate their model. Comparing experimental data to their simulations showed that they correctly predicted that the error was not statistically significant. While this scale of computing remains limited to only a few machines, their computing power makes quantifying errors in models possible, and as technology continues to mature, the accessibility by more researchers to such machines should increase. Further, once a limited number of highly complex simulations are executed, it is possible to develop and use simpler, surrogate approaches. While new models are being developed constantly, standardizing how the data are created and stored is important. One way that this can be conducted is by using the software DREAM3D. Created as a data pipeline to support the reconstruction of serial-section experiments in the earliest days of 3D materials characterization, DREAM3D expanded into software that may be used to build synthetic microstructures, generate surface meshes, and help with microstructural quantification [274]. There are other methods that store microstructural data, such as electron backscatter diffraction (EBSD) data and the statistical analysis of grains and their size and texture. There is extensive software available for visualizing EBSD data, such as the MTEX toolbox for MATLAB [275].

Regarding the foundational details of AM microstructures, there is much to build upon, as primary solidification microstructures have been studied for many decades (Figure 3b). There is a solid basis for knowledge related to solidification structures and their relation to processing parameters for traditional manufacturing methods such as casting and welding. A portion of this knowledge can be readily applied to AM. For example, one crucial aspect that determines the microstructure, as mentioned in the Monitoring subsection, is the G/R ratio. At low G/R ratios, equiaxed grain growth is often observed, whereas at high G/R ratios, columnar grain growth is found [167,276]. The G/R ratio can change the overall microstructure of a part, as seen by Dehoff et al. [271]. Solidification speeds can also change the microstructure within layers, known as the columnar to equiaxed transition (CET). The CET is a phenomenon where the solidification structure starts as columnar growth and ends as epitaxial growth, forming equiaxed grains [277]. Another important factor that governs the microstructure is the thermodynamic stability of phases. Some common AM alloys are inherently two-phase alloys, such as α/β titanium alloys and nickel superalloys [16,138]. Other alloys can form deleterious phases at intermediate temperatures, such as stainless steels [278]. Therefore, it is common to see CALPHAD and solidification models, such as Scheil-Gulliver models, intertwined with other microstructural models to serve as a basis for how a microstructure evolves in a dynamic process [14,279].

There are several ways that the solidification and microstructure of AM can be modeled. The most common types of models are those based upon kinetic Monte Carlo (KMC), cellular automata (CA), or phase field approaches (PF) (Figure 8) [18,175,280]. In general, PF has the best ability to connect to thermodynamics and kinetics in real life but is also the most computationally intensive of the three models. KMC uses randomness to model microstructural growth and is not very computationally intensive but is hard to connect back to physical representations. CA offers a compromise on both fronts, as it can connect to real life through its model formulation and is often not as costly as PF models. This compromise is the reason that CA has become one of the most popular methods for solidification modeling in recent years [165,281–283]. Texture is commonly modeled using CA and PF for AM, with recent introductions of texture modeling with KMC [175]. As with many models, it is common to see experimental validation of models that are created and reiteration of experiments with data that is gleaned from simulations. One example is the modeling, simulation, and experimental validation of a Ti-xW binary alloy and its solidification structure [284,285].

Several codes have been created to model and simulate AM microstructures from ExaAM. These microstructure codes are ExaCA, AMPE, Tusas, and MEUMAPPS-SS. These codes use a variety of model types and have unique microstructural aspects that they are designed for. ExaCA is a CA code that simulates grain growth [160]. CA is a relatively simple model to implement and has relatively low computational cost using one of several mathematical formulations, such as finite element (FE) and finite difference (FD) [175]. AMPE and Tusas are PF codes that are designed for simulating subgrain microstructural features during solidification by solving systems of partial differential equations [161–163]. MEUMAPPS-SS (Microstructure Evolution Using Massively Parallel Phase-field Simulation for Solid State) is a PF code written originally in Fortran that was converted into C++ and focuses on simulating solid-state transformations that occur as a result of the complex thermal histories seen in AM [164]. The US Naval Research Laboratory (NRL) has also created AMCAFE, a CAFE (Cellular Automata Finite Element) model to simulate the solidification in AM [165]. The model has been validated using a laser PBF 316L build.

Kinetic Monte Carlo microstructure models have also been made for AM [159]. SP-PARKS (Stochastic Parallel PARticle Kinetic Simulator) was created at Sandia National Laboratories to pioneer a way to model microstructure. SPPARKS has modules specifically for AM applications and uses the spin-based Potts model by introducing probabilities to change the spin [286,287]. While KMC can simulate texture, SPPARKS cannot. As AM parts are highly textured, leading to anisotropic properties [288], the modeling and simulation of texture formation and evolution becomes important. Of the above codes that are mentioned, ExaCA and AMCAFE also are capable of simulating texture in AM. However, there is ongoing research to extend texture modeling to more methods, such as Monte Carlo (MC). For example, Pauza et al. [166] used an MC Potts model to model texture in additive manufacturing using Inconel 718 [165].

3.3.3. Microstructural Statistics

Predicting the microstructure of any as-built part relies on having detailed knowledge of relevant aspects of the selected AM process conditions, as well as the material used [111]. An aspirational goal is to integrate monitoring data and modeling to predict local microstructure, its evolution, and, consequently, local and global properties, and performance (Figure 4b). Such a goal has been realized to different degrees of success in which researchers have integrated thermal models and physical processes such as evaporation and microstructural evolution to predict microstructure and have used sensor data to calibrate the models [115,213,289,290]. An example of a more fully integrated workflow is the ExaAM project that has produced multiple different models targeted towards metal AM processes, one of which is aimed towards predicting local microstructure; however, the problem that exists in all models, i.e., verification, remains reliant on the use of statistics, thus requiring additional computational resources. One open area of potential research is the use of statistical methods to understand the uncertainty and fidelity of difference monitoring approaches when implemented into AM systems.

3.4. Residual Stresses and Distortion

Residual stresses and distortion are inherent in AM processes (Figure 1b). The complex thermal histories that exist spatially and temporally within and among layers can lead to complex stress states. Distortion is a direct cause of these residual stresses, leading to the deformation of the material from the intended shape of a part due to the thermal stresses that exist during the manufacturing process.

Classic approaches to residual stress measurements use techniques such as XRD, neutron diffraction, ultrasonic techniques, and measuring relaxed strain through the hole drilling method [291]. Traditional measurement techniques use a powdered sample or a polished surface on bulk material or are semi-destructive to the sample. Depending on the level of stress within a sample, measuring geometric features or final dimensions can show distortion.

3.4.1. Monitoring of Distortion and Residual Stress

The line-by-line, layer-by-layer nature of metallic-based AM necessarily means that the material experiences gyrating thermal gradients and cyclic heating and cooling. Such gyrating thermal gradients and cyclic thermal cycles result in complex residual stresses that impact the mechanical properties of the final part (Figure 2b), such as strength and fatigue life, and can lead to cracking or distortion [225,226,229]. Efforts to minimize or control residual stresses in AM usually involve preventing steep temperature gradients by preheating the build/deposition/powder bed or applying different scan strategies [225]. Residual stress can be characterized using both destructive and NDE techniques. Some NDE techniques involve diffraction-based methods such as XRD, by measuring beam broadening that indicates lattice distortion of the crystal structure of the material and determining dilation in any direction. However, the need for a perfect stress-free control sample to obtain baseline lattice spacing is difficult during an AM process leading to diffraction-based methods being less suitable than other residual stress monitoring techniques, such as ultrasonic measurements [127]. Ultrasonic monitoring includes a variety of methods in multiple systems [119] and is viewed by many as a more viable approach to monitoring residual stresses than diffraction-based methods. If the residual stresses are large enough, the part can experience significant distortion, and under extreme cases can fracture or fail while still attached to a build plate. This correlation between distortion and residual stress provides a basis for monitoring distortion using approaches such as digital image correlation (DIC) [292]. Distortion measurements can use single-point laser displacement sensors (LDS) or full-field (2D or 3D DIC) imaging to monitor the distortion of the build or build substrate (Figure 13) [120–122,128]. Distortion measurements based on DIC use multiple points on a sample, tracking their relative displacements to provide distortion information in at least two dimensions, though more advanced 3D systems have been used to monitor the build plate during steel cladding tests [123]. Further, a system has been developed to show the full-field displacement of a laser cladding system using a partially fixed build substrate allowing distortion [124]. Distortion has been evaluated, as each new layer has been deposited [125]. Importantly, the natural inconsistencies on the surface of a DED build were used to show the full-field strain of a thin wall [126], as opposed to the typical DIC approaches, which require the application of artificial, stochastic speckle patterns. For monitoring, residual stress distortion-based systems are more widely studied and are the most feasible of the current sensing technologies (Table 3).

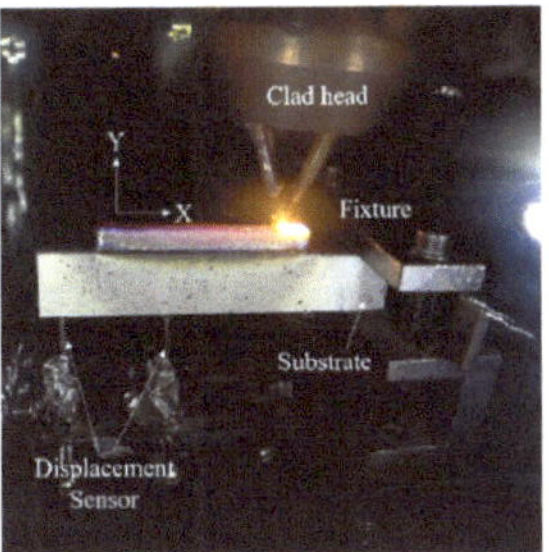

Figure 13. Experimental distortion detection setup used in a laser AM setup. Figure is reprinted from [120] under Creative Commons Attribution License (CC BY).

3.4.2. Modeling and Understanding Residual Stress and Distortion

There are a multitude of reasons to model and predict distortion and residual stress, ranging from (i) the desire to produce a part in as near to net shape as possible, (ii) the desire to avoid post-deposition stress relief [293] treatments, and (iii) the need to mitigate/eliminate risks associated with sudden fracture/failure events.

There has been significant research in the last decade to create models that predict residual stress and distortion (Figures 3b and 14) [227]. FEM is commonly used to model residual stress and distortion due to its common use for mechanical models. Since residual stress is a result of the large thermal gradients in AM, models require thermal information derived from heat source models [128,179,294], which were discussed in Section 2.2 [225–227,229,295]. Over time, these models have developed more complexity, considering phase transformations that can negate significant amounts of residual stress [178,296]. There are several commercially available software packages that estimate the residual stress and distortion of AM parts (Table 5) [297–304].

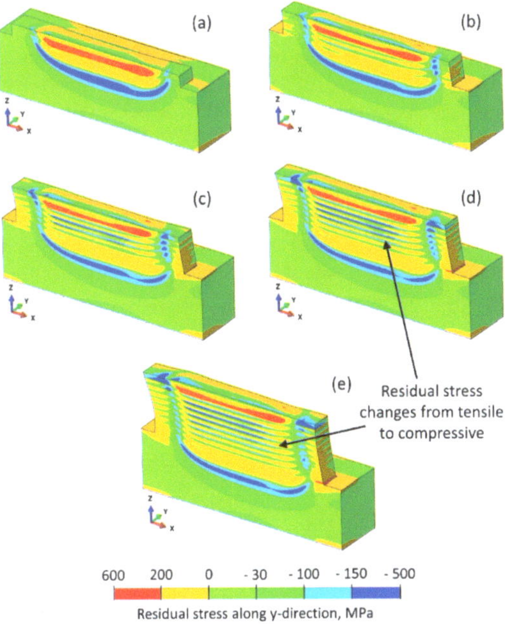

Figure 14. Modeling of residual stress in an additively manufactured sample at different layers. Figure is reprinted from [227] under Creative Commons Attribution License (CC BY).

Table 5. Commercially available AM modeling software packages in alphabetic order.

Name	Company	Modeled Features	References
3DEXPERIENCE	Dassault	Thermal profile and stresses	[297]
Additive Print	ANSYS	Distortion and residual stress	[298]
Amphyon	Oqton	Thermal profiles, distortion, and residual stress	[299]
COMSOL Multiphysics	COMSOL	Coupled thermomechanical models	[300]
GENOA 3DP	AlphaSTAR	Defects and distortion	[301]
Netfabb	Autodesk	Thermal profiles and distortion	[302]
Simufact Additive	Hexagon	Thermal profiles, distortion, and residual stress	[303]
Sunata	Atlas 3D	Thermal profiles and distortion	[304]

3.4.3. Statistics of Distortion and Failure Prediction

The statistical approaches available for predicting distortion are very immature (Figure 4b). Such immaturity is due, in large part, to the nature of distortion and its strong dependence on not only process parameters and thermal gradients but also how those thermal gradients couple with the part topology through the full 3D stress and strain tensors [129,296,305]. It is essential to recall that residual stress is not an average singular value but depends upon lattice distortion gradients in the object. Thus, residual stress is not even directly tied to dislocation densities, though this is erroneously assumed at times. A part can have a very high dislocation density (and thus strength) but have few if any gradients in the dislocation densities. Conversely, steep gradients can exist in parts with lower average dislocation densities. Thus, residual stress is the descriptor of a spatially and directionally varying metric, and statistical approaches are less well developed, resulting in an increase in the quantity of data and complexity of the analysis [125,129,130].

With these difficulties noted, there are opportunities. Residual stress can often be reduced to an "average" or "maximum" residual stress within a part. Under such treatment, the application of many existing statistical approaches is appropriate. It is likely that the incorporation of modeling, including full FEA analysis with stress triaxiality and Lode parameter angle, will result in the most useful information that might then be analyzed using various statistical approaches. Conversely, a new statistical method has been recently leveraged that relies on building a Bayesian model from small sample builds to inform the in-plane deviation from distortion of new parts [306]. As most AM builds can be abstracted to simpler two-dimensional cross-sections, this model studied the distortion seen in differing cross-sections to inform a predictive model that could be applied to more geometrically complex shapes to see how much in-plane deviation would be found in final builds. This method is limited to individual cross-section analyses, so distortion in the build direction can go unseen but can be a powerful tool when paired with other techniques. In principle, if the training data are large enough, eventually there could be a machine learning algorithm made that takes in that data and suggests how the part's geometry should be altered ahead of time so that the distortions are either minimized or lead to correct geometry in the final build.

3.5. Mechanical Properties

One of the important performance metrics for metallic parts made using fusion-based AM processes is the mechanical properties. Metal AM parts often have a vastly different material state and therefore different properties than wrought or cast parts. Some of the important mechanical properties are the (i) tensile properties, including yield strength, ultimate tensile strength, and ductility; (ii) fatigue properties, which are related to both the rough surfaces produced by AM processes as well as the residual stress; and (iii) fracture toughness.

Measuring the mechanical properties of metal AM samples is commonplace in materials characterization. The ex situ testing of AM samples can be conducted to obtain elastic and plastic properties, strength, hardness, toughness, and impact resistance. There are many standards for measuring the mechanical properties of metal samples with a wide range of instruments [307–315]. Most classical testing methods are destructive or alter the sample in some way, so most measurement techniques are not suitable for in situ monitoring.

3.5.1. Monitoring of Mechanical Properties

A primary objective in the manufacturing of any arbitrary part is that the mechanical properties meet the design properties. Given the complexity and multi-scale nature of the process, it may be necessary to use a variety of in situ monitoring techniques to develop sufficient information to predict properties. Most of these systems monitor an aspect of the build that will affect the final properties (e.g., the melt pool stability, thermal gradients, defects, feedstock, and composition). Ultimately, monitoring information should be used to inform models to predict mechanical properties. Intriguingly, some mechanical properties would be possible to monitor during a build. For example, acoustic methods may be used to measure elastic modulus. However, even the use of acoustics for determining elasticity is not without difficulties, as the existence of thermal gradients during deposition necessarily means the elastic properties vary significantly throughout the part and are time-dependent. Similarly, hybrid additive/subtractive approaches have the potential to infer other mechanical properties by measuring machining forces [135,136]. However, this too is likely to be, at best, a time- and temperature-dependent surrogate measurement technique. Thus, in the main, there are no techniques to directly measure properties during deposition (Table 3).

3.5.2. Modeling of Mechanical Properties

As noted previously, research from the welding community has provided a critical foundation for many models, including, in some cases, the mechanical behavior. While these models are generally empirical in nature, the fact remains that the mechanical properties often important to welding are also important to AM applications, such as yield and tensile strength, ductility, and impact toughness. Many of these models are based on the microstructure of the samples. These models are generally highly dependent on the material. For example, duplex stainless steel's impact toughness depends on the phase fraction of austenite, which is dependent on the thermal history [288,316]. Other microstructural features influencing properties are composition, inclusion content, and acicular ferrite content in high-strength low alloy (HSLA) steels [276]. Similar to welding research, empirical relationships have been investigated for AM materials, such as Ti-6Al-4V [138]. These relations often consider the composition and microstructure of each material class.

The finite element method (FEM) has emerged as a preferred method for modeling solid mechanics. Therefore, many codes developed for predicting and modeling mechanical properties are based on FEM. One type of FEM that is prominent to local properties is crystal plasticity FEM (CPFE). CPFE uses slip as the primary mechanism for stress compensation, taking the form of dislocations. CPFE has been applied to many different microstructures and materials, and there are a variety of constitutive models that can be used in CPFE [317].

As a part of ExaAM, two codes, developed specifically for use in AM, are ExaConstit and Diablo [318,319]. ExaConstit is a crystal plasticity FEM code using a Newton–Raphson scheme. ExaConstit takes the microstructure data from the codes mentioned in Section 3.3 and thermal history data from the codes mentioned in Section 2.2. Diablo is a code intended to be the first step in the ExaAM workflow and provides the initial thermal history with additions of various constitutive models for mechanical behavior [210,305,319,320].

Since there are many features of the microstructure spanning many length scales that influence the properties of a material (grain size, porosity, phases, precipitates, and composition) [7,17,138], the integration of several models is necessary to gain an understanding of how the process influences many of these aspects. Model integration can help build a picture of the part to inform material property predictions [17–23]. This is, in essence, the reason for the advent of ICME, which assists with manufacturing parts with well-understood properties by taking several models that work on several length scales and can be used to create an informed simulation of a build and its properties. See Tables 4 and 5.

3.5.3. Statistics of Mechanical Property Models

Modeling and predicting mechanical properties, such as yield strength, often involves the creation of a constitutive equation, as has been the case for AM Ti-6Al-4V for example [25,26,47,138]. The variables that compose such constitutive equations are, by definition, variables associated with the material state, such as composition, texture, phase fractions, or dislocation densities, and continuing to improve monitoring efforts will only better inform models such as these. One novel approach to determining statistical information, such as "design allowables", used these constitutive equations and their statistical distributions to predict, probabilistically, cumulative distribution functions, not breaking the constitutive model. This methodology, called distribution translation and rotation (DTR), has been shown to be effective in calibrating yield strength models for AM Ti-6Al-4V [25], and, as long as constitutive equations are made for their corresponding properties, could be extended for use in calibrating models for other mechanical properties built via other manufacturing methods.

4. Summary

This paper has reviewed the monitoring, modeling, and statistics associated with additive manufacturing of metallic systems. Throughout, the case has been made to consider multiple interrelationships, including (i) the importance of considering monitoring, modeling, and statistics concurrently; (ii) the importance of considering additive manufacturing from the perspective of both the process and the material state; and, (iii) the complex interdependencies that exist between and among aspects of the process (e.g., energy, feedstock, plume) and material state (e.g., composition, phase fraction). The three pillars of this paper—monitoring, modeling, and statistical analysis—are summarized briefly below.

Monitoring can be used to (i) help determine the presence of defects that would make the part unusable for operating conditions, (ii) provide data to optimize future parts, and (iii) provide information to create models that represent a specific aspect or characteristic of the part. By carefully considering the signals that may be naturally generated from the location being monitored, one can not only determine which technique is most appropriate to deploy but also suggest integrating techniques that are either used in other manufacturing sectors or new techniques entirely. New computational techniques used for monitoring in AM, based on machine learning or genetic algorithms, are proving to be valuable tools to analyze large data sets and provide quality assessments of the builds as well as inform statistical models. While some monitoring techniques have a level of maturity that can be easily applied to both laboratory and commercial systems (e.g., optical monitoring of melt pool and defects, or heat flow), others are considered emerging technologies (e.g., feedstock or distortion-focused monitoring), or only suitable for specific experiments and development efforts (e.g., synchrotron).

Modeling offers a way to understand the complex mechanisms that exist within AM processes. There are many models that have been used for AM. The first models that were created that are applicable to AM are moving heat source models. These have increased in complexity over the years and now can be extremely complex and thorough finite element models. The other part of AM processing that is modeled is the melt pool to understand how the energy source interacts with the feedstock and base material. These processing models are useful to understand how processing influences the material state. Therefore, to understand the material state better, models have been created that demonstrate the formation of defects, compositional variations, microstructure, residual stress, and distortion. Understanding how defects form from melt pool interactions and thermal histories that result in high residual stress is an important part of AM research, and models can assist with this research. Microstructure is influenced by thermal history. Residual stresses and distortion have been consistently researched. However, more research is needed to create better-informed models for future builds. Commercial software is available that models residual stress and distortion, especially for production manufacturing. Modeling is often performed with the intention that the information gathered can be used to understand the properties and performance of the material. Therefore, with validation, these models are useful to understand how properties are impacted, leading to the leverage of integrated computational materials engineering (ICME) to create well-informed builds quickly.

The statistical treatment of data, including the emerging in situ monitoring data associated with AM processes, is not only necessary to better understand the process and develop a fundamental understanding of the physics in operation but is also improved by integrating data into and from models. The process of better collection and better use of collected data creates a positive feedback loop where improved monitoring techniques help improve models, and improved models suggest what else needs monitoring. Data from monitoring processes can be used to better understand the system, as well as establish statistical methods such as confidence intervals to estimate how closely the monitoring data could be representative of the ground truth. As the real-time data analysis of monitoring increases, the use of machine learning to identify failed prints prematurely presents itself as both a time and material cost savings. For the modeling side, statistics can be used to bolster mathematical models to calibrate mechanical properties models of additively manufactured parts. The efforts behind model-informed qualification are pointing the way to approaches to accelerate the adoption of AM processes to manufacture parts for a variety of applications.

Each of these different threads has been treated relative to the process and the material state. However, there are only a few large research activities that seek to integrate subject matter expertise across the three domains covered by this paper. This gap arises largely due to the competition between requisite technical depth to advance each discrete approach (e.g., new sensor packages integrated into physical systems, computational algorithms, and physical architecture for on-board modeling incorporating real-time signals, accelerating AI/ML approaches) and the need to think broadly about the fundamental interdisciplinary nature of the problem. New research will be achieved by teams containing subject matter expertise that is only possible by working across research organizations.

Author Contributions: Conceptualization, G.A.J., M.M.D., J.T.Z., M.J.Q. and P.C.C.; methodology, G.A.J., M.M.D., J.T.Z., M.J.Q. and P.C.C.; investigation, G.A.J., M.M.D., J.T.Z., M.J.Q. and P.C.C.; resources, G.A.J., M.M.D., J.T.Z., M.J.Q. and P.C.C.; writing—original draft preparation, G.A.J., M.M.D., J.T.Z. and M.J.Q.; writing—review and editing, G.A.J., M.M.D., J.T.Z., M.J.Q. and P.C.C.; supervision, M.J.Q. and P.C.C.; project administration, P.C.C.; funding acquisition, P.C.C. All authors have read and agreed to the published version of the manuscript.

Funding: The authors gratefully acknowledge the support of the Department of Defense, specifically the Office of Naval Research (ONR) under Grant Numbers FA9550-22-1-0385 and N000142312721.

Institutional Review Board Statement: Not applicable.

Informed Consent Statement: Not applicable.

Data Availability Statement: No new data were created or analyzed in this study. Data sharing is not applicable to this article.

Acknowledgments: The authors are also grateful to Sierra Kruger for her work on designing Figures 1–4 and Jaydeep Saha for his previous work in establishing the basis for Figure 6.

Conflicts of Interest: The authors declare no conflicts of interest.

References

1. Buynak, C.F.; Blackshire, J.; Lindgren, E.A.; Jata, K.V. Challenges and Opportunities in NDE, ISHM and Material State Awareness for Aircraft Structures: US Air Force Perspective. *Proc. AIP Conf. Proc.* **2008**, *975*, 1789–1801.
2. Aldrin, J.C.; Lindgren, E.A. The Need and Approach for Characterization—U.S. Air Force Perspectives on Materials State Awareness. In *Proceedings of the AIP Conference Proceedings*; American Institute of Physics Inc.: College Park, MA, USA, 2018; Volume 1949.
3. Jacobs, L.J. Nonlinear Ultrasonics for Material State Awareness. In *Proceedings of the AIP Conference Proceedings*; American Institute of Physics Inc.: College Park, MA, USA, 2014; Volume 1581, pp. 13–20.
4. Farret, J.; Witherell, P. Data Formats in Additive Manufacturing. In *ASM Handbook*; ASM International: Detroit, MI, USA, 2024; Volume 24A, pp. 184–194.
5. Frazier, W.E. Metal Additive Manufacturing: A Review. *J. Mater. Eng. Perform.* **2014**, *23*, 1917–1928. [CrossRef]
6. Srivastava, M.; Rathee, S.; Tiwari, A.; Dongre, M. Wire Arc Additive Manufacturing of Metals: A Review on Processes, Materials and Their Behaviour. *Mater. Chem. Phys.* **2023**, *294*, 126988. [CrossRef]
7. Collins, P.C.; Brice, D.A.; Samimi, P.; Ghamarian, I.; Fraser, H.L. Microstructural Control of Additively Manufactured Metallic Materials. *Annu. Rev. Mater. Res.* **2016**, *46*, 63–91. [CrossRef]
8. Gorsse, S.; Hutchinson, C.; Gouné, M.; Banerjee, R. Additive Manufacturing of Metals: A Brief Review of the Characteristic Microstructures and Properties of Steels, Ti-6Al-4V and High-Entropy Alloys. *Sci. Technol. Adv. Mater.* **2017**, *18*, 584–610. [CrossRef]
9. Zhang, D.; Liu, A.; Yin, B.; Wen, P. Additive Manufacturing of Duplex Stainless Steels—A Critical Review. *J. Manuf. Process* **2022**, *73*, 496–517. [CrossRef]
10. Mukherjee, T.; Zuback, J.S.; De, A.; DebRoy, T. Printability of Alloys for Additive Manufacturing. *Sci. Rep.* **2016**, *6*, 19717. [CrossRef]
11. Johnson, L.; Mahmoudi, M.; Zhang, B.; Seede, R.; Huang, X.; Maier, J.T.; Maier, H.J.; Karaman, I.; Elwany, A.; Arróyave, R. Assessing Printability Maps in Additive Manufacturing of Metal Alloys. *Acta Mater.* **2019**, *176*, 199–210. [CrossRef]
12. Sun, Z.; Ma, Y.; Ponge, D.; Zaefferer, S.; Jägle, E.A.; Gault, B.; Rollett, A.D.; Raabe, D. Thermodynamics-Guided Alloy and Process Design for Additive Manufacturing. *Nat. Commun.* **2022**, *13*, 4361. [CrossRef]
13. Martin, J.H.; Yahata, B.D.; Hundley, J.M.; Mayer, J.A.; Schaedler, T.A.; Pollock, T.M. 3D Printing of High-Strength Aluminium Alloys. *Nature* **2017**, *549*, 365–369. [CrossRef]
14. Bocklund, B.; Bobbio, L.D.; Otis, R.A.; Beese, A.M.; Liu, Z.K. Experimental Validation of Scheil–Gulliver Simulations for Gradient Path Planning in Additively Manufactured Functionally Graded Materials. *Materialia* **2020**, *11*, 100689. [CrossRef]
15. Gockel, J.; Sheridan, L.; Koerper, B.; Whip, B. The Influence of Additive Manufacturing Processing Parameters on Surface Roughness and Fatigue Life. *Int. J. Fatigue* **2019**, *124*, 380–388. [CrossRef]
16. Kwabena, N.A.; Haghdadi, N.; Primig, S. Electron and Laser-Based Additive Manufacturing of Ni-Based Superalloys: A Review of Heterogeneities in Microstructure and Mechanical Properties. *Mater. Des.* **2022**, *223*, 111245. [CrossRef]
17. Allison, J.; Li, M.; Wolverton, C.; Su, X. Enhanced for the Web Virtual Aluminum Castings: An Industrial Application of ICME. *JOM* **2006**, *58*, 28–35. [CrossRef]
18. Shi, R.; Khairallah, S.; Heo, T.W.; Rolchigo, M.; McKeown, J.T.; Matthews, M.J. Integrated Simulation Framework for Additively Manufactured Ti-6Al-4V: Melt Pool Dynamics, Microstructure, Solid-State Phase Transformation, and Microelastic Response. *JOM* **2019**, *71*, 3640–3655. [CrossRef]
19. Martin, B.W.; Ales, T.K.; Rolchigo, M.R.; Collins, P.C. Developing and Applying ICME + Modeling Tools to Predict Performance of Additively Manufactured Aerospace Parts. In *Additive Manufacturing for the Aerospace Industry*; Elsevier Inc.: Amsterdam, The Netherlands, 2019; pp. 375–400, ISBN 9780128140635.
20. Herriott, C.; Li, X.; Kouraytem, N.; Tari, V.; Tan, W.; Anglin, B.; Rollett, A.D.; Spear, A.D. A Multi-Scale, Multi-Physics Modeling Framework to Predict Spatial Variation of Properties in Additive-Manufactured Metals. *Model. Simul. Mat. Sci. Eng.* **2019**, *27*, 025009. [CrossRef]
21. Yi Wang, W.; Li, J.; Liu, W.; Liu, Z.K. Integrated Computational Materials Engineering for Advanced Materials: A Brief Review. *Comput. Mater. Sci.* **2019**, *158*, 42–48. [CrossRef]
22. Martukanitz, R.; Michaleris, P.; Palmer, T.; DebRoy, T.; Liu, Z.K.; Otis, R.; Heo, T.W.; Chen, L.Q. Toward an Integrated Computational System for Describing the Additive Manufacturing Process for Metallic Materials. *Addit. Manuf.* **2014**, *1*, 52–63. [CrossRef]
23. Witherell, P.; Feng, S.; Simpson, T.W.; Saint John, D.B.; Michaleris, P.; Liu, Z.K.; Chen, L.Q.; Martukanitz, R. Toward Metamodels for Composable and Reusable Additive Manufacturing Process Models. *J. Manuf. Sci. Eng. Trans. ASME* **2014**, *136*, 061025. [CrossRef]
24. King, W.E.; Anderson, A.T.; Ferencz, R.M.; Hodge, N.E.; Kamath, C.; Khairallah, S.A.; Rubenchik, A.M. Laser Powder Bed Fusion Additive Manufacturing of Metals; Physics, Computational, and Materials Challenges. *Appl. Phys. Rev.* **2015**, *2*, 041304. [CrossRef]
25. Collins, P.C.; Harlow, D.G. Probability and Statistical Modeling: Ti-6Al-4V Produced Via Directed Energy Deposition. *J. Mater. Eng. Perform.* **2021**, *30*, 6905–6912. [CrossRef]
26. Haden, C.V.; Collins, P.C.; Harlow, D.G. Yield Strength Prediction of Titanium Alloys. *JOM* **2015**, *67*, 1357–1361. [CrossRef]

27. *ASTM B214-22*; Standard Test Method for Sieve Analysis of Metal Powders. ASTM International: West Conshohocken, PA, USA, 2022.

28. *ASTM B330-20*; Standard Test Methods for Estimating Average Particle Size of Metal Powders and Related Compounds Using Air Permeability. ASTM International: West Conshohocken, PA, USA, 2020.

29. *ASTM B213-20*; Standard Test Methods for Flow Rate of Metal Powders Using the Hall Flowmeter Funnel. ASTM International: West Conshohocken, PA, USA, 2020.

30. Slotwinski, J.A.; Garboczi, E.J.; Stutzman, P.E.; Ferraris, C.F.; Watson, S.S.; Peltz, M.A. Characterization of Metal Powders Used for Additive Manufacturing. *J. Res. Natl. Inst. Stand. Technol.* **2014**, *119*, 460–493. [CrossRef] [PubMed]

31. Zegzulka, J.; Gelnar, D.; Jezerska, L.; Prokes, R.; Rozbroj, J. Characterization and Flowability Methods for Metal Powders. *Sci. Rep.* **2020**, *10*, 21004. [CrossRef] [PubMed]

32. Spierings, A.B.; Voegtlin, M.; Bauer, T.; Wegener, K. Powder Flowability Characterisation Methodology for Powder-Bed-Based Metal Additive Manufacturing. *Prog. Addit. Manuf.* **2016**, *1*, 9–20. [CrossRef]

33. Muñiz-Lerma, J.A.; Nommeots-Nomm, A.; Waters, K.E.; Brochu, M. A Comprehensive Approach to Powder Feedstock Characterization for Powder Bed Fusion Additive Manufacturing: A Case Study on AlSi7Mg. *Materials* **2018**, *11*, 2386. [CrossRef]

34. Le, T.P.; Wang, X.; Seita, M. An Optical-Based Method to Estimate the Oxygen Content in Recycled Metal Powders for Additive Manufacturing. *Addit. Manuf.* **2022**, *59*, 103127. [CrossRef]

35. Montazeri, M.; Yavari, R.; Rao, P.; Boulware, P. In-Process Monitoring of Material Cross-Contamination Defects in Laser Powder Bed Fusion. *J. Manuf. Sci. Eng. Trans. ASME* **2018**, *140*, 111001. [CrossRef]

36. Grasso, M.; Remani, A.; Dickins, A.; Colosimo, B.M.; Leach, R.K. In-Situ Measurement and Monitoring Methods for Metal Powder Bed Fusion: An Updated Review. *Meas. Sci. Technol.* **2021**, *32*, 112001. [CrossRef]

37. Tan, J.H.; Wong, W.L.E.; Dalgarno, K.W. An Overview of Powder Granulometry on Feedstock and Part Performance in the Selective Laser Melting Process. *Addit. Manuf.* **2017**, *18*, 228–255. [CrossRef]

38. Tan, Y.; Zhang, J.; Li, X.; Xu, Y.; Wu, C.Y. Comprehensive Evaluation of Powder Flowability for Additive Manufacturing Using Principal Component Analysis. *Powder Technol.* **2021**, *393*, 154–164. [CrossRef]

39. Todorov, E.I.; Boulware, P.; Gaah, K. Demonstration of Array Eddy Current Technology for Real-Time Monitoring of Laser Powder Bed Fusion Additive Manufacturing Process. In Proceedings of the SPIE, San Diego, CA, USA, 27 March 2018; p. 40.

40. Baere, D.D.; Hinderdael, M.; Jardon, Z.; Sanchez-Medina, J.; Helsen, J.; Powell, J. Experimental Evaluation of the Metal Powder Particle Flow on the Melt Pool during Directed Energy Deposition. *J. Laser Appl.* **2023**, *35*, 022008. [CrossRef]

41. GranuDrum. Available online: https://www.granutools.com/en/granudrum (accessed on 4 September 2024).

42. Whiting, J.; Springer, A.; Sciammarella, F. Real-Time Acoustic Emission Monitoring of Powder Mass Flow Rate for Directed Energy Deposition. *Addit. Manuf.* **2018**, *23*, 312–318. [CrossRef] [PubMed]

43. Scipioni Bertoli, U.; Guss, G.; Wu, S.; Matthews, M.J.; Schoenung, J.M. In-Situ Characterization of Laser-Powder Interaction and Cooling Rates through High-Speed Imaging of Powder Bed Fusion Additive Manufacturing. *Mater. Des.* **2017**, *135*, 385–396. [CrossRef]

44. Xia, C.; Pan, Z.; Polden, J.; Li, H.; Xu, Y.; Chen, S.; Zhang, Y. A Review on Wire Arc Additive Manufacturing: Monitoring, Control and a Framework of Automated System. *J. Manuf. Syst.* **2020**, *57*, 31–45. [CrossRef]

45. Ng Beng, C.-H. Wire Arc Additive Manufacturing of Metastable β-Ti Alloys. Ph.D. Thesis, University of Queensland, Brisbane, QLD, Australia, 2023.

46. Ghamarian, I.; Samimi, P.; Dixit, V.; Collins, P.C. A Constitutive Equation Relating Composition and Microstructure to Properties in Ti-6Al-4V: As Derived Using a Novel Integrated Computational Approach. *Metall. Mater. Trans. A Phys. Metall. Mater. Sci.* **2015**, *46*, 5021–5037. [CrossRef]

47. Quintana, M.J.; Temple, A.J.; Harlow, D.G.; Collins, P.C. On the Prediction of Uniaxial Tensile Behavior Beyond the Yield Point of Wrought and Additively Manufactured Ti-6Al-4V. *Integr. Mater. Manuf. Innov.* **2022**, *11*, 327–338. [CrossRef]

48. Samimi, P.; Liu, Y.; Ghamarian, I.; Collins, P.C. A Novel Tool to Assess the Influence of Alloy Composition on the Oxidation Behavior and Concurrent Oxygen-Induced Phase Transformations for Binary Ti-XMo Alloys at 650 °C. *Corros. Sci.* **2014**, *89*, 295–306. [CrossRef]

49. Amano, H.; Ishimoto, T.; Nakano, T. Importance of Atmospheric Gas Selection in Metal Additive Manufacturing: Effects on Spatter, Microstructure, and Mechanical Properties. *Mater. Trans.* **2023**, *64*, 2–9. [CrossRef]

50. Whiting, J.; Fox, J. Characterization of Feedstock in the Powder Bed Fusion Process: Sources of Variation in Particle Size Distribution and the Factors That Influence Them. In Proceedings of the 2016 International Solid Freeform Fabrication Symposium, Austin, TX, USA, 8–10 August 2016.

51. Spierings, A.B.; Herres, N.; Levy, G.; Spierings, M.A.B. Influence of the Particle Size Distribution on Surface Quality and Mechanical Properties in Additive Manufactured Stainless Steel Parts. In Proceedings of the 2010 International Solid Freeform Fabrication Symposium, University of Texas at Austin, Austin, TX, USA, 23 September 2010.

52. Besterci, M.; Käerdi, H.; Kulu, P.; Mikli, V. Characterization of Powder Particle Morphology. *Proc. Est. Acad. Sciences. Eng.* **2001**, *7*, 22. [CrossRef]

53. Koester, L.W.; Taheri, H.; Bigelow, T.A.; Bond, L.J.; Faierson, E.J. In-Situ Acoustic Signature Monitoring in Additive Manufacturing Processes. In *Proceedings of the AIP Conference Proceedings*; American Institute of Physics Inc.: College Park, MA, USA, 2018; Volume 1949.

54. Pringle, A.M.; Oberloier, S.; Petsiuk, A.L.; Sanders, P.G.; Pearce, J.M. Open Source Arc Analyzer: Multi-Sensor Monitoring of Wire Arc Additive Manufacturing. *HardwareX* **2020**, *8*, e00137. [CrossRef] [PubMed]
55. RoboWAAM, Wire Arc Additive Manfuacturing Machine. Available online: https://www.waam3d.com/hardware/robowaam (accessed on 4 September 2024).
56. Tapia, G.; Elwany, A. A Review on Process Monitoring and Control in Metal-Based Additive Manufacturing. *J. Manuf. Sci. Eng. Trans. ASME* **2014**, *136*, 060801. [CrossRef]
57. Cordero, P.M.; Mireles, J.; Ridwan, S.; Wicker, R.B. Evaluation of Monitoring Methods for Electron Beam Melting Powder Bed Fusion Additive Manufacturing Technology. *Prog. Addit. Manuf.* **2017**, *2*, 1–10. [CrossRef]
58. Vest, A.M.; St-Pierre, D.R.; Rock, S.; Maniatty, A.M.; Lewis, D.J.; Hocker, S.J.A. *Thermocouple Temperature Measurements in Selective Laser Melting Additive Manufacturing*; National Aeronautics and Space Administration, Langley Research Center: Hampton, VA, USA, 2022.
59. Haley, J.; Leach, C.; Jordan, B.; Dehoff, R.; Paquit, V. In-Situ Digital Image Correlation and Thermal Monitoring in Directed Energy Deposition Additive Manufacturing. *Opt. Express* **2021**, *29*, 9927. [CrossRef] [PubMed]
60. Murphy, R.D.; Forrest, E.C. A Review of In-Situ Temperature Measurements for Additive Manufacturing Technologies. In Proceedings of the NCSL International Workshop & Symposium, Saint Paul, MN, USA, 20 May 2016.
61. Van Elsen, M.; Al-Bender, F.; Kruth, J.P. Application of Dimensional Analysis to Selective Laser Melting. *Rapid Prototyp. J.* **2008**, *14*, 15–22. [CrossRef]
62. Mukherjee, T.; Manvatkar, V.; De, A.; DebRoy, T. Dimensionless Numbers in Additive Manufacturing. *J. Appl. Phys.* **2017**, *121*, 064904. [CrossRef]
63. Gu, H.; Gong, H.; Pal, D.; Rafi, K.; Starr, T.; Stucker, B. Influences of Energy Density on Porosity and Microstructure of Selective Laser Melted 17-4PH Stainless Steel. In Proceedings of the 24th Annual International Solids Freeform Fabrication Symposium, Austin, TX, USA, 12–14 August 2013.
64. Ciurana, J.; Hernandez, L.; Delgado, J. Energy Density Analysis on Single Tracks Formed by Selective Laser Melting with CoCrMo Powder Material. *Int. J. Adv. Manuf. Technol.* **2013**, *68*, 1103–1110. [CrossRef]
65. Kempen, K.; Thijs, L.; Van Humbeeck, J.; Kruth, J.P. Processing AlSi10Mg by Selective Laser Melting: Parameter Optimisation and Material Characterisation. *Mater. Sci. Technol.* **2015**, *31*, 917–923. [CrossRef]
66. Le, T.N.; Lee, M.H.; Lin, Z.H.; Tran, H.C.; Lo, Y.L. Vision-Based in-Situ Monitoring System for Melt-Pool Detection in Laser Powder Bed Fusion Process. *J. Manuf. Process* **2021**, *68*, 1735–1745. [CrossRef]
67. Clijsters, S.; Craeghs, T.; Buls, S.; Kempen, K.; Kruth, J.P. In Situ Quality Control of the Selective Laser Melting Process Using a High-Speed, Real-Time Melt Pool Monitoring System. *Int. J. Adv. Manuf. Technol.* **2014**, *75*, 1089–1101. [CrossRef]
68. Xia, C.; Pan, Z.; Li, Y.; Chen, J.; Li, H. Vision-Based Melt Pool Monitoring for Wire-Arc Additive Manufacturing Using Deep Learning Method. *Int. J. Adv. Manuf. Technol.* **2022**, *120*, 551–562. [CrossRef]
69. Da Silva, A.; Frostevarg, J.; Kaplan, A.F.H. Melt Pool Monitoring and Process Optimisation of Directed Energy Deposition via Coaxial Thermal Imaging. *J. Manuf. Process* **2023**, *107*, 126–133. [CrossRef]
70. Chen, X.; Zhang, H.; Hu, J.; Xiao, Y. A Passive On-Line Defect Detection Method for Wire and Arc Additive Manufacturing Based on Infrared Thermography. In Proceedings of the Solid Freeform Fabrication 2019: Proceedings of the 30th Annual International Solid Freeform Fabrication Symposium—An Additive Manufacturing Conference, SFF, Austin, TX, USA, 12–14 August 2019.
71. Cheng, B.; Lydon, J.; Cooper, K.; Cole, V.; Northrop, P.; Chou, K. Melt Pool Sensing and Size Analysis in Laser Powder-Bed Metal Additive Manufacturing. *J. Manuf. Process* **2018**, *32*, 744–753. [CrossRef]
72. Akbari, P.; Ogoke, F.; Kao, N.Y.; Meidani, K.; Yeh, C.Y.; Lee, W.; Barati Farimani, A. MeltpoolNet: Melt Pool Characteristic Prediction in Metal Additive Manufacturing Using Machine Learning. *Addit. Manuf.* **2022**, *55*, 102817. [CrossRef]
73. Leung, C.L.A.; Marussi, S.; Atwood, R.C.; Towrie, M.; Withers, P.J.; Lee, P.D. In Situ X-Ray Imaging of Defect and Molten Pool Dynamics in Laser Additive Manufacturing. *Nat. Commun.* **2018**, *9*, 1355. [CrossRef]
74. Wolff, S.J.; Webster, S.; Parab, N.D.; Aronson, B.; Gould, B.; Greco, A.; Sun, T. In-Situ Observations of Directed Energy Deposition Additive Manufacturing Using High-Speed X-Ray Imaging. *JOM* **2021**, *73*, 189–200. [CrossRef]
75. Ozel, T. A Review on In-Situ Process Sensing and Monitoring Systems for Fusion-Based Additive Manufacturing. *Int. J. Mechatron. Manuf. Syst.* **2023**, *16*, 115–154. [CrossRef]
76. Simonds, B.J.; Sowards, J.; Hadler, J.; Pfeif, E.; Wilthan, B.; Tanner, J.; Harris, C.; Williams, P.; Lehman, J. Time-Resolved Absorptance and Melt Pool Dynamics during Intense Laser Irradiation of a Metal. *Phys. Rev. Appl.* **2018**, *10*, 044061. [CrossRef]
77. Kube, C.M.; Shu, Y.; Lew, A.J.; Galles, D. Real-Time Characterization of Laser-Generated Melt Pools Using Ultrasound. *Mater. Eval.* **2018**, *76*, 525–534.
78. Grasso, M.; Demir, A.G.; Previtali, B.; Colosimo, B.M. In Situ Monitoring of Selective Laser Melting of Zinc Powder via Infrared Imaging of the Process Plume. *Robot. Comput. Integr. Manuf.* **2018**, *49*, 229–239. [CrossRef]
79. Ren, W.; Zhang, Z.; Lu, Y.; Wen, G.; Mazumder, J. In-Situ Monitoring of Laser Additive Manufacturing for Al7075 Alloy Using Emission Spectroscopy and Plume Imaging. *IEEE Access* **2021**, *9*, 61671–61679. [CrossRef]
80. Stutzman, C.B.; Mitchell, W.F.; Nassar, A.R. Optical Emission Sensing for Laser-Based Additive Manufacturing—What Are We Actually Measuring? *J. Laser Appl.* **2021**, *33*, 012010. [CrossRef]
81. Zhang, Y.; Hong, G.S.; Ye, D.; Zhu, K.; Fuh, J.Y.H. Extraction and Evaluation of Melt Pool, Plume and Spatter Information for Powder-Bed Fusion AM Process Monitoring. *Mater. Des.* **2018**, *156*, 458–469. [CrossRef]

82. Lednev, V.N.; Sdvizhenskii, P.A.; Asyutin, R.D.; Tretyakov, R.S.; Grishin, M.Y.; Stavertiy, A.Y.; Pershin, S.M. In Situ Multi-Elemental Analysis by Laser Induced Breakdown Spectroscopy in Additive Manufacturing. *Addit. Manuf.* **2019**, *25*, 64–70. [CrossRef]

83. Lednev, V.N.; Sdvizhenskii, P.A.; Asyutin, R.D.; Tretyakov, R.S.; Grishin, M.Y.; Stavertiy, A.Y.; Fedorov, A.N.; Pershin, S.M. In Situ Elemental Analysis and Failures Detection during Additive Manufacturing Process Utilizing Laser Induced Breakdown Spectroscopy. *Opt. Express* **2019**, *27*, 4612. [CrossRef]

84. Song, L.; Huang, W.; Han, X.; Mazumder, J. Real-Time Composition Monitoring Using Support Vector Regression of Laser-Induced Plasma for Laser Additive Manufacturing. *IEEE Trans. Ind. Electron.* **2017**, *64*, 633–642. [CrossRef]

85. Lin, J.; Yang, J.; Huang, Y.; Lin, X. Defect Identification of Metal Additive Manufacturing Parts Based on Laser-Induced Breakdown Spectroscopy and Machine Learning. *Appl. Phys. B* **2021**, *127*, 173. [CrossRef]

86. Nassar, A.R.; Spurgeon, T.J.; Reutzel, E.W. Sensing Defects during Directed-Energy Additive Manufacturing of Metal Parts Using Optical Emissions Spectroscopy. In Proceedings of the 25th Annual International Solid Freeform Fabrication Symposium: An Additive Manufacturing Conference, SFF 2014, Austin, TX, USA, 4–6 August 2014.

87. Repossini, G.; Laguzza, V.; Grasso, M.; Colosimo, B.M. On the Use of Spatter Signature for In-Situ Monitoring of Laser Powder Bed Fusion. *Addit. Manuf.* **2017**, *16*, 35–48. [CrossRef]

88. Fu, Y.; Priddy, B.; Downey, A.; Yuan, L. Real-Time Splatter Tracking in Laser Powder Bed Fusion Additive Manufacturing. In Proceedings of the SPIE, Long Beach, CA, USA, 10 March 2023.

89. Young, Z.A.; Guo, Q.; Parab, N.D.; Zhao, C.; Qu, M.; Escano, L.I.; Fezzaa, K.; Everhart, W.; Sun, T.; Chen, L. Types of Spatter and Their Features and Formation Mechanisms in Laser Powder Bed Fusion Additive Manufacturing Process. *Addit. Manuf.* **2020**, *36*, 101438. [CrossRef]

90. Guo, Q.; Zhao, C.; Escano, L.I.; Young, Z.; Xiong, L.; Fezzaa, K.; Everhart, W.; Brown, B.; Sun, T.; Chen, L. Transient Dynamics of Powder Spattering in Laser Powder Bed Fusion Additive Manufacturing Process Revealed by In-Situ High-Speed High-Energy X-Ray Imaging. *Acta Mater.* **2018**, *151*, 169–180. [CrossRef]

91. Ahmadi, F.; Song, J.; Zoughi, R. Electromagnetic Scattering of Metal Powder Spatter in Laser Powder Bed Fusion Additive Manufacturing Process. In Proceedings of the IEEE Conference on Antenna Measurements and Applications, CAMA, Genoa, Italy, 15–17 November 2023.

92. Li, Y.; Gu, D. Parametric Analysis of Thermal Behavior during Selective Laser Melting Additive Manufacturing of Aluminum Alloy Powder. *Mater. Des.* **2014**, *63*, 856–867. [CrossRef]

93. Gong, H.; Gu, H.; Zeng, K.; Dilip, J.J.S.; Pal, D.; Stucker, B.; Christiansen, D.; Beuth, J.; Lewandowski, J.J. Melt Pool Characterization for Selective Laser Melting of Ti-6Al-4V Pre-Alloyed Powder. In Proceedings of the 2014 International Solid Freeform Fabrication Symposium, Austin, TX, USA, 4–6 August 2014.

94. Gusarov, A.V.; Smurov, I. Modeling the Interaction of Laser Radiation with Powder Bed at Selective Laser Melting. In *Proceedings of the Physics Procedia*; Elsevier B.V.: Amsterdam, The Netherlands, 2010; Volume 5, pp. 381–394.

95. Romano, J.; Ladani, L.; Razmi, J.; Sadowski, M. Temperature Distribution and Melt Geometry in Laser and Electron-Beam Melting Processes—A Comparison among Common Materials. *Addit. Manuf.* **2015**, *8*, 1–11. [CrossRef]

96. Zhang, B.; Liu, S.; Shin, Y.C. In-Process Monitoring of Porosity during Laser Additive Manufacturing Process. *Addit. Manuf.* **2019**, *28*, 497–505. [CrossRef]

97. Scime, L.; Beuth, J. Using Machine Learning to Identify In-Situ Melt Pool Signatures Indicative of Flaw Formation in a Laser Powder Bed Fusion Additive Manufacturing Process. *Addit. Manuf.* **2019**, *25*, 151–165. [CrossRef]

98. Everton, S.K.; Hirsch, M.; Stavroulakis, P.I.; Leach, R.K.; Clare, A.T. Review of In-Situ Process Monitoring and in-Situ Metrology for Metal Additive Manufacturing. *Mater. Des.* **2016**, *95*, 431–445. [CrossRef]

99. Cannizzaro, D.; Varrella, A.G.; Paradiso, S.; Sampieri, R.; Chen, Y.; Macii, A.; Patti, E.; Cataldo, S. Di In-Situ Defect Detection of Metal Additive Manufacturing: An Integrated Framework. *IEEE Trans. Emerg. Top. Comput.* **2022**, *10*, 74–86. [CrossRef]

100. Sun, W.; Zhang, Z.; Ren, W.; Mazumder, J.; Jin, J.J. In Situ Monitoring of Optical Emission Spectra for Microscopic Pores in Metal Additive Manufacturing. *J. Manuf. Sci. Eng. Trans. ASME* **2022**, *144*, 011006. [CrossRef]

101. Cunningham, R.; Zhao, C.; Parab, N.; Kantzos, C.; Pauza, J.; Fezzaa, K.; Sun, T.; Rollett, A.D. Keyhole Threshold and Morphology in Laser Melting Revealed by Ultrahigh-Speed x-Ray Imaging. *Science (1979)* **2019**, *363*, 849–852. [CrossRef]

102. Ren, Z.; Gao, L.; Clark, S.J.; Fezzaa, K.; Shevchenko, P.; Choi, A.; Everhart, W.; Rollett, A.D.; Chen, L.; Sun, T. Machine Learning-Aided Real-Time Detection of Keyhole Pore Generation in Laser Powder Bed Fusion. *Science (1979)* **2023**, *379*, 89–94. [CrossRef]

103. Liu, Z.; Yang, Y.; Han, C.; Zhou, H.; Zhou, H.; Wang, M.; Liu, L.; Wang, H.; Bai, Y.; Wang, D. Effects of Gas Flow Parameters on Droplet Spatter Features and Dynamics during Large-Scale Laser Powder Bed Fusion. *Mater. Des.* **2023**, *225*, 111534. [CrossRef]

104. Gaja, H.; Liou, F. Depth of Cut Monitoring for Hybrid Manufacturing Using Acoustic Emission Sensor. In Proceedings of the International Solid Freeform Fabrication Symposium, Austin, TX, USA, 10–12 August 2015; pp. 422–436.

105. Kononenko, D.Y.; Nikonova, V.; Seleznev, M.; van den Brink, J.; Chernyavsky, D. An in Situ Crack Detection Approach in Additive Manufacturing Based on Acoustic Emission and Machine Learning. *Addit. Manuf. Lett.* **2023**, *5*, 100130. [CrossRef]

106. Kim, F.H.; Moylan, S.P.; Garboczi, E.J.; Slotwinski, J.A. Investigation of Pore Structure in Cobalt Chrome Additively Manufactured Parts Using X-Ray Computed Tomography and Three-Dimensional Image Analysis. *Addit. Manuf.* **2017**, *17*, 23–38. [CrossRef] [PubMed]

107. Tang, M.; Pistorius, P.C.; Beuth, J.L. Prediction of Lack-of-Fusion Porosity for Powder Bed Fusion. *Addit. Manuf.* **2017**, *14*, 39–48. [CrossRef]
108. Min, Y.; Shen, S.; Li, H.; Liu, S.; Mi, J.; Zhou, J.; Mai, Z.; Chen, J. Online Monitoring of an Additive Manufacturing Environment Using a Time-of-Flight Mass Spectrometer. *Measurement* **2022**, *189*, 110473. [CrossRef]
109. Stolidi, A.; Touron, A.; Toulemonde, L.; Paradis, H.; Lapouge, P.; Coste, F.; Garandet, J.P. Towards In-Situ Fumes Composition Monitoring during an Additive Manufacturing Process Using Energy Dispersive X-Ray Fluorescence Spectrometry. *Addit. Manuf. Lett.* **2023**, *6*, 100153. [CrossRef]
110. Wang, S.; Liu, C. Real-Time Monitoring of Chemical Composition in Nickel-Based Laser Cladding Layer by Emission Spectroscopy Analysis. *Materials* **2019**, *12*, 2637. [CrossRef]
111. Kimura, T.; Nakamoto, T. Microstructures and Mechanical Properties of A356 (AlSi7Mg0.3) Aluminum Alloy Fabricated by Selective Laser Melting. *Mater. Des.* **2016**, *89*, 1294–1301. [CrossRef]
112. Aboulkhair, N.T.; Maskery, I.; Tuck, C.; Ashcroft, I.; Everitt, N.M. On the Formation of AlSi10Mg Single Tracks and Layers in Selective Laser Melting: Microstructure and Nano-Mechanical Properties. *J. Mater. Process Technol.* **2016**, *230*, 88–98. [CrossRef]
113. Raplee, J.; Plotkowski, A.; Kirka, M.M.; Dinwiddie, R.; Okello, A.; Dehoff, R.R.; Babu, S.S. Thermographic Microstructure Monitoring in Electron Beam Additive Manufacturing. *Sci. Rep.* **2017**, *7*, srep43554. [CrossRef]
114. Farshidianfar, M.H.; Khajepour, A.; Gerlich, A. Real-Time Control of Microstructure in Laser Additive Manufacturing. *Int. J. Adv. Manuf. Technol.* **2016**, *82*, 1173–1186. [CrossRef]
115. Jamnikar, N.D.; Liu, S.; Brice, C.; Zhang, X. In Situ Microstructure Property Prediction by Modeling Molten Pool-Quality Relations for Wire-Feed Laser Additive Manufacturing. *J. Manuf. Process* **2022**, *79*, 803–814. [CrossRef]
116. Grayeli, N.; Ilic, D.B.; Stanke, F.; Chou, C.H.; Shyne, J.C. Studies of Steel Microstructure by Acoustical Methods. In Proceedings of the Ultrasonics Symposium, New Orleans, LA, USA, 26–28 September 1979.
117. Sojiphan, K.; Wangsupangkul, P.; Chailampangsuksakul, T. Application of Ultrasonic Inspection for Microstructure Analysis of Stainless Steel Grade 304L. In *Proceedings of the Key Engineering Materials*; Trans Tech Publications Ltd.: Wollerau, Switzerland, 2019; Volume 798 KEM, pp. 32–37.
118. Evans, R.; Walker, J.; Middendorf, J.; Gockel, J. Modeling and Monitoring of the Effect of Scan Strategy on Microstructure in Additive Manufacturing. *Metall. Mater. Trans. A Phys. Metall. Mater. Sci.* **2020**, *51*, 4123–4129. [CrossRef]
119. Acevedo, R.; Sedlak, P.; Kolman, R.; Fredel, M. Residual Stress Analysis of Additive Manufacturing of Metallic Parts Using Ultrasonic Waves: State of the Art Review. *J. Mater. Res. Technol.* **2020**, *9*, 9457–9477. [CrossRef]
120. He, W.; Shi, W.; Li, J.; Xie, H. In-Situ Monitoring and Deformation Characterization by Optical Techniques; Part I: Laser-Aided Direct Metal Deposition for Additive Manufacturing. *Opt. Lasers Eng.* **2019**, *122*, 74–88. [CrossRef]
121. Heigel, J.C. Thermo-Mechanical Model Development and Experimental Validation for Directed Energy Deposition. Ph.D. Thesis, The Pennsylvania State University, University Park, PA, USA, 2015.
122. Corbin, D.J.; Nassar, A.R.; Reutzel, E.W.; Kistler, N.A.; Beese, A.M.; Michaleris, P. Impact of Directed Energy Deposition Parameters on Mechanical Distortion of Laser Deposited Ti-6Al-4V. In Proceedings of the Solid Freeform Fabrication 2016: Proceedings of the 27th Annual International Solid Freeform Fabrication Symposium—An Additive Manufacturing Conference, SFF 2016, Austin, TX, USA, 8–10 August 2016.
123. Ocelík, V.; Bosgra, J.; de Hosson, J.T.M. In-Situ Strain Observation in High Power Laser Cladding. *Surf. Coat. Technol.* **2009**, *203*, 3189–3196. [CrossRef]
124. Lu, X.; Lin, X.; Chiumenti, M.; Cervera, M.; Hu, Y.; Ji, X.; Ma, L.; Huang, W. In Situ Measurements and Thermo-Mechanical Simulation of Ti–6Al–4V Laser Solid Forming Processes. *Int. J. Mech. Sci.* **2019**, *153–154*, 119–130. [CrossRef]
125. Biegler, M.; Graf, B.; Rethmeier, M. In-Situ Distortions in LMD Additive Manufacturing Walls Can Be Measured with Digital Image Correlation and Predicted Using Numerical Simulations. *Addit. Manuf.* **2018**, *20*, 101–110. [CrossRef]
126. Xie, R.; Zhao, Y.; Chen, G.; Lin, X.; Zhang, S.; Fan, S.; Shi, Q. The Full-Field Strain Distribution and the Evolution Behavior during Additive Manufacturing through in-Situ Observation. *Mater. Des.* **2018**, *150*, 49–54. [CrossRef]
127. Xu, C.; Song, W.; Pan, Q.; Li, H.; Liu, S. Nondestructive Testing Residual Stress Using Ultrasonic Critical Refracted Longitudinal Wave. In *Proceedings of the Physics Procedia*; Elsevier: Amsterdam, The Netherlands, 2015; Volume 70, pp. 594–598.
128. Denlinger, E.R.; Heigel, J.C.; Michaleris, P. Residual Stress and Distortion Modeling of Electron Beam Direct Manufacturing Ti-6Al-4V. *Proc. Inst. Mech. Eng. B J. Eng. Manuf.* **2015**, *229*, 1803–1813. [CrossRef]
129. Mercelis, P.; Kruth, J.P. Residual Stresses in Selective Laser Sintering and Selective Laser Melting. *Rapid Prototyp. J.* **2006**, *12*, 254–265. [CrossRef]
130. Vasinonta, A.; Beuth, J.; Griffith, M. Process Maps for Controlling Residual Stress and Melt Pool Size in Laser-Based SFF Processes. In Proceedings of the 2000 International Solid Freeform Fabrication Symposium, Austin, TX, USA, 8–10 August 2000.
131. Sharples, S.D.; Clark, M.; Somekh, M.G. Spatially Resolved Acoustic Spectroscopy for Fast Noncontact Imaging of Material Microstructure. *Opt. Express* **2006**, *14*, 10435–10440. [CrossRef] [PubMed]
132. Smith, R.J.; Li, W.; Coulson, J.; Clark, M.; Somekh, M.G.; Sharples, S.D. Spatially Resolved Acoustic Spectroscopy for Rapid Imaging of Material Microstructure and Grain Orientation. *Meas. Sci. Technol.* **2014**, *25*, 055902. [CrossRef]
133. Li, W.; Sharples, S.D.; Smith, R.J.; Clark, M.; Somekh, M.G. Determination of Crystallographic Orientation of Large Grain Metals with Surface Acoustic Waves. *J. Acoust. Soc. Am.* **2012**, *132*, 738–745. [CrossRef] [PubMed]

134. Dryburgh, P.; Li, W.; Pieris, D.; Fuentes-Domínguez, R.; Patel, R.; Smith, R.J.; Clark, M. Measurement of the Single Crystal Elasticity Matrix of Polycrystalline Materials. *Acta Mater.* **2022**, *225*, 117551. [CrossRef]
135. Suárez Fernández, D.; Wynne, B.P.; Crawforth, P.; Jackson, M. Titanium Alloy Microstructure Fingerprint Plots from In-Process Machining. *Mater. Sci. Eng. A* **2021**, *811*, 141074. [CrossRef]
136. Suárez Fernández, D.; Jackson, M.; Crawforth, P.; Fox, K.; Wynne, B.P. Using Machining Force Feedback to Quantify Grain Size in Beta Titanium. *Materialia* **2020**, *13*, 100856. [CrossRef]
137. Blanch, O.L.; Fernández, D.S.; Graves, A.; Jackson, M. MulTi-FAST: A Machinability Assessment of Functionally Graded Titanium Billets Produced from Multiple Alloy Powders. *Materials* **2022**, *15*, 3237. [CrossRef]
138. Hayes, B.J.; Martin, B.W.; Welk, B.; Kuhr, S.J.; Ales, T.K.; Brice, D.A.; Ghamarian, I.; Baker, A.H.; Haden, C.V.; Harlow, D.G.; et al. Predicting Tensile Properties of Ti-6Al-4V Produced via Directed Energy Deposition. *Acta Mater.* **2017**, *133*, 120–133. [CrossRef]
139. Thompson, A.P.; Aktulga, H.M.; Berger, R.; Bolintineanu, D.S.; Brown, W.M.; Crozier, P.S.; in 't Veld, P.J.; Kohlmeyer, A.; Moore, S.G.; Nguyen, T.D.; et al. LAMMPS—A Flexible Simulation Tool for Particle-Based Materials Modeling at the Atomic, Meso, and Continuum Scales. *Comput. Phys. Commun.* **2022**, *271*, 108171. [CrossRef]
140. Kloss, C.; Goniva, C.; Hager, A.; Amberger, S.; Pirker, S. Models, Algorithms and Validation for Opensource DEM and CFD-DEM. *Prog. Comput. Fluid. Dyn.* **2012**, *12*, 140–152. [CrossRef]
141. Parteli, E.J.R. DEM Simulation of Particles of Complex Shapes Using the Multisphere Method: Application for Additive Manufacturing. In *Proceedings of the AIP Conference Proceedings*; American Institute of Physics Inc.: College Park, MA, USA, 2013; Volume 1542, pp. 185–188.
142. Johnson, K.L.; Kendall, K.; Roberts, A.D. Surface Energy and the Contact of Elastic Solids. *Proc. R. Soc. Lond.* **1971**, *324*, 301–313.
143. Derjaguin, B.V.; Muller, V.M.; Toporov, Y.P. Effect of Contact Deformations on the Adhesion of Particles. *J. Colloid. Interface Sci.* **1975**, *53*, 314–326. [CrossRef]
144. Shenouda, S.; Hoff, A. *Discrete Element Method Analysis for Metal Powders Used in Additive Manufacturing, and DEM Simulation Tutorial Using LIGGGHTS-PUBLIC*; Lawrence Livermore National Lab. (LLNL): Livermore, CA, USA, 2020.
145. Yee, I. Powder Bed Surface Quality and Particle Size Distribution for Metal Additive Manufacturing and Comparison with Discrete Element Model. Master's Thesis, California Polytechnic State University, San Luis Obispo, CA, USA, 2018.
146. Shi, C.; Zou, Y.; Zou, Z.; Wu, D. Twin-Wire Indirect Arc Welding by Modeling and Experiment. *J. Mater. Process Technol.* **2014**, *214*, 2292–2299. [CrossRef]
147. Stützer, J.; Totzauer, T.; Wittig, B.; Zinke, M.; Jüttner, S. GMAW Cold Wire Technology for Adjusting the Ferrite–Austenite Ratio of Wire and Arc Additive Manufactured Duplex Stainless Steel Components. *Metals* **2019**, *9*, 564. [CrossRef]
148. Rosenthal, D. Mathematical Theory of Heat Distribution during Welding and Cutting. *Weld. J.* **1941**, *20*, 220–234.
149. Eagar, T.W.; Tsai, N.-S. Temperature Fields Produced by Traveling Distributed Heat Sources. *Weld. J.* **1983**, *162*, 346–355.
150. Chen, F.F.; Xiang, J.; Thomas, D.G.; Murphy, A.B. Model-Based Parameter Optimization for Arc Welding Process Simulation. *Appl. Math. Model.* **2020**, *81*, 386–400. [CrossRef]
151. Coleman, J.; Kincaid, K.; Knapp, G.L.; Stump, B.; Plotkowski, A.J. *AdditiveFOAM*; Oak Ridge National Laboratory (ORNL): Oak Ridge, TN, USA, 2023.
152. Korzekwa, D.A. Truchas—A Multi-Physics Tool for Casting Simulation. *Int. J. Cast. Met. Res.* **2009**, *22*, 178–191. [CrossRef]
153. Slattery, S.R. ExaMPM 2017.
154. Weller, H.G.; Tabor, G.; Jasak, H.; Fureby, C. A Tensorial Approach to Computational Continuum Mechanics Using Object-Oriented Techniques. *Comput. Phys.* **1998**, *12*, 620–631. [CrossRef]
155. Bayat, M.; Thanki, A.; Mohanty, S.; Witvrouw, A.; Yang, S.; Thorborg, J.; Tiedje, N.S.; Hattel, J.H. Keyhole-Induced Porosities in Laser-Based Powder Bed Fusion (L-PBF) of Ti6Al4V: High-Fidelity Modelling and Experimental Validation. *Addit. Manuf.* **2019**, *30*, 100835. [CrossRef]
156. Langmuir, I. The Vapor Pressure of Metallic Tungsten. *Phys. Rev.* **1913**, *2*, 329–342. [CrossRef]
157. Klassen, A.; Scharowsky, T.; Körner, C. Evaporation Model for Beam Based Additive Manufacturing Using Free Surface Lattice Boltzmann Methods. *J. Phys. D Appl. Phys.* **2014**, *47*, 275303. [CrossRef]
158. Klassen, A.; Forster, V.E.; Körner, C. A Multi-Component Evaporation Model for Beam Melting Processes. *Model. Simul. Mat. Sci. Eng.* **2017**, *25*, 025003. [CrossRef]
159. Rodgers, T.M.; Madison, J.D.; Tikare, V. Simulation of Metal Additive Manufacturing Microstructures Using Kinetic Monte Carlo. *Comput. Mater. Sci.* **2017**, *135*, 78–89. [CrossRef]
160. Rolchigo, M.; Reeve, S.T.; Stump, B.; Knapp, G.L.; Coleman, J.; Plotkowski, A.; Belak, J. ExaCA: A Performance Portable Exascale Cellular Automata Application for Alloy Solidification Modeling. *Comput. Mater. Sci.* **2022**, *214*, 111692. [CrossRef]
161. Dorr, M.R.; Fattebert, J.L.; Wickett, M.E.; Belak, J.F.; Turchi, P.E.A. A Numerical Algorithm for the Solution of a Phase-Field Model of Polycrystalline Materials. *J. Comput. Phys.* **2010**, *229*, 626–641. [CrossRef]
162. Fattebert, J.L.; Wickett, M.E.; Turchi, P.E.A. Phase-Field Modeling of Coring during Solidification of Au-Ni Alloy Using Quaternions and CALPHAD Input. *Acta Mater.* **2014**, *62*, 89–104. [CrossRef]
163. The Tusas Development Team Tusas Project Website 2017. Available online: https://github.com/chrisknewman/tusas (accessed on 15 November 2024).
164. Radhakrishnan, B.; Gorti, S.B.; Song, Y. *MEUMAPPS (Microstructure Evolution Using Massively Parallel Phase-Field Simulations)*; Oak Ridge National Laboratory (ORNL): Oak Ridge, TN, USA, 2020.

165. Teferra, K.; Rowenhorst, D.J. Optimizing the Cellular Automata Finite Element Model for Additive Manufacturing to Simulate Large Microstructures. *Acta Mater.* **2021**, *213*, 116930. [CrossRef]
166. Pauza, J.G.; Tayon, W.A.; Rollett, A.D. Computer Simulation of Microstructure Development in Powder-Bed Additive Manufacturing with Crystallographic Texture. *Model. Simul. Mat. Sci. Eng.* **2021**, *29*, 055019. [CrossRef]
167. Kurz, W.; Fisher, D.J. *Fundamentals of Solidification*, 4th ed.; Trans Tech Publications: Wollerau, Switzerland, 1998.
168. Bae, C.J.; Halloran, J.W. A Segregation Model Study of Suspension-Based Additive Manufacturing. *J. Eur. Ceram. Soc.* **2018**, *38*, 5160–5166. [CrossRef]
169. Tolochko, N.K.; Laoui, T.; Khlopkov, Y.V.; Mozzharov, S.E.; Titov, V.I.; Ignatiev, M.B. Absorptance of Powder Materials Suitable for Laser Sintering. *Rapid Prototyp. J.* **2000**, *6*, 155–160. [CrossRef]
170. Negi, S.; Nambolan, A.A.; Kapil, S.; Joshi, P.S.; Manivannan, R.; Karunakaran, K.P.; Bhargava, P. Review on Electron Beam Based Additive Manufacturing. *Rapid Prototyp. J.* **2020**, *26*, 485–498. [CrossRef]
171. Drstvenšek, I.; Zupanič, F.; Bončina, T.; Brajlih, T.; Pal, S. Influence of Local Heat Flow Variations on Geometrical Deflections, Microstructure, and Tensile Properties of Ti-6Al-4 V Products in Powder Bed Fusion Systems. *J. Manuf. Process* **2021**, *65*, 382–396. [CrossRef]
172. Baier, D.; Wolf, F.; Weckenmann, T.; Lehmann, M.; Zaeh, M.F. Thermal Process Monitoring and Control for a Near-Net-Shape Wire and Arc Additive Manufacturing. *Prod. Eng.* **2022**, *16*, 811–822. [CrossRef]
173. Sugden, A.A.B. Towards the Prediction of Weld Metal Properties. Ph.D. Thesis, University of Cambridge, Cambridge, UK, 1989.
174. Nuñez, L.; Sabharwall, P.; van Rooyen, I.J. In Situ Embedment of Type K Sheathed Thermocouples with Directed Energy Deposition. *Int. J. Adv. Manuf. Technol.* **2023**, *127*, 3611–3623. [CrossRef]
175. Gatsos, T.; Elsayed, K.A.; Zhai, Y.; Lados, D.A. Review on Computational Modeling of Process–Microstructure–Property Relationships in Metal Additive Manufacturing. *JOM* **2020**, *72*, 403–419. [CrossRef]
176. Du, H.; Wei, Y.; Wang, W.; Lin, W.; Fan, D. Numerical Simulation of Temperature and Fluid in GTAW-Arc under Changing Process Conditions. *J. Mater. Process Technol.* **2009**, *209*, 3752–3765. [CrossRef]
177. Ushio, M.; Matsuda, F. Mathematical Modelling of Heat Transfer of Welding Arc. *Trans. JWRI* **1982**, *11*, 7–15. [CrossRef]
178. Denlinger, E.R.; Gouge, M.; Irwin, J.; Michaleris, P. Thermomechanical Model Development and in Situ Experimental Validation of the Laser Powder-Bed Fusion Process. *Addit. Manuf.* **2017**, *16*, 73–80. [CrossRef]
179. Ning, J.; Praniewicz, M.; Wang, W.; Dobbs, J.R.; Liang, S.Y. Analytical Modeling of Part Distortion in Metal Additive Manufacturing. *Int. J. Adv. Manuf. Technol.* **2020**, *107*, 49–57. [CrossRef]
180. Gusarov, A.V.; Grigoriev, S.N.; Volosova, M.A.; Melnik, Y.A.; Laskin, A.; Kotoban, D.V.; Okunkova, A.A. On Productivity of Laser Additive Manufacturing. *J. Mater. Process Technol.* **2018**, *261*, 213–232. [CrossRef]
181. Kohser, R.; Collins, P.; Black, J. *DeGarmo's Materials and Processes in Manufacturing*, 14th ed.; John Wiley & Sons: Hoboken, NJ, USA, 2025; *in preparation for printing*.
182. Kamath, C.; Fan, Y.J. Regression with Small Data Sets: A Case Study Using Code Surrogates in Additive Manufacturing. *Knowl. Inf. Syst.* **2018**, *57*, 475–493. [CrossRef]
183. Trivedi, R.; David, S.A.; Eshelman, M.A.; Vitek, J.M.; Babu, S.S.; Hong, T.; DebRoy, T. In Situ Observations of Weld Pool Solidification Using Transparent Metal-Analog Systems. *J. Appl. Phys.* **2003**, *93*, 4885–4895. [CrossRef]
184. Grugel, R.N.; Well, A.H. Alloy Solidification in Systems Containing a Liquid Miscibility Gap. *Metall. Trans. A* **1981**, *12*, 669–681. [CrossRef]
185. Steube, R.; Hellawell, A. The Use of a Transparent Aqueous Analogue to Demonstrate the Development of Segregation Channels During Alloy Solidification, Vertically Upwards. *Video J. Engng. Res.* **1993**, *3*, 1–16.
186. Kledwig, C.; Perfahl, H.; Reisacher, M.; Brückner, F.; Bliedtner, J.; Leyens, C. Analysis of Melt Pool Characteristics and Process Parameters Using a Coaxial Monitoring System during Directed Energy Deposition in Additive Manufacturing. *Materials* **2019**, *12*, 308. [CrossRef]
187. Wu, B.; Ji, X.Y.; Zhou, J.X.; Yang, H.Q.; Peng, D.J.; Wang, Z.M.; Wu, Y.J.; Yin, Y.J. In Situ Monitoring Methods for Selective Laser Melting Additive Manufacturing Process Based on Images—A Review. *China Foundry* **2021**, *18*, 265–285. [CrossRef]
188. Zheng, H.; Li, H.; Lang, L.; Gong, S.; Ge, Y. Effects of Scan Speed on Vapor Plume Behavior and Spatter Generation in Laser Powder Bed Fusion Additive Manufacturing. *J. Manuf. Process* **2018**, *36*, 60–67. [CrossRef]
189. McCann, R.; Obeidi, M.A.; Hughes, C.; McCarthy, É.; Egan, D.S.; Vijayaraghavan, R.K.; Joshi, A.M.; Acinas Garzon, V.; Dowling, D.P.; McNally, P.J.; et al. In-Situ Sensing, Process Monitoring and Machine Control in Laser Powder Bed Fusion: A Review. *Addit. Manuf.* **2021**, *45*, 102058. [CrossRef]
190. Harbig, J.; Wenzler, D.L.; Baehr, S.; Kick, M.K.; Merschroth, H.; Wimmer, A.; Weigold, M.; Zaeh, M.F. Methodology to Determine Melt Pool Anomalies in Powder Bed Fusion of Metals Using a Laser Beam by Means of Process Monitoring and Sensor Data Fusion. *Materials* **2022**, *15*, 1265. [CrossRef]
191. Craeghs, T.; Clijsters, S.; Yasa, E.; Kruth, J.-P. Online Quality Control of Selective Laser Melting. In Proceedings of the Solid Freeform Fabrication Symposium, Austin, TX, USA, 8–10 August 2011.
192. Kamath, R.R.; Choo, H.; Fezzaa, K.; Babu, S.S. Estimation of Spatio-Temporal Temperature Evolution During Laser Spot Melting Using In Situ Dynamic X-Ray Radiography. *Metall. Mater. Trans. A Phys. Metall. Mater. Sci.* **2024**, *55*, 983–991. [CrossRef]
193. Wolff, S.J.; Wu, H.; Parab, N.; Zhao, C.; Ehmann, K.F.; Sun, T.; Cao, J. In-Situ High-Speed X-Ray Imaging of Piezo-Driven Directed Energy Deposition Additive Manufacturing. *Sci. Rep.* **2019**, *9*, 962. [CrossRef] [PubMed]

194. Ye, D.; Zhu, K.; Fuh, J.Y.H.; Zhang, Y.; Soon, H.G. The Investigation of Plume and Spatter Signatures on Melted States in Selective Laser Melting. *Opt. Laser Technol.* **2019**, *111*, 395–406. [CrossRef]
195. Baehr, S.; Melzig, L.; Bauer, D.; Ammann, T.; Zaeh, M.F. Investigations of Process By-Products by Means of Schlieren Imaging during the Powder Bed Fusion of Metals Using a Laser Beam. *J. Laser Appl.* **2022**, *34*, 042045. [CrossRef]
196. Martin, A.A.; Calta, N.P.; Khairallah, S.A.; Wang, J.; Depond, P.J.; Fong, A.Y.; Thampy, V.; Guss, G.M.; Kiss, A.M.; Stone, K.H.; et al. Dynamics of Pore Formation during Laser Powder Bed Fusion Additive Manufacturing. *Nat. Commun.* **2019**, *10*, 1987. [CrossRef]
197. Foster, S.J.; Carver, K.; Dinwiddie, R.B.; List, F.; Unocic, K.A.; Chaudhary, A.; Babu, S.S. Process-Defect-Structure-Property Correlations During Laser Powder Bed Fusion of Alloy 718: Role of In Situ and Ex Situ Characterizations. *Metall. Mater. Trans. A Phys. Metall. Mater. Sci.* **2018**, *49*, 5775–5798. [CrossRef]
198. Yang, Z.; Lu, Y.; Yeung, H.; Krishnamurty, S. Investigation of Deep Learning for Real-Time Melt Pool Classification in Additive Manufacturing. In Proceedings of the IEEE International Conference on Automation Science and Engineering, Vancouver, BC, Canada, 22–26 August 2019; Volume 2019.
199. Sala, V.; Vandone, A.; Mazzucato, F.; Banfi, M.; Baraldo, S.; Valente, A. AI-Aided Thermal Imaging with Multispectral Camera for Direct Energy Deposition. In Proceedings of the 2024 IEEE International Workshop on Metrology for Industry 4.0 and IoT, MetroInd4.0 and IoT 2024—Proceedings, Firenze, Italy, 29–31 May 2024; Institute of Electrical and Electronics Engineers Inc.: Piscataway, NJ, USA, 2024; pp. 150–155.
200. Mochi, V.H.; Núñez, H.H.L.; Ribeiro, K.S.B.; Venter, G.S. Real-Time Prediction of Deposited Bead Width in L-DED Using Semi-Supervised Transfer Learning. *Int. J. Adv. Manuf. Technol.* **2023**, *129*, 5643–5654. [CrossRef]
201. Jasien, C.; Saville, A.; Becker, C.G.; Klemm-Toole, J.; Fezzaa, K.; Sun, T.; Pollock, T.; Clarke, A.J. In Situ X-Ray Radiography and Computational Modeling to Predict Grain Morphology in β-Titanium during Simulated Additive Manufacturing. *Metals* **2022**, *12*, 1217. [CrossRef]
202. Lamb, J.; Ochoa, R.; Eres-Castellanos, A.; Klemm-Toole, J.; Echlin, M.P.; Sun, T.; Fezzaa, K.; Clarke, A.; Pollock, T.M. Quantification of Melt Pool Dynamics and Microstructure during Simulated Additive Manufacturing. *Scr. Mater.* **2024**, *245*, 116036. [CrossRef]
203. *FLOW-3D®, version 2023R1*; Flow Science, Inc.: Santa Fe, NM, USA, 2023.
204. Tong, M.; Browne, D.J. Smoothed Particle Hydrodynamics Modelling of the Fluid Flow and Heat Transfer in the Weld Pool during Laser Spot Welding. *Proc. IOP Conf. Ser. Mater. Sci. Eng.* **2011**, *27*, 012080. [CrossRef]
205. Olleak, A.; Adcock, E.; Hinnebusch, S.; Dugast, F.; Rollett, A.D.; To, A.C. Understanding the Role of Geometry and Interlayer Cooling Time on Microstructure Variations in LPBF Ti6Al4V through Part-Scale Scan-Resolved Thermal Modeling. *Addit. Manuf. Lett.* **2024**, *9*, 100197. [CrossRef]
206. Sikan, F.; Wanjara, P.; Gholipour, J.; Kumar, A.; Brochu, M. Thermo-Mechanical Modeling of Wire-Fed Electron Beam Additive Manufacturing. *Materials* **2021**, *14*, 911. [CrossRef] [PubMed]
207. Sharma, S.; Krishna, K.V.M.; Joshi, S.S.; Radhakrishnan, M.; Palaniappan, S.; Dussa, S.; Banerjee, R.; Dahotre, N.B. Laser Based Additive Manufacturing of Tungsten: Multi-Scale Thermo-Kinetic and Thermo-Mechanical Computational Model and Experiments. *Acta Mater.* **2023**, *259*, 119244. [CrossRef]
208. Mazumder, S.; Palaniappan, S.; Pantawane, M.V.; Radhakrishnan, M.; Patil, S.M.; Dowden, S.; Mahajan, C.; Mukherjee, S.; Dahotre, N.B. Electrochemical Response of Heterogeneous Microstructure of Laser Directed Energy Deposited CoCrMo in Physiological Medium. *Appl. Phys. A Mater. Sci. Process* **2023**, *129*, 332. [CrossRef]
209. Sharma, S.; Mani Krishna, K.V.; Radhakrishnan, M.; Pantawane, M.V.; Patil, S.M.; Joshi, S.S.; Banerjee, R.; Dahotre, N.B. A Pseudo Thermo-Mechanical Model Linking Process Parameters to Microstructural Evolution in Multilayer Additive Friction Stir Deposition of Magnesium Alloy. *Mater. Des.* **2022**, *224*, 111412. [CrossRef]
210. Turner, J.A.; Belak, J.; Barton, N.; Bement, M.; Carlson, N.; Carson, R.; DeWitt, S.; Fattebert, J.L.; Hodge, N.; Jibben, Z.; et al. ExaAM: Metal Additive Manufacturing Simulation at the Fidelity of the Microstructure. *Int. J. High. Perform. Comput. Appl.* **2022**, *36*, 13–39. [CrossRef]
211. Mniszewski, S.M.; Belak, J.; Fattebert, J.L.; Negre, C.F.A.; Slattery, S.R.; Adedoyin, A.A.; Bird, R.F.; Chang, C.; Chen, G.; Ethier, S.; et al. Enabling Particle Applications for Exascale Computing Platforms. *Int. J. High. Perform. Comput. Appl.* **2021**, *35*, 572–597. [CrossRef]
212. Zhao, C.; Shi, B.; Chen, S.; Du, D.; Sun, T.; Simonds, B.J.; Fezzaa, K.; Rollett, A.D. Laser Melting Modes in Metal Powder Bed Fusion Additive Manufacturing. *Rev. Mod. Phys.* **2022**, *94*, 045002. [CrossRef]
213. Ales, T.K. An Integrated Model for the Probabilistic Prediction of Yield Strength in Electron-Beam Additively Manufactured Ti-6Al-4V. Master's Thesis, Iowa State University, Ames, IA, USA, 2018.
214. Semiatin, S.L.; Ivanchenko, V.G.; Akhonin, S.V.; Ivasishin, O.M. Diffusion Models for Evaporation Losses during Electron-Beam Melting of Alpha/Beta-Titanium Alloys. *Metall. Mater. Trans. B* **2004**, *35*, 235–245. [CrossRef]
215. Babu, S.; DebRoy, T. Liquid Metal Expulsion during Laser Irradiation. *J. Appl. Phys.* **1992**, *72*, 3317–3322. [CrossRef]
216. Stokes, M.A.; Khairallah, S.A.; Volkov, A.N.; Rubenchik, A.M. Fundamental Physics Effects of Background Gas Species and Pressure on Vapor Plume Structure and Spatter Entrainment in Laser Melting. *Addit. Manuf.* **2022**, *55*, 102819. [CrossRef]
217. Liu, J.; Wen, P. Metal Vaporization and Its Influence during Laser Powder Bed Fusion Process. *Mater. Des.* **2022**, *215*, 110505. [CrossRef]
218. O'Donnell, K. The Use of Defects and Compositional Variations to Elucidate Physical Phenomena in Electron Beam Melted Ti-6Al-4V across Scan Strategies. Ph.D. Thesis, Iowa State University, Ames, IA, USA, 2023.

219. O'Donnell, K.; Quintana, M.J.; Collins, P.C. Understanding the Effect of Electron Beam Melting Scanning Strategies on the Aluminum Content and Materials State of Single Ti-6Al-4V Feedstock. *Materials* **2023**, *16*, 6366. [CrossRef] [PubMed]
220. Sharma, A.; Chen, J.; Diewald, E.; Imanian, A.; Beuth, J.; Liu, Y. Data-Driven Sensitivity Analysis for Static Mechanical Properties of Additively Manufactured Ti–6Al–4V. *ASCE-ASME J. Risk Uncertain. Eng. Syst. Part. B Mech. Eng.* **2022**, *8*, 011108. [CrossRef]
221. Grasso, M.; Colosimo, B.M. A Statistical Learning Method for Image-Based Monitoring of the Plume Signature in Laser Powder Bed Fusion. *Robot. Comput. Integr. Manuf.* **2019**, *57*, 103–115. [CrossRef]
222. Yang, D.; Li, H.; Liu, S.; Song, C.; Yang, Y.; Shen, S.; Lu, J.; Liu, Z.; Zhu, Y. In Situ Capture of Spatter Signature of SLM Process Using Maximum Entropy Double Threshold Image Processing Method Based on Genetic Algorithm. *Opt. Laser Technol.* **2020**, *131*, 106371. [CrossRef]
223. Maslow, A.H. A Theory of Human Motivation. *Psychol. Rev.* **1943**, *50*, 370–396. [CrossRef]
224. Ferro, P.; Meneghello, R.; Savio, G.; Berto, F. A Modified Volumetric Energy Density-Based Approach for Porosity Assessment in Additive Manufacturing Process Design. *Int. J. Adv. Manuf. Technol.* **2020**, *110*, 1911–1921. [CrossRef]
225. Li, C.; Liu, Z.Y.; Fang, X.Y.; Guo, Y.B. Residual Stress in Metal Additive Manufacturing. In *Proceedings of the Procedia CIRP*; Elsevier B.V.: Amsterdam, The Netherlands, 2018; Volume 71, pp. 348–353.
226. Fergani, O.; Berto, F.; Welo, T.; Liang, S.Y. Analytical Modelling of Residual Stress in Additive Manufacturing. *Fatigue Fract. Eng. Mater. Struct.* **2017**, *40*, 971–978. [CrossRef]
227. Mukherjee, T.; Zhang, W.; DebRoy, T. An Improved Prediction of Residual Stresses and Distortion in Additive Manufacturing. *Comput. Mater. Sci.* **2017**, *126*, 360–372. [CrossRef]
228. Biswal, R.; Zhang, X.; Shamir, M.; Al Mamun, A.; Awd, M.; Walther, F.; Khadar Syed, A. Interrupted Fatigue Testing with Periodic Tomography to Monitor Porosity Defects in Wire + Arc Additive Manufactured Ti-6Al-4V. *Addit. Manuf.* **2019**, *28*, 517–527. [CrossRef]
229. Vastola, G.; Zhang, G.; Pei, Q.X.; Zhang, Y.W. Controlling of Residual Stress in Additive Manufacturing of Ti6Al4V by Finite Element Modeling. *Addit. Manuf.* **2016**, *12*, 231–239. [CrossRef]
230. Körner, C.; Markl, M.; Koepf, J.A. Modeling and Simulation of Microstructure Evolution for Additive Manufacturing of Metals: A Critical Review. *Metall. Mater. Trans. A Phys. Metall. Mater. Sci.* **2020**, *51*, 4970–4983. [CrossRef]
231. Bellini, C.; Borrelli, R.; Cocco, V.D.; Franchitti, S.; Iacoviello, F.; Maletta, C.; Mocanu, L.P. Fatigue Crack Growth in Ti-6Al-4V EBMed Samples: Impact of Powder Recycling. In *Proceedings of the Procedia Structural Integrity*; Elsevier B.V.: Amsterdam, The Netherlands, 2024; Volume 53, pp. 129–135.
232. Quintana, M.J.; Ji, Y.; Collins, P.C. A Perspective of the Needs and Opportunities for Coupling Materials Science and Nondestructive Evaluation for Metals-Based Additive Manufacturing. *Mater. Eval.* **2022**, *80*, 45–63. [CrossRef]
233. *ISO/ASTMTR52905-EB*; Additive Manufacturing of Metals—Nondestructive Testing and Evaluation—Defect Detection in Parts. ASTM International: West Conshohocken, PA, USA, 2023.
234. Baumgartl, H.; Tomas, J.; Buettner, R.; Merkel, M. A Deep Learning-Based Model for Defect Detection in Laser-Powder Bed Fusion Using in-Situ Thermographic Monitoring. *Prog. Addit. Manuf.* **2020**, *5*, 277–285. [CrossRef]
235. D'Accardi, E.; Krankenhagen, R.; Ulbricht, A.; Pelkner, M.; Pohl, R.; Palumbo, D.; Galietti, U. Capability to Detect and Localize Typical Defects of Laser Powder Bed Fusion (L-PBF) Process: An Experimental Investigation with Different Non-Destructive Techniques. *Prog. Addit. Manuf.* **2022**, *7*, 1239–1256. [CrossRef]
236. Remani, A.; Williams, R.; Thompson, A.; Dardis, J.; Jones, N.; Hooper, P.; Leach, R. Design of a Multi-Sensor Measurement System for in-Situ Defect Identification in Metal Additive Manufacturing. In Proceedings of the Joint Special Interest Group Meeting Between euspen and ASPE Advancing Precision in Additive Manufacturing Inspire AG, St. Gallen, Switzerland, 20–22 September 2021.
237. Paulson, N.H.; Gould, B.; Wolff, S.J.; Stan, M.; Greco, A.C. Correlations between Thermal History and Keyhole Porosity in Laser Powder Bed Fusion. *Addit. Manuf.* **2020**, *34*, 101213. [CrossRef]
238. Petrich, J.; Snow, Z.; Corbin, D.; Reutzel, E.W. Multi-Modal Sensor Fusion with Machine Learning for Data-Driven Process Monitoring for Additive Manufacturing. *Addit. Manuf.* **2021**, *48*, 102364. [CrossRef]
239. Snow, Z.; Diehl, B.; Reutzel, E.W.; Nassar, A. Toward In-Situ Flaw Detection in Laser Powder Bed Fusion Additive Manufacturing through Layerwise Imagery and Machine Learning. *J. Manuf. Syst.* **2021**, *59*, 12–26. [CrossRef]
240. Bugatti, M.; Colosimo, B.M. Towards Real-Time in-Situ Monitoring of Hot-Spot Defects in L-PBF: A New Classification-Based Method for Fast Video-Imaging Data Analysis. *J. Intell. Manuf.* **2022**, *33*, 293–309. [CrossRef]
241. Mitchell, J.A.; Ivanoff, T.A.; Dagel, D.; Madison, J.D.; Jared, B. Linking Pyrometry to Porosity in Additively Manufactured Metals. *Addit. Manuf.* **2020**, *31*, 100946. [CrossRef]
242. Wu, Q.; Yang, F.; Lv, C.; Liu, C.; Tang, W.; Yang, J. In-Situ Quality Intelligent Classification of Additively Manufactured Parts Using a Multi-Sensor Fusion Based Melt Pool Monitoring System. *Addit. Manuf. Front.* **2024**, *3*, 200153. [CrossRef]
243. Ye, J.; Patel, M.; Alam, N.; Vargas-Uscategui, A.; Cole, I. A Dimensionless Group-Incorporating Artificial Neural Network (DI-ANN) Model for Single-Track Depth Prediction of SS316L for Laser-Directed Energy Deposition (L-DED). *Int. J. Adv. Manuf. Technol.* **2024**, *135*, 3529–3545. [CrossRef]
244. Liu, H.; Gobert, C.; Ferguson, K.; Abranovic, B.; Chen, H.; Beuth, J.L.; Rollett, A.D.; Kara, L.B. Inference of Highly Time-Resolved Melt Pool Visual Characteristics and Spatially-Dependent Lack-of-Fusion Defects in Laser Powder Bed Fusion Using Acoustic and Thermal Emission Data. *Addit. Manuf.* **2024**, *83*, 104057. [CrossRef]

245. Carter, L.N.; Wang, X.; Read, N.; Khan, R.; Aristizabal, M.; Essa, K.; Attallah, M.M. Process Optimisation of Selective Laser Melting Using Energy Density Model for Nickel Based Superalloys. *Mater. Sci. Technol.* **2016**, *32*, 657–661. [CrossRef]
246. Aidibe, A.; Tahan, A.; Brailovski, V. Metrological Investigation of a Selective Laser Melting Additive Manufacturing System: A Case Study. *IFAC-PapersOnLine* **2016**, *49*, 25–29. [CrossRef]
247. Boyce, B.L.; Salzbrenner, B.C.; Rodelas, J.M.; Swiler, L.P.; Madison, J.D.; Jared, B.H.; Shen, Y. Extreme-Value Statistics Reveal Rare Failure-Critical Defects in Additive Manufacturing. *Adv. Eng. Mater.* **2017**, *19*, 1700102. [CrossRef]
248. Wilson-Heid, A.E.; Beese, A.M. Combined Effects of Porosity and Stress State on the Failure Behavior of Laser Powder Bed Fusion Stainless Steel 316L. *Addit. Manuf.* **2021**, *39*, 101862. [CrossRef]
249. Wilson-Heid, A.E.; Beese, A.M. Fracture of Laser Powder Bed Fusion Additively Manufactured Ti–6Al–4V under Multiaxial Loading: Calibration and Comparison of Fracture Models. *Mater. Sci. Eng. A* **2019**, *761*, 137967. [CrossRef]
250. Wilson-Heid, A.E.; Furton, E.T.; Beese, A.M. Contrasting the Role of Pores on the Stress State Dependent Fracture Behavior of Additively Manufactured Low and High Ductility Metals. *Materials* **2021**, *14*, 3657. [CrossRef]
251. Furton, E.T.; Wilson-Heid, A.E.; Beese, A.M. Effect of Stress Triaxiality and Penny-Shaped Pores on Tensile Properties of Laser Powder Bed Fusion Ti-6Al-4V. *Addit. Manuf.* **2021**, *48*, 102414. [CrossRef]
252. du Plessis, A.; Yadroitsava, I.; Yadroitsev, I. Effects of Defects on Mechanical Properties in Metal Additive Manufacturing: A Review Focusing on X-Ray Tomography Insights. *Mater. Des.* **2020**, *187*, 108385. [CrossRef]
253. Sigworth, G.K. Fundamentals of Solidification in Aluminum Castings. *Int. J. Met.* **2014**, *8*, 7–20. [CrossRef]
254. Luo, Q.; Shimanek, J.D.; Simpson, T.W.; Beese, A.M. An Image-Based Transfer Learning Approach for Using In Situ Processing Data to Predict Laser Powder Bed Fusion Additively Manufactured Ti-6Al-4V Mechanical Properties. *3D Print. Addit. Manuf.* **2024**. [CrossRef]
255. Luo, Q.; Huang, N.; Fu, T.; Wang, J.; Bartles, D.L.; Simpson, T.W.; Beese, A.M. New Insight into the Multivariate Relationships among Process, Structure, and Properties in Laser Powder Bed Fusion AlSi10Mg. *Addit. Manuf.* **2023**, *77*, 103804. [CrossRef]
256. Luo, Q.; Yin, L.; Simpson, T.W.; Beese, A.M. Effect of Processing Parameters on Pore Structures, Grain Features, and Mechanical Properties in Ti-6Al-4V by Laser Powder Bed Fusion. *Addit. Manuf.* **2022**, *56*, 102915. [CrossRef]
257. Oliveira, J.P.; LaLonde, A.D.; Ma, J. Processing Parameters in Laser Powder Bed Fusion Metal Additive Manufacturing. *Mater. Des.* **2020**, *193*, 108762. [CrossRef]
258. Song, L.; Mazumder, J. Real Time Cr Measurement Using Optical Emission Spectroscopy during Direct Metal Deposition Process. *IEEE Sens. J.* **2012**, *12*, 958–964. [CrossRef]
259. Kim, C.S. *Thermophysical Properties of Stainless Steels*; Argonne National Lab.: Lemont, IL, USA, 1975.
260. Bogaard, R.H.; Desai, P.D.; Li, H.H.; Ho, C.Y. Thermophysical Properties of Stainless Steels. *Thermochim. Acta* **1993**, *218*, 373–393. [CrossRef]
261. Montero-Sistiaga, M.L.; Mertens, R.; Vrancken, B.; Wang, X.; Van Hooreweder, B.; Kruth, J.-P.; Van Humbeeck, J. Changing the Alloy Composition of Al7075 for Better Processability by Selective Laser Melting. *J. Mater. Process Technol.* **2016**, *238*, 437–445. [CrossRef]
262. Towfighi, S.; Romilly, D.P.; Olson, J.A. Elevated Temperature Material Characteristics of AISI 304L Stainless Steel. *Mater. High. Temp.* **2013**, *30*, 151–155. [CrossRef]
263. Squires, B.; Flannery, D.; Bivens, T.; Banerjee, R.; McWilliams, B.; Cho, K.; Neogi, A.; Dahotre, N.B.; Voevodin, A.A. Laser-Induced Breakdown Spectroscopy for Composition Monitoring during Directed Energy Deposition of Graded Fe-Ni Alloys. *Int. J. Adv. Manuf. Technol.* **2024**, *132*, 3877–3888. [CrossRef]
264. Shassere, B.; Nycz, A.; Noakes, M.W.; Masuo, C.; Sridharan, N. Correlation of Microstructure and Mechanical Properties of Metal Big Area Additive Manufacturing. *Appl. Sci.* **2019**, *9*, 787. [CrossRef]
265. Wang, L.; Xue, J.; Wang, Q. Correlation between Arc Mode, Microstructure, and Mechanical Properties during Wire Arc Additive Manufacturing of 316L Stainless Steel. *Mater. Sci. Eng. A* **2019**, *751*, 183–190. [CrossRef]
266. Köhnen, P.; Létang, M.; Voshage, M.; Schleifenbaum, J.H.; Haase, C. Understanding the Process-Microstructure Correlations for Tailoring the Mechanical Properties of L-PBF Produced Austenitic Advanced High Strength Steel. *Addit. Manuf.* **2019**, *30*, 100914. [CrossRef]
267. Lee, D.; Park, S.; Lee, C.H.; Hong, H.U.; Oh, J.; So, T.Y.; Kim, W.S.; Seo, D.; Han, J.; Ko, S.H.; et al. Correlation between Microstructure and Mechanical Properties in Additively Manufactured Inconel 718 Superalloys with Low and High Electron Beam Currents. *J. Mater. Res. Technol.* **2024**, *28*, 2410–2419. [CrossRef]
268. Kobryn, P.A.; Semiatin, S.L. Microstructure and Texture Evolution during Solidification Processing of Ti-6Al-4V. *J. Mater. Process Technol.* **2003**, *135*, 330–339. [CrossRef]
269. Bontha, S.; Klingbeil, N.W.; Kobryn, P.A.; Fraser, H.L. Effects of Process Variables and Size-Scale on Solidification Microstructure in Beam-Based Fabrication of Bulky 3D Structures. *Mater. Sci. Eng. A* **2009**, *513–514*, 311–318. [CrossRef]
270. Dehoff, R.R.; Kirka, M.M.; List, F.A.; Unocic, K.A.; Sames, W.J. Crystallographic Texture Engineering through Novel Melt Strategies via Electron Beam Melting: Inconel 718. *Mater. Sci. Technol.* **2015**, *31*, 939–944. [CrossRef]
271. Dehoff, R.R.; Kirka, M.; Sames, W.J.; Bilheux, H.; Tremsin, A.S.; Lowe, L.E.; Babu, S.S. Site Specific Control of Crystallographic Grain Orientation through Electron Beam Additive Manufacturing. *Mater. Sci. Technol.* **2015**, *31*, 931–938. [CrossRef]
272. Li, W.; Dryburgh, P.; Pieris, D.; Patel, R.; Clark, M.; Smith, R.J. Imaging Microstructure on Optically Rough Surfaces Using Spatially Resolved Acoustic Spectroscopy. *Appl. Sci.* **2023**, *13*, 3424. [CrossRef]

273. Carson, R.; Rolchigo, M.; Coleman, J.; Titov, M.; Belak, J.; Bement, M. Uncertainty Quantification of Metal Additive Manufacturing Processing Conditions Through the Use of Exascale Computing. In Proceedings of the ACM International Conference Proceeding Series, Denver, CO, USA, 12 November 2023; Association for Computing Machinery: New York, NY, USA, 2023; pp. 380–383.

274. Donegan, S.P.; Tucker, J.C.; Rollett, A.D.; Barmak, K.; Groeber, M. Extreme Value Analysis of Tail Departure from Log-Normality in Experimental and Simulated Grain Size Distributions. *Acta Mater.* **2013**, *61*, 5595–5604. [CrossRef]

275. Hielscher, R.; Schaeben, H. A Novel Pole Figure Inversion Method: Specification of the MTEX Algorithm. *J. Appl. Crystallogr.* **2008**, *41*, 1024–1037. [CrossRef]

276. Grong, Ø. *Metallurgical Modelling of Welding*, 2nd ed.; The Institute of Materials: London, UK, 1997.

277. Liu, P.; Wang, Z.; Xiao, Y.; Horstemeyer, M.F.; Cui, X.; Chen, L. Insight into the Mechanisms of Columnar to Equiaxed Grain Transition during Metallic Additive Manufacturing. *Addit. Manuf.* **2019**, *26*, 22–29. [CrossRef]

278. Hosseini, V.; Karlsson, L.; Engelberg, D.; Wessman, S. Time-Temperature-Precipitation and Property Diagrams for Super Duplex Stainless Steel Weld Metals. *Weld. World* **2018**, *62*, 517–533. [CrossRef]

279. Ji, Y.; Chen, L.; Chen, L.Q. Understanding Microstructure Evolution During Additive Manufacturing of Metallic Alloys Using Phase-Field Modeling. In *Thermo-Mechanical Modeling of Additive Manufacturing*; Elsevier: Amsterdam, The Netherlands, 2017; pp. 93–116, ISBN 9780128118207.

280. Sharma, S.; Joshi, S.S.; Pantawane, M.V.; Radhakrishnan, M.; Mazumder, S.; Dahotre, N.B. Multiphysics Multi-Scale Computational Framework for Linking Process–Structure–Property Relationships in Metal Additive Manufacturing: A Critical Review. *Int. Mater. Rev.* **2023**, *68*, 943–1009. [CrossRef]

281. Rolchigo, M.R.; LeSar, R. Modeling of Binary Alloy Solidification under Conditions Representative of Additive Manufacturing. *Comput. Mater. Sci.* **2018**, *150*, 535–545. [CrossRef]

282. Rolchigo, M.R.; LeSar, R. Application of Alloy Solidification Theory to Cellular Automata Modeling of Near-Rapid Constrained Solidification. *Comput. Mater. Sci.* **2019**, *163*, 148–161. [CrossRef]

283. Rolchigo, M.; Stump, B.; Belak, J.; Plotkowski, A. Sparse Thermal Data for Cellular Automata Modeling of Grain Structure in Additive Manufacturing. *Model. Simul. Mat. Sci. Eng.* **2020**, *28*, 065003. [CrossRef]

284. Mendoza, M.Y.; Samimi, P.; Brice, D.A.; Martin, B.W.; Rolchigo, M.R.; LeSar, R.; Collins, P.C. Microstructures and Grain Refinement of Additive-Manufactured Ti-XW Alloys. *Metall. Mater. Trans. A Phys. Metall. Mater. Sci.* **2017**, *48*, 3594–3605. [CrossRef]

285. Rolchigo, M.R.; Mendoza, M.Y.; Samimi, P.; Brice, D.A.; Martin, B.; Collins, P.C.; LeSar, R. Modeling of Ti-W Solidification Microstructures Under Additive Manufacturing Conditions. *Metall. Mater. Trans. A Phys. Metall. Mater. Sci.* **2017**, *48*, 3606–3622. [CrossRef]

286. Rodgers, T.M.; Lim, H.; Brown, J.A. Three-Dimensional Additively Manufactured Microstructures and Their Mechanical Properties. *JOM* **2020**, *72*, 75–82. [CrossRef]

287. Trageser, J.E.; Mitchell, J.A.; Johnson, K.L.; Rodgers, T.M. A Bézier Curve Fit to Melt Pool Geometry for Modeling Additive Manufacturing Microstructures. *Comput. Methods Appl. Mech. Eng.* **2023**, *415*, 116208. [CrossRef]

288. Ngo, T.D.; Kashani, A.; Imbalzano, G.; Nguyen, K.T.Q.; Hui, D. Additive Manufacturing (3D Printing): A Review of Materials, Methods, Applications and Challenges. *Compos. B Eng.* **2018**, *143*, 172–196. [CrossRef]

289. Collins, P.C.; Haden, C.V.; Ghamarian, I.; Hayes, B.J.; Ales, T.; Penso, G.; Dixit, V.; Harlow, G. Progress toward an Integration of Process-Structure-Property-Performance Models for "Three-Dimensional (3-D) Printing" of Titanium Alloys. *JOM* **2014**, *66*, 1299–1309. [CrossRef]

290. Halder, R.; Pistorius, P.C.; Blazanin, S.; Sardey, R.P.; Quintana, M.J.; Pierson, E.A.; Verma, A.K.; Collins, P.C.; Rollett, A.D. The Effect of Interlayer Delay on the Heat Accumulation, Microstructures, and Properties in Laser Hot Wire Directed Energy Deposition of Ti-6Al-4V Single-Wall. *Materials* **2024**, *17*, 3307. [CrossRef]

291. Kandil, F.A.; Lord, J.D.; Fry, A.T.; Grant, P.V. *A Review of Residual Stress Measurement Methods—A Guide to Technical Selection*; NPL Materials Centre: Teddington, UK, 2001.

292. Cunha, F.G.; Santos, T.G.; Xavier, J. In Situ Monitoring of Additive Manufacturing Using Digital Image Correlation: A Review. *Materials* **2021**, *14*, 1511. [CrossRef]

293. Baker, A.H.; Collins, P.C.; Williams, J.C. New Nomenclatures for Heat Treatments of Additively Manufactured Titanium Alloys. *JOM* **2017**, *69*, 1221–1227. [CrossRef]

294. Denlinger, E.R.; Irwin, J.; Michaleris, P. Thermomechanical Modeling of Additive Manufacturing Large Parts. *J. Manuf. Sci. Eng. Trans. ASME* **2014**, *136*, 061007. [CrossRef]

295. Megahed, M.; Mindt, H.W.; N'Dri, N.; Duan, H.; Desmaison, O. Metal Additive-Manufacturing Process and Residual Stress Modeling. *Integr. Mater. Manuf. Innov.* **2016**, *5*, 61–93. [CrossRef]

296. Denlinger, E.R.; Michaleris, P. Effect of Stress Relaxation on Distortion in Additive Manufacturing Process Modeling. *Addit. Manuf.* **2016**, *12*, 51–59. [CrossRef]

297. van der Velden, W. Simulation for Additive Manufacturing of a Marine Propeller. Available online: https://technews.shvtech.com/2024/05/simulation-for-additive-manufacturing-of-a-marine-propeller/ (accessed on 12 May 2024).

298. Ansys Additive Print. Available online: https://www.ansys.com/products/additive/ansys-additive-print (accessed on 12 May 2024).

299. Amphyon: Metal Additive Manufacturing Software. Available online: https://oqton.com/amphyon/ (accessed on 12 May 2024).

300. Leung, K.; Imanian, A. Simulation of Laser Powder-Bed Fusion Additive Manufacturing Process Using the COMSOL Multiphysics Software. In Proceedings of the COMSOL Conference, Boston, MA, USA, 11 October 2018.
301. GENOA 3DP Simulation. Available online: https://alphastarcorp.com/genoa-3dp-simulation/ (accessed on 12 May 2024).
302. Autodesk Fusion with Netfabb: Additive Manufacturing, Design, and Simulation. Available online: https://www.autodesk.com/products/netfabb (accessed on 12 May 2024).
303. Simufact Additive. Available online: https://hexagon.com/products/simufact-additive (accessed on 12 May 2024).
304. Metal Additive Manufacturing: A Flawed Process. Available online: https://atlas3d.xyz/features/ (accessed on 12 May 2024).
305. Hodge, N.E.; Ferencz, R.M.; Vignes, R.M. Experimental Comparison of Residual Stresses for a Thermomechanical Model for the Simulation of Selective Laser Melting. *Addit. Manuf.* **2016**, *12*, 159–168. [CrossRef]
306. Sabbaghi, A.; Huang, Q.; Dasgupta, T. Bayesian Model Building From Small Samples of Disparate Data for Capturing In-Plane Deviation in Additive Manufacturing. *Technometrics* **2018**, *60*, 532–544. [CrossRef]
307. *ASTM E647-24*; Standard Test Method for Measurement of Fatigue Crack Growth Rates. ASTM International: West Conshohocken, PA, USA, 2024.
308. *ASTM E606-21*; Standard Test Method for Strain-Controlled Fatigue Testing 1. ASTM International: West Conshohocken, PA, USA, 2021.
309. *ASTM E92-23*; Standard Test Methods for Vickers Hardness and Knoop Hardness of Metallic Materials. ASTM International: West Conshohocken, PA, USA, 2023.
310. *ASTM E10-23*; Standard Test Method for Brinell Hardness of Metallic Materials. ASTM International: West Conshohocken, PA, USA, 2023.
311. *ASTM E18-24*; Standard Test Methods for Rockwell Hardness of Metallic Materials. ASTM International: West Conshohocken, PA, USA, 2024.
312. *ASTM E8-24*; Standard Test Methods for Tension Testing of Metallic Materials. ASTM International: West Conshohocken, PA, USA, 2024.
313. *ASTM E1304-97*; Standard Test Method for Plane-Strain (Chevron-Notch) Fracture Toughness of Metallic Materials. ASTM International: West Conshohocken, PA, USA, 1997.
314. *ASTM E466-21*; Standard Practice for Conducting Force Controlled Constant Amplitude Axial Fatigue Tests of Metallic Materials. ASTM International: West Conshohocken, PA, USA, 2021.
315. *ASTM E837-20*; Standard Test Method for Determining Residual Stresses by the Hole-Drilling Strain-Gage Method. ASTM International: West Conshohocken, PA, USA, 2020.
316. Cederberg, E.; A Hosseini, V.; Kumara, C.; Karlsson, L. Physical Simulation of Additively Manufactured Super Duplex Stainless Steels—Microstructure and Properties. *Addit. Manuf.* **2020**, *34*, 101269. [CrossRef]
317. Roters, F.; Eisenlohr, P.; Bieler, T.; Raabe, D. *Crystal Plasticity Finite Element Methods in Materials Science and Engineering*; Wiley-VCH: Hoboken, NJ, USA, 2010; ISBN 9783527324477.
318. Carson, R.A.; Wopschall, S.R.; Bramwell, J.A. ExaConstit 2019.
319. Hodge, N.E.; Ferencz, R.M.; Solberg, J.M. Implementation of a Thermomechanical Model for the Simulation of Selective Laser Melting. *Comput. Mech.* **2014**, *54*, 33–51. [CrossRef]
320. Hodge, N.E. Towards Improved Speed and Accuracy of Laser Powder Bed Fusion Simulations via Representation of Multiple Time Scales. *Addit. Manuf.* **2021**, *37*, 101600. [CrossRef]

 *materials*

Article

Physics-Informed Online Learning for Temperature Prediction in Metal AM

Pouyan Sajadi, Mostafa Rahmani Dehaghani, Yifan Tang and G. Gary Wang *

Product Design and Optimization Laboratory, Simon Fraser University, Surrey, BC V3T 0A3, Canada;
sps11@sfu.ca (P.S.); mra91@sfu.ca (M.R.D.); yta88@sfu.ca (Y.T.)
* Correspondence: gary_wang@sfu.ca

Abstract: In metal additive manufacturing (AM), precise temperature field prediction is crucial for process monitoring, automation, control, and optimization. Traditional methods, primarily offline and data-driven, struggle with adapting to real-time changes and new process scenarios, which limits their applicability for effective AM process control. To address these challenges, this paper introduces the first physics-informed (PI) online learning framework specifically designed for temperature prediction in metal AM. Utilizing a physics-informed neural network (PINN), this framework integrates a neural network architecture with physics-informed inputs and loss functions. Pretrained on a known process to establish a baseline, the PINN transitions to an online learning phase, dynamically updating its weights in response to new, unseen data. This adaptation allows the model to continuously refine its predictions in real-time. By integrating physics-informed components, the PINN leverages prior knowledge about the manufacturing processes, enabling rapid adjustments to process parameters, geometries, deposition patterns, and materials. Empirical results confirm the robust performance of this PI online learning framework in accurately predicting temperature fields for unseen processes across various conditions. It notably surpasses traditional data-driven models, especially in critical areas like the Heat Affected Zone (HAZ) and melt pool. The PINN's use of physical laws and prior knowledge not only provides a significant advantage over conventional models but also ensures more accurate predictions under diverse conditions. Furthermore, our analysis of key hyperparameters—the learning rate and batch size of the online learning phase—highlights their roles in optimizing the learning process and enhancing the framework's overall effectiveness. This approach demonstrates significant potential to improve the online control and optimization of metal AM processes.

Keywords: physics-informed neural networks; metal additive manufacturing; online learning; real-time modeling; temperature field prediction

Citation: Sajadi, P.; Rahmani Dehaghani, M.; Tang, Y.; Wang, G.G. Physics-Informed Online Learning for Temperature Prediction in Metal AM. *Materials* **2024**, *17*, 3306. https://doi.org/10.3390/ma17133306

Academic Editor: Tomasz Trzepieciński

Received: 5 June 2024
Revised: 27 June 2024
Accepted: 2 July 2024
Published: 4 July 2024

1. Introduction

Metal additive manufacturing (AM) introduces a paradigm shift in the field of manufacturing technologies, significantly enhancing adaptability across a range of sectors such as aerospace, biomedical engineering, and defense [1]. The capacity of this technology to construct detailed, tailor-made 3D configurations through sequential layer deposition not only enables the realization of mass customization but also facilitates the fabrication of components with reduced mass, optimized material consumption, and expedited prototype development [2].

Central to understanding and optimizing metal AM processes is the Parameter–Signature–Quality (PSQ) model, which explains the complex relationships between the settings of the manufacturing process (Parameters), the observable effects during the process (Signatures), and the characteristics of the final product (Qualities). The PSQ model serves as a critical framework, clarifying how variations in process parameters influence the physical and mechanical properties of the final product through changes in process signatures [3–5]. This relationship is key to ensuring that the manufacturing process yields parts that conform to predefined quality and performance standards, allowing for precise control over the variables to achieve the desired outcomes [6].

Temperature emerges as a vital signature in this framework, with its management and prediction being fundamental to the integrity and quality of AM parts [7,8]. In metal AM, the rapid heating and cooling cycles lead to significant temperature fluctuations within the substrate and deposited layers, affecting part quality and integrity through stress, distortion, and microstructural changes [9,10]. Temperature variations can also cause defects such as porosity due to trapped gas bubbles, a lack of fusion from insufficient melting, and surface rugosity from thermal gradients. Non-uniform cooling rates can lead to heterogeneous microstructures and low ductility, compromising mechanical properties. Additionally, trapped inert gas bubbles can form due to improper shielding gas flow or high scanning speeds, further degrading part quality. These defects have been widely reported in the literature [11–14]. In addressing these challenges, real-time or near-real-time temperature prediction models are essential. They enable the dynamic adjustment of process parameters based on thermal feedback, thus enhancing thermal management and reducing defects for improved precision and quality of the final product [15].

Therefore, the capability to monitor and control temperature directly influences PSQ dynamics, impacting thermal signatures and, consequently, the structural and material properties of manufactured components. In improving control over the AM process, these developments not only fulfill specific application requirements but also push the limits of manufacturing efficiency and product innovation in AM.

There has been extensive study on offline, data-driven approaches for thermal modeling in metal AM. Offline, or batch learning, is a method where models are trained on a complete dataset before deployment, without updating or learning from new data during operation. These studies have explored various methodologies for predicting temperature distributions and their effects on the final product's quality, relying on historical data and computational simulations to inform their predictions. For instance, research by Pham et al. [16] developed a feed-forward neural network surrogate model to accurately predict temperature evolution and melting pool sizes in metal bulk samples fabricated using the directed energy deposition (DED) process. In another work, Mozaffar et al. [17] developed a recurrent neural network (RNN)-based model to predict the thermal history of manufactured parts. Utilizing a considerable amount of data produced by the Finite Element Method (FEM), this model is adept at forecasting temperature fields both on the surface and within the interior of fabricated parts. Moreover, in [18], Le et al. used FEM data from five processes with different currents and voltages to train a neural network for temperature prediction at mesh points, using coordinates, travel speed, and current as inputs. The model, demonstrating over 99% accuracy, can predict temperature histories in new cases.

Adding to these conventional approaches, physics-informed neural networks (PINNs), as introduced by Raissi et al. [19], have emerged as a novel machine learning paradigm by integrating physical laws, typically described by partial differential equations (PDEs), directly into the neural network architecture. Notable implementations include the study by Zhu et al. [20], which utilized PINNs for temperature and melt pool dynamics in Laser Powder Bed Fusion (LPBF), and the research by Xie et al. [21], focusing on temperature prediction in DED. Additionally, Jiang et al. [22] demonstrated the effectiveness of PINNs in melt pool temperature predictions with limited training data. These instances underscore the potential of PINNs to improve predictive accuracy and computational efficiency in metal AM by employing physical principles, especially when faced with sparse data.

However, both traditional offline models and PINNs face shortcomings that limit their practicality within the evolving domain of metal AM. A shared challenge is their limited adaptability to real-time manufacturing variations, frequently leading to discrepancies between predicted outcomes and the dynamics of actual processes. These approaches struggle to generalize across the diverse AM processes characterized by different materials, geometries, and process parameters. Additionally, data-driven models require large training datasets, which are often not available in manufacturing. Meanwhile, PINNs, despite leveraging physical laws for prediction, typically focus on narrow segments of the AM process, which curtails their overall utility. Lastly, these models often require extensive computational resources and processing time for training, which can impede swift decision making critical for optimizing AM processes in real time. For example, the processing time reported in [17] reached 40 h, rendering it impractical for scenarios requiring real-time control.

Other approaches, such as analytical or numerical methods, can address some of these limitations but still have their challenges. Recently, Yang et al. [23] implemented an analytical model with three different heat sources to predict melt pool dimensions and temperature distributions. Their study effectively analyzed the effects of laser processing parameters and material thermophysical properties on the temperature field and melt pool size, offering valuable insights into the intricate thermal behavior in metal additive manufacturing. However, analytical models lack real-time adaptability and cannot incorporate dynamic process variations.

Online learning, a branch of machine learning, updates model parameters dynamically with the arrival of new data, presenting a flexible alternative to conventional offline approaches. This method ensures that the model stays relevant and accurate through real-time adaptation and is more efficient due to its reduced memory storage needs [24]. Unlike batch learning, which requires access to the entire dataset for training, online learning processes data incrementally, removing the need for substantial data storage. The benefits of online learning are evident across various fields. For example, Yang et al. [25] designed an online deep learning model for the continuous monitoring of train traction motor temperatures, adjusting the model's structure in response to new data. Wang et al. [26] proposed a combined method that leverages offline learning's predictive capabilities with online learning's adaptability for the real-time control of deformable objects.

In the realm of metal AM, the application of online learning is emerging. Ouidadi et al. [27] applied online learning for real-time defect detection in Laser Metal Deposition (LMD), using transfer learning and adaptive models like K-means and self-organizing maps to enhance quality control by updating predictions with incoming data. Mu et al. [28] developed an online simulation model for Wire Arc Additive Manufacturing (WAAM) that employs neural network techniques to adapt predictions based on live data, showcasing an improvement over conventional models. Despite these advances, a significant gap exists in the specific application of online learning for thermal modeling within metal AM. In one of the first attempts at surrogate modeling in this context, our recent study [29] proposed an online thermal field prediction method using artificial neural networks for thermal field mapping and a reduced-order model (ROM) for thermal field reconstruction, demonstrating its effectiveness in various experiments and simulations. However, this study is limited to thin wall structures, which restricts its applicability to more complex geometries in metal AM. Additionally, the method requires multiple pyrometers that need to be moved as new layers are added, increasing the complexity and potential for measurement errors.

To the authors' best knowledge, this study is the first to explore physics-informed online learning for thermal modeling in metal AM, marking a pioneering step into this research area. Our approach improves upon the existing state-of-the-art approaches by offering dynamic adaptability to real-time data, enhanced predictive accuracy, and greater flexibility and generalizability across different settings. Furthermore, it reduces the need for extensive sensor setups, addressing practical challenges in the field.

To this end, this paper introduces a framework that integrates a physics-informed neural network (PINN) with transfer learning and online learning to address the challenges of thermal modeling in metal AM. At its core, this methodology leverages real-time temperature field data alongside heat boundary conditions to accurately predict 2D temperature fields at future timestamps for processes previously unseen and for which no prior data are available before training. This novel approach is distinguished from prior methods by its dynamic adaptability to a wide array of AM scenarios, including variations in geometries, deposition patterns, and process parameters. This adaptability significantly improves the precision and utility of thermal field modeling in metal AM. Our PINN approach offers several unique advantages, including the following:

1. **Real-Time Adaptability:** The PINN approach can quickly adapt to new data in real time, allowing for immediate updates and predictions during the manufacturing process. This is vital for applications requiring rapid decision making and adjustments.
2. **Faster Computation Times:** Neural networks, once trained, can perform predictions much faster than traditional numerical simulations. This speed advantage is crucial for the real-time monitoring and control of the AM process.
3. **Continuous Learning:** The PINN framework can continuously learn and improve from new data, enhancing its predictive accuracy over time. This capability allows the model to become more robust and reliable with ongoing use.
4. **Utilizing Process Data:** The PINN model leverages data gathered directly from the manufacturing process, allowing for more accurate and context-specific predictions. This use of real-time process data helps tailor the model to the specific conditions of and variations in the ongoing AM process, further enhancing its applicability and precision.

The organization of the rest of this paper is as follows. Section 2 introduces the physics-informed online learning framework tailored for real-time temperature field prediction. Section 3 outlines the data generation and model implementation strategies. Section 4 discusses the results, emphasizing the framework's capabilities and exploring potential limitations. Finally, Section 5 concludes this paper, summarizing the contributions and suggesting future directions for enhancing the adaptability and precision of thermal modeling in metal AM.

2. Methodology

In this section, we delve into the development of the proposed physics-informed online learning framework aimed at the prediction of 2D temperature field in metal AM processes. The framework consists of two distinct phases. Initially, an offline learning phase is conducted, where a PINN, incorporating a neural network with physics-informed modifications, is trained on a dataset from a previously conducted metal AM process. This initial phase enables the PINN to grasp the fundamental patterns and dynamics inherent in AM thermal processes, creating a robust foundation of knowledge for subsequent real-time data exposure.

In the online learning phase, this pre-trained PINN serves as the base model for dynamic adaptation to an unseen process, for which data are acquired in real time. The integration of Synaptic Intelligence (SI) [30] and an adaptive learning rate helps maintain and apply previously acquired knowledge as the PINN adapts to new data from different AM processes. This setup enables the PINN to dynamically update its weights through online gradient descent [31] in response to new information, demonstrating its flexibility and prompt response to temperature changes in the new process.

The PINN's architecture consists of three core components: the neural network, physics-informed (PI) input, and a PI loss function. It employs a series of thermal images to predict the 2D temperature field at future time steps, which capture the 2D temperature fields of the currently deposited layer of the manufactured part, along with a PI input that includes heat input characteristics. More precisely, the model inputs a sequence of w thermal images spanning from timestamps $(n - w)$ to (n) and the process' heat input characteristics. It uses these data to forecast the thermal image (i.e., 2D temperature field) at the timestamp $(n + i)$. Here, w represents the window size of input data, capturing a specific range of thermal imaging, and i denotes the hyperparameter indicating the future timestamp targeted for prediction, focusing on the evolving thermal conditions of the currently deposited layer. Figure 1 presents an overview of the proposed framework.

It is worth mentioning that while thermal imaging provides valuable temperature distribution data, its absolute values can be unreliable due to the non-linearity of gray body emissions in high-temperature metals. This challenge can be mitigated by incorporating a pyrometer for calibration or using two synchronized sensing devices at different wavelengths to calculate temperature based on emissions using Planck's Law [32]. Additionally, hyperspectral thermal imaging and two-wavelength pyrometry can offer accurate temperature distributions without needing emissivity adjustments [33].

We will further elaborate on the proposed framework in the following sections: Section 2.1 delves into the components of the PINN; Section 2.2 discusses the pretraining phase, highlighting how the model is prepared using historical AM data; finally, Section 2.3 covers the online learning phase, detailing the adaptation of the pre-trained PINN to new, real-time AM processes.

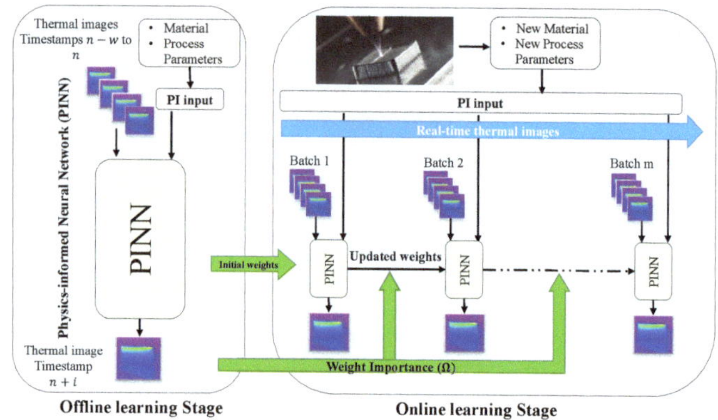

Figure 1. Schematic of the two-stage framework for online 2D temperature prediction in metal AM. Both stages leverage physics-informed components based on process parameters and material to guide the PINN: initially through training on previous AM thermal patterns, and subsequently through updates with real-time thermal images.

2.1. Proposed Physics-Informed Neural Network

The PINN used in our framework is designed to integrate real-time data with physics-based constraints, enabling accurate predictions of thermal fields in metal additive manufacturing This specialized neural network incorporates convolutional long short-term memory (ConvLSTM) layers and convolutional layers into the model's architecture. It also includes an auxiliary input for process heat flux information, represented as a 2D matrix (i.e., PI input), and integrates a boundary condition (BC) loss term into the overall loss function (i.e., PI loss). Each component is essential for accurately predicting the thermal field by leveraging both data-driven insights and physics-informed constraints. In Figure 2, the proposed PINN—which comprises three key components, the neural network, PI loss, and PI input—is presented.

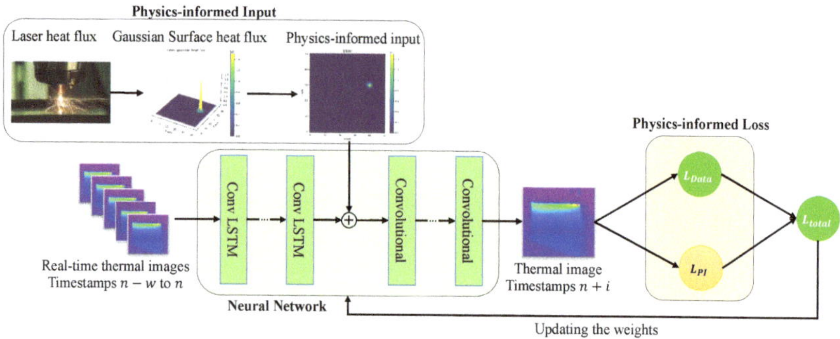

Figure 2. PINN with its components: the neural network, PI input, and PI loss.

2.1.1. Neural Network Architecture

The architecture of the neural network is designed to capture the complexities of thermal field prediction within the domain of metal additive manufacturing. Central to this architecture are the convolutional long short-term memory (ConvLSTM) layers, which can simultaneously process spatial and temporal information, making them an ideal choice for tasks addressed in this paper, where 2D thermal images serve as inputs, and the goal is to predict a 2D temperature distribution evolving over time. The choice of CNNs and RNNs over ANNs is due to the need to handle both spatial and temporal dependencies in image data, which is critical for accurate thermal field prediction. To complement the ConvLSTM layers, the architecture also includes traditional convolutional layers. These layers specialize in extracting spatial features from each thermal image, identifying intricate patterns and temperature distributions essential for constructing an accurate predictive model. This architecture is superior because it leverages the strengths of both CNNs for spatial feature extraction and RNNs for temporal sequence modeling, leading to more precise and reliable predictions. Together, these neural network components form a powerful and efficient architecture designed to tackle the challenges of predicting thermal fields in the dynamic environment of metal AM processes.

2.1.2. Physics-Informed Input

In our framework, the Physics-Informed (PI) input is incorporated to infuse the model with physics-based information regarding the heat input of the manufacturing process, thereby enabling the model to capture the intricate relationship between process parameters and the resultant temperature field. This addition ensures that the neural network can access a richer set of data that reflect the underlying physical processes, enhancing its ability to make better predictions. We chose the laser heat flux as the PI input. This parameter serves as a quantifiable measure of the energy intensity and distribution as the laser interacts with the material, which is vital for understanding the thermal dynamics involved in the process. The laser heat flux represents a critical factor that influences key thermal phenomena, including the melting and solidification processes and the creation of thermal gradients within the layer currently under fabrication.

The process of accurately estimating the laser heat flux is a cornerstone in enhancing our model's predictive accuracy. In the literature, laser heat flux is modeled in different modes, namely point heat source, Gaussian surface heat source, and Gaussian body heat source [23]. Building on this foundation, we utilized a Gaussian surface heat flux model [34] to facilitate a nuanced representation of how the laser's energy is dispersed across the material layer. This model incorporates critical factors such as laser power (P), the radius of the laser beam (r_{beam}), and the material's absorptivity (η) to create a comprehensive picture of the energy input:

$$q_{\text{laser}}(x, y) = -\frac{2\eta P}{\pi r_{\text{beam}}^2} \exp\left(\frac{2d^2}{r_{\text{beam}}^2}\right) \tag{1}$$

This equation calculates the heat flux (i.e., q_{laser}) at each point (x, y) on the material's surface, based on the distance d from the laser center, effectively mapping out the spatial energy profile imposed by the laser. The precision in capturing this energy distribution is critical for simulating the thermal dynamics during the additive manufacturing process accurately. It allows our model to predict the resulting temperature fields with sufficient fidelity, considering how variations in laser settings or material properties could impact the thermal environment within the layer being printed. Figure 3 illustrates an instance of laser application on a surface, modeled using the Gaussian surface heat flux, along with its corresponding q_{laser} matrix.

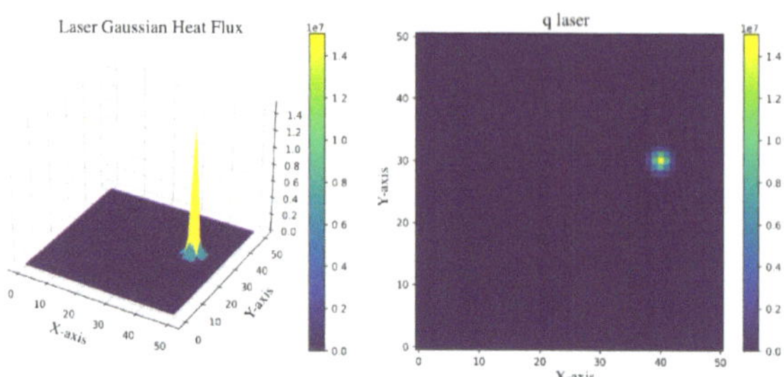

Figure 3. An example of Gaussian surface heat flux on a surface (**left**) and its corresponding q_{laser} matrix (**right**).

2.1.3. Physics-Informed Loss

In our PINN, the PI loss function is essential to reflecting the physical laws governing the system. The PI loss is designed to specifically enforce boundary conditions, which are critical for accurately modeling thermal processes in metal additive manufacturing.

Boundary conditions dictate how surfaces of a part interact with their environment, influencing heat transfer mechanisms such as conduction, convection, and radiation, which are crucial for predicting temperature fields during manufacturing. Figure 4 showcases the three heat transfer mechanisms considered in our problem. The PI loss minimizes the residual of the physical equations at these boundaries, improving the network's adherence to physical constraints and ensuring that predictions are both statistically accurate and physically plausible.

By penalizing deviations from these physical laws, our model's reliability is bolstered. For the problem of predicting the 2D temperature field of the currently deposited layer, the specific boundary condition for the top surface is mathematically expressed as follows:

$$-k\frac{\partial T}{\partial n} = h_c(T - T_{\text{amb}}) + \sigma\varepsilon(T^4 - T_{\text{amb}}^4) + Q_{\text{laser}} \tag{2}$$

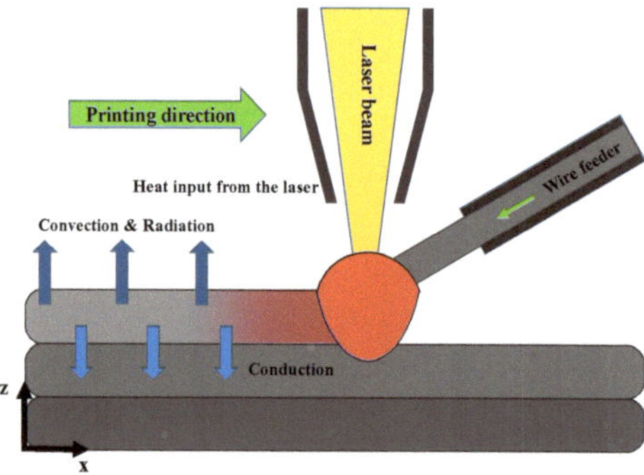

Figure 4. The boundary conditions used in the physics-informed loss function for a DED process, illustrating the heat transfer mechanisms, including heat from the laser, heat conduction to the layer below, convective heat loss, and radiative heat loss.

In this equation, T denotes the matrix of the 2D temperature fields, representing the temperature distribution during the manufacturing process captured by thermal images. This equation encompasses several heat transfer mechanisms: The term $-k\frac{\partial T}{\partial \vec{n}}$ calculates the conductive heat flux through the surface, with $\vec{n}$ denoting the normal direction outward from the surface. This term accounts for interlayer heat conduction in the normal direction (z-axis or interlayer direction). h_c is the convective heat transfer coefficient, modeling heat loss caused by air or fluid motion over the surface. Radiative heat loss is given by

$\sigma\epsilon(T^4 - T_{\text{amb}}^4)$, where σ is the Stefan–Boltzmann constant, and ϵ is the emissivity, indicating energy lost as radiation based on the fourth power of the temperature difference between the surface and ambient air. Q_{laser} represents the heat input from the laser, crucial for the melting and fusing of material layers. Figure 4 illustrates the boundary conditions used in the physics-informed loss function. Constants k, h_c, σ, and ϵ are set based on the values for the materials used in the simulation, derived from the material data in the simulation software.

Given the boundary condition, the residual for the boundary condition can be defined as

$$R_{PI}(T) = k\frac{\partial T}{\partial \vec{n}} + h_c(T + T_{\text{amb}}) + \sigma\epsilon(T^4 - T_{\text{amb}}^4) + Q_{\text{laser}} \tag{3}$$

The PI loss, L_{PI}, is computed as the mean squared error of the residuals across N thermal images:

$$L_{PI} = \frac{1}{N}\sum_{i=1}^{N} R_{PI}^2(T_i) \tag{4}$$

The total loss function for the PINN, which includes both the data-based loss (L_{Data}) and the physics-informed loss (L_{PI}), is weighted to balance the contribution from each component. The weights for each loss term, w_{PI} for L_{PI} and w_D for L_{Data}, are established during the training process. This involves observing and adjusting the impact of each loss term to ensure that they contribute equally to the overall loss. The combined loss function is expressed as follows:

$$L_{Total} = w_{PI}L_{PI} + w_D L_{Data} \tag{5}$$

Here, w_{PI} and w_D denote the weights assigned to the PI loss and data loss, respectively. These weights are proportionally set to balance the scale between two terms in the loss function, contributing to the improved training robustness of the model.

It is worth mentioning that the physical principle in Equation (2) cannot capture all the complex phenomena involved. The physics incorporated have some limitations. Firstly, it does not account for the intricate nature of heat transfer in the porous material ahead of the heat source, or fully capture the dynamics of the initial layers. Additionally, the assumption of a semi-infinite plate may not accurately reflect the physical realities at the metal powder size resolution. However, our approach is open to incorporating more detailed physics equations, which will strengthen the technology as a general method in future studies.

2.2. Offline Learning Stage

In the offline learning stage, our framework trains a PINN on data from a completed metal AM process. This training integrates physics-based inputs and loss functions tailored to the specific material properties and heat input characteristics of the process, equipping the model with an understanding of the underlying thermal dynamics. This comprehensive preparation sets a solid foundation for the model's subsequent application in the online learning stage.

The PI loss function in our framework incorporates material properties—the thermal conductivity (k) and convection heat transfer coefficient (h_c), determined by the materials used in the process. These properties are crucial for enforcing realistic boundary behaviors for accurate thermal modeling. Additionally, heat input parameters such as laser power (P), beam radius (r_{beam}), and material absorptivity (η), are integrated into a physics-informed input that characterizes the laser heat flux.

As the model transitions to the online learning phase, it requires the careful management of model updates to ensure that new data do not drastically alter the neural network's established weight configurations. It is essential to maintain the network's fundamental knowledge from the offline learning phase, enabling it to adjust to new data while preserving its accuracy and ability to predict temperature fields as taught by historical data.

Synaptic Intelligence addresses this by evaluating the significance of each weight relative to the tasks mastered previously. This evaluation is captured mathematically by omega values (Ω), which highlight weights that are key to the model's prior tasks and should therefore be conservatively adjusted with incoming data.

$$\Omega_i = \frac{\sum_t g_{i,t} \Delta w_{i,t}}{(\Delta w_i)^2} \tag{6}$$

Here, Ω_i is the omega value for weight i, $g_{i,t}$ is the gradient of the loss with respect to weight i at time t, $\Delta w_{i,t}$ is the change in weight i at time t, and Δw_i is the total change in weight i over the training period. To streamline computation and avoid potential issues like division by zero, an approximation is used:

$$\Omega_i \approx \sum_t g_{i,t}^2 \tag{7}$$

This approach posits that the importance of weight is indicated by the sum of its gradients' magnitudes over time, implying that weights with consistently significant gradients are more important for the model's function than otherwise.

Incorporating these omega values into the online learning phase allows the model to adjust weight importance flexibly, integrating new data while maintaining the core insights gained previously. The careful balance between incorporating new information and preserving valuable existing knowledge improves the model's adaptability and performance across varied additive manufacturing scenarios.

2.3. Online Learning Stage

In the online learning phase, the primary objective is to predict the temperature field for a new, previously unseen metal AM process that may exhibit characteristics different from the process used during the offline learning stage. This phase is crucial for allowing the model to transition smoothly from utilizing foundational knowledge to integrating new insights, thus maintaining accuracy and adaptability during the dynamic conditions of AM processes. A critical aspect of this phase involves updating the PI input and PI loss function to incorporate new process characteristics as they are encountered. For instance, modifications in laser power, material properties, or beam radius require adjustments to the PI input that captures the laser heat flux and the PI loss terms that enforce adherence to new boundary conditions and material behaviors. By dynamically adapting these PI components to reflect new process characteristics, the model effectively incorporates updated knowledge, enhancing its accuracy and adaptability across different AM setups.

In this phase, the PINN, previously trained during the offline learning stage, serves as the initial model. Its weights are updated in real time as new data from the ongoing process are received. The PINN employs online gradient descent (OGD) to dynamically adapt to new data. This method allows for immediate adjustments to the model's parameters, enhancing its ability to refine predictions continuously as new information becomes available. The mathematical expression for updating the weights through OGD is shown below:

$$w_{t+1} = w_t - \alpha \nabla L(w_t, x_t, y_t) \tag{8}$$

where w_t and w_{t+1} represent the weights before and after processing the new data point, respectively. The learning rate, α, determines the step size for the update, and $\nabla L(w_t, x_t, y_t)$ is the gradient of the loss function relative to the weights for the current data point x_t and its target value y_t.

The loss function used during this phase includes components for data loss (L_{data}), PI loss (L_{PI}), and Synaptic Intelligence (SI) loss (L_{SI}):

$$L_{\text{total}} = w_{\text{Data}} L_{\text{data}} + w_{\text{PI}} L_{\text{PI}} + w_{\text{SI}} L_{\text{SI}} \tag{9}$$

where L_{SI} is specifically formulated as

$$L_{SI} = \sum_i \Omega_i (\Delta w_i)^2 \tag{10}$$

Here, w_{Data}, w_{PI}, and w_{SI} are the weights that are determined during the training process to maintain a balanced scale among the various terms in the loss function. Ω_i quantifies the importance of each weight i in the neural network, reflecting its role in prior tasks learned by the model. The term Δw_i measures the change in weight i due to new data, and squaring this change $(\Delta w_i)^2$ aims to minimize large shifts in critical weights, thus preserving essential knowledge and mitigating catastrophic forgetting. This strategic formulation ensures that the model not only adapts to new data but also retains accuracy and stability in predictions by balancing novel learning with the preservation of previously acquired knowledge.

Additionally, we adjust the learning rate dynamically, starting with a lower value to prevent drastic parameter shifts when limited data are available. This conservative approach maintains model stability. As more data are integrated and the model adapts, we increase the learning rate to accelerate learning and enhance adaptability, ensuring that the model remains responsive to new information while preserving its accuracy.

3. Data Generation and Model Implementation

In this section, we outline the dataset generation using simulations, emphasizing the distinct characteristics of these simulations. We utilized finite element simulations with ANSYS software, specifically the Workbench AM DED 2022 R2 module, to create the training and testing datasets for our online learning framework. We conducted a total of four simulations, varying materials, geometries, deposition patterns, and process parameters to evaluate the framework's generalizability under diverse conditions.

These simulations standardized the pass width and layer thickness at a consistent 1 mm. Materials such as 17–4PH stainless steel and Inconel 625 were chosen for their prevalent use and unique properties pertinent to our study. Geometrically, cylinders and cubes were explored for their common industrial applications, and three distinct deposition patterns were investigated to further assess the framework's flexibility.

For process parameters, we defined two distinct scenarios: one with a deposition speed of 10 mm/s and a higher laser power, and another at 6 mm/s with a lower laser power. Our thermal analysis incorporated factors such as thermal conductivity, convection, and radiation, maintaining substrate and ambient temperatures at a constant 23 °C. Activation temperatures were set at 2000 °C for the lower laser power and 2400 °C for the higher laser power, noting that the simulation abstracts from directly modeling the laser's heat flux. Our study focused on multi-layer fabrications, where each layer follows the same printing pattern and geometry. This approach was chosen to systematically investigate the thermal behavior and ensure consistency across layers. The number of layers for these processes were 10, 10, 9, and 8. An overview of different simulated processes A–D is illustrated in Table 1.

Although only four simulations were conducted, they were deliberately designed to encompass a range of characteristics, including variations in material types, geometries, deposition patterns, and laser parameters. This diversity was intended to validate the transferability of knowledge from one process to another. The boundary conditions for these simulations were carefully chosen to replicate realistic metal AM processes. Specifically, we considered factors such as thermal conductivity, convection, and radiation, while maintaining constant ambient conditions. Despite these efforts, simulation data cannot fully capture the complexities and variabilities of actual AM processes, such as unforeseen environmental factors and material inconsistencies. Future work should extend the number of simulations and incorporate experimental data to enhance the reliability and applicability of the findings.

Table 1. Illustrations of four simulated processes. In the geometry and deposition patterns, the solid lines indicate the deposition geometry, and the dashed lines indicate the direction of the laser scanning.

	Process A	Process B	Process C	Process D
Material	17-4PH Stainless Steel	17-4PH Stainless Steel	17-4PH Stainless Steel	Inconel 625
Process Temperature	2400 °C	2000 °C	2000 °C	2000 °C
Travel Speed	10 mm/s	6 mm/s	6 mm/s	20 mm/s
Number of Layers	10	10	9	8
Geometry and Deposition Pattern				

Each simulation produced datasets capturing transient temperature values at each timestamp, from which approximately 16,000 input–output pairs per simulation were extracted for training and validation. These pairs consist of sequences of the 2D temperature field for the currently deposited layer as inputs, with the outputs representing the 2D temperature field of the same layer for the subsequent timestamp.

The model was implemented in TensorFlow. We constructed a neural network with six ConvLSTM layers and four convolutional layers, each utilizing 20 filters. This network was pre-trained using data from the simulated processes, with a learning rate set to 10^{-5}. The pretraining phase employed three previous timestamps ($w = 3$) to predict the temperature field five seconds ahead ($i = 50$). Upon completing pretraining, both the model's parameters and the omegas for each parameter were saved. These omegas indicate the importance of the weights in retaining learned knowledge, which is vital for the subsequent online learning phase.

In the online learning stage, the architecture remains unchanged, and the pre-trained model, equipped with the saved weights, is introduced to streaming data from a new process simulation.

Based on the insights from Section 4, the learning rate is dynamically adjusted throughout the phase, starting at 5×10^{-8} and incrementally increasing to 5×10^{-5} to better accommodate learning from the dynamic, real-time data. While setting w_{Data} to 1, the regularization parameter w_{SI} is set to 10^{-5} to balance the scales of the data loss (L_{Data}) and the Synaptic Intelligence loss (L_{SI}), ensuring harmony between adapting to new information and preserving essential insights from previous learning. Similarly, w_{PI} is set to 10^{-15} to align the physics-informed loss (L_{PI}) with the other terms in the loss function, maintaining consistency across the model's evaluation criteria. This calibration supports the neural network to be effectively trained on the fly with the new process data, ensuring the model's continuous adaptation and robustness in predicting temperature fields across varied additive manufacturing scenarios.

4. Results and Discussion

To evaluate the real-time adaptability of our online learning framework to new data, we carried out a sequence of experiments. We designated Process A (Table 1) as the baseline dataset for the offline learning phase, creating a foundational knowledge base for the model to understand the standard thermal patterns of metal AM processes. We then transitioned to the online learning phase, progressively introducing datasets from Processes B, C, and D (Table 1). Each dataset represented a unique scenario and varied incrementally from the parameters of Process A.

The experimental processes were strategically designed to evaluate the performance of the online learning framework under various scenarios: Process B mirrored Process A, differing only in specific operational parameters to test the model's sensitivity to such changes under controlled conditions. Process C presented a greater challenge by varying not only the process parameters but also the deposition patterns, assessing the model's adaptability to both thermal and process changes. Process D, introducing a change in geometry and material, represented the most significant departure from the initial setup, testing the model's ability to adapt predictions to new structural contexts.

To evaluate the model's performance, we used the Mean Absolute Error (MAE) and Mean Absolute Percentage Error (MAPE). The mathematical representations for these metrics are as follows:

$$\text{MAE} = \frac{1}{n \times m} \sum_{i=1}^{n} \sum_{j=1}^{m} |Y_{ij} - \hat{Y}_{ij}|, \tag{11}$$

$$\text{MAPE} = \frac{100\%}{n \times m} \sum_{i=1}^{n} \sum_{j=1}^{m} \left| \frac{Y_{ij} - \hat{Y}_{ij}}{Y_{ij}} \right|. \tag{12}$$

In these equations, Y_{ij} and $\hat{Y}_{ij}$ denote the actual and predicted temperature values for each element in the 2D temperature field, respectively, with n and m representing the number of rows and columns in the temperature matrix. The individual errors calculated using these metrics provide insights into the precision at each data point, while the overall error for the validation dataset, obtained by averaging these individual errors, offers an evaluation across the entire dataset.

4.1. Performance of Physics-Informed Online Learning

The performance evaluation of the physics-informed (PI) model during the online learning phase involves a structured training procedure. The dataset from each process—B, C, and D—was split such that the first 80% of data were used incrementally as the training set, allowing the model to continuously update and refine its predictions. The remaining 20% of the data serve as the validation set, used to assess the model's predictive accuracy and to validate its generalization capability on unseen data.

The model's ability to process new batches of data swiftly is a significant advantage, particularly in real-time applications. On average, updating the model with a new batch of data takes just 0.21 s on Compute Canada's infrastructure using a single NVIDIA Tesla T4 GPU. This demonstrates the framework's efficiency and practical utility in scenarios where rapid data processing and immediate decision making are crucial.

In Figure 5, the MAPE and MAE are illustrated, providing clear visual indicators of the model's performance across various stages of the learning process. These metrics are plotted against the percentage of process data that were incrementally introduced to the model, which are represented on the x-axis. The y-axis, meanwhile, displays the values of the MAPE or MAE, indicating the model's error rate at each stage of data integration.

For Process B, the initial spike in the MAPE suggests an adjustment period as the model adapts to changes in process parameters. As more data are processed, a steady decrease in error is observed, highlighting the framework's capacity to learn and refine its predictions based on closely related baseline conditions. In contrast, Process C, with its introduction of new deposition patterns in addition to varied process parameters, starts with a higher MAPE. This underscores the complexity introduced by the new material properties and deposition strategies. However, the subsequent decline in the MAPE indicates effective model adaptation to these complexities, showcasing its ability to manage multi-faceted changes in AM processes.

Process D shows an initially high MAPE, reflecting the challenge of accommodating a new material and geometric configuration, but this quickly improves as the model assimilates more data and fine-tunes its predictions to the altered geometry. Despite Process B's closer resemblance to Process A, the modeling for Process D results in lower errors than for Process B. This could be attributed to the circular deposition patterns in Processes A and B, which may complicate thermal management due to surface curvature affecting heat conduction and convection, thereby posing greater modeling challenges than the uniform cubic geometry of Process D. We utilized multi-layer simulations to capture the cumulative thermal effects and inter-layer interactions, which are crucial for accurate temperature predictions and process optimization. However, we recognize the importance of analyzing more complex geometries and their impact on thermal predictions, which will be considered in future work.

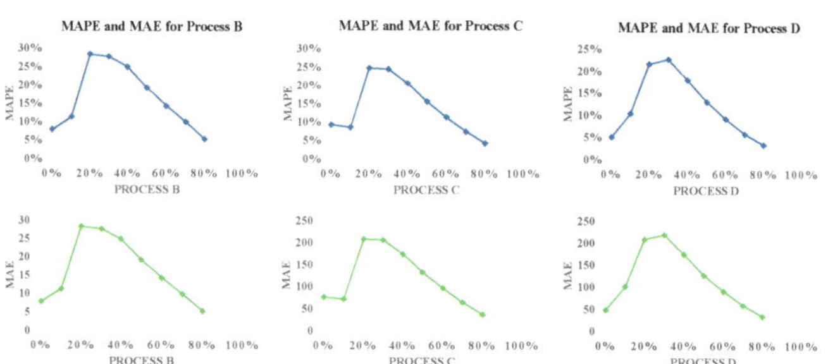

Figure 5. Comparative analysis of MAPE and MAE across processes B, C, and D.

4.2. Comparison of Proposed Framework with Machine Learning Framework

In this study, we conducted a comparative analysis between the physics-informed (PI) online learning framework and a data-driven approach. As outlined in Section 2.1, the PINN model that is used in the PI framework incorporates three main components: the neural network architecture, physics-informed (PI) inputs, and physics-informed (PI) loss function. In contrast, the data-driven model used for comparison utilizes the same neural architecture but excludes the PI components, focusing solely on data-centric learning methods.

To ensure a comprehensive evaluation, we assessed the performance of both the PI and data-driven frameworks across two critical areas: the entire temperature field of the layer being deposited and specifically within the Heat-Affected Zone (HAZ) and melt pool area. The HAZ is a crucial region in metal AM processes, characterized by the surrounding material of the weld or melt pool that experiences thermal cycling without melting. This zone is drastically influential in microstructural changes due to thermal exposure, which can significantly impact the mechanical properties and integrity of the final product [35]. The accurate identification and management of the HAZ and melt pool are essential for ensuring the quality of manufactured components.

The literature indicates that the identification of the HAZ in AM processes typically involves sophisticated methods such as thermal imaging, metallurgical analysis, and computational modeling. In our study, the HAZ is defined as the region where the temperature exceeds a specific threshold that modifies the microstructure. For materials like 17–4PH stainless steel and Inconel 625, the HAZ temperature thresholds are set at 1050 °C [36] and 960 °C [37], respectively, reflecting their unique thermal characteristics. These established thresholds facilitate the analysis of both the HAZ and melt pool, which is crucial for the comparative evaluation of the frameworks in our study.

Figure 6 presents the MAPEs of both the PI and data-driven frameworks as they predict temperature fields across the entire deposited layer (upper row), the melt pool, and the surrounding HAZ (lower row). In Process B, the PI framework consistently outperforms the data-driven model, exhibiting lower errors when predicting the temperature field for future timestamps as it continuously integrates new data. However, for Processes C and D, despite the PI model initially showing lower error rates—thanks to the incorporation of physics-based knowledge during training—the data-driven model achieves a slightly lower MAPE in the latter half of the process. This shift can be attributed to the additional requirements imposed by the physics-informed constraints in the PI model, which slightly decelerates error reduction.

Regarding the specific areas of the HAZ and melt pool, the PI framework maintains superior performance throughout the online learning process, consistently presenting lower error rates compared to its data-driven counterpart. Notably, the disparity between the PI and data-driven models is more noticeable at the beginning of the online learning phase when less data are available. This underscores the significant advantage of incorporating prior physical knowledge into the neural network, which enhances initial model guidance and prediction accuracy at the early stages. It is important to note that the error rates for both models are generally higher in the HAZ and melt pool areas. This increased error is due to the more complex thermal behavior in these zones, influenced by phenomena such as steep thermal gradients, rapid solidification rates, and varied material properties at high temperatures. These factors complicate the thermal dynamics, making accurate predictions more challenging [20].

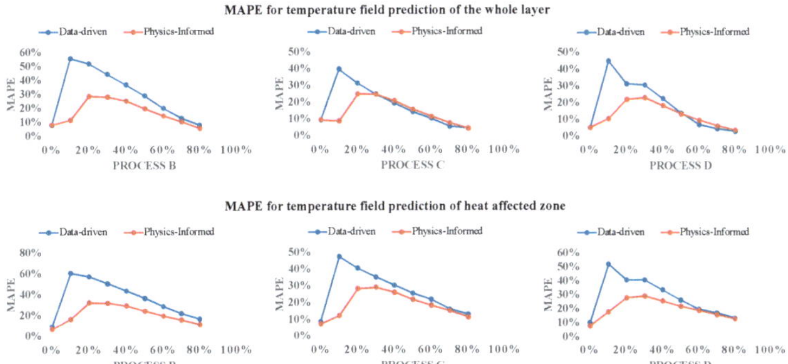

Figure 6. Comparative MAPE of physics-informed and data-driven models across Processes B, C, and D.

The comparative analysis between the PI and the data-driven frameworks is further elucidated by examining their respective outputs against the ground truth provided by simulations. Figure 7 illustrates the predicted temperature fields for both the PINN and data-driven models alongside the simulated "true" temperature distributions for process C.

In the provided examples, both the PI and data-driven frameworks align variably with the simulation results, with most predicted temperatures differing by less than 20 °C from the simulated temperature. Near the melt pool, discrepancies increase, though the PI model generally approximates simulated values more closely, especially around critical areas like the HAZ and melt pool itself. This precision is indicative of the PINN's ability to incorporate physical laws into its predictions of complex phenomena in metal AM processes.

The figure also displays the absolute difference panels, which quantify the discrepancies between the predictions and the simulations. These discrepancies highlight areas where the models struggle to capture the exact thermal behaviors, potentially due to the intricate dynamics within the melt pool and HAZ that are challenging to model precisely with data-driven approaches alone.

The "HAZ + Melt Pool MAPE" images further provide a focused view of the error distribution within the HAZ, emphasizing the regions where the predictions deviate most significantly from the observed data. This detailed error analysis is critical for refining the models and for understanding the specific conditions under which each model may require further tuning or additional data to enhance the accuracy.

In these comparisons, the PI framework consistently outperforms the data-driven framework, particularly in critical areas such as the HAZ and melt pool, where precise temperature knowledge is crucial for ensuring the quality and integrity of the manufactured parts. This superiority of the PI framework is especially significant when considering that the provided examples are outputs at a stage where only 80% of the data from the new process have been integrated into the models, as indicated in Figure 6. It is noteworthy that the PI framework's performance advantage becomes even more pronounced when less data are available, underscoring its robust capability to effectively utilize physical laws to predict complex phenomena with limited input data. This makes the PI model particularly valuable in the early stages of new process integration, where data scarcity can often hinder accurate modeling. Additionally, the PI framework's integration of physical principles allows it to maintain a high prediction accuracy across various conditions and complexities of the AM processes, providing a reliable and efficient tool for process optimization and control. This reliability is crucial for applications where obtaining extensive training data is impractical or time-consuming, thus ensuring consistent quality and performance in real-world manufacturing scenarios.

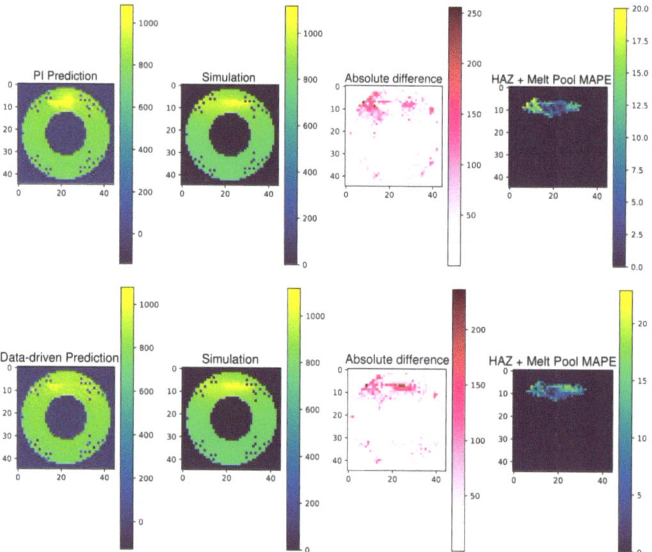

Figure 7. Comparative top view of temperature fields for Process C, showing framework predictions alongside simulation results with physics-informed predictions on top and data-driven predictions below.

4.3. Effect of Varying Learning Rates

This section evaluates the impact of different learning rate strategies on the performance of the PI framework during its online learning stage. As the model transitions from pretraining on a previous process's data to incremental data from a new process, selecting an optimal learning rate strategy becomes critical for managing adaptability and accuracy.

We explore three main learning rate strategies: constant, linear increasing, and linear decreasing. A constant learning rate strategy provides a stable update mechanism throughout the learning process, suitable for environments where data properties do not vary significantly. A linear increasing learning rate strategy allows the model to start with cautious adjustments and progressively increase its responsiveness as it adapts to the new data. Conversely, a linear decreasing learning rate strategy enables the model to initially make broad updates and gradually refine these adjustments to focus on detailed patterns.

Figure 8 illustrates the MAPEs for these strategies, showcasing how each one impacts model accuracy over the training period.

The constant learning rate was set at 5×10^{-5}, a value carried over from the offline learning stage to provide a baseline of stability. In the increasing learning rate strategy, the learning rate began at a much lower value, 5×10^{-8}, intentionally chosen to demonstrate the effect of gradually adapting the learning rate on the model's performance. This strategy proved particularly beneficial as it allowed the model to adjust more significantly as its confidence in the new data increased, leading to a consistent reduction in error rates throughout the training process. By gradually increasing the learning rate, this approach helps prevent stagnation, ensuring that the model remains dynamic and responsive as it encounters new and increasingly complex data. This calibration of learning rate adjustments fosters a more gradual adaptation to new data, enabling the model to evolve its learning strategy in sync with the unfolding complexities of the process. On the other hand, the decreasing learning rate started at 5×10^{-5}, aiming to quickly assimilate broad patterns before reducing the rate to 5×10^{-8}. However, this approach occasionally resulted in higher error rates in training, indicating that a high initial rate might compromise the model's ability to adapt to new complexities as they arise.

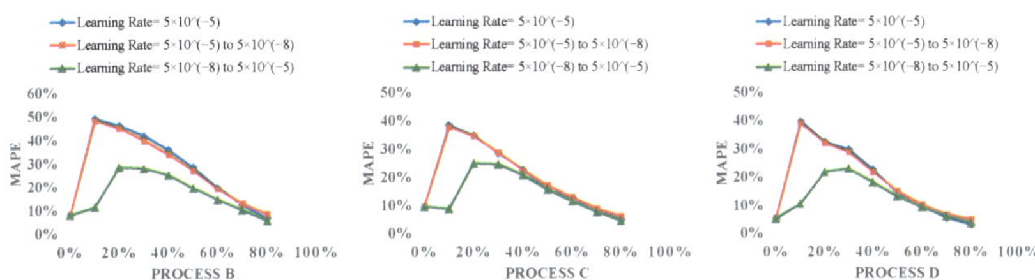

Figure 8. Comparison of learning rate strategies on PINN performance.

The analysis indicates that the choice of learning rate strategy and its initial setting plays an important role in the model's ability to adapt to new data. Among the strategies examined, the increasing learning rate strategy not only facilitated better initial learning with minimal risk of error escalation but also allowed for enhanced adaptability and precision as the complexity of the data increased. This strategy's success underscores the importance of a dynamic learning rate adjustment in environments where data characteristics are expected to evolve substantially, such as in metal AM.

4.4. Effect of Varying Batch Sizes

In this section, we explore the impact of varying the batch sizes of new online data on the performance of the PI online learning framework, specifically focusing on batch sizes of 2, 4, 8, 16, and 32. The batch size is a critical hyperparameter in machine learning that determines the number of training examples used in one iteration to calculate the gradient during the model training process. This parameter significantly influences both the computational efficiency and the convergence behavior of the training algorithm.

Figure 9 illustrates the MAPE for different batch sizes across Processes B, C, and D. This visualization helps assess how the batch size impacts model accuracy and learning dynamics during the online training phase. The results reveal that as the batch size increases, there is a noticeable stabilization in error reduction throughout the online learning process. Larger batch sizes tend to smooth out the learning updates due to the averaging of gradients across more data points. This aggregation diminishes the influence of outliers and reduces the variability of weight updates, leading to a more consistent and gradual decrease in error rates.

Moreover, larger batch sizes offer computational advantages, particularly for real-time applications. Processing larger batches can utilize computational resources more efficiently, potentially speeding up the training process since fewer updates are required per epoch. However, this comes with the caveat that the model must wait for the entire batch of data to be collected and processed before proceeding with an update. This requirement can introduce delays in scenarios where data are being gathered incrementally, such as in real-time monitoring or streaming applications. Hence, there is a trade-off between computational speed and update latency that needs to be carefully managed to optimize real-time performance.

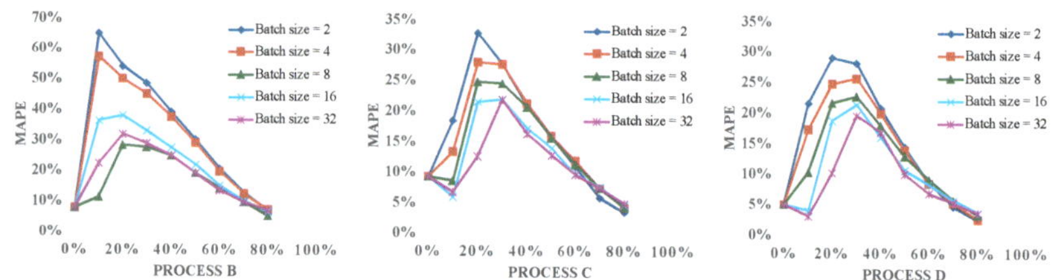

Figure 9. Impact of batch size on model performance across Processes B, C, and D.

In conclusion, the choice of the batch size is a decision that balances several factors, including error stability, computational efficiency, and responsiveness to new data. Larger batches may be preferable for scenarios where computational speed is crucial and data are abundant, but they require careful consideration of the delay in model updates.

4.5. Discussion on Geometry and Multi-Layer Fabrications

We utilized multi-layer simulations with consistent geometries across layers to capture the cumulative thermal effects and inter-layer interactions, which are important for accurate temperature predictions and process optimization. This approach allowed us to systematically investigate thermal behavior and ensure consistency in our predictions.

Geometrical variations, such as changes in the shape and size of the printed part, significantly impact the thermal distribution due to differences in heat conduction, convection, and radiation pathways. Complex geometries can result in non-uniform thermal fields, leading to localized areas of higher or lower temperatures that can affect material properties and structural integrity. The accumulation of heat in certain areas can cause thermal stresses and deformations, making it important to predict these thermal fields accurately. Understanding and predicting these thermal variations are important for refining and optimizing the printing path to ensure uniform temperature distribution, reduce thermal stresses, and improve the overall quality of the final product.

Analyzing more complex geometries and varying geometries across layers is valuable for fully understanding their impact on thermal behavior and inter-layer interactions. This knowledge will enhance the model's applicability and provide a more comprehensive understanding of metal AM processes.

5. Conclusions

In this paper, we introduce the first physics-informed (PI) online learning framework specifically designed for temperature field prediction in metal additive manufacturing (AM), utilizing a physics-informed neural network (PINN). This innovative framework integrates a PINN that includes three main components—a neural network architecture, physics-informed inputs, and physics-informed loss functions. Initially, the PINN is pretrained on a known process during the offline learning stage to establish a foundational model. It then transitions to the online learning stage where it continuously adapts to new, unseen process data by dynamically updating its weights.

Our results demonstrate the robust performance of the PI online learning framework in predicting temperature fields for unseen processes, effectively handling various manufacturing conditions. Particularly notable is its superior performance over data-driven counterparts in predicting temperatures in critical areas such as the Heat-Affected Zone (HAZ) and melt pool. These regions are vital for the overall quality and structural integrity of the manufactured parts, highlighting the importance of precise temperature predictions in these areas. The PINN's integration of physical laws and prior knowledge provides a distinct advantage, enabling more accurate predictions under diverse conditions. Additionally, our analysis of key operational parameters—including the learning rate and batch size of the online learning process—reveals their roles in optimizing the learning process, further enhancing the framework's effectiveness.

The key contributions of this study are as follows:

1. Online Learning and Prediction: This study is potentially the first attempt to apply online learning for the real-time modeling and prediction of temperature fields in previously unseen AM processes. This pioneering effort represents an advancement toward adaptable manufacturing technologies.
2. Physics-Informed Integration: We incorporated heat boundary conditions into our framework as the physics-informed loss function, and heat input characteristics as physics-informed input within the neural network. This integration significantly increases the prediction accuracy and reliability.
3. Framework Generality: Our methodology proves highly versatile, demonstrating effectiveness across a diverse range of AM conditions. It can accommodate changes in process parameters, materials, geometries, and deposition patterns, showcasing an essential step toward a universally adaptable AM framework.
4. Improvement in Predictive Accuracy and Process Adaptability: By integrating real-time data with PINNs, this research enhances the predictive accuracy and adaptability of thermal models in metal AM. The framework's dynamic adaptation to new data and varying conditions ensures precise temperature predictions, improving quality and consistency in AM processes. This advancement over existing methods enables more accurate and reliable thermal modeling, supporting the development of adaptable and efficient AM technologies.

In conclusions, this study represents a pioneering effort in applying physics-informed online learning to metal AM, offering significant improvements in predictive accuracy and operational efficiency. Looking ahead, experimental data from actual metal AM processes can be integrated to enhance the physics-informed machine learning framework. The incorporation of real-world data is expected to improve the accuracy of the model's predictions by capturing the complex and nuanced behaviors of AM environments. We will also consider more complex geometries and varying geometries across layers in future work to further refine our predictive capabilities.

Author Contributions: Conceptualization, P.S., Y.T., M.R.D. and G.G.W.; methodology, P.S.; validation, P.S., Y.T. and M.R.D.; formal analysis, P.S.; investigation, P.S.; resources, G.G.W.; data curation, P.S.; writing—original draft preparation, P.S.; writing—review and editing, P.S., Y.T., M.R.D. and G.G.W.; visualization, P.S.; supervision, G.G.W.; project administration, G.G.W.; funding acquisition, G.G.W. All authors have read and agreed to the published version of the manuscript. Additionally, we acknowledge the use of generative AI tools, specifically ChatGPT, for light editing assistance, including spelling and grammar corrections.

Funding: This work was supported by the Natural Sciences and Engineering Research Council (NSERC) of Canada, under Grant No. RGPIN-2019-06601.

Institutional Review Board Statement: Not applicable.

Informed Consent Statement: Not applicable.

Data Availability Statement: The original contributions presented in the study are included in the article, further inquiries can be directed to the corresponding author.

Conflicts of Interest: The authors declare no conflicts of interest. The funders had no role in the design of the study; in the collection, analyses, or interpretation of data; in the writing of the manuscript; or in the decision to publish the results.

References

1. Ford, S.; Despeisse, M. Additive manufacturing and sustainability: An exploratory study of the advantages and challenges. *J. Clean. Prod.* **2016**, *137*, 1573–1587. [CrossRef]
2. Cooke, S.; Ahmadi, K.; Willerth, S.; Herring, R. Metal additive manufacturing: Technology, metallurgy and modelling. *J. Manuf. Process.* **2020**, *57*, 978–1003. [CrossRef]
3. Zhang, J.; Yin, C.; Xu, Y.; Sing, S.L. Machine learning applications for quality improvement in laser powder bed fusion: A state-of-the-art review. *Int. J. Mater. Des.* **2024**, *1*, 2301. [CrossRef]
4. Shrestha, R.; Shamsaei, N.; Seifi, M.; Phan, N. An investigation into specimen property to part performance relationships for laser beam powder bed fusion additive manufacturing. *Addit. Manuf.* **2019**, *29*, 100807. [CrossRef]
5. Mani, M.; Lane, B.M.; Donmez, M.A.; Feng, S.C.; Moylan, S.P. A review on measurement science needs for real-time control of additive manufacturing metal powder bed fusion processes. *Int. J. Prod. Res.* **2017**, *55*, 1400–1418. [CrossRef]
6. Dehaghani, M.R.; Sahraeidolatkhaneh, A.; Nilsen, M.; Sikström, F.; Sajadi, P.; Tang, Y.; Wang, G.G. System identification and closed-loop control of laser hot-wire directed energy deposition using the parameter-signature-quality modeling scheme. *J. Manuf. Process.* **2024**, *112*, 1–13. [CrossRef]
7. Thompson, S.M.; Bian, L.; Shamsaei, N.; Yadollahi, A. An overview of Direct Laser Deposition for additive manufacturing; Part I: Transport phenomena, modeling and diagnostics. *Addit. Manuf.* **2015**, *8*, 36–62. [CrossRef]
8. Chua, C.; Liu, Y.; Williams, R.J.; Chua, C.K.; Sing, S.L. In-process and post-process strategies for part quality assessment in metal powder bed fusion: A review. *J. Manuf. Syst.* **2024**, *73*, 75–105. [CrossRef]
9. Bontha, S.; Klingbeil, N.W.; Kobryn, P.A.; Fraser, H.L. Thermal process maps for predicting solidification microstructure in laser fabrication of thin-wall structures. *J. Mater. Process. Technol.* **2006**, *178*, 135–142. [CrossRef]
10. Barua, S.; Liou, F.; Newkirk, J.; Sparks, T. Vision-based defect detection in laser metal deposition process. *Rapid Prototyp. J.* **2014**, *20*, 77–85. [CrossRef]
11. Vilaro, T.; Colin, C.; Bartout, J.D. As-fabricated and heat-treated microstructures of the Ti-6Al-4V alloy processed by selective laser melting. *Metall. Mater. Trans.* **2011**, *42*, 3190–3199. [CrossRef]
12. Chern, A.H.; Nandwana, P.; Yuan, T.; Kirka, M.M.; Dehoff, R.R.; Liaw, P.K.; Duty, C.E. A review on the fatigue behavior of Ti-6Al-4V fabricated by electron beam melting additive manufacturing. *Int. J. Fatigue* **2019**, *119*, 173–184. [CrossRef]
13. Kasperovich, G.; Hausmann, J. Improvement of fatigue resistance and ductility of TiAl6V4 processed by selective laser melting. *J. Mater. Process. Technol.* **2015**, *220*, 202–214. [CrossRef]
14. Liu, S.; Shin, Y.C. Additive manufacturing of Ti6Al4V alloy: A review. *Mater. Des.* **2019**, *164*, 107552. [CrossRef]
15. Yan, Z.; Liu, W.; Tang, Z.; Liu, X.; Zhang, N.; Li, M.; Zhang, H. Review on thermal analysis in laser-based additive manufacturing. *Opt. Laser Technol.* **2018**, *106*, 427–441. [CrossRef]
16. Pham, T.Q.D.; Hoang, T.V.; Van Tran, X.; Pham, Q.T.; Fetni, S.; Duchêne, L.; Tran, H.S.; Habraken, A.M. Fast and accurate prediction of temperature evolutions in additive manufacturing process using deep learning. *J. Intell. Manuf.* **2023**, *34*, 1701–1719. [CrossRef]
17. Mozaffar, M.; Paul, A.; Al-Bahrani, R.; Wolff, S.; Choudhary, A.; Agrawal, A.; Ehmann, K.; Cao, J. Data-driven prediction of the high-dimensional thermal history in directed energy deposition processes via recurrent neural networks. *Manuf. Lett.* **2018**, *18*, 35–39. [CrossRef]
18. Le, V.T.; Bui, M.C.; Pham, T.Q.D.; Tran, H.S.; Van Tran, X. Efficient prediction of thermal history in wire and arc additive manufacturing combining machine learning and numerical simulation. *Int. J. Adv. Manuf. Technol.* **2023**, *126*, 4651–4663. [CrossRef]
19. Raissi, M.; Perdikaris, P.; Karniadakis, G.E. Physics-informed neural networks: A deep learning framework for solving forward and inverse problems involving nonlinear partial differential equations. *J. Comput. Phys.* **2019**, *378*, 686–707. [CrossRef]
20. Zhu, Q.; Liu, Z.; Yan, J. Machine learning for metal additive manufacturing: Predicting temperature and melt pool fluid dynamics using physics-informed neural networks. *Comput. Mech.* **2021**, *67*, 619–635. [CrossRef]
21. Xie, J.; Chai, Z.; Xu, L.; Ren, X.; Liu, S.; Chen, X. 3D temperature field prediction in direct energy deposition of metals using physics informed neural network. *Int. J. Adv. Manuf. Technol.* **2022**, *119*, 3449–3468. [CrossRef]
22. Jiang, F.; Xia, M.; Hu, Y. Physics-Informed Machine Learning for Accurate Prediction of Temperature and Melt Pool Dimension in Metal Additive Manufacturing. *Print. Addit. Manuf.* **2023**, *ahead of print*.
23. Yang, Z.; Zhang, S.; Ji, X.; Liang, S.Y. Model-Based Sensitivity Analysis of the Temperature in Laser Powder Bed Fusion. *Materials* **2024**, *17*, 2565. [CrossRef] [PubMed]
24. Hoi, S.C.; Sahoo, D.; Lu, J.; Zhao, P. Online learning: A comprehensive survey. *Neurocomputing* **2021**, *459*, 249–289. [CrossRef]
25. Yang, Z.; Dong, H.; Man, J.; Jia, L.; Qin, Y.; Bi, J. Online Deep Learning for High-Speed Train Traction Motor Temperature Prediction. *IEEE Trans. Transp. Electrif.* **2023**, *10*, 608–622. [CrossRef]
26. Wang, C.; Zhang, Y.; Zhang, X.; Wu, Z.; Zhu, X.; Jin, S.; Tang, T.; Tomizuka, M. Offline-online learning of deformation model for cable manipulation with graph neural networks. *IEEE Robot. Autom. Lett.* **2022**, *7*, 5544–5551. [CrossRef]
27. Ouidadi, H.; Guo, S.; Zamiela, C.; Bian, L. Real-time defect detection using online learning for laser metal deposition. *J. Manuf. Process.* **2023**, *99*, 898–910. [CrossRef]
28. Mu, H.; He, F.; Yuan, L.; Hatamian, H.; Commins, P.; Pan, Z. Online distortion simulation using generative machine learning models: A step toward digital twin of metallic additive manufacturing. *J. Ind. Inf. Integr.* **2024**, *38*, 100563. [CrossRef]

29. Tang, Y.; Dehaghani, M.R.; Sajadi, P.; Balani, S.B.; Dhalpe, A.; Panicker, S.; Wu, D.; Coatanea, E.; Wang, G.G. Online thermal field prediction for metal additive manufacturing of thin walls. *J. Manuf. Process.* **2023**, *108*, 529–550. [CrossRef]
30. Zenke, F.; Poole, B.; Ganguli, S. Continual learning through synaptic intelligence. In Proceedings of the 34th International Conference on Machine Learning, PMLR, Sydney, Australia, 6–11 August 2017; pp. 3987–3995.
31. Zinkevich, M. Online convex programming and generalized infinitesimal gradient ascent. In Proceedings of the 20th International Conference on Machine Learning (Icml-03), Washington, DC, USA, 21–24 August 2003; pp. 928–936.
32. Doubenskaia, M.; Pavlov, M.; Chivel, Y. Optical system for on-line monitoring and temperature control in selective laser melting technology. *Key Eng. Mater.* **2010**, *437*, 458–461. [CrossRef]
33. Lane, B.; Lane, B.; Jacquemetton, L.; Piltch, M.; Beckett, D. *Thermal Calibration of Commercial Melt Pool Monitoring Sensors on a Laser Powder Bed Fusion System*; US Department of Commerce, National Institute of Standards and Technology: Gaithersburg, MD, USA, 2020.
34. Al Hamahmy, M.I.; Deiab, I. Review and analysis of heat source models for additive manufacturing. *Int. J. Adv. Manuf. Technol.* **2020**, *106*, 1223–1238. [CrossRef]
35. Saboori, A.; Aversa, A.; Marchese, G.; Biamino, S.; Lombardi, M.; Fino, P. Microstructure and mechanical properties of AISI 316L produced by directed energy deposition-based additive manufacturing: A review. *Appl. Sci.* **2020**, *10*, 3310. [CrossRef]
36. Cheruvathur, S.; Lass, E.A.; Campbell, C.E. Additive manufacturing of 17-4 PH stainless steel: Post-processing heat treatment to achieve uniform reproducible microstructure. *JOM* **2016**, *68*, 930–942. [CrossRef]
37. Wang, S.; Gu, H.; Wang, W.; Li, C.; Ren, L.; Wang, Z.; Zhai, Y.; Ma, P. The influence of heat input on the microstructure and properties of wire-arc-additive-manufactured Al-Cu-Sn alloy deposits. *Metals* **2020**, *10*, 79. [CrossRef]

Article

Transferability of Temperature Evolution of Dissimilar Wire-Arc Additively Manufactured Components by Machine Learning

Håvard Mo Fagersand, David Morin, Kjell Magne Mathisen, Jianying Heand Zhiliang Zhang *

Department of Structural Engineering, Norwegian University of Science and Technology (NTNU), 7491 Trondheim, Norway; havard.m.fagersand@ntnu.no (H.M.F.); david.morin@ntnu.no (D.M.); kjell.mathisen@ntnu.no (K.M.M.); jianying.he@ntnu.no (J.H.)
* Correspondence: zhiliang.zhang@ntnu.no; Tel.: +47-735-92-530

Abstract: Wire-arc additive manufacturing (WAAM) is a promising industrial production technique. Without optimization, inherent temperature gradients can cause powerful residual stresses and microstructural defects. There is therefore a need for data-driven methods allowing real-time process optimization for WAAM. This study focuses on machine learning (ML)-based prediction of temperature history for WAAM-produced aluminum bars with different geometries and process parameters, including bar length, number of deposition layers, and heat source movement speed. Finite element (FE) simulations are used to provide training and prediction data. The ML models are based on a simple multilayer perceptron (MLP) and performed well during baseline training and testing, giving a testing mean absolute percentage error (MAPE) of less than 0.7% with an 80/20 train–test split, with low variation in model performance. When using the trained models to predict results from FE simulations with greater length or number of layers, the MAPE increased to an average of 3.22% or less, with greater variability. In the cases of greatest difference, some models still returned a MAPE of less than 1%. For different scanning speeds, the performance was worse, with some outlier models giving a MAPE of up to 14.91%. This study demonstrates the transferability of temperature history for WAAM with a simple MLP approach.

Keywords: additive manufacturing; WAAM; machine learning; neural networks; temperature history; finite element method

Citation: Fagersand, H.M.; Morin, D.; Mathisen, K.M.; He, J.; Zhang, Z. Transferability of Temperature Evolution of Dissimilar Wire-Arc Additively Manufactured Components by Machine Learning. *Materials* **2024**, *17*, 742. https://doi.org/10.3390/ma17030742

Academic Editors: Tuhin Mukherjee, Guozheng Quan and Qianru Wu

Received: 30 November 2023
Revised: 26 January 2024
Accepted: 1 February 2024
Published: 3 February 2024

1. Introduction

Additive manufacturing (AM) of metals is known as a promising technology for a wide range of industrial applications. In general, AM involves the production of components through the melting of precursor material, called the feedstock, by a powerful, localized heat source based on information from a computer-aided design (CAD) file [1]. AM has certain advantages over conventional subtractive manufacturing techniques such as grinding or milling; it produces less wasted material, can create more complex shapes, and offers reduction in lead time [2,3]. Customization of the manufacturing process is much easier with AM, as there is no need to create custom tooling or molds [4]. Wire-arc AM (WAAM) is a particular form of AM derived from welding, where the feedstock is a metal wire and the heat source is a welding arc [5–7]. Compared with other AM techniques like powder-bed-based methods, WAAM offers faster deposition rates and larger part size, at the expense of a reduced ability to create complex designs and reduced surface quality [7]. There are many potential industrial applications of WAAM, such as for aluminum parts for aerospace applications, steel bars for construction, or large parts like propellers and rudders for use in shipbuilding [3,8–11].

Today, there are still challenges which limit the spread of the AM of metals for industrial application. The powerful localized heat source used causes steep temperature gradients, which in turn cause significant residual stresses in the material. These stresses can further lead to cracking and distortion, reducing the quality of the manufactured parts [12–14]. The rapid heating and cooling can also result in pore defects, which result in parts with lower tensile strength and greater susceptibility to fatigue [15]. To mitigate this, postprocessing with heat treatment to reduce the residual stress and reduce porosity is necessary [15,16]. The cycles of rapid heating, cooling, and reheating is unique to metal AM, and since it is a young technology compared with conventional manufacturing techniques, the exact relationship between process parameters and the resulting mechanical properties is not yet known [14,17]. However, it is well established that the temperature distribution during AM affects the residual stresses, with a more uniform distribution resulting in stresses of lower magnitude [17–21]. A greater understanding of the WAAM process in general and the temperature distribution specifically is therefore needed if widespread industrial application is to become a reality. A potential aid here is a digital twin (DT), which is defined as "a digital representation of a production system or service" [22]. A DT could be used to predict properties like temperature and residual stress distribution before the AM procedure is performed, integrate measured data for real-time prediction, and even alter the manufacturing process based on results from the digital model [23,24]. DT models are believed to greatly increase the viability of metal AM for industry [24,25].

One tool which can be used to model AM is finite element (FE) simulation. FE simulations can model various AM processes with a high degree of accuracy and give insight into the temperature evolution during production as well as the resulting residual stress profile [16,26,27]. However, FE simulation has a major drawback: it is slow and computationally expensive, often requiring hours or days to obtain results. Real-time prediction of temperature evolution is therefore impossible with an FE model alone. For this reason, surrogate models, simplified alternatives to numerical simulations, are being explored as an option [28]. Many researchers have investigated the possibility of using machine learning (ML) to create surrogate models, with the ultimate goal of achieving real-time prediction while retaining good accuracy [29,30]. Many different ML methods have been explored, with varying degrees of complexity.

Mozaffar et al. [31] developed a surrogate method for determining the thermal history during directed energy deposition (DED), a subset of AM processes which includes WAAM, as a function of process parameters. Their surrogate model used a recurrent neural network (RNN), with its inputs being an engineered feature for the tool path, deposition time, laser intensity, and current layer height. To aid with process optimization for powder-bed-based methods, another major subgroup of AM, Stathatos et al. [32] created a multilayer perceptron (MLP)-based model for predicting temperature evolution and density. The model consisted of several neural networks in sequence—one main network for determining the temperature and additional "rider" networks for predicting other properties that depend on the temperature. As input, the main network in their model took in information about the laser path and past temperatures, and they applied it to data from a simulation of a laser following a random path.

More recently, there have been many more studies performed using a variety of different ML methods. Ness et al. [33] created data-driven models for temperature prediction using the extra trees algorithm. Extra trees is in the decision-tree family of algorithms, and their model utilized several engineered features, including the distance between the node and the heat source and the power influence of the heat source [34]. They applied their models to FE-simulated AM depositions of aluminum alloy, with varying deposition patterns and power intensities, to investigate how well the models would perform when applied to another system with a different pattern or intensity. Their results showed that while the models worked well for prediction on the training system, with a mean absolute percentage error (MAPE) of less than 5%, transferability to different systems gave worse results, with the MAPE ranging from around 5 to 25% [33]. Le et al. [35] applied an MLP-based method to model temperature distribution in the WAAM of 316L steel using FE simulations. As input features, they used the heat input, node coordinates, and time. Even with these relatively simple inputs, they were able to obtain good prediction accuracy with their model for WAAM-deposited multilayer bars. Xie et al. [36] used a hybrid physical/data-driven method, a physics-informed neural network (PINN), to model the three-dimensional temperature field during the DED of bars of nickel–chromium alloy. Their model used an approximation of the partial differential equation for heat conduction during the process for the physics-informed part and the laser power and scanning speed, time, and coordinates as input for the data-driven part. The data were obtained from FE simulations of deposition of single-layer and multilayer bars. They found that they were able to achieve high-accuracy prediction with a smaller amount of data than purely data-driven models require. Wacker et al. [37] used two kinds of neural network models to predict the resulting accuracy and distortion for multilayer parts produced by WAAM. The training data were obtained experimentally through WAAM performed on steel. The first data-driven model used only the welding parameters as inputs, while the second model also included recursive parameters from the output.

Our overarching goal is to construct a comprehensive dataset for WAAM. This dataset will facilitate real-time predictions of the thermal history of WAAM-produced components, regardless of their shape or size, once the component design is finalized by CAD and fundamental process parameters are selected. We refer to this predictive capability as transferability, meaning that we can transfer the thermal history from known components to new and unknown ones. In this study, we used MLP models to test the transferability for WAAM production of different thin rectangular bars. MLP models were chosen over RNNs, which are by definition more appropriate for a time-series prediction [38]. We wanted to evaluate the feasibility of simple MLPs as an alternative, since they are very simple to implement and train. Additionally, if simple MLPs are found to perform well at predicting WAAM temperature history, more advanced RNN-based models will be expected to perform even better.

We performed 40 FE simulations of different WAAM processes with varying bar lengths, numbers of deposition layers, and scanning speeds. Unlike the MLP models used by Le et al. [35], our MLP models used only the current time and past temperatures of the node as input. The MLPs were trained on data from one of the FE simulations and were tested on data from other simulations with different parameters. In Section 2, we discuss how the FE simulations were conducted and how the MLPs were set up. Then, in Section 3 we present the results of testing the transferability using the MLP models. Finally, in Section 4 we sum up our conclusions and present some future perspectives.

2. Materials and Methods

2.1. FE Simulation

The overall flow of this study is illustrated in Figure 1. First, FE simulations of WAAM deposition of aluminum bars were performed using the FE software Abaqus 2019 [39]. The output data were then postprocessed by taking data from every fourth timestep; this was done to limit data size. For each node in the deposited bar, the current elapsed time, current temperature, and temperature at the last five timesteps were recorded. A total of 40 different simulations were performed. Each FE simulation consisted of a rectangular substrate onto which a rectangular bar was deposited. A transient heat transfer analysis was used, with the Abaqus AM module used to simulate the deposition of material and movement of the heat source. The heat source was modelled as a Goldak heat source, also called a double ellipsoid heat source [40,41]. We used the Abaqus implicit solver, Abaqus/Standard, with linear 8-node heat transfer elements (DC3D8). Most of the simulations were performed with a fixed timestep of 0.025 s. For the systems with a scanning speed different from 0.015 m/s or with 7 to 9 layers, automatic timesteps were used instead, with a maximum and starting value of 0.025 s and a minimum of 10^{-6} s. The initial as well as the ambient temperature was 20 °C.

The deposited bar varied both in length and in number of layers, while the substrate length was equal to the length of the deposited bar plus an additional 5 cm at both ends. The substrate height and width, as well as the width and height of each deposited layer, were the same across all simulations. The parameters are summed up in the Supplementary Material in Table S1. Figure 2a shows a visualization of the FE model for a one-layer system, and Figure 2b depicts one of the four-layer systems. The deposited material was 2319 aluminum alloy. The properties of this material, such as melting point, density, etc., and the process parameters used were taken from FE simulations performed by Ness et al. [33].

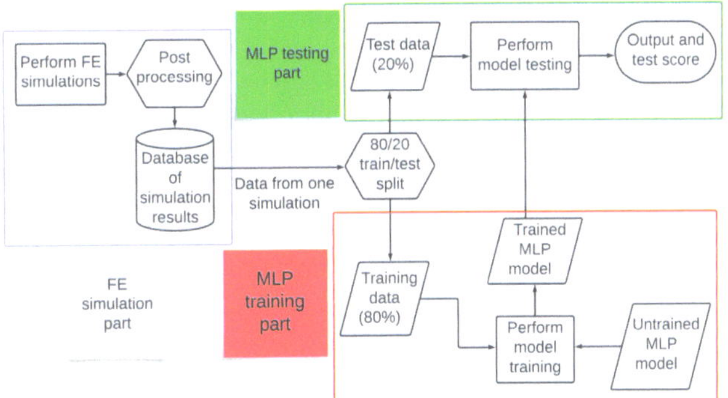

Figure 1. A flowchart of the work performed in this study. FE simulations were performed to generate data, and the postprocessed data were used to train and test MLP models.

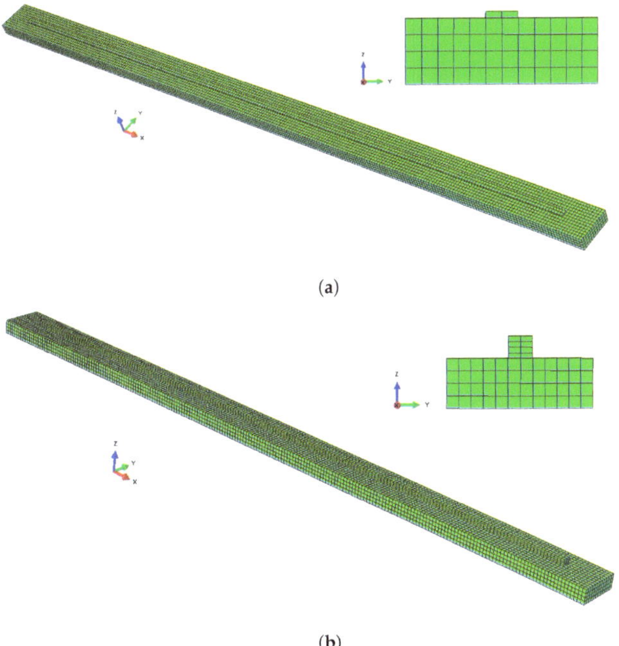

(**a**)

(**b**)

Figure 2. A visualization of two of the FE models: (**a**) a one-layer model, and (**b**) a four-layer model.

2.2. Neural Networks

The postprocessed data obtained from FE simulation were then used to train a number of ML models. The models were based on an MLP, a relatively simple model in the neural network family [42]. The MLPs were created and trained using the Python package

Pytorch [43]. For the loss function, we employed a modified version of $L1$ loss, called smooth $L1$ loss. It is defined as follows:

$$\text{Smooth } L1 \text{ loss} = \begin{cases} \frac{1}{2\beta}(y_i - x_i)^2 & \text{if } |y_i - x_i| < \beta, \\ |y_i - x_i| - \frac{\beta}{2} & \text{otherwise.} \end{cases} \tag{1}$$

Here, $y_i - x_i$ is the difference between the prediction and the true value at timestep i, and β is a parameter. We choose a value of $\beta = 1$. Smooth $L1$ loss is more robust than standard $L1$ loss while avoiding the greater sensitivity to outliers inherent to $L2$ loss [44]. The MLPs had six input values: The current time and the node temperature at the last five timesteps. For output, they returned a single value, the current temperature. They consisted of two hidden layers, each containing 64 nodes, with a rectified linear unit (ReLU) activation function [42]. Minibatch gradient descent was used, with the batch size initially set to 64. The number of epochs was initially set to 5. An overview of all the parameters used is shown in Table 1. The chosen values for the hyperparameters were determined through trial and error. Extensive optimization of the MLP hyperparameters is beyond the scope of this study.

Table 1. A list of the parameters and functions for the MLP models used in this study.

Parameter	Chosen Value
Loss function	Smooth $L1$ loss
β	1
Activation function	ReLU
Learning rate	0.002
Nodes in hidden layer	64
Batch size	64–960
Number of hidden layers	2
Number of epochs	5–8

2.3. Baseline Model Performance

To obtain a baseline performance for the MLP models, training and testing was performed on data from each of the FE simulations. The size of each dataset was large, ranging from 6.5×10^5 to 2.5×10^7 data points. A train/test split of 0.8/0.2 was used, i.e., 80% of data were used for training, and the remaining 20% were reserved for testing. Due to the large size of the datasets used, we deemed it sufficient to use an 0.8/0.2 split instead of cross-validation. The dataset was randomly shuffled before applying the split. The metric used to evaluate the performance was the MAPE, which is defined as follows:

$$M = \frac{100\%}{n} \sum_{i=1}^{n} \left| \frac{y_i - x_i}{x_i} \right|. \tag{2}$$

Here, M is the MAPE, n the sample size, y_i the ith predicted value, and x_i the corresponding true value. Four groups of FE simulations were considered: one consisted of one-layer bars of varying lengths; one of four-layer bars of varying lengths; one with bars of length 0.96 m with different numbers of layers; and one with four-layer bars of length 0.96 m with different scanning speeds. An illustration of the four groups and how the FE simulations in each group differ is shown in Figure 3. From here, we will refer to them as Groups 1 through 4. The data were rescaled between 0 and 1 before training, based on the highest value observed in each group of the FE simulations, so that the data for each model trained on FE simulations in the same group were rescaled the same way. Data from one node in the center of the bottom layer were also excluded from the training dataset; the data from this node were instead used to test model prediction on sequential data from a single node. To begin with, the viability of the MLP models was tested by using trained models to predict the temperature evolution of the one-layer systems in Group 1. Figure 4a shows the training and testing MAPEs for MLP models applied to each of the nine Group 1 systems. An overview of the systems and their bar lengths is shown in Table 2; while the testing MAPE shows more variation than the training MAPE, it is lower than 0.4% in all cases and still close to the training MAPE. This suggests that overfitting is not a significant issue in the one-layer case.

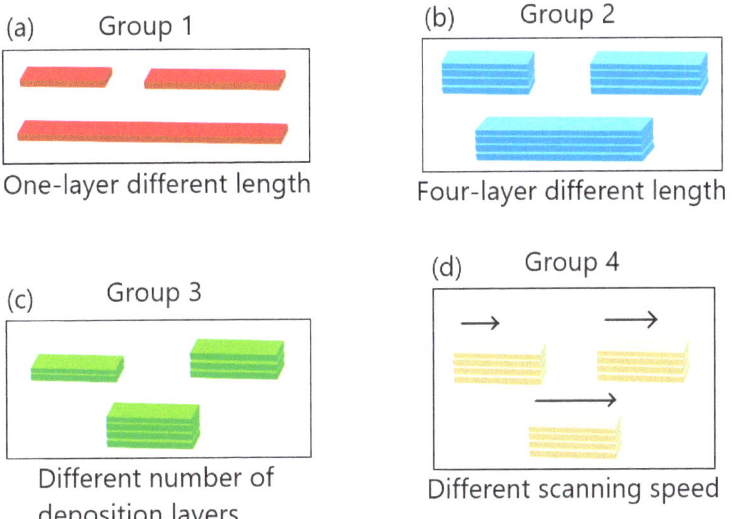

Figure 3. An illustration of the four groups of FE systems considered in this study. (**a**) illustrates the group with one-layer bars of different length, (**b**) the group with four-layer bars of different length, (**c**) the group with bars of length 0.96 m with different numbers of layers, and (**d**) the group with four-layer bars of length 0.96 m with different scanning speeds.

Next, to examine the effect of randomness on model performance, four additional sets of nine MLP models were created. The models in each set were trained and tested on one of the Group 1 systems in the same way as before, except the models in each additional set had a different initial seed. Figure 4b shows the resulting testing MAPEs, where the Model Set A models are the same as used in Figure 4a, and the Model Sets B through E are the new models. As can be seen, the randomness does have an effect on the resulting test accuracy, though the MAPE is below 0.5% in all the cases shown.

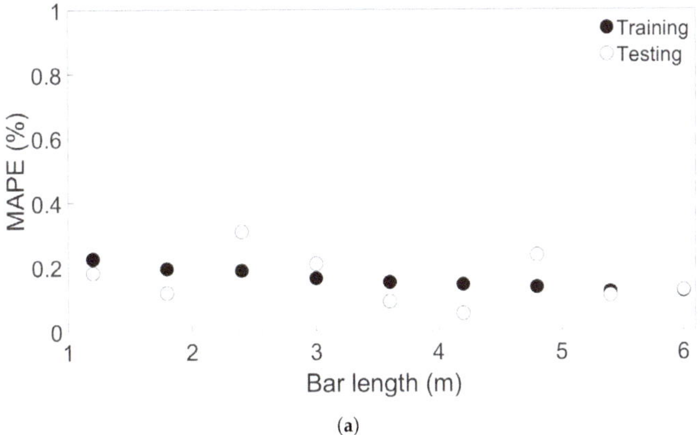

Figure 4. *Cont.*

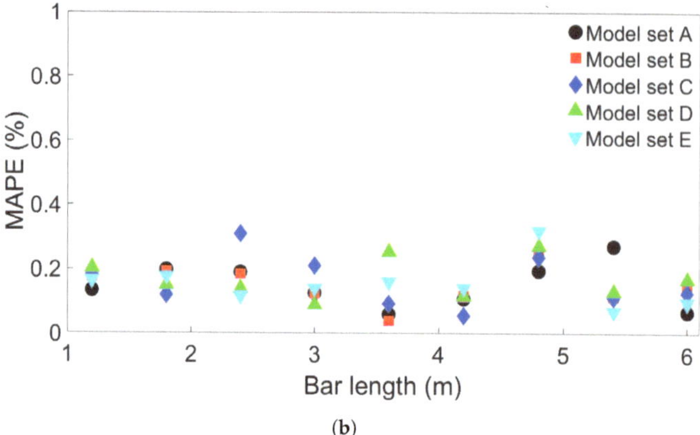

(**b**)

Figure 4. (**a**) The training and testing MAPEs for nine MLP models, each trained and tested on one of the FE systems in Group 1. (**b**) The testing MAPEs for five sets of nine MLP models, each trained and tested on one of the Group 1 systems. The models in each set had the same initial seed.

Table 2. The bar length of each system considered in Group 1.

System	Bar Length (m)
1	1.2
2	1.8
3	2.4
4	3.0
5	3.6
6	4.2
7	4.8
8	5.4
9	6.0

When the MLP models are used to predict the temperature evolution in another system, in principle, the whole dataset can be used for testing. To check how the MLP performance changes when fewer testing data are used, tests were performed with an MLP model which was trained on System 1. We used the model to test System 2 and System 9, using progressively fewer data, going from 100% to 10% of the dataset. For each test, the data were randomly shuffled in five different ways. In both cases, we found that there was a slight deviation in model performance when only 10% of the test data were used. When using 20% or more of the test data, the difference in performance from the 100% case was negligible. We chose to use 20% of the dataset for cross-system testing as well as same-system testing for the sake of consistency.

Next, the MLPs were tested on the three groups of multilayer systems to explore if their performance would be different in the multilayer cases. Training and testing MAPEs were compared in the same manner as for Group 1. The results are shown in the Supplementary Material, in Figure S1 for Group 2, Figure S2 for Group 3, and Figure S3 for Group 4. The testing MAPE is generally close to the training MAPE, with one exception in the varying-scanning-speed case, as seen in Figure S3a. Here, for scanning speed 0.035 m/s, the training MAPE is 0.17% and the testing MAPE is 0.61%. Still, the overall error is low, at less than 0.7% for all the models considered. We conclude that the MLP models show good performance in the multilayer cases as well.

3. Results and Discussion

As stated in the Introduction, the ultimate purpose of our ongoing efforts is to develop a digital platform that is able to predict the temperature history of WAAM-produced components with any geometry (shape and size) for a given material and a set of process parameters. Three key parameters were identified and studied independently, namely, the bar length, the number of deposited layers, and the scanning speed. The effects of these three parameters on the transferability are reported first, in Sections 3.1–3.3, respectively. In Section 3.4, we take a closer look at the temperature evolution of single nodes in systems from Groups 2 and 4 to check how the MLP temperature predictions compare to the real temperatures from the FE simulations. Finally, in Section 3.5, we show the results of further study on the effect of batch size in order to investigate the source of a particularly large error found in the different-scanning-speed case.

3.1. Transferability for Bars with Different Lengths

As observed in Section 2.3, the MLP models have a good baseline performance. The next step is to use the MLP models to test the transferability of the thermal history among FE systems in the four groups. This was undertaken by using trained MLP models to test data from different FE systems in the same group. Here, we present the results for transferability in Groups 1 and 2, which consist of FE systems with one-layer and four-layer bars of different lengths, respectively.

The bar lengths of each of the Group 1 systems are shown in Table 2. Three sets of MLP models were created—one set was trained on System 1, the second trained on System 5, and the third trained on System 9. Each set consists of 5 models, and each of the five models has a different random seed that was set before the training process, for a total of 15 MLP models. This process is illustrated in Figure 5. Figure 6a shows the resulting testing MAPEs for the models trained on System 1, Figure 6b the MAPEs for models trained on System 5, and Figure 6c the MAPEs for models trained on System 9. The MLP models in each set are labelled A through E. A negative length difference means the training system is longer than the test system, while a positive length difference means the opposite, as illustrated in Figure 6d. As expected, the test accuracy is good when the difference in length between the training and testing systems is small. A small growth in error is observed when the bar length of the testing system becomes progressively shorter than that of the training system. When the test system bar length becomes progressively longer than that of the training system, the error growth is much greater. For the models trained on System 1, shown in Figure 6a, the average MAPE for testing on System 1 is 0.17%. This increases to an average of 3.22% for testing on System 9. The difference in MAPE for the models in each set increases as well—the standard deviation in MAPE is 0.024% for System 1 and increases to 1.43% for System 9. The models trained on System 1 are able to achieve good prediction, with a MAPE of less than 3% when the length difference is 3 m or less. We created more MLP models to train on System 1, each with a different initial seed, to see if any of them would exceed this value. A total of 15 additional models were made for each set, giving each set a total of 20 models. These extra models were only used to test data from Systems 6 and 9. The MAPEs from these tests are shown in the Supplementary Material, in Table S2 for System 6 and Table S3 for System 9. Including the five models shown in Figure 6a, this resulted in an average MAPE of 1.02% with a standard deviation of 0.70%; the highest MAPE found was 2.60%. For System 9, an average MAPE of 2.95% and standard deviation 1.59% were obtained. In addition, two of the additional models were found to yield a MAPE of less than 1%.

The models trained on the other Group 1 systems do not show as high an error. Figure 6b shows that for models trained on System 5, the largest positive length difference of 2.4 m results in a low MAPE, at less than 2% for all Models A through E. The average MAPE increases from 0.12% for baseline testing to 1.05% for testing on System 9. The difference between the baseline case and testing on System 1 is very small, with testing on System 1 giving an average MAPE of 0.14%. Models trained on System 9, the system with the longest bar length, show a similar small increase in error as length difference increases. The results are shown in Figure 6c; the average MAPE goes from 0.12% with a standard deviation of 0.04% in the baseline case to 0.25% with a standard deviation of 0.11% when tested on System 1.

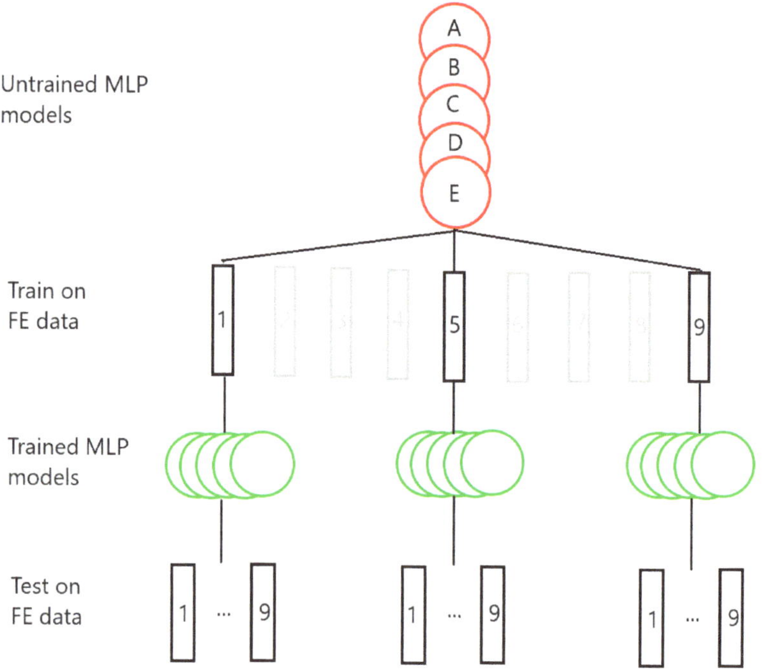

Figure 5. An illustration of the transferability testing process. Groups of five MLP models are trained on data from selected FE simulations, and the trained models are subsequently used to test data from each of the simulations.

Next, the Group 2 systems were examined. The eleven systems and the associated bar lengths are shown in Table 3. Three sets of five trained MLP models were tested on data from each of the eleven systems, like in the Group 1 case. Figure 7a shows the resulting MAPEs for models trained on System 1 from Group 2, Figure 7b the MAPEs for models trained on System 6, and Figure 7c the MAPEs for models trained on System 11. Cross-testing with the other Group 2 systems shows a similar trend as for Group 1. When the length difference is small, or when the training system is larger, the error remains low. We observe that a small length difference leads to a much greater difference in accuracy than what was found in the Group 1 case. For instance, a length increase of 1.2 m in the Group 1 case results in an average MAPE of 0.24%, while for Group 2, the same increase gives an average MAPE of 3.18% with a standard deviation of 2.74%. Taking a closer look at Figure 7a, the MAPE remains below 3% for the five models shown for training on System 1 and testing on System 7, with a length difference of 0.72 m. Like in the Group 1 case, more examples were obtained by training 15 additional models with different seeds on System 1 and applying them to System 7. The results are shown in Table S4 in the Supplementary Material. Here, one of the models returned a MAPE of 3.25%, though all the others gave a MAPE of less than 3%. The average MAPE in this case was 0.88%. The additional models were also tested on System 11, with results shown in Table S5; this gives an average MAPE of 2.27% and a standard deviation of 2.2465%, with nine of the models returning an error of less than 1%. In Figure 7b, the MAPE shows a slight increase as the test system length decreases and a larger increase as the test system length increases compared to the training system. This is comparable to what is observed in Figure 6b. With System 11 as the training system, shown in Figure 7c, the increase in error as test system length shrinks is very small. The average MAPE goes from a baseline value of 0.11% to 0.14% for testing on System 1.

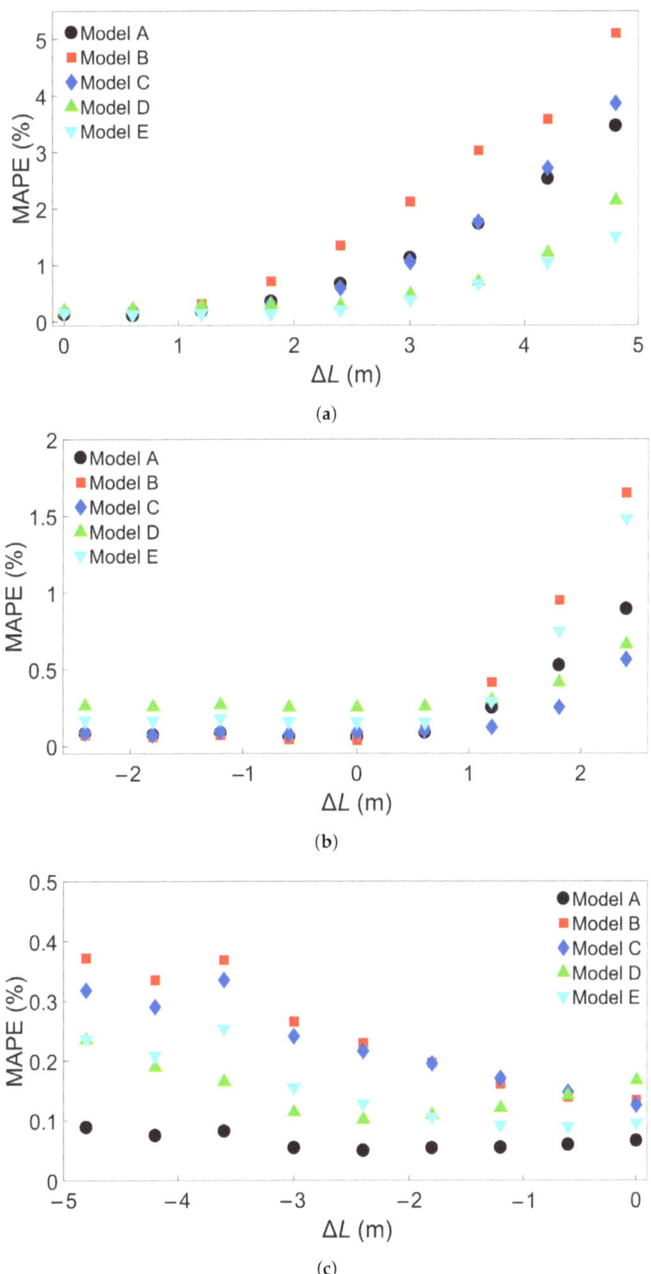

Figure 6. *Cont.*

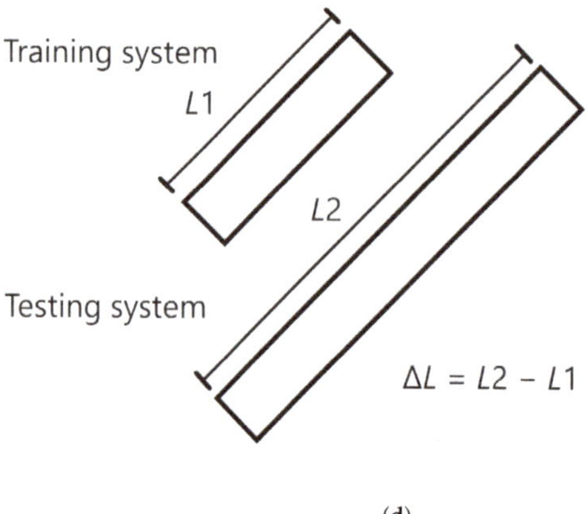

(**d**)

Figure 6. The testing MAPE for MLP models applied to all nine Group 1 systems. (**a**) Models were trained on System 1. (**b**) Models were trained on System 5. (**c**) Models were trained on System 9. Models A through E differ in initial seed before training. (**d**) Definition of the difference in length, ΔL.

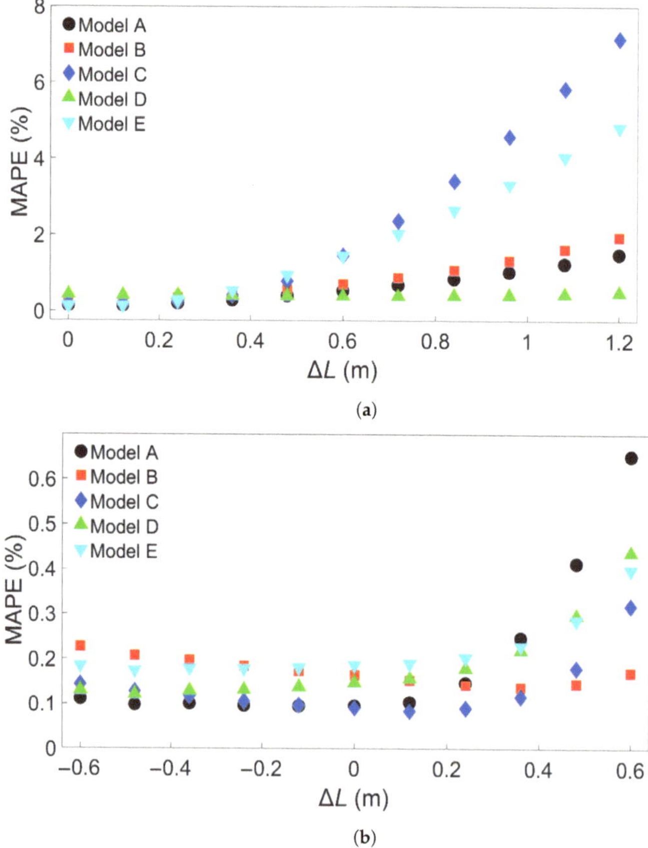

Figure 7. *Cont.*

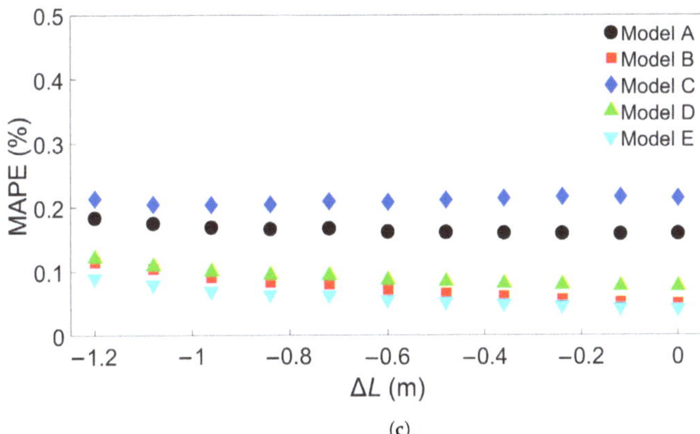

(c)

Figure 7. The testing MAPE for MLP models applied to each of the eleven Group 2 systems. (**a**) The models were trained on System 1. (**b**) The models were trained on System 6. (**c**) The models were trained on System 11. The models A through E in each case differ only in the initial seed set before training.

Table 3. The bar length of each system considered in Group 2.

System	Bar Length (m)
1	0.48
2	0.60
3	0.72
4	0.84
5	0.96
6	1.08
7	1.20
8	1.32
9	1.44
10	1.56
11	1.68

To better understand how the models perform during tests with both low and high error rates, testing was performed on temperature data from single nodes using Model B from Figure 6a and Model C from Figure 7a. The nodes were located in the middle of the first deposition layer, in the position labelled Node B in Figure 8, and were taken from Systems 1 and 9 from Group 1 and Systems 1 and 11 from Group 2. All of the temperature data for each node were used, and the data were ordered by timestep instead of being shuffled. The resulting temperature predictions for the Group 1 systems are shown in Figure 9, compared with the real temperatures as taken from the FE data. Temperature predictions for the Group 2 systems compared with the real temperatures are shown in Figure 10. For both situations, the difference in predicted temperature evolution between the baseline and cross-system predictions is clear to see: the predicted temperatures in both Figures 9b and 10b deviate from the real temperature during the later parts of the simulation. For all the temperature predictions, there is also noticeable error at the peaks, particularly the first peak.

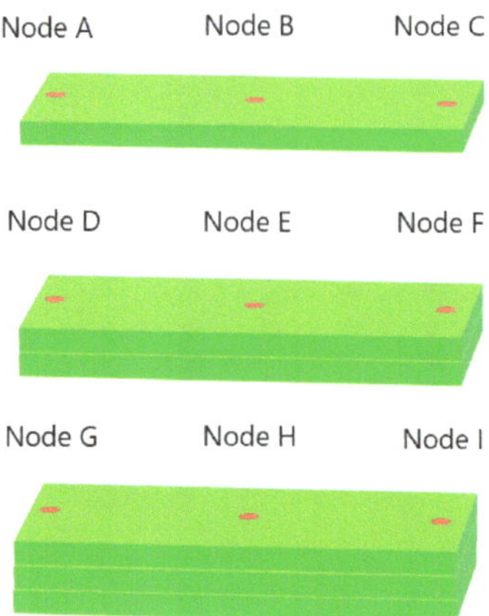

Figure 8. The locations of the single nodes examined in this study. The substrate is not pictured.

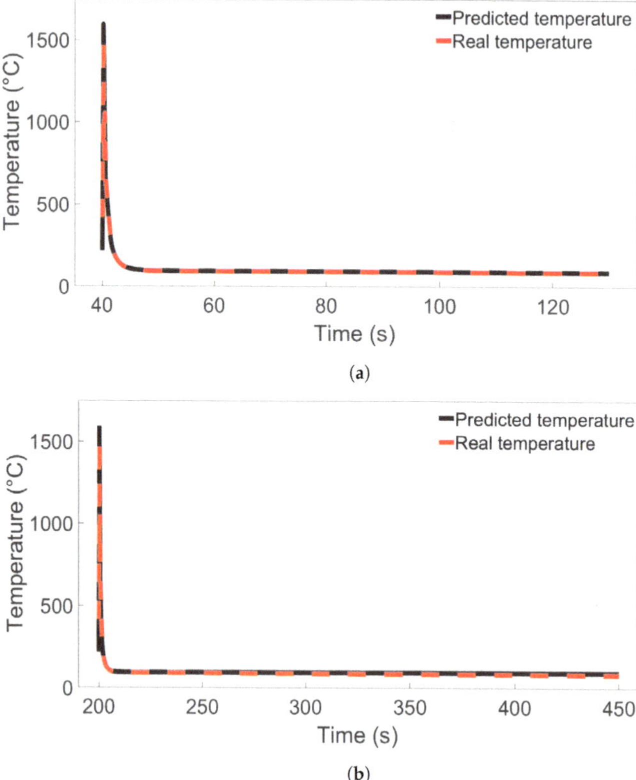

Figure 9. (**a**) Comparison of the real temperature and the temperature predicted by Model B from Figure 6a for a single node in System 1 from Group 1. (**b**) Comparison for a single node in System 9 from Group 1.

In summary, for FE simulations with different bar lengths, the increase in MAPE with length is greater for four-layer systems than single-layer systems. The standard deviation is also consistently greater for Group 2. In both Figures 6a and 7a, it can be seen that MLP models which show good performance at moderate length difference tend to perform better at greater length difference as well. Next, we test transferability for the other groups of FE systems.

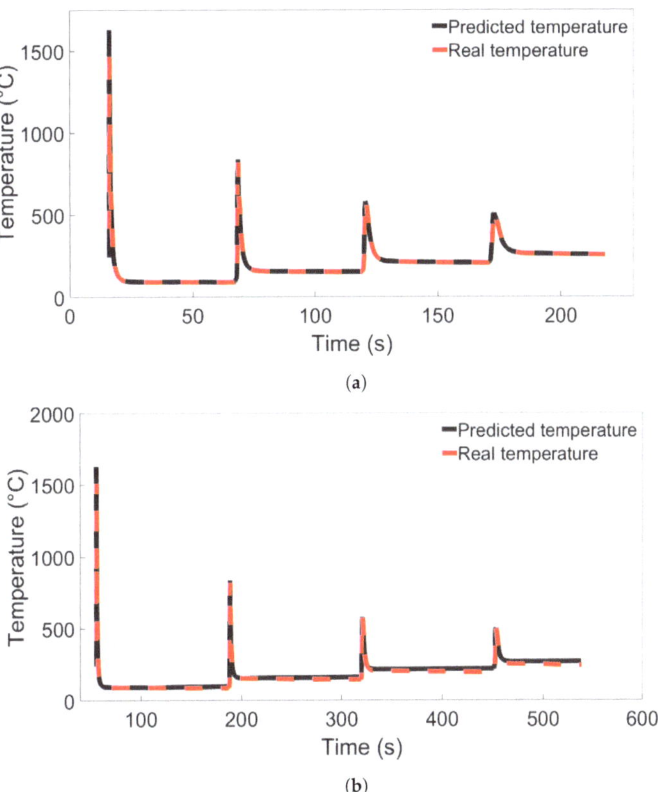

Figure 10. (**a**) Comparison of the real temperature and the temperature predicted by Model C from Figure 7a for a single node in System 1 from Group 2. (**b**) Comparison for a single node in System 11 from Group 2.

3.2. Transferability for Bars with Different Numbers of Layers

The third group of systems investigated was Group 3, which consisted of multilayer bars with different numbers of layers, ranging from 2 to 10. For the Group 3 systems, the length of each deposited layer was kept constant at 0.96 m. A total of nine such systems were simulated, including one which was also part of Group 2. A list of the Group 3 systems is shown in Table 4. Figure 11a shows the results from cross-system testing using models trained on System 1, Figure 11b for models trained on System 5, and Figure 11c for System 9. Similar trends to the ones observed in Figure 6 and Figure 7 can also be seen here, though the increase in MAPE is much smaller, reaching a maximum value of around 1.2% when one of the models trained on System 1 is used to test System 9. Two of the models show limited error growth, while Model B shows a slightly decreasing error. A total of 15 additional models were trained on System 1 and tested on System 9 to further examine model performance. The results are shown in Table S6 in the Supplementary Material. With these results included, the average MAPE for prediction of System 9 is 0.66%, with a maximum value of 4.02%. This is better than the average MAPEs that were found for prediction of System 9 from Group 1 using models trained on System 1 from Group 1 and for prediction of System 11 from Group 2 using models trained on System 1 from Group 2. Figure 11b shows that Models B and E perform worse as the number of layers of the test system increases, while the other models do not. In Figure 11c, we observe a marginal increase in error as the difference in number of layers increases. For both Figure 11b,c, the MAPE remains very small at less than 0.5%.

Table 4. The FE systems in Group 3 and the number of deposited layers in each.

System	Number of Layers
1	2
2	3
3	4
4	5
5	6
6	7
7	8
8	9
9	10

Single-node prediction was also performed. Model A from Figure 11a was used to predict temperature evolution in a single node from System 1 and System 9 from Group 3, located in the Node B position as shown in Figure 8. The resulting temperature evolutions are compared with the real temperatures in Figure 12. The error in predicted temperature when the model is tested on the 10-layer System 9 is lower than was observed in Figure 10b, but a slight error is again observed during the later parts of the deposition process.

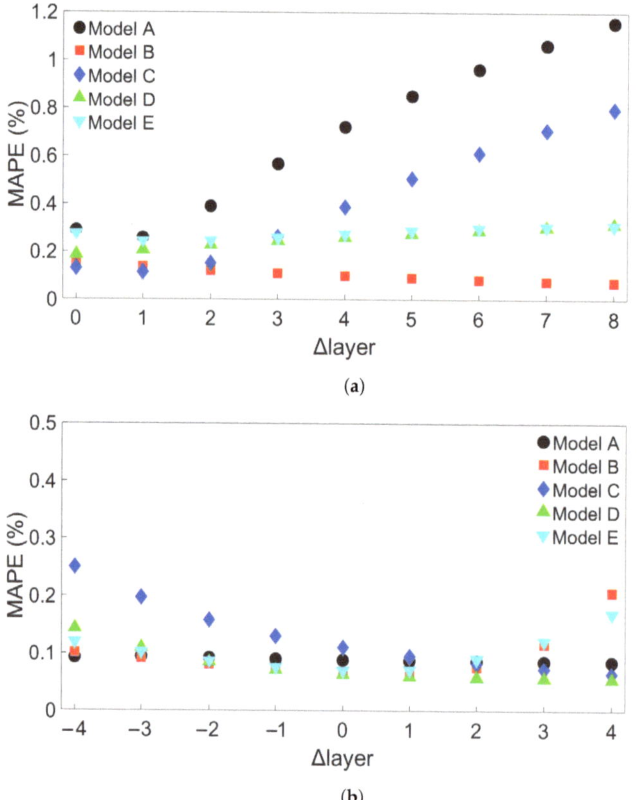

(a)

(b)

Figure 11. *Cont.*

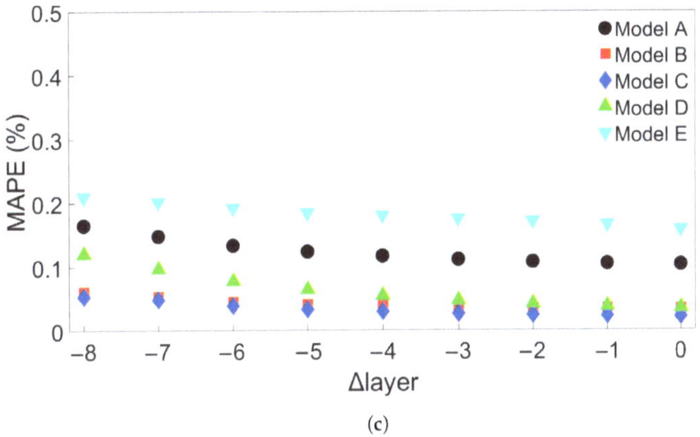

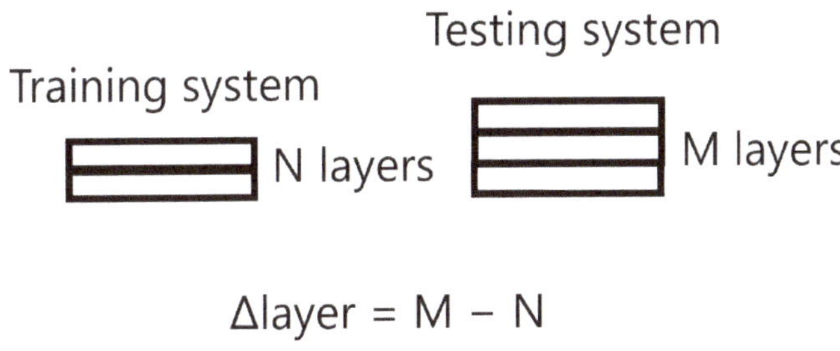

(**d**)

Figure 11. The testing MAPEs for MLP models applied to each of the nine Group 3 systems. (**a**) Models were trained on System 1. (**b**) Models were trained on System 5. (**c**) Models were trained on System 9. Models A through E in each case differ only in the initial seed set before training. (**d**) Definition of the difference in number of layers, Δlayer.

3.3. Transferability for Bars with Different Scanning Speeds

The last group of FE simulations considered was Group 4, which consisted of 13 four-layer bars with the same length of 0.96 m but with different scanning speeds. Table 5 shows the scanning speed for each of the Group 4 systems. Figure 13a shows the MAPE for cross-system testing with Group 4 systems when models are trained on System 1, Figure 13b the MAPE for models trained on System 4, and Figure 13c the MAPE for models trained on System 13. Here, we observe a greater increase in MAPE when a model trained on a system with a slower scanning speed is tested on a system with a faster speed; the MAPE still remains good when tested on a system with a similar scanning speed. Figure 13a shows that the models all give a MAPE of less than 1% at a difference in scanning speed of 0.015 m/s. When used to test System 13, which has a 0.035 m/s faster scanning speed than the training system, the errors ranges from 2 to 5%. In Figure 13b, however, it can be seen that one of the models has a radically different error growth than the others, namely, Model C. The other models all have approximately the same error growth, giving a MAPE of between 2 and 3% when testing on System 13. Figure 13c shows a general decrease in error up to a difference of about 0.02 m/s; further differences lead to a slight increase in error, though the error still remains lower than 0.5%. In order to see if this trend would hold, 15 additional models with different initial random seeds were trained on System 4 and tested on System 13. The MAPEs are shown in Table S7 in the Supplementary Material. None of these additional models gave a higher MAPE than Model E's error of 3.03%, leaving Model C's MAPE of almost 15% a clear outlier. The average MAPE for all 20 models is 2.88%.

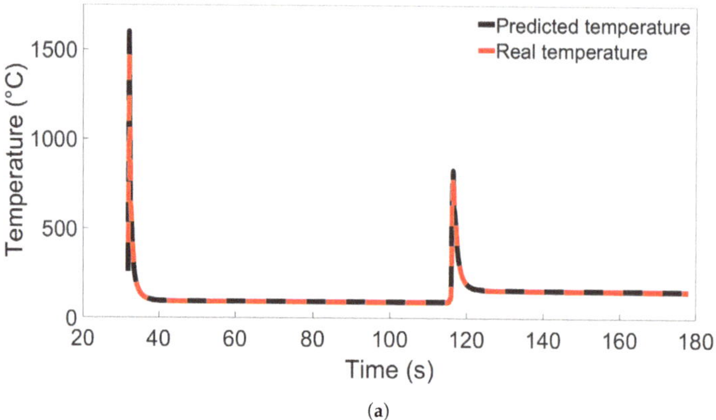

(a)

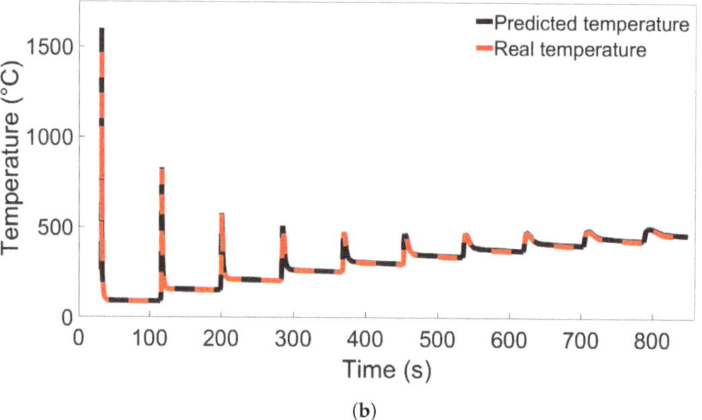

(b)

Figure 12. The temperature evolution of single nodes predicted by Model A from Figure 11a compared with the real temperature evolution from the FE simulation. In (**a**), the node is from System 1 from Group 3; in (**b**), the node is from System 9 from Group 3.

When examining the temperature evolution of single nodes from Group 4 systems using Model C from Figure 13b, we found a different trend than was observed in single-node predictions for Groups 1 through 3. Like in the previous cases, the nodes were taken from the Node B position shown in Figure 8. The results are shown in Figure 14a for a single node from System 4 and Figure 14b for a node from System 13, and in the latter case, the temperature evolution shows a large divergence between the first and second temperature peaks. Interestingly, this error disappears after the second peak, and there are no visible errors during the later parts of the simulation. Additional single-node tests for Model C on System 13 were performed; these are presented in Section 3.4.

Table 5. The FE systems in Group 4 and the scanning speed of each.

System	Scanning Speed (m/s)
1	0.01
2	0.01125
3	0.01375
4	0.015
5	0.01625
6	0.0175
7	0.01875
8	0.02
9	0.025
10	0.03
11	0.035
12	0.04
13	0.045

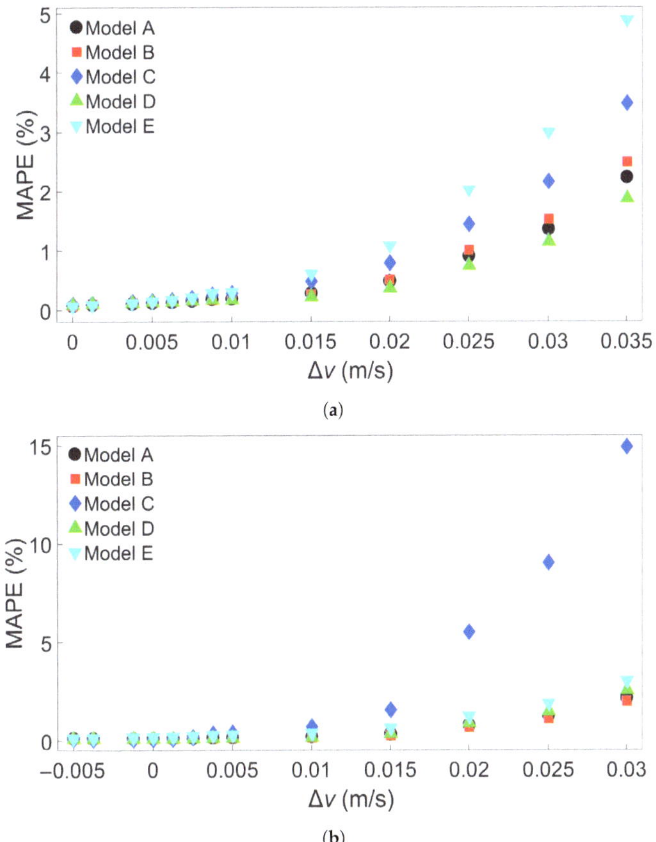

Figure 13. *Cont.*

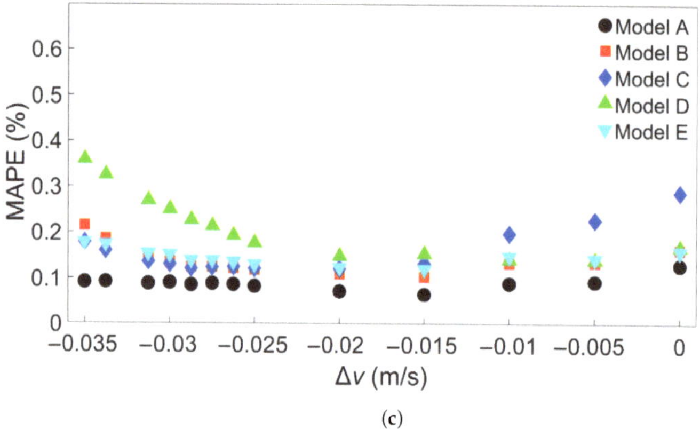

(c)

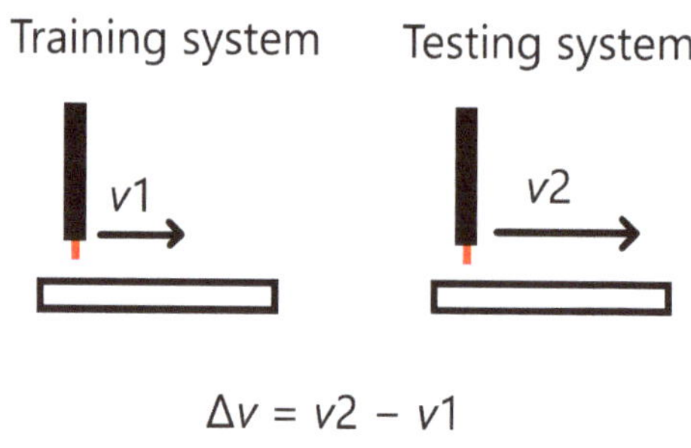

(d)

Figure 13. The MAPEs for MLP models trained on (**a**) System 1, (**b**) System 4, and (**c**) System 13 from Group 4 and tested on each system in Group 4. Models A through E differ only in initial seed before training. (**d**) Definition of the difference in scanning speed, Δv.

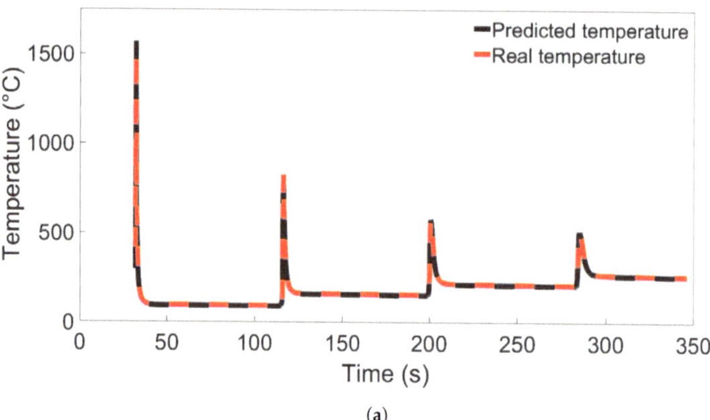

(a)

Figure 14. *Cont.*

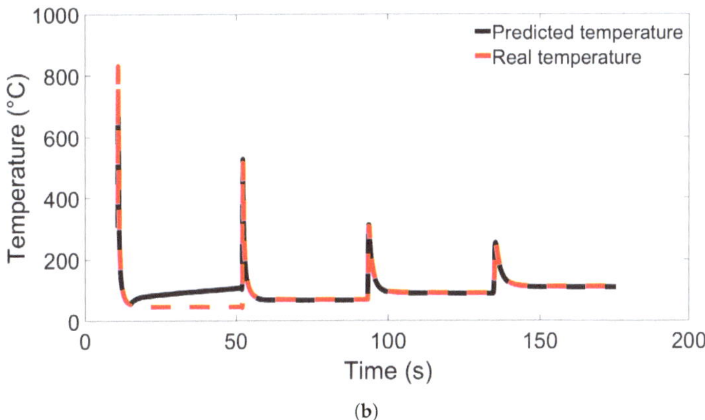

Figure 14. Comparison of real temperature and temperature predicted by Model C from Figure 13b for (**a**) a single node from System 4 from Group 4, and (**b**) a single node from System 13 from Group 4.

3.4. Temperature Evolution of Single Nodes

To investigate the error observed in Section 3.3 further, we performed additional single-node temperature predictions for System 13 from Group 4. Nine different nodes from System 13 were used, labelled Nodes A through I, in the positions illustrated in Figure 8. Nodes A and C were positioned 0.025 m from each end of the bar, and Node B was positioned in the center, all in the first layer. Node B's position was the same as the single nodes examined in the previous sections. Nodes D through F and G through I were positioned at those same horizontal locations, but in layers 2 and 3, respectively. Models A and C from Figure 13b were used to test the single-node data to compare the poorly performing Model C with a model that performs better. The results for testing on data from Node B are shown in Figure 15a, while the results for Node A and Nodes C through I are shown in Figures S4 through S11 in the Supplementary Material. From the figure, we observe that the temperature evolution predicted by Model A shows the same trend as Model C but with a smaller error between the first two temperature peaks. In both cases, the error disappears after the second peak. In Figures S4–S11, showing the results from prediction on Nodes A and C through I, this error is present for all predictions between the first and second peaks. For the nodes near the bar edges, there is also a prediction error between the second and third peaks. This error is present regardless of whether the node is in the first or second layer. We see that the prediction during the first peak varies between overshooting and undershooting the actual temperature.

Additional single-node predictions were examined in System 11 from Group 2 as well to compare the results with those found for Group 4. The nodes were placed and labelled as in the previous case, illustrated in Figure 8. Models A and C from Figure 7a were used. The resulting temperature predictions for Node B are shown in Figure 15b, where Model A shows the same trend as Model C but with a lower error. Results for Nodes A and C through I are shown in Figures S12 through S19 in the Supplementary Material. These figures all show the same trend of predicted temperature diverging from the real temperature during later parts of the simulation, with Model C displaying greater divergence than Model A.

From these results, we observe that the MLP model predictions are consistent for different nodes in the same FE system. There is a clear difference in how the prediction errors are distributed in models used on the Group 4 system compared to models used on the systems from Groups 1 through 3. This could be the reason the particularly high error observed in Figure 13b does not appear for any of the tests performed on the Groups 1 through 3 systems. We also observe that for Group 4, the discrepancy between the predicted and real temperature disappears after peak two or three.

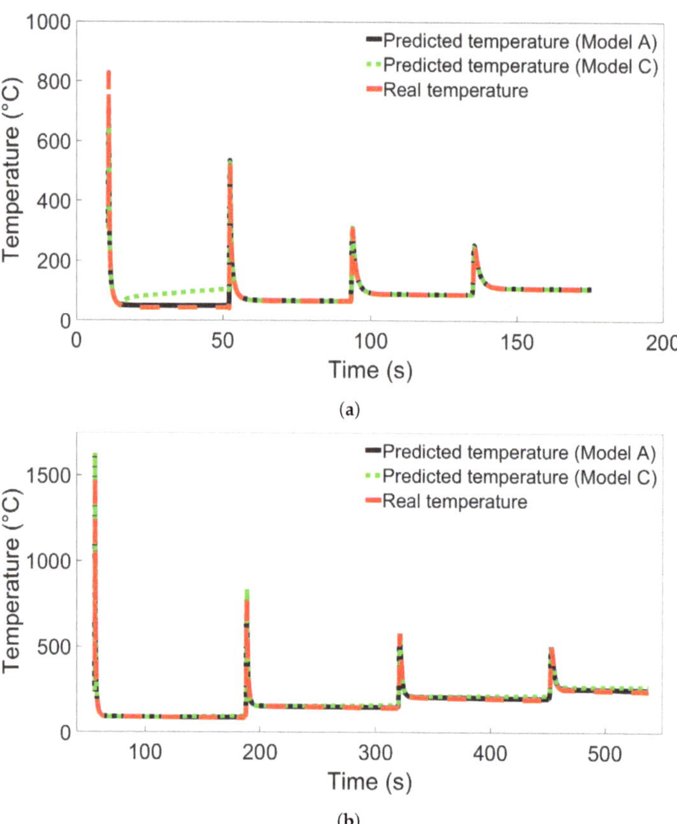

Figure 15. (**a**) The temperature evolution over time for Node B in System 13 from Group 4 as predicted by Model C from Figure 13b compared with the real temperature evolution. (**b**) The temperature evolution over time for Node B in System 11 from Group 2 as predicted by Model C from Figure 7a compared with the real temperature evolution.

3.5. Effects of Batch Size

One potential cause of the particularly high error observed in Section 3.3 is overfitting; while the baseline tests in Section 2.3 do not show a high error, it is possible that a small degree of overfitting in a model could become more prominent when the model is used to test data from substantially different FE simulations. Thus far, the batch size was set to 64. One way to reduce overfitting is to increase the batch size, which lowers the amount of times the MLP weights are updated during training. To explore the effect of batch size on the models used thus far, additional MLP models were created with the same properties as previously described, except for batch size and number of epochs. The batch size for these additional models was set to 640, ten times larger than the batch size of 64 used for the previous models. The number of epochs for the additional models was determined through preliminary testing. For the models trained on Group 1 systems, the number of epochs was kept at five, while for the models trained on other systems, it was increased to eight. Figure 16a shows the MAPE when batch-size-640 models trained on System 1 from Group 1 are tested on the other Group 1 systems, similar to Figure 6a. Figure 16b shows the MAPE for batch-size-640 models trained on System 1 from Group 2 and tested on other Group 2 systems, as in Figure 7a. Comparing each of the figures, the batch-size-640 models appear to perform better overall than the previously examined batch-size-64 models. It is also clear from the figures that there is no guarantee that any given model which performs well at a given seed and low batch size will give good performance with a higher batch size, as seen with the two Model Es trained on Group 1, Model 1, in Figures 6a and 16a. Additional tests were performed with 15 additional batch-size-640 models with different random seeds for both the Group 1 and the Group 2 systems. The models were trained on System 1 from Group 1 or Group 2 and then used to test System 9 from Group 1 or System 11 from Group 2, respectively. The resulting MAPEs are shown in the Supplementary Material, the Group 1 results in Table S8 and the Group 4 results in Table S9. From these tables, we see that the performances of Model E in Figure 16a and Model B in Figure 16b, while particularly high, are not outliers. In the Group 1 batch-size-640 case, including results from the additional models led to an average MAPE of 1.83% with standard deviation 1.11% for testing on System 9. For the Group 2 batch-size-640 case, an average of 1.51% with standard deviation 1.54% was found for testing on System 11. For both sets, this is an improvement over the average errors obtained using batch size 64. In the case of the largest length differences, for the models trained on System 1 from Group 1, there were five models with a MAPE of less than 1%, while for models trained on System 1 from Group 2 there were nine, though there were also several models with a MAPE just above 1%.

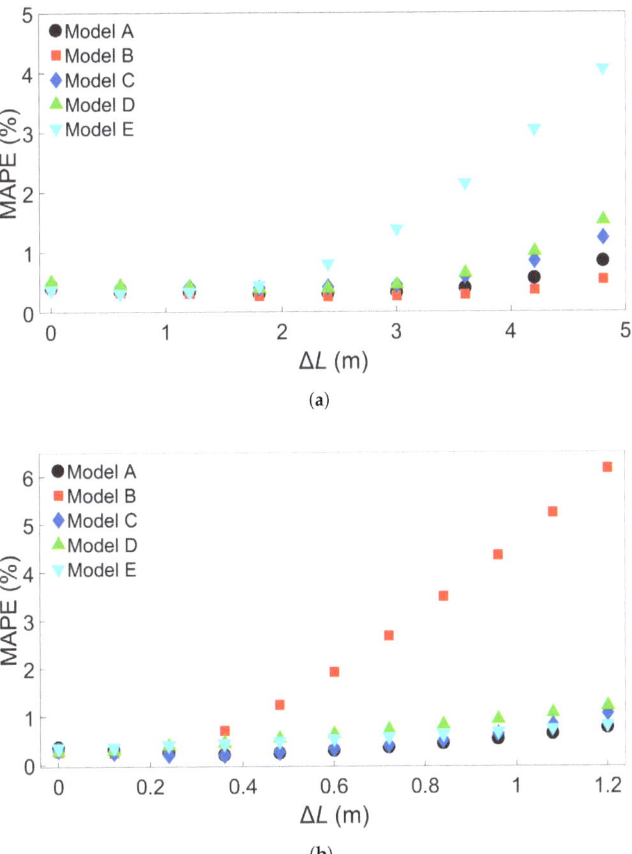

(a)

(b)

Figure 16. (**a**) The MAPEs for MLP models which were trained on System 1 from Group 1 and used to test each of the Group 1 systems. (**b**) The MAPEs for MLP models which were trained on System 1 from Group 2 and used to test each of the Group 2 systems. The models have a batch size of 640, and Models A through E differ only in initial seed before training.

We examined the Group 3 and Group 4 cases in the same fashion. Figure 17a shows the MAPE for five batch-size-640 models trained on System 1 from Group 3 and tested on each of the other Group 3 systems. Here, all the models show decreasing error as exhibited by Model B in Figure 11a. Figure 17b shows the MAPE for five models trained on System 4 from Group 4. All of these models show error growth comparable to the nonoutlier results in the batch-size-64 case. A total of 15 additional batch-size-640 models were trained on System 1 from Group 3 and 15 more on System 4 from Group 4 to further examine performance when testing on System 9 from Group 3 and System 13 from Group 4, respectively. The results are shown in Tables S10 and S11 in the Supplementary Material. Including these additional results, the average error for testing on System 9 from Group 3 is 0.32%, with a maximum error of 2.32%, showing improved results compared to those obtained using models with batch size 64. The MAPEs for prediction on System 13 from Group 4 do not show any outlier behavior. The average MAPE with batch-size-640 models is 1.96%, again lower than for the batch-size-64 models. Even if the outlier, batch-size-64 Model C, is removed, the average MAPE is still largest for the batch-size-64 models at 2.25%.

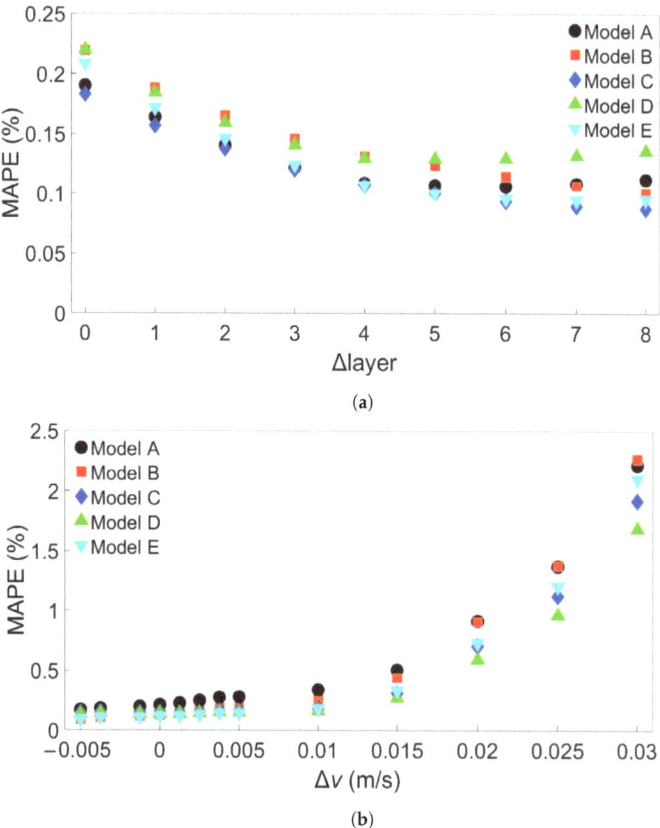

Figure 17. (**a**) The MAPEs for MLP models which were trained on System 1 from Group 3 and used to test each of the Group 3 systems. (**b**) The MAPEs for MLP models which were trained on System 4 from Group 4 and used to test each of the Group 4 systems. The models have a batch size of 640, and Models A through E differ only in initial seed before training.

We observe that for all four groups of FE systems, batch-size-640 models return a lower average MAPE than the batch-size-64 models. For Groups 1 and 3, batch size

640 resulted in more models with a highest MAPE of less than 1% than with batch size 64. The improvement in performance, as well as the lack of any outliers among the batch-size-640 models trained on System 4 from Group 4, could be due to reduced overfitting. To further explore the effect of increasing batch size, additional tests were performed using models with a range of different batch sizes. A total of 25 sets of four models with batch sizes 120, 240, 480, and 960 were trained on System 4 from Group 4, and tested on System 13 from Group 4. Each set had a different initial seed for its models. The resulting MAPEs are shown in Figure 18a. A total of 2 of the 100 models do indeed result in much higher errors than the rest—1 model with batch size 120 and 1 with batch size 960, showing that higher-batch-size models can also give unusually high errors. Table 6 shows the maximum MAPE, average MAPE, and variance for each batch size. For batch sizes 120 and 960, two variations of the results are shown, both including and excluding the outlier values. The variance is naturally much higher for the groups containing these outliers. When they are excluded, a slight decrease in both average and maximum MAPE is observed.

Additional MLP models were also created for the Group 1 systems to compare with the results for Group 4. A total of 25 similar sets of four models with the same batch sizes as for the Group 4 case were trained on System 1 from Group 1 and tested on System 9 from Group 1. The results are shown in Figure 18b. Here, there is a much greater spread of values compared to what is observed for Group 4. There are also no extreme outliers observed compared to what is seen in Figure 18a. Table 7 shows the mean and variance of the MAPE as well as the maximum MAPE for the models by batch size. For batch sizes 480 and 960, all of these values are lower than for the groups of lower-batch-size models.

Table 6. The average MAPE, standard deviation of the MAPE, and maximum MAPE for the models shown in Figure 18a, by batch size. For batch sizes 120 and 960, the results are shown both with and without the outlier value.

Batch Size	Average MAPE (%)	Std. Dev. (%)	Max MAPE (%)
120	2.42	1.40	9.00
120 (without outlier)	2.15	0.29	2.77
240	2.10	0.34	2.72
480	2.00	0.25	2.64
960	2.42	2.20	12.93
960 (without outlier)	1.98	0.25	2.37

Table 7. The average MAPE, standard deviation for the MAPE, and maximum MAPE for the models shown in Figure 18b, by batch size.

Batch Size	Average MAPE (%)	Std. Dev. (%)	Max MAPE (%)
120	3.19	1.66	6.38
240	3.55	1.60	7.09
480	2.27	1.20	4.79
960	1.71	0.87	4.05

It appears that the Group 4 case is particularly vulnerable to extreme spikes in error, and these can occur unpredictably—models with different number of batches or a different initial seed behave consistently. The Groups 1, 2, and 3 cases, in contrast, do not seem to exhibit this—at least not in the cases studied. The error vulnerability appears to be related to how the MLP models perform when predicting results in the first and second valleys between peaks in temperature, as described in Section 3.4. Though the models tend to perform well in the Group 4 case, care should be taken to check each particular model before it is used, even if the batch size is set to an optimal value.

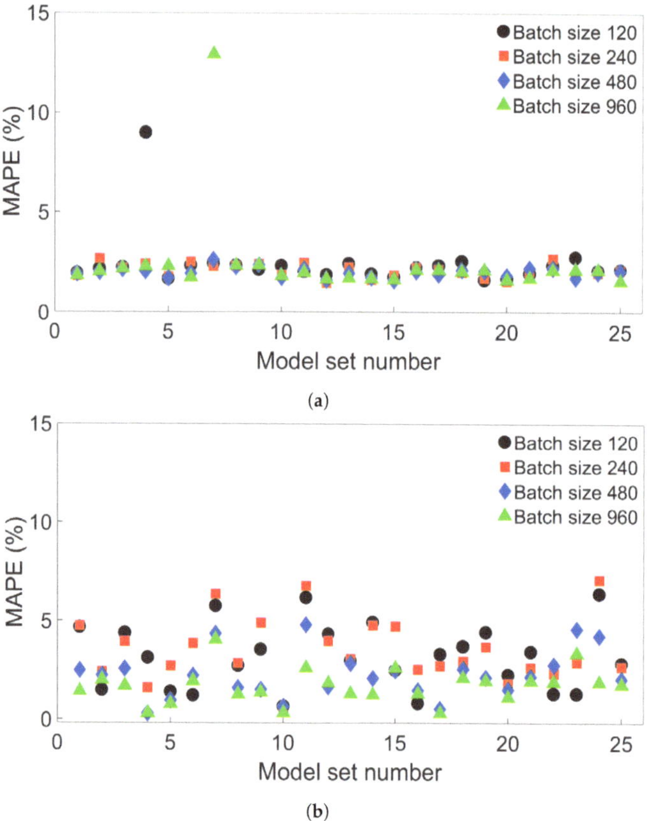

Figure 18. The MAPEs for different sets of four MLP models with different batch sizes. (**a**) The models were trained on System 4 from Group 4 and tested on System 13 from Group 4. (**b**) The models were trained on System 1 from Group 1 and tested on System 9 from Group 1.

4. Conclusions

In this study, we performed 40 FE simulations and used the simulation data to train and test multilayer perceptron (MLP) models for predicting temperature evolution. Our goal for this study was to investigate the transferability of thermal history in wire-arc additive manufacturing (WAAM) by using a simple MLP model which takes past temperature history and time as input. Particular attention was paid to situations where high error was found to occur, and additional MLP models were created for further testing when necessary. We have shown that, with proper precautions taken, the simple MLP-based model explored in this study is able to predict temperature evolution in finite element (FE) simulations with different bar lengths, numbers of layers, or scanning speeds. When a trained model is used to test data from FE systems with shorter bar lengths, fewer layers, or slower scanning speeds, the MLP models consistently result in a small error. A slight growth in error is observed as the difference from the simulation used for training increases, but the mean absolute percentage error (MAPE) remains less than 0.5% for one-layer systems of different lengths and four-layer systems with different scanning speeds, and less than 0.3% for four-layer systems of different lengths and systems with different numbers of layers and a bar length of 0.96 m.

When a trained model is used to test simulations with longer bar lengths, more layers, or faster scanning speeds, low error is still observed when the difference in length, number of layers, or speed is small. As the difference increases, the average MAPE increases as well, as does the variation in model performance due to randomness. In the groups of one-layer and four-layer bars of different lengths and bars with different numbers of layers and a length of 0.96 m, we were still able to identify models which yielded a MAPE of less than 1% in the cases of greatest difference. For four-layer bars with different scanning speeds, testing on data from a system with a faster scanning speed can in a few cases result in unusually high error. The difference in error shows more natural variation in the variable-length and variable-layer cases.

The performance of a given model is observed to be consistent for an increasing difference in length, number of layers, or scanning speed. Models which show a relatively low error growth when tested on an FE system with a certain difference from the training system tend to also give relatively low error when tested on an FE system with a greater difference. Similarly, if the model shows a relatively high error when applied to a system with a certain difference, it tends to give a high error when applied to systems with greater difference. Therefore, it is possible that the performance of a given model can be predicted through preliminary testing before the model is applied.

Increasing the batch size from the initial value of 64 to 640 was found to result in overall better results, though increasing the batch size of a particular model does not guarantee it will perform better. Of course, the optimal values for the batch size and number of epochs depends on how many data points are used for training. We found that the particularly high errors obtained with some of the models trained on four-layer systems with different scanning speeds could be found in models with both small and large batch sizes, suggesting that these errors are not caused by overfitting. From single-node considerations, we observe that this increase in error comes from the model performance between the first and second passes of the heat source; this is not the case for the other multilayer systems.

In the future, we plan to continue the systematic study by creating recurrent neural network (RNN)- and physics-informed neural network (PINN)-based models and comparing their performance to the MLP models studied in the current work. We will examine whether the same patterns that were found in this work also appear when different models are used. We will also perform simulations of WAAM deposition of two-dimensional plates and once again compare how different models perform for these cases. Another possible candidate for further study is the effect of varying other process parameters, like laser power.

Supplementary Materials: The following are available online at https://www.mdpi.com/article/10.3390/ma17030742/s1, Figures S1–S3: Comparison of training and testing MAPE as well as testing MAPE for models with different initial seeds for System 1 from Group 2, System 1 from Group 3, and System 1 from Group 4, respectively; Figures S4–S11: Single-node predictions on System 13 from Group 4 performed using Models A and C from Figure 13b; Figures S12–S19: Single-node predictions on System 11 from Group 2 performed using Models A and C from Figure 7a; Table S1: FE model parameters, Tables S2–S11: MAPEs for tests performed with additional models.

Author Contributions: Conceptualization, Z.Z. and H.M.F.; methodology, H.M.F., Z.Z., K.M.M. and J.H.; investigation, H.M.F.; writing—original draft preparation, H.M.F.; writing—review & editing, H.M.F., D.M., K.M.M., J.H. and Z.Z.; supervision, Z.Z., K.M.M., J.H. and D.M. All authors have read and agreed to the published version of the manuscript.

Funding: This research received no external funding.

Institutional Review Board Statement: Not applicable.

Informed Consent Statement: Not applicable.

Data Availability Statement: Datasets used in this study are available upon request.

Acknowledgments: Supercomputer CPU hours used for training MLP models were provided by the Norwegian Metacenter for Computational Science (NRIS) (Grant No. NN9391K).

Conflicts of Interest: The authors declare no conflict of interest.

Abbreviations

The following abbreviations are used in this manuscript:

AM	Additive manufacturing
WAAM	Wire-arc additive manufacturing
DED	Directed energy deposition
CAD	Computer-aided design
DT	Digital twin
ML	Machine learning
MLP	Multilayer perceptron
ReLU	Rectified linear unit
RNN	Recurrent neural network
PINN	Physics-informed neural network
FE	Finite element
MAPE	Mean absolute percentage error

References

1. Beaman, J.J.; Bourell, D.L.; Seepersad, C.C.; Kovar, D. Additive Manufacturing Review: Early Past to Current Practice. *J. Manuf. Sci. Eng.* **2020**, *142*, 110812. [CrossRef]
2. Blakey-Milner, B.; Gradl, P.; Snedden, G.; Brooks, M.; Pitot, J.; Lopez, E.; Leary, M.; Berto, F.; du Plessis, A. Metal additive manufacturing in aerospace: A review. *Mater. Des.* **2021**, *209*, 110008. [CrossRef]
3. Li, J.L.Z.; Alkahari, M.R.; Rosli, N.A.B.; Hasan, R.; Sudin, M.N.; Ramli, F.R. Review of Wire Arc Additive Manufacturing for 3D Metal Printing. *Int. J. Autom. Technol.* **2019**, *13*, 346–353. [CrossRef]
4. Durakovic, B. Design for Additive Manufacturing: Benefits, Trends and Challenges. *Period. Eng. Nat. Sci.* **2018**, *6*, 179–191. [CrossRef]
5. Rodrigues, T.A.; Duarte, V.; Miranda, R.M.; Santos, T.G.; Oliveira, J.P. Current Status and Perspectives on Wire and Arc Additive Manufacturing (WAAM). *Materials* **2019**, *12*, 1121. [CrossRef]
6. Singh, S.R.; Khanna, P. Wire arc additive manufacturing (WAAM): A new process to shape engineering materials. *Mater. Today Proc.* **2021**, *44*, 118–128. [CrossRef]
7. Treutler, K.; Wesling, V. The Current State of Research of Wire Arc Additive Manufacturing (WAAM): A Review. *Appl. Sci.* **2021**, *11*, 8619. [CrossRef]
8. Müller, J.; Grabowski, M.; Müller, C.; Hensel, J.; Unglaub, J.; Thiele, K.; Kloft, H.; Dilger, K. Design and Parameter Identification of Wire and Arc Additively Manufactured (WAAM) Steel Bars for Use in Construction. *Metals* **2019**, *9*, 725. [CrossRef]
9. Brandl, E.; Baufeld, B.; Leyens, C.; Gault, R. Additive manufactured Ti-6Al-4V using welding wire: Comparison of laser and arc beam deposition and evaluation with respect to aerospace material specifications. *Phys. Procedia* **2010**, *5*, 595–606. [CrossRef]
10. Omiyale, B.O.; Olugbade, T.O.; Abioye, T.E.; Farayibi, P.K. Wire arc additive manufacturing of aluminium alloys for aerospace and automotive applications: A review. *Mater. Sci. Technol.* **2022**, *38*, 391–408. [CrossRef]
11. Taşdemir, A.; Nohut, S. An overview of wire arc additive manufacturing (WAAM) in shipbuilding industry. *Ships Offshore Struct.* **2021**, *16*, 797–814. [CrossRef]
12. Machirori, T.; Liu, F.Q.; Yin, Q.Y.; Wei, H.L. Spatiotemporal variations of residual stresses during multi-track and multi-layer deposition for laser powder bed fusion of Ti-6Al-4V. *Comput. Mater. Sci.* **2021**, *195*, 110462. [CrossRef]
13. Wu, B.; Pan, Z.; Ding, D.; Cuiuri, D.; Li, H.; Xu, J.; Norrish, J. A review of the wire arc additive manufacturing of metals: properties, defects and quality improvement. *J. Manuf. Process.* **2018**, *35*, 127–139. [CrossRef]
14. Bartlett, J.L.; Croom, B.P.; Burdick, J.; Henkel, D.; Li, X. Revealing mechanisms of residual stress development in additive manufacturing via digital image correlation. *Addit. Manuf.* **2018**, *22*, 1–12. [CrossRef]
15. Du, C.; Zhao, Y.; Jiang, J.; Wang, Q.; Wang, H.; Li, N.; Sun, J. Pore defects in Laser Powder Bed Fusion: Formation mechanism, control method, and perspectives. *J. Alloys Compd.* **2023**, *944*, 169215. [CrossRef]
16. Kumar, M.D.B.; Manikandan, M. Assessment of Process, Parameters, Residual Stress Mitigation, Post Treatments and Finite Element Analysis Simulations of Wire Arc Additive Manufacturing Technique. *Met. Mater. Int.* **2022**, *28*, 54–111. [CrossRef]
17. Li, C.; Liu, Z.Y.; Fang, X.Y.; Guo, Y.B. Residual Stress in Metal Additive Manufacturing. *Procedia CIRP* **2018**, *71*, 348–353. [CrossRef]
18. Sun, L.; Ren, X.; He, J.; Zhang, Z. Numerical investigation of a novel pattern for reducing residual stress in metal additive manufacturing. *J. Mater. Sci. Technol.* **2021**, *67*, 11–22. [CrossRef]
19. Sun, L.; Ren, X.; He, J.; Olsen, J.S.; Pallaspuro, S.; Zhang, Z. A new method to estimate the residual stresses in additive manufacturing characterized by point heat source. *Int. J. Adv. Manuf. Technol.* **2019**, *105*, 2415–2429. [CrossRef]
20. Sun, L.; Ren, X.; He, J.; Zhang, Z. A bead sequence-driven deposition pattern evaluation criterion for lowering residual stresses in additive manufacturing. *Addit. Manuf.* **2021**, *48*, 102424. [CrossRef]

21. Zhao, Y.; Sun, W.; Wang, Q.; Sun, Y.; Chen, J.; Du, C.; Xing, H.; Li, N.; Tian, W. Effect of beam energy density characteristics on microstructure and mechanical properties of Nickel-based alloys manufactured by laser directed energy deposition. *J. Mater. Process. Technol.* **2023**, *319*, 118074. [CrossRef]

22. Zhang, L.; Chen, X.; Zhou, W.; Cheng, T.; Chen, L.; Guo, Z.; Han, B.; Lu, L. Digital Twins for Additive Manufacturing: A State-of-the-Art Review. *Appl. Sci.* **2020**, *10*, 8350. [CrossRef]

23. Phua, A.; Davies, C.H.J.; Delaney, G.W. A digital twin hierarchy for metal additive manufacturing. *Comput. Ind.* **2022**, *140*, 103667. [CrossRef]

24. Knapp, G.L.; Mukherjee, T.; Zuback, J.S.; Wei, H.L.; Palmer, T.A.; De, A.; DebRoy, T. Building blocks for a digital twin of additive manufacturing. *Acta Mater.* **2017**, *135*, 390–399. [CrossRef]

25. Bartsch, K.; Pettke, A.; Hübert, A.; Lakämper, J.; Lange, F. On the digital twin application and the role of artificial intelligence in additive manufacturing: A systematic review. *J. Phys. Mater.* **2021**, *4*, 032005. [CrossRef]

26. Vastola, G.; Zhang, G.; Pei, Q.X.; Zhang, Y.W. Controlling of residual stress in additive manufacturing of Ti6Al4V by finite element modeling. *Addit. Manuf.* **2016**, *12*, 231–239. [CrossRef]

27. Cook, P.S.; Murphy, A.B. Simulation of melt pool behaviour during additive manufacturing: Underlying physics and progress. *Addit. Manuf.* **2020**, *31*, 100909. [CrossRef]

28. Alizadeh, R.; Allen, J.K.; Mistree, F. Managing computational complexity using surrogate models: A critical review. *Res. Eng. Des.* **2020**, *31*, 275–298. [CrossRef]

29. Jiang, J.; Xiong, Y.; Zhang, Z.; Rosen, D.W. Machine learning integrated design for additive manufacturing. *J. Intell. Manuf.* **2022**, *33*, 1073–1086. [CrossRef]

30. Zhu, Q.; Liu, Z.; Yan, J. Machine learning for metal additive manufacturing: Predicting temperature and melt pool fluid dynamics using physics-informed neural networks. *Comput. Mech.* **2021**, *67*, 619–635. [CrossRef]

31. Mozaffar, M.; Paul, A.; Al-Bahrani, R.; Wolff, S.; Choudhary, A.; Agrawal, A.; Ehmann, K.; Cao, J. Data-driven prediction of the high-dimensional thermal history in directed energy deposition processes via recurrent neural networks. *Manuf. Lett.* **2018**, *18*, 35–39. [CrossRef]

32. Stathatos, E.; Vosniakos, G. Real-time simulation for long paths in laser-based additive manufacturing: A machine learning approach. *Int. J. Avd. Manuf. Technol.* **2019**, *104*, 1967–1984. [CrossRef]

33. Ness, K.L.; Paul, A.; Sun, L.; Zhang, Z. Towards a generic physics-based machine learning model for geometry invariant thermal history prediction in additive manufacturing. *J. Mater. Process. Technol.* **2022**, *302*, 117472. [CrossRef]

34. Geurts, P.; Ernst, D.; Wehenkel, L. Extremely randomized trees. *Mach. Learn.* **2006**, *63*, 3–42. [CrossRef]

35. Le, V.T.; Nguyen, H.D.; Bui, M.C.; Pham, T.Q.D.; Le, H.T.; Tran, V.X.; Tran, H.S. Rapid and accurate prediction of temperature evolution in wire plus arc additive manufacturing using feedforward neural network. *Manuf. Lett.* **2022**, *32*, 28–31. [CrossRef]

36. Xie, J.; Chai, Z.; Xu, L.; Ren, X.; Liu, S.; Chen, X. 3D temperature field prediction in direct energy deposition of metals using physics informed neural network. *Int. J. Adv. Manuf. Technol.* **2022**, *119*, 3449–3468. [CrossRef]

37. Wacker, C.; Köhler, M.; David, M.; Aschersleben, F.; Gabriel, F.; Hensel, J.; Dilger, K.; Dröder, K. Geometry and Distortion Prediction of Multiple Layers for Wire Arc Additive Manufacturing with Artificial Neural Networks. *Appl. Sci.* **2021**, *11*, 4694. [CrossRef]

38. Bonassi, F.; Farina, M.; Xie, J.; Scattolini, R. On Recurrent Neural Networks for learning-based control: Recent results and ideas for future developments. *J. Process Control* **2022**, *114*, 92–104. [CrossRef]

39. Abaqus–Finite Element Analysis for Mechanical Engineering and Civil Engineering. Available online: https://www.3ds.com/products-services/simulia/products/abaqus/ (accessed on 27 November 2023).

40. Goldak, J.; Chakravarti, A.; Bibby, M. A new finite element model for welding heat sources. *Metall. Trans. B* **1984**, *15*, 299–305. [CrossRef]

41. Chujutalli, J.H.; Lourenço, M.I.; Estefen, S.F. Experimental-based methodology for the double ellipsoidal heat source parameters in welding simulations. *Mar. Syst. Ocen Technol.* **2020**, *15*, 110–123. [CrossRef]

42. Patterson, J.; Gibson, A. *Deep Learning: A Practitioner's Approach*, 1st ed.; O'Reilly Media: Sebastopol, CA, USA, 2017; pp. 41–80.

43. Paszke, A.; Gross, S.; Massa, F.; Lerer, A.; Bradbury, J.; Chanan, G.; Killeen, T.; Lin, Z.; Gimelshein, N.; L., A.; et al. PyTorch: An Imperative Style, High-Performance Deep Learning Library. In Proceedings of the Advances in Neural Information Processing Systems 32, Vancouver, Canada, 8–14 December 2019. [CrossRef]

44. Girshick, R. Fast R-CNN. In Proceedings of the 2015 IEEE International Conference on Computer Vision (ICCV 2015), Santiago, Chile, 13–16 December 2015; pp. 1440–1448. [CrossRef]

Article

A Robust Recurrent Neural Networks-Based Surrogate Model for Thermal History and Melt Pool Characteristics in Directed Energy Deposition

Sung-Heng Wu [1], Usman Tariq [1], Ranjit Joy [1], Muhammad Arif Mahmood [2]

[1] Department of Mechanical Engineering, Missouri University of Science and Technology, Rolla, MO 65409, USA; swm54@umsystem.edu (S.-H.W.)
[2] Intelligent Systems Center, Missouri University of Science and Technology, Rolla, MO 65409, USA
[3] National Strategic Planning and Analysis Research Center (NSPARC), Department of Electrical and Computer Engineering, Mississippi State University, Starkville, MS 39759, USA
*   Correspondence: liou@mst.edu; Tel.: +1-573-341-4908

Abstract: In directed energy deposition (DED), accurately controlling and predicting melt pool characteristics is essential for ensuring desired material qualities and geometric accuracies. This paper introduces a robust surrogate model based on recurrent neural network (RNN) architectures—Long Short-Term Memory (LSTM), Bidirectional LSTM (Bi-LSTM), and Gated Recurrent Unit (GRU). Leveraging a time series dataset from multi-physics simulations and a three-factor, three-level experimental design, the model accurately predicts melt pool peak temperatures, lengths, widths, and depths under varying conditions. RNN algorithms, particularly Bi-LSTM, demonstrate high predictive accuracy, with an R-square of 0.983 for melt pool peak temperatures. For melt pool geometry, the GRU-based model excels, achieving R-square values above 0.88 and reducing computation time by at least 29%, showcasing its accuracy and efficiency. The RNN-based surrogate model built in this research enhances understanding of melt pool dynamics and supports precise DED system setups.

Keywords: directed energy deposition; surrogate model; recurrent neural network; melt pool characterization; thermal history

Citation: Wu, S.-H.; Tariq, U.; Joy, R.; Mahmood, M.A.; Malik, A.W.; Liou, F. A Robust Recurrent Neural Networks-Based Surrogate Model for Thermal History and Melt Pool Characteristics in Directed Energy Deposition. *Materials* **2024**, *17*, 4363. https://doi.org/10.3390/ma17174363

Academic Editors: Tuhin Mukherjee and Qianru Wu

Received: 30 July 2024
Revised: 23 August 2024
Accepted: 27 August 2024
Published: 3 September 2024

1. Introduction

Directed energy deposition (DED) is an additive manufacturing (AM) technique for metals that creates parts by melting metal feedstocks with concentrated thermal energy [1,2]. Compared to the laser powder bed fusion process, DED is more cost-efficient and capable of producing parts with greater efficiency and adaptability [3]. These remarkable characteristics make DED an attractive option for rapid prototyping, manufacturing functionally graded materials, and repairing high-value components [4]. Specifically, DED excels in repairing worn or damaged components, thereby extending the service life of industrial and aerospace equipment by restoring structural integrity and functionality [5]. Over the last decade, DED's usage has expanded in the defense, manufacturing, and automotive industries [6]. For instance, DED has been employed to repair airfoils in airplane engines [7]. The DED market size is projected to reach more than USD 700 million by 2025 [8]. Despite DED's advantages over other AM techniques, challenges remain in minimizing defects during printing. Factors contributing to defect generation include gas entrapment, insufficient melting, and unstable melt pool generation [9,10]. Comprehending the thermal behavior and melt pool generation in relation to process parameters is essential for reducing defects during DED printing [11].

In DED, the melt pool is defined as the regime where metal particles are melted during laser–material interaction, generating an orbicular droplet [2,12]. Within the molten pool, the thermal distribution plays a crucial role in defining the microstructure and defects of the manufactured part [13]. In the case of a small molten pool, a relatively reduced thermal distribution can result in inadequate adjacent melt pools' overlap, leading to the lack of fusion defects [14]. Additionally, an irregular molten pool caused by elevated energy density can cause keyhole formation, leading to substantial material vaporization [15,16]. The dense plasma plume results in a recoil force on the molten material which leads to gas entrapment, creating defects [17,18]. Attaining and monitoring optimal thermal distribution is essential for an appropriate melting flow within the molten pool [19]. DED often encounters non-uniform thermal distribution along with rapid heating and slow cooling cycles, developing anisotropic microstructures, characterized by porosity and uneven grains [20]. The uneven grains affect the mechanical properties negatively [21,22]. For the DED process, the thermal distribution within the molten pool can be monitored using sensors such as thermocouples, IR cameras, and pyrometers. IR camera, in combination with image processing, was applied to observe thermal distribution within the melt pool. Comparatively reliable results were obtained at a 100 kHz sampling rate as well as 20 μm resolution [23]. In addition, the IR camera and pyrometers can monitor radiation from moving bodies and capture thermal distribution without surface contact, thus assisting in situ monitoring of the DED process [24]. On the other hand, thermocouples are flexible and resource-effective compared to other sensing devices. However, direct contact is required for thermocouples, which limits their usage [25].

To predict molten pool thermal distribution, researchers have explored multi-physics and machine learning-based approaches [26,27]. In multi-physics techniques, FEM and analytical methods have been elaborated. On the one hand, an extensive multi-physics FEM model may provide reliable results on the verge of computational [28]. On the other hand, a simplified FEM model faces limitations owing to incomplete multi-physics involved in simulation analysis [29,30]. Furthermore, the accuracy of the FEM model is also affected by factors such as element type, initial and boundary conditions, and meshing size [29]. In addition, the analytical techniques utilize multi-physics equations solved based on the initial and boundary conditions, simulating the thermal distribution and melt pool formation in the DED process [31]. These methods are unreliable due to mass and volume variation with time and uncertainties involved in DED processes.

Machine learning (ML)-based approaches have demonstrated significant advantages in modeling the intricate thermal distributions and melt pool formations essential to Directed Energy Deposition (DED), achieving solid accuracy and efficiency [32]. These approaches significantly reduce the high costs associated with extensive experimental procedures in research and development and alleviate the burden of lengthy computational times typically required by traditional simulation methods [33]. ML-based models are fundamentally data-driven, analyzing the relationship between each process parameter, like laser power, scanning speed, powder feed rate, and its outputs, such as thermal distribution and mechanical properties [34]. The data for training these models are usually collected from experiments or simulations, and the predictive insights provided by ML models greatly enhance the scalability of applications across various scenarios [35]. Various ML algorithms, such as SVM, clustering, and artificial neural networks, have been utilized to predict melt pool characteristics [36,37]. In addition, the defects of printed parts can be detected by predicting the melt pool dimension [38]. Despite these advancements, the dynamics of melt pools pose complex challenges. Primarily, the acquisition of large, robust datasets necessary for training these models is prohibitively expensive and time-intensive [39]. Additionally, current research inadequately addresses the sequential nature of melt pool dynamics, highlighting a critical need and understanding for more sophisticated applications of recurrent neural network (RNN) algorithms. Furthermore, the computational demand and memory requirements of these models also need optimization to enhance their reliability and robustness.

To address these challenges, this research introduces a pioneering RNN-based surrogate model designed specifically to predict both the thermal history and the geometric characteristics of melt pools in DED. The comprehensive framework that incorporates a factorial design of experiments, multi-physics modeling, refined data processing, and rigorous surrogate model training, evaluation, and comparison are proposed for this research. This innovative approach deepens the understanding of complex melt pool dynamics and significantly advances the operational capabilities of DED systems. It marks a substantial progression in the field, enhancing the precision and efficiency of ML-based surrogate models and facilitating their practical application in optimizing DED processes.

2. Methodology

The method used to develop the robust machine learning-based surrogate model for predicting melt pool thermal history and characteristics is presented in Figure 1. The process begins with the design of experiments, focusing on various parameters such as geometry, material, laser power, scanning speed, and hatch spacing. This is followed by multi-physics modeling, which includes finite element (FE) simulations and thermal modeling with temperature-dependent material properties. Key data points such as melt pool peak temperature and dimensions are extracted for building the surrogate model. In this research, the surrogate model is machine learning-based, employing multiple machine learning algorithms including Extreme Gradient Boosting (XGBoost), Long Short-Term Memory (LSTM), Bidirectional Long Short-Term Memory (Bi-LSTM), and Gated Recurrent Unit (GRU) to ensure accurate predictions of melt pool thermal history and dimensions. The evaluation and comparison of each algorithm are based on R-square values, Root Mean Square Error (RMSE), and Mean Absolute Error (MAE), ensuring robust model performance. A detailed description of each section is provided in the following content.

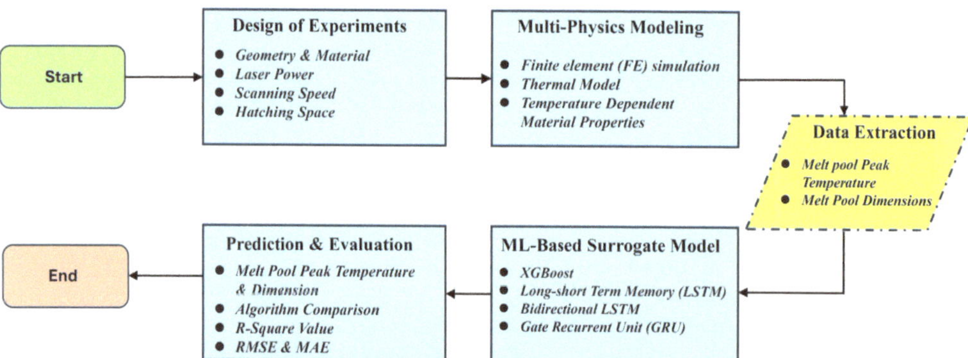

Figure 1. Proposed flow chart of current research.

2.1. Design of Experiments

In this research, a factorial design of experiments (DOE) is employed, involving three factors, each at three different levels. This methodical approach is designed to thoroughly investigate the interactions and effects of the variables on the outcomes. The chosen factors, critical to the Directed Energy Deposition (DED) process, include laser power (W), scanning speed (mm/s), and hatching space (%). Specifically, the laser power varies between 600 and 1000 watts, the scanning speed ranges from 2 to 6 mm per second, and the hatching space is adjusted from 40% to 60%. These parameters are selected based on their significant influence on the melt pool thermal distribution. A total of 27 experimental runs are conducted to explore the full factorial space, providing a comprehensive understanding of the process dynamics. The schematic detailing these experiments and their configurations is depicted in Figure 2.

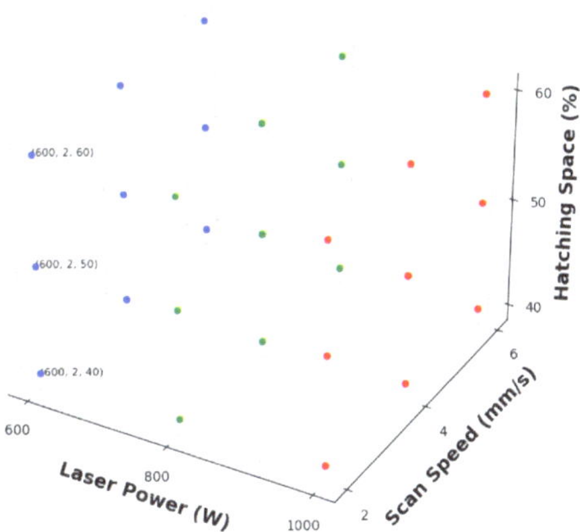

Figure 2. Factorial design of experiments.

In this research, Ti-6Al-4V is utilized as both the substrate material and the powder. Figure 3 depicts the simulation setup and the laser tool path, featuring a substrate thickness of 6.35 mm. This design incorporates four vertical single laser tracks that run from top to bottom. The total width of the deposit varies from 4.4 mm to 5.6 mm depending on the hatching space, with a length of 15 mm and a thickness of 0.5 mm. The red-colored line indicates that the laser is active, while the purple dashed line signifies that the gantry is moving to the next track and the laser is turned off. In this setup, cantilever clamping, shown in green, extends from the left end to 20 mm. Table 1 details the process parameters for the factorial design of experiments, while Table 2 presents the complete design used for the subsequent multi-physics simulation analysis.

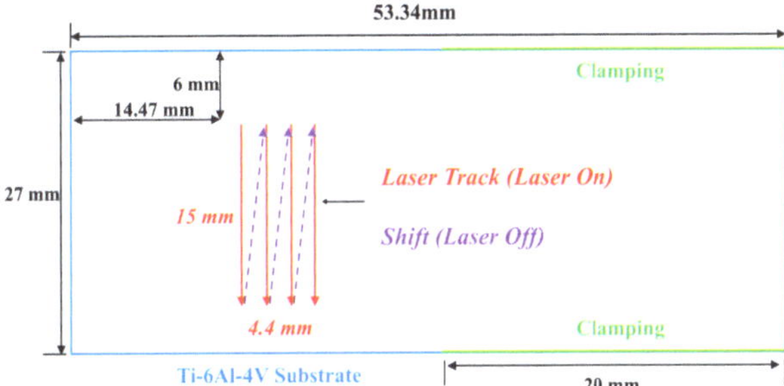

Figure 3. Tool path and simulation setup.

Table 1. Summary of process parameters.

Process Parameters (Unit)	Values
Laser Power (W)	600, 800, 1000
Scanning Speed (mm/s)	2, 4, 6
Hatching Space (%)	40, 50, 60
Laser Beam Size (mm)	2
Layer Thickness (mm)	0.5
Thermal Properties	Shown in Figure 4

Table 2. The twenty-seven-run design of experiment for multi-physics simulation.

Run	Laser Power (W)	Scanning Speed (mm/s)	Hatch Space (%)
1	600	2	60
2	600	2	50
3	600	2	40
4	600	4	60
5	600	4	50
6	600	4	40
7	600	6	60
8	600	6	50
9	600	6	40
10	800	2	60
11	800	2	50
12	800	2	40
13	800	4	60
14	800	4	50
15	800	4	40
16	800	6	60
17	800	6	50
18	800	6	40
19	1000	2	60
20	1000	2	50
21	1000	2	40
22	1000	4	60
23	1000	4	50
24	1000	4	40
25	1000	6	60
26	1000	6	50
27	1000	6	40

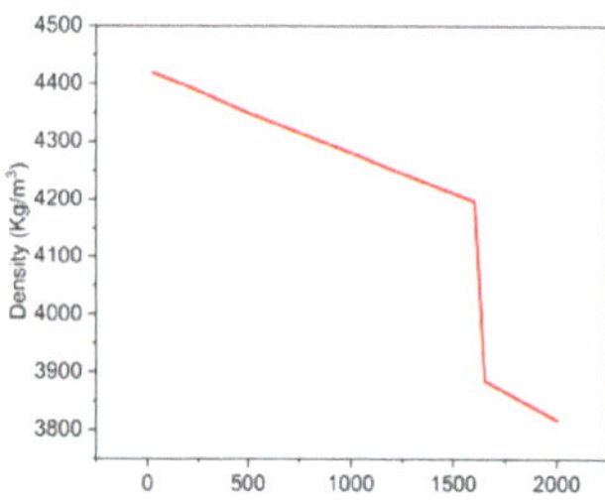 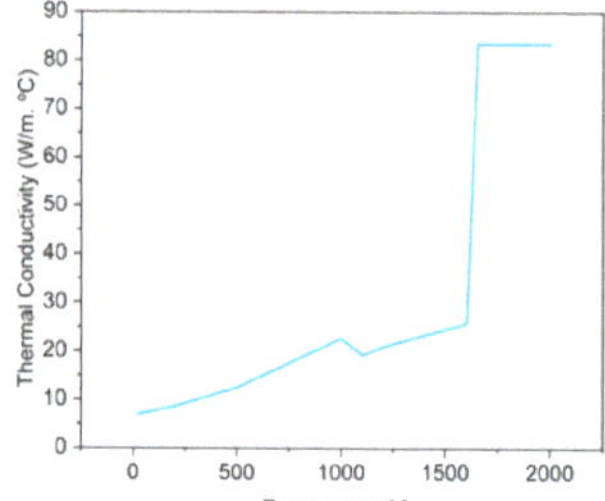 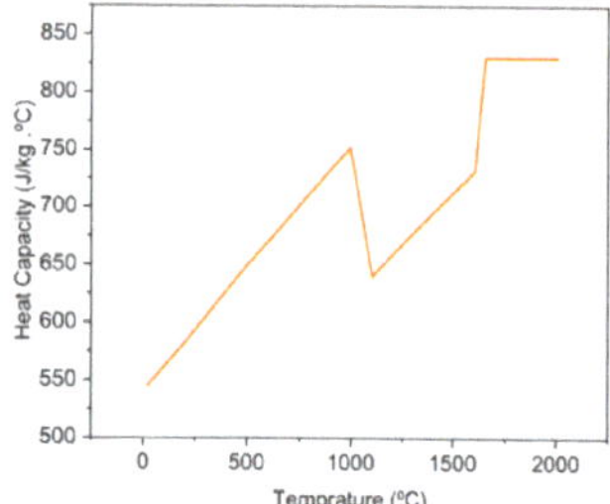

Figure 4. Thermal properties of Ti6Al4V [40].

2.2. Multi-Physics Simulation

After designing the experiments, each of the 27 runs was simulated in Abaqus CAE using the AM Modeler plug-in. For the thermal simulation, temperature-dependent material properties of Ti6Al4V were used, as shown in Figure 4.

To perform the calculation of thermal distribution during laser and material deposition, 3D heat conduction equation was employed over the domain shown as $T(x, y, z, t)$ while incorporating appropriate initial and boundary conditions as shown in Equation (1) [41,42].

$$\rho C \frac{\partial T}{\partial t} = \frac{\partial}{\partial x}\left(k \frac{\partial T}{\partial x}\right) + \frac{\partial}{\partial y}\left(k \frac{\partial T}{\partial y}\right) + \frac{\partial}{\partial z}\left(k \frac{\partial T}{\partial z}\right) + Q \tag{1}$$

where ρ is density, C is specific heat, T is temperature, t is time, k is thermal conductivity, and Q is heat flux in the form of laser heat source. To calculate heat loss due to convection, Newton's law cooling was employed as shown in Equation (2) [41,42].

$$q_{conv} = h(T - T_{env}) \tag{2}$$

where h is convective coefficient which is 30.0 (W/m^2·K^4), T is shown as temperature at any given time on the surface of the substrate, and T_{env} is room temperature which is 25.0 °C. Heat loss due to radiation is calculated using the Stephen–Boltzmann radiation law as shown in Equation (3) [41,42].

$$q_{rad} = \epsilon\sigma\left(T^4 - T_{env}^4\right) \tag{3}$$

where ϵ is known as emissivity and its value is taken as 0.8, σ represents the Stephen–Boltzmann constant with a value of 5.67×10^{-8} W/m^2·K^4. For body heat flux, Goldaks's double ellipsoid heat distribution is used as shown in Equation (4) [42,43].

$$Q = \frac{6\sqrt{3}P\eta}{abc\sqrt{\pi}}exp\left(-\frac{3x^2}{a^2} - \frac{3y^2}{b^2} - \frac{3(z + V_s t)^2}{c^2}\right) \tag{4}$$

where P is power in Watts, η is the efficiency of laser absorption and taken as 0.6, a and b values are taken as 1 and 2 mm, c for both front and back are taken 1 and 2 mm, respectively. V_s is the scan speed with which the laser moves in the z-direction.

2.3. Data Generation and Extraction

After solving all the given designs of experiments, data were extracted from each of the ODB files using a Python script. Two types of data were extracted: maximum temperature and melt pool dimensions. Therefore, separate scripts were used for each type. For example, run number 27 is shown in Figure 5a during material deposition and analysis.

For the maximum temperature, the highest temperature value was extracted for each increment from each frame as shown in Figure 5b during Run27 simulation. The same concept was applied to extract and calculate the melt pool dimensions. For each successful increment solved during the analysis, all nodal locations in all directions with values equal to or above 1605 °C were extracted. Once extracted for the specific increment, the location with the highest value in length was subtracted from the location with the lowest value of length, essentially providing the relevant dimension. The same method was employed for extracting the melt pool length, width, and depth as shown in Figure 5c.

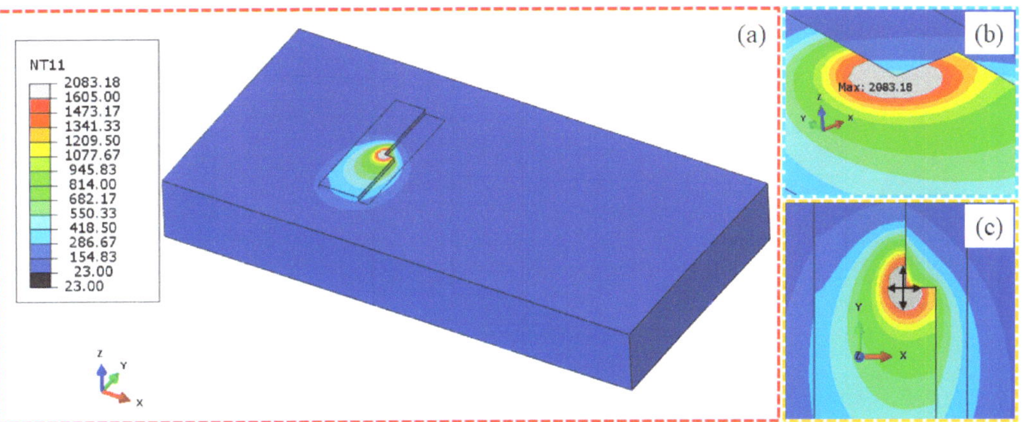

Figure 5. Thermal simulation during material deposition of Run27 as shown in (**a**), maximum temperature value extraction (**b**) and melt pool dimension (**c**).

2.4. Machine Learning Models

After extracting data from finite element simulations, four machine learning algorithms—Extreme Gradient Boosting (XGBoost), Long Short-Term Memory (LSTM), Bidirectional Long Short-Term Memory (Bi-LSTM), and Gated Recurrent Units (GRUs)—are prepared to build a surrogate model for predicting the thermal history and dimensions of the melt pool. Both the accuracy and computational time of these algorithms are considered to construct a robust machine learning-based surrogate model. The subsequent sections describe each algorithm's advantages and mathematical concepts.

2.4.1. Extreme Gradient Boosting (XGBoost)

XGBoost is recognized as one of the most effective applications of gradient-boosted decision trees [44]. Explicitly proposed to augment memory utilization and leverage hardware computational power, XGBoost significantly reduces accomplishment time while enhancing performance compared to other ML algorithms. The core concept of boosting involves sequentially constructing sub-trees from an original one, where each successive tree aims to lessen the errors of the preceding one. This iterative method updates the prior residuals, thereby minimizing the error of the cost function. Let us assume a dataset illustrated as [44]

$$D = \{(x_i, y_i) \mid x_i \in \mathbb{R}^m, y_i \in \mathbb{R}\}. \tag{5}$$

Here, m, x_i, and y_i are the feature dimensions and the samples' (i) responses, respectively. In addition, n represents the sample number ($|D| = n$). The forecasted output (y_i) for an entry (i) is as follows [44]:

$$y^i = \sum_{k=1}^{K} f_k(x_i), \quad f_k \in F. \tag{6}$$

In the above Equation, f_k represents a standalone tree within F, and $f_k(x_i)$ indicates the projected result from the ith trial and kth tree. The objective function ($\mathcal{L}$) is written as [44]

$$\mathcal{L} = \sum_{i=1}^{n} l(y_i, \hat{y}^i) + \sum_{k=1}^{K} \Omega(f_k). \tag{7}$$

By minimizing the actual function ($\mathcal{L}$), the regression tree model functions (f_k) are attained. The loss function ($l(y_i, \hat{y}_i)$) assesses the differentiation between estimated ($\hat{y}_i$) and real outputs (y_i). So, the term Ω is applied to prevent the overfitting issue by correcting the model intricacy, explained as [44]

$$\Omega(f_k) = \gamma T + \frac{1}{2}\lambda\|w\|^2. \tag{8}$$

Here, γ as well as λ are regularization factors, T and w are designated as the number and score of the leaf, respectively. A Taylor series expansion with the second degree can be applied to estimate the target function. We assume that $I_j = \{i \mid q(x_i) = j\}$ is an insistence set of leaf j having $q(x)$ as a permanent configuration. The optimum weights w_j^* of j and the subsequent quantity are estimated as [44]

$$w_j^* = -\frac{g_j}{h_j + \lambda}. \tag{9}$$

$$L^* = -\frac{1}{2}\sum_{j=1}^{T} \frac{\left(\sum_{i \in I_j} g_i\right)^2}{\sum_{i \in I_j} h_i + \lambda} + \lambda T. \tag{10}$$

Here, the first- and second-order gradients for $\mathcal{L}$ are represented by g_i and h_i, respectively. $\mathcal{L}$ can be applied as a quality index of the tree (q) so that the model is outstanding if the score is lower. It is not possible to consider the whole tree structure at a time. An excellent algorithm should resolve the challenge by initiating from an individual leaf and iteratively increasing branches. We assume that the right and left instance nodes are represented by I_R as well as I_L, respectively. Considering $I = I_R \cup I_L$, the loss reduction can be written as following the split [44]:

$$L_{\text{split}} = \frac{1}{2} \left[\frac{\left(\sum_{i \in I_L} g_i \right)^2}{\sum_{i \in I_L} h_i + \lambda} + \frac{\left(\sum_{i \in I_R} g_i \right)^2}{\sum_{i \in I_R} h_i + \lambda} - \frac{\left(\sum_{i \in I} g_i \right)^2}{\sum_{i \in I} h_i + \lambda} \right] - \gamma. \tag{11}$$

The XGBoost model employs numerous simple trees and assigns scores to leaf nodes during the splitting process.

2.4.2. Long Short-Term Memory (LSTM)

LSTM networks, an advanced type of recurrent neural networks, effectively address long-range dependencies in sequence data, crucial in scenarios like directed energy deposition processes. Characterized by three distinct gates—input, forget, and output—LSTMs manage information flow, selectively retaining or discarding data to precisely learn dependencies. The input state (i_t) decides which new information to incorporate into the cell state (c_t) and candidate state ($\tilde{c}_t$), enabling the model to update its memory with relevant data. The forget gate (f_t) selectively removes irrelevant information from the cell state to maintain the model's focus on pertinent data through time. The output gate (o_t) controls the flow of information from the cell state to the next layer or time step, determining what part of the hidden state (h_t) is used to compute the output and pass to next iteration.

This architecture mitigates gradient vanishing and exploding issues, enhancing robustness and accuracy in predictive models and making LSTM ideal for capturing complex thermal and mechanical interactions in additive manufacturing. The LSTM architecture is shown in Figure 6. The operator '×' denotes pointwise multiplication, and '+' denotes pointwise addition. The mathematical framework of LSTMs is presented in [45].

Forget gate:

$$f_t = \sigma(W_{fh}h_{t-1} + W_{fx}x_t + P_f \cdot c_{t-1} + b_f) \tag{12}$$

Input gate:

$$i_t = \sigma(W_{ih}h_{t-1} + W_{ix}x_t + P_i \cdot c_{t-1} + b_i) \tag{13}$$

$$\tilde{c}_t = \tanh(W_{ch}h_{t-1} + W_{cx}x_t + b_{\tilde{c}}) \tag{14}$$

$$c_t = f_t \cdot c_{t-1} + i_t \cdot \tilde{c}_t \tag{15}$$

Output gate:

$$o_t = \sigma(W_{oh}h_{t-1} + W_{ox}x_t + P_o \cdot c_t + b_o) \tag{16}$$

$$h_t = o_t \cdot \tanh(c_t) \tag{17}$$

Here, W_f, W_i, W_c, and W_o are the weights of each input. The x_t, h_t, and y_t are represented as input, hidden state (recurrent information), and output concerning time. Furthermore, the f_t is the forget cell starting from 0, P_f, P_i, and P_o are the peephole weights for f_t, input, and output gates. The c_t denotes the LSTM cell state, and b_i, b_f, $b_{\tilde{c}}$, and b_o are the biases. Figure 7 shows the architecture of the series of LSTM structures.

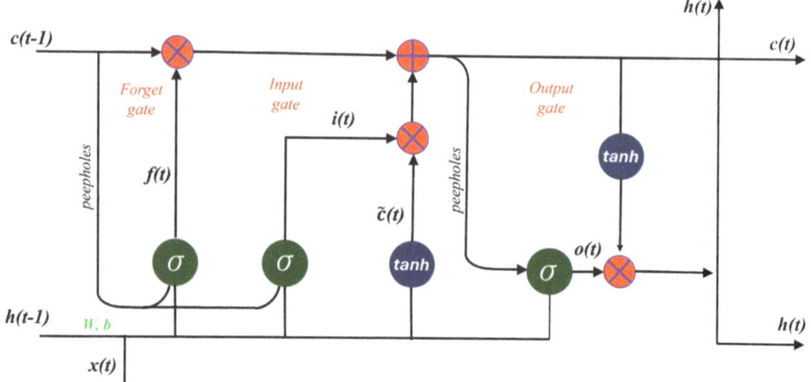

Figure 6. Architecture of LSTM algorithm.

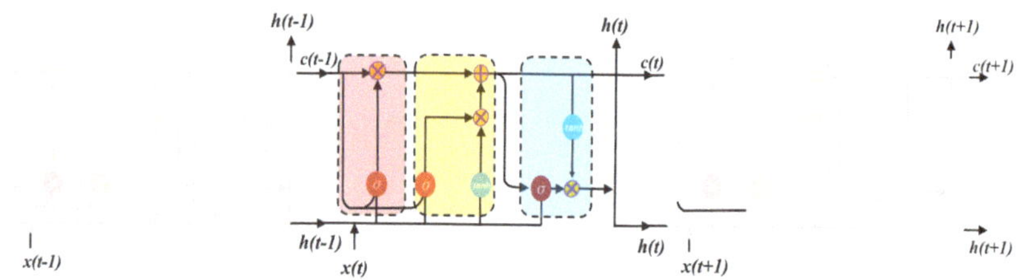

Figure 7. Series of LSTM architecture.

2.4.3. Bidirectional Long Short-Term Memory (Bi-LSTM)

Bi-LSTM networks enhance traditional LSTM by processing data both forwards and backwards, enriching sequence context understanding. This dual-path approach not only boosts predictive accuracy in tasks like outcome prediction in directed energy deposition but also captures nuanced temporal dynamics from both past and future contexts. Despite their increased computational demands and potential for overfitting with small datasets, Bi-LSTMs remain valuable for thoroughly analyzing thermal and mechanical properties in AM. Leveraging LSTM strengths, they effectively manage long-term dependencies and mitigate gradient issues, providing a robust model for complex material behaviors. The Bi-LSTM architecture is defined in Figure 8, and the LSTM block within this architecture follows the structure shown in Figure 6. The mathematical expression is given in [45].

$$f_t^L = \sigma(W_{fh}^L h_{t-1}^L + W_{fx}^L h_t^{L-1} + b_f^L) \tag{18}$$

$$i_t^L = \sigma(W_{ih}^L h_{t-1}^L + W_{ix}^L h_t^{L-1} + b_i^L), \tag{19}$$

$$\tilde{c}_t^L = \tanh(W_{\tilde{c}h}^L h_{t-1}^L + W_{\tilde{c}x}^L h_{t-1}^L + b_{\tilde{c}}^L), \tag{20}$$

$$c_t^L = f_t^L \cdot c_{t-1}^L + i_t^L \cdot \tilde{c}_t^L, \tag{21}$$

$$o_t^L = \sigma(W_{oh}^L h_{t-1}^L + W_{ox}^L h_{t-1}^L + b_o^L) \tag{22}$$

$$h_t^L = o_t^L \cdot \tanh(c_t^L). \tag{23}$$

$$y_t = W_{hy}^{\rightarrow} h_t + W_{hy}^{\leftarrow} h_t + b_y \tag{24}$$

Here, h_t^L represents the output of the hidden state in the (L)th layer at time t. Equation (24) shows the output of architecture, where $W_{hy}^{\rightarrow}$ denotes the weight of the forward pass, $W_{hy}^{\leftarrow}$ indicates the weight of the backward pass, and b_y signifies the bias of the output.

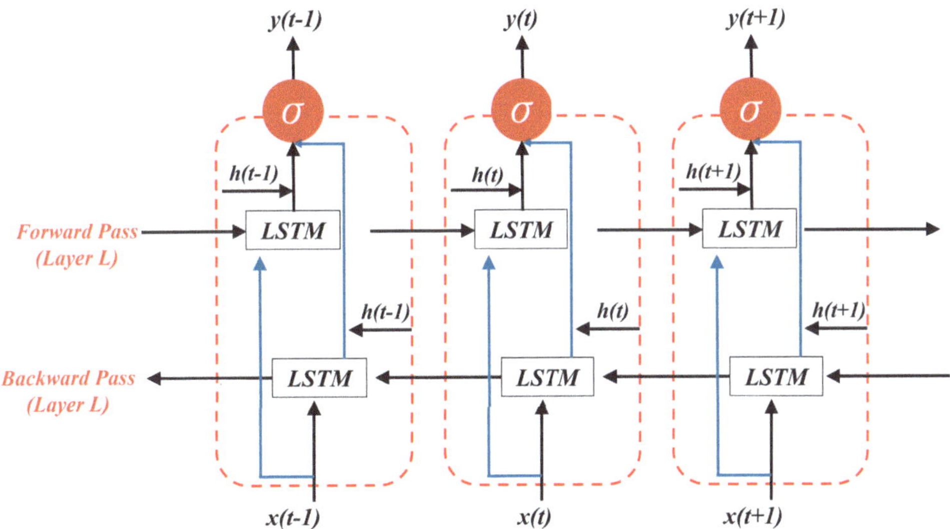

Figure 8. Architecture of Bi-LSTM algorithm.

2.4.4. Gated Recurrent Units (GRUs)

GRUs offer a streamlined alternative to LSTMs and Bi-LSTMs, ideal for modeling thermal histories in directed energy deposition. By employing just two gates—the reset and update gates—GRUs enhance computational efficiency and reduce model complexity, making them well-suited for scenarios with limited data or computational resources. The reset gate determines how much past information to forget. In contrast, the update gate decides how much of the current input should be incorporated, allowing the model to handle time dependencies dynamically. Although GRUs may struggle with extremely long dependencies, their ability to efficiently process sequential data without significant computational overhead keeps them highly relevant for improving predictive models in AM. Based on Figure 9, the following mathematical model has been proposed for GRU [45]:

Reset gate:

$$r_t = \sigma(W_{rh} h_{t-1} + W_{rx} x_t + b_r), \tag{25}$$

Update gate:

$$z_t = \sigma(W_{zh} h_{t-1} + W_{zx} x_t + b_z), \tag{26}$$

$$\tilde{h}_t = \tanh(W_{\tilde{h}h}(r_t \cdot h_{t-1}) + W_{\tilde{h}x} x_t + b_{\tilde{h}}), \tag{27}$$

$$h_t = (1 - z_t) \cdot h_{t-1} + z_t \cdot \tilde{h}_t. \tag{28}$$

Here, W_r, W_z, and $W_{\tilde{h}h}$ are the weights, and b_r and b_z are the biases.

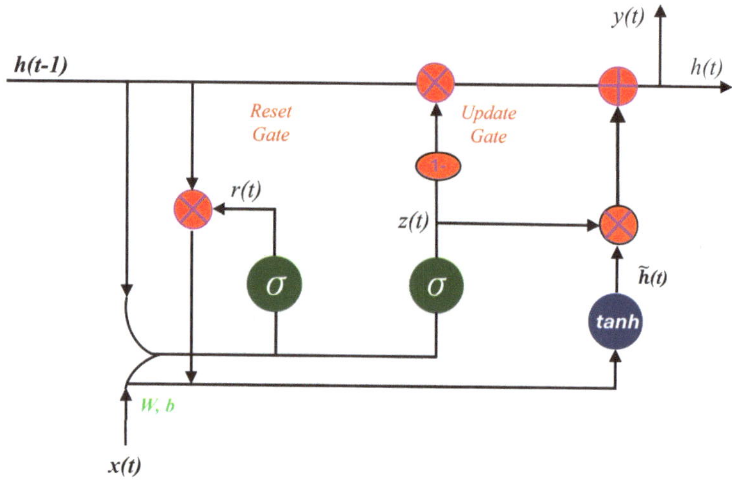

Figure 9. Architecture of GRU algorithm.

2.4.5. Model Evaluation

In the model evaluation part, the performance of the surrogate model is assessed using several statistical metrics to ensure accuracy and reliability. These metrics include R-squared (R^2), which measures the proportion of variance in the dependent variable that is predictable from the independent variables; Root Mean Square Error (RMSE), which provides the standard deviation of the prediction errors or residuals; and Mean Absolute Error (MAE), which represents the average magnitude of the errors in a set of predictions, without considering their direction. The mathematical formulas are presented as follows:

$$R^2 = 1 - \frac{\sum_{i=1}^{n}(y_i - \hat{y}_i)^2}{\sum_{i=1}^{n}(y_i - \bar{y})^2} \tag{29}$$

$$RMSE = \sqrt{\frac{1}{n}\sum_{i=1}^{n}(y_i - \hat{y}_i)^2} \tag{30}$$

$$MAE = \frac{1}{n}\sum_{i=1}^{n}|y_i - \hat{y}_i| \tag{31}$$

Here, y_i are the observed values, $\hat{y}_i$ are the predicted values, and $\bar{y}$ is the mean of the observed values. Additionally, the computation time of the training model is also considered a factor in evaluating model performance.

3. Results and Discussion

3.1. Data Pre-Processing and Model Training

The data used to build the surrogate models for melt pool peak temperature and melt pool dimension in this research originated from 27 runs of multi-physics modeling, employing a three-level, three-factor factorial design of experiments. A total of 54,956 data points were extracted. For the melt pool peak temperature model, data points that did not reach the melting point of Ti-6Al-4V (1605 °C) or exceeded (3200 °C) were excluded. The vaporization point of Ti-6Al-4V is 3040 °C, but melt pool peak temperatures occasionally exceed this threshold. To accommodate most conditions during deposition, temperatures above the vaporization point were also considered. After cleaning, the dataset contained 38,867 peak temperature points, with 28,683 allocated for training and 10,184 for testing. The training and testing sets accounted for 73.8% and 26.2%, respectively. Figure 10 displays the training features: time, x position, y position, z position, laser power, scanning speed, and hatch space. Figure 11 depicts the training label for melt pool peak temperature. A detailed and clear description of the training label is shown in Figure 12. The peak temperature dramatically increases when the laser is on and drops when it is off. Each run consists of four tracks, and fluctuations occur during the movement in each track.

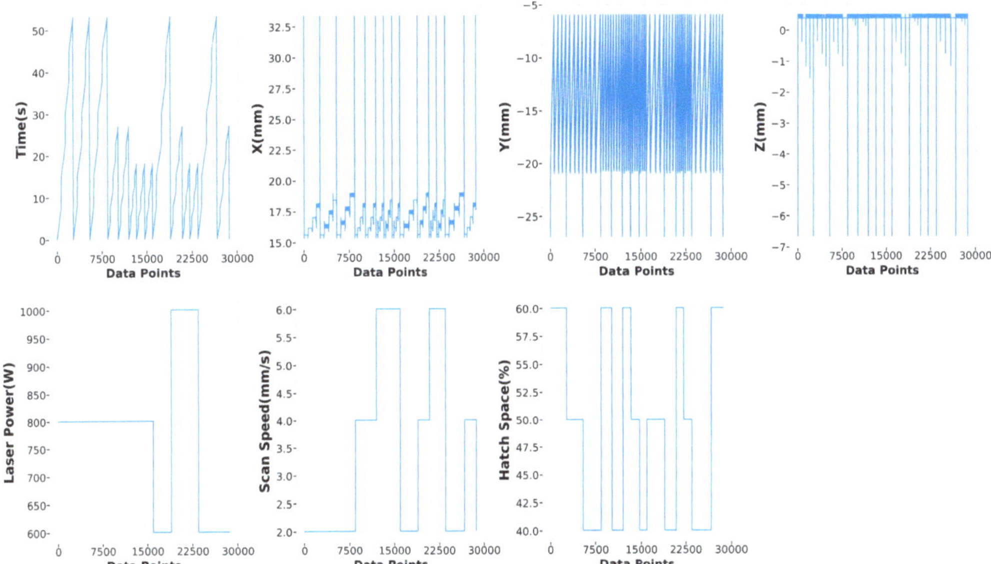

Figure 10. Training Features of melt pool peak temperature model.

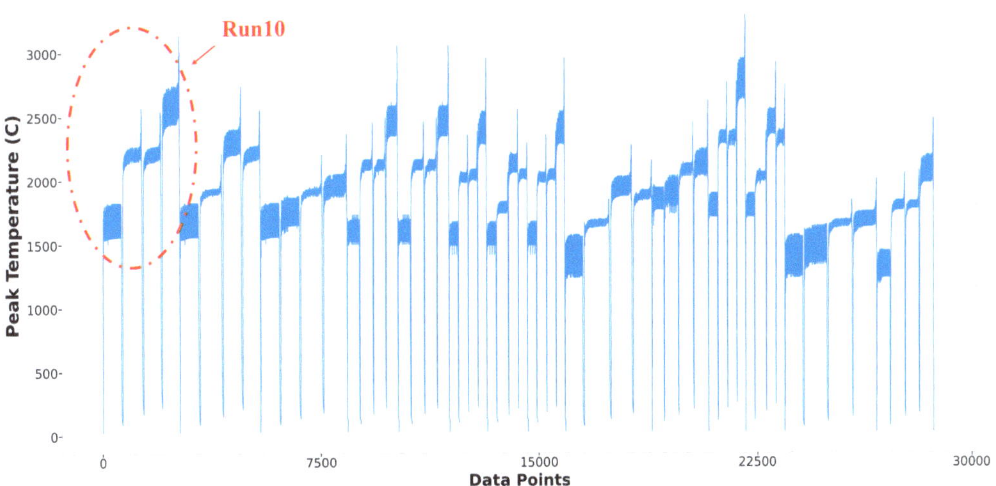

Figure 11. Training Label of melt pool peak temperature model.

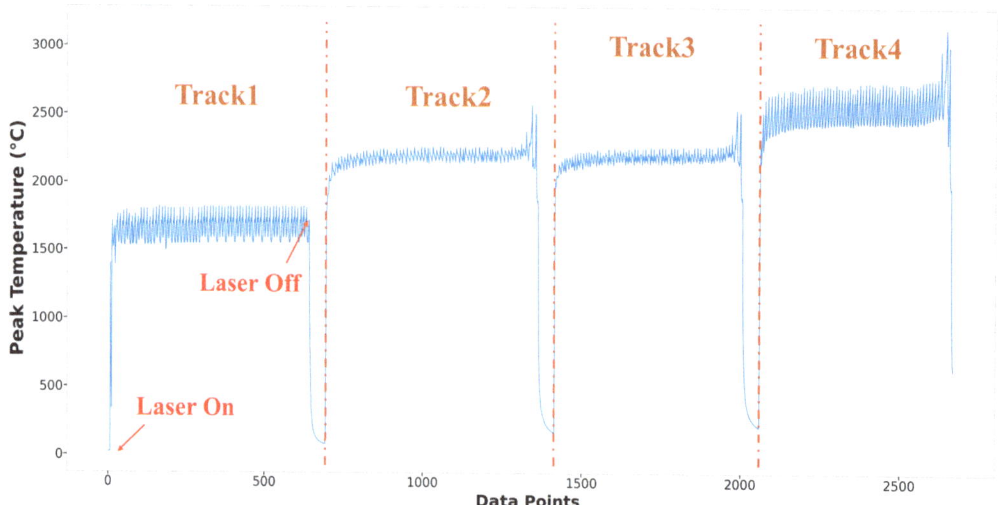

Figure 12. Training label of Run10 in melt pool peak temperature model.

Regarding the melt pool dimension model, 27,772 data points were collected because data on melt pool dimensions are extracted only when the border of the melt pool exceeds 1605 °C, as described in Section 2. These points are divided into 20,182 (72.6%) for training and 7590 (27.4%) for testing. The features and labels are shown in Figure 13. In the melt pool dimension model, time (s), laser power (W), scanning speed (mm/s), hatching space (%), and peak temperature (°C) are considered as features, while melt pool length, width, and depth (mm) are considered as labels. After removing outliers, such as extremely high and low thermal histories, 19 runs of data remain: 14 runs are designated for training and 5 runs for testing. To mitigate the impact of disproportionately large values among the process parameters and training features, normalization is applied in the data pre-processing stage. The details of the data for the two surrogate models, the melt pool peak temperature model, and the melt pool dimension model are described in Table 3.

With regard to model training, the grid search method is applied to find the proper hyperparameters. For the XGBoost algorithm, the tree depth is set to five to avoid overfitting, with a learning rate of 0.01 to ensure steady convergence. The objective is defined as 'reg:squarederror' to minimize squared errors in regression tasks. L1 regularization (reg_alpha) is applied at 0.01 to promote parameter sparsity, and L2 regularization (reg_lambda) is set at 1 to reduce weight extremes. Both subsample and colsample_bytree are maintained at 0.8, allowing the model to learn from 80% of data and features, respectively, to prevent overfitting. The evaluation metric used is 'rmse', measuring prediction accuracy. Training involves 10,000 rounds, optimizing learning against computational demands. In terms of RNN algorithms, the hyperparameters for all LSTM, Bi-LSTM, and GRU algorithms are unified to ensure a fair comparison among the models. The sequence length of data is set to 10, batch size to 64, dropout rate to 0.25, hidden dimension to 100, number of layers to 2, learning rate to 0.001, and number of epochs to 100.

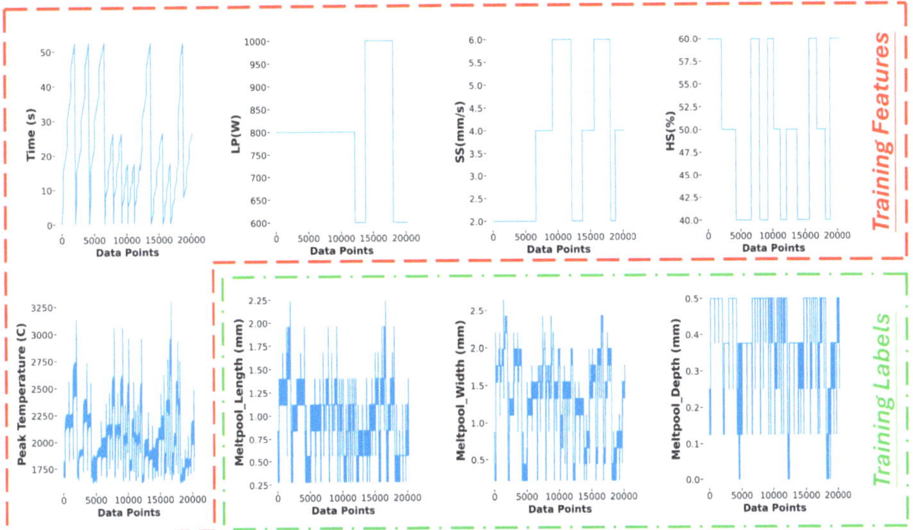

Figure 13. Training features and labels of melt pool dimension model.

Table 3. Summary of training and testing data of surrogate models.

Model	Training Data	Testing Data	Training Size	Testing Size	Features	Labels
Melt Pool Peak Temperature	Run2-4, Run10-13, Run15-18, Run24-26	Run1, Run5, Run14, Run23, Run27	28,683	10,184	Time, Position X, Y, Z, Laser Power, Scanning Speed, Hatch Space	Melt Pool Peak Temperature
Melt Pool Dimension	Run2-4, Run10-13, Run15-18, Run24-26	Run1, Run5, Run14, Run23, Run27	20,182	7590	Time, Peak Temperature, Laser Power, Scanning Speed, Hatch Space	Melt Pool Length, Melt Pool Width, Melt Pool Depth

3.2. Melt Pool Peak Temperature Model

In this section, four different algorithms—XGBoost, Bi-LSTM, LSTM, and GRU—are applied to this research. To compare the pros and cons of tree-based versus RNN algorithms, the predicted results by XGBoost and Bi-LSTM are presented together in one figure. In terms of comparing the complexity of RNN algorithms, the results of LSTM and GRU are displayed together in another figure. Additionally, two specific runs, Run1 (Laser Power: 600W, Scanning Speed: 2 mm/s, Hatching Space: 60%) and Run27 (Laser Power: 1000 W, Scanning Speed: 6 mm/s, Hatching Space: 40%), are extracted and analyzed to facilitate a detailed comparison and enhance clarity. A comprehensive comparison of predictions by four algorithms is also included in this section.

Figure 14 depicts the comparison of Run1 among actual values and predicted values by Bi-LSTM and XGBoost. It shows that the melt pool peak temperatures predicted by Bi-LSTM closely match the actual peak temperatures. The results from XGBoost also demonstrate reasonably good prediction performance. However, in Run27, the predictions by XGBoost significantly deviate from the actual peak temperatures, especially in the second and third tracks, where the predictions have more fluctuation and are higher than the actual values. In contrast, the results from Bi-LSTM closely align with the actual values, demonstrating the robustness of the model built using the Bi-LSTM algorithm, as shown in Figure 15. In terms of the other two algorithms, LSTM and GRU, both achieve good predictions that closely fit the actual values. In the first track of Run1, both predictions are slightly lower than the actual values, yet the remaining predictions demonstrate good performance, as depicted in Figure 16. In Run27, shown in Figure 17, except for the fourth track, where the predictions are slightly lower than the actual values, most of the results closely match the actual values.

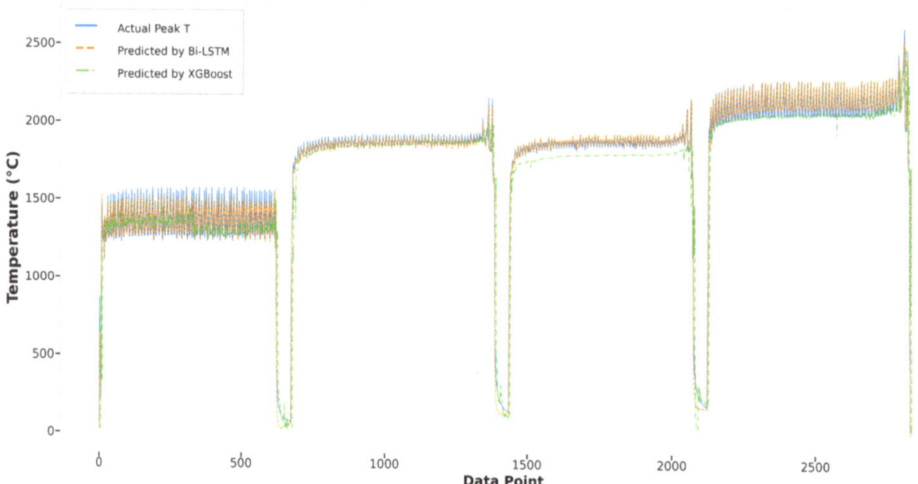

Figure 14. Run1: actual peak temperature versus prediction by Bi-LSTM and XGBoost.

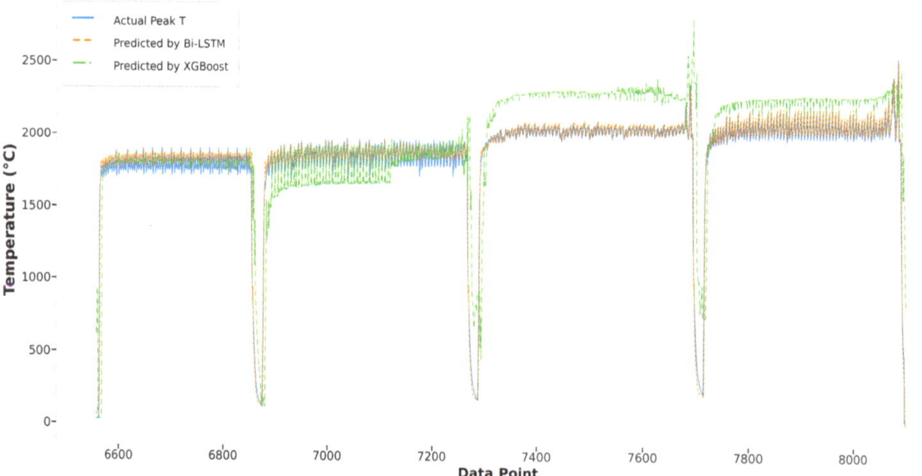

Figure 15. Run27: actual peak temperature versus prediction by Bi-LSTM and XGBoost.

All predicted temperatures versus actual temperatures scatter plots are shown in Figure 18. It demonstrates that the predicted values by XGBoost are relatively less accurate than those produced by RNN algorithms. Most of the predicted results by RNN algorithms closely match the red line, which has a slope of one, indicating that the predictions are both accurate and robust. To compare the performance of the four algorithms, Table 4 reveals that the Bi-LSTM model has the highest accuracy, longest computational time, and greatest memory usage. Although XGBoost performs well in terms of computational time and memory usage, its accuracy is not robust enough to predict melt pool peak temperatures reliably. The accuracy of the LSTM and GRU models is similar; however, the computational time and memory usage of the GRU model are lower than those of the LSTM model by 20.7% and 5.4%, respectively. In conclusion, the Bi-LSTM model provides the most accurate results, while the GRU model offers comparable accuracy with lower computational time and memory usage.

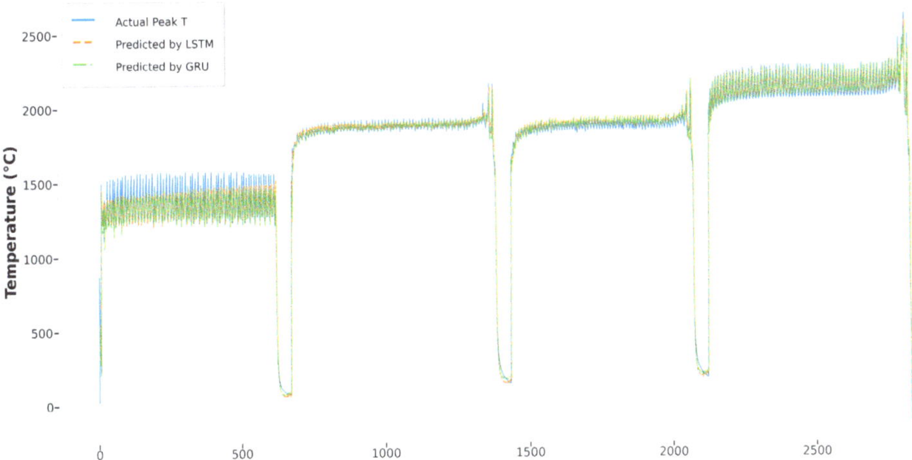

Figure 16. Run1: actual peak temperature versus prediction by LSTM and GRU.

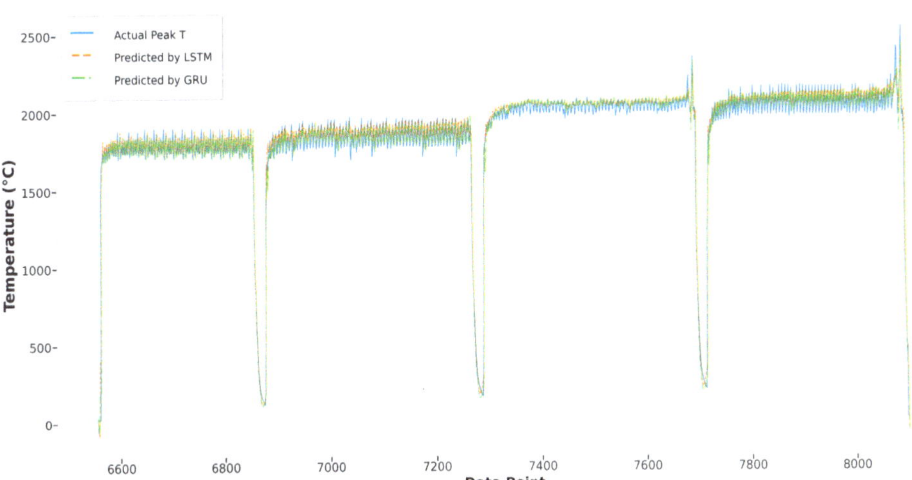

Figure 17. Run27: Actual peak temperature versus prediction by LSTM and GRU.

Table 4. Evaluation and comparative analysis: melt pool peak temperature model.

Algorithms	R-Square	RMSE	MAE	Computation Time (s)	Memory Usage (GB)
XGBoost	0.852	0.0550	0.0382	16.67	0.747
LSTM	0.979	0.0178	0.0126	238.60	2.41
Bi-LSTM	0.983	0.0153	0.0101	290.25	5.24
GRU	0.978	0.0179	0.0129	189.30	2.28

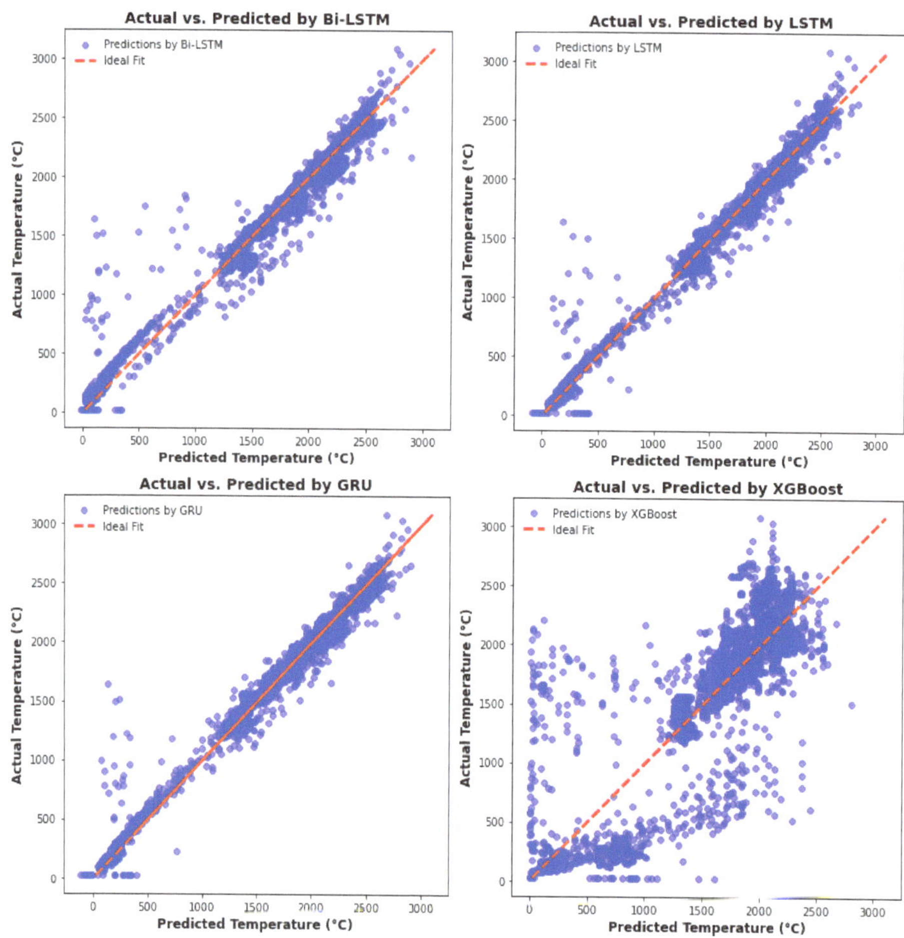

Figure 18. Actual peak temperature with predictions from four algorithms.

3.3. Melt Pool Geometry Model

In this section, three surrogate models are presented: melt pool length, width, and depth, respectively. To clarify the comparison, results from Run23 and Run14 are extracted for discussion. Additionally, the overall results of the four algorithms are compared using scatter plots and a comprehensive table in the subsequent contents.

3.3.1. Melt Pool Length Model

In Run14 and Run23, the Bi-LSTM model consistently outperforms the XGBoost model in predicting melt pool length. As depicted in Figure 19, the Bi-LSTM model demonstrates superior accuracy in predicting higher melt pool lengths, particularly for data points from 5700 to 5800. Moreover, towards the end of Run14, from data points 6300 to 6700, the Bi-LSTM model shows significantly less fluctuation compared to the XGBoost model, indicating its enhanced stability under varying conditions. In Figure 20, although the XGBoost model accurately predicts the melt pool lengths for data points from 3400 to 3750, the overall performance of the Bi-LSTM model remains more consistent and aligned with the actual length.

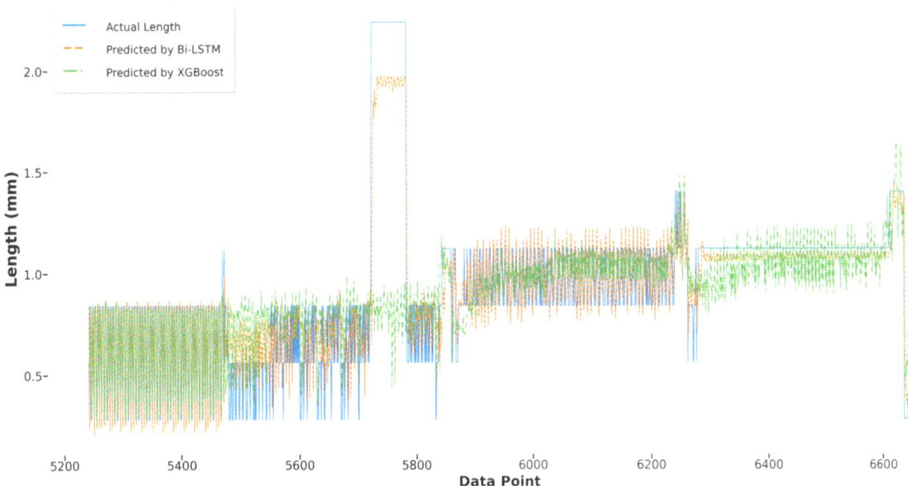

Figure 19. Run14: Actual length versus prediction by Bi-LSTM and XGBoost.

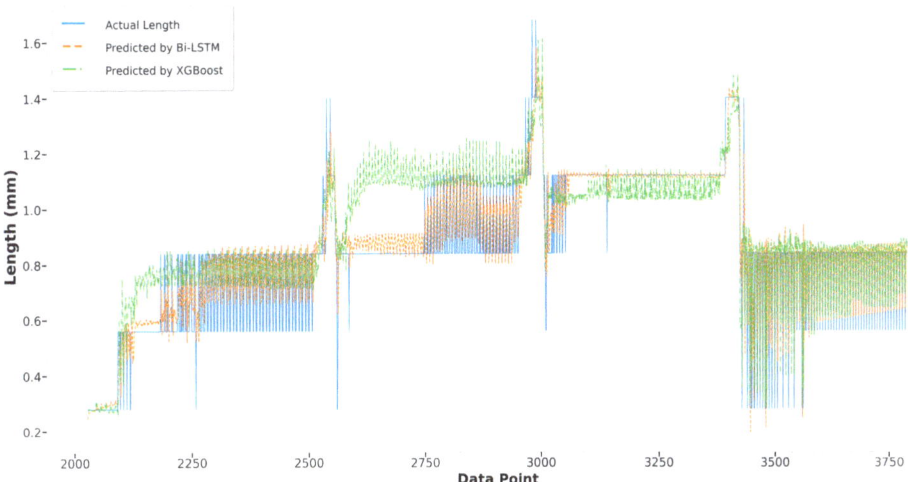

Figure 20. Run23: Actual length versus prediction by Bi-LSTM and XGBoost.

Regarding the LSTM and GRU models, the GRU model exhibits less error in predicting longer melt pool lengths, as evident in Figure 21 for data points from 5700 to 5800. Throughout the remainder of Run14, both models achieve commendable accuracy in fitting the actual length. In Run23, despite both models displaying a similar trend in capturing the actual values, the LSTM model exhibits greater deviations from the actual lengths compared to the GRU model, as illustrated in Figure 22. This result suggests that while the LSTM model is generally reliable, the GRU model may offer better consistency and precision under certain conditions.

The comparison of overall predictions among four algorithms is presented in a scatter plot. Figure 23 demonstrates that the melt pool lengths predicted by the RNN algorithms are more accurate than those predicted by the XGBoost algorithm. Notably, when the melt pool length exceeds 2 mm, predictions from the XGBoost model deviate significantly from the ideal fit, resulting in decreased accuracy. Table 5 summarizes the evaluation and comparative analysis of the melt pool length models. Although the XGBoost algorithm impresses with its computation time and memory usage, its accuracy needs improvement. In terms of RNN algorithms, the GRU and Bi-LSTM models perform better. In particular, the GRU model not only achieves the highest R-square value but also requires the least computation time and memory usage among all RNN algorithms. Compared to the Bi-LSTM model, the GRU model's computation time and memory usage are lower by 44% and 51%, respectively, making it the most suitable candidate for predicting melt pool length in this research.

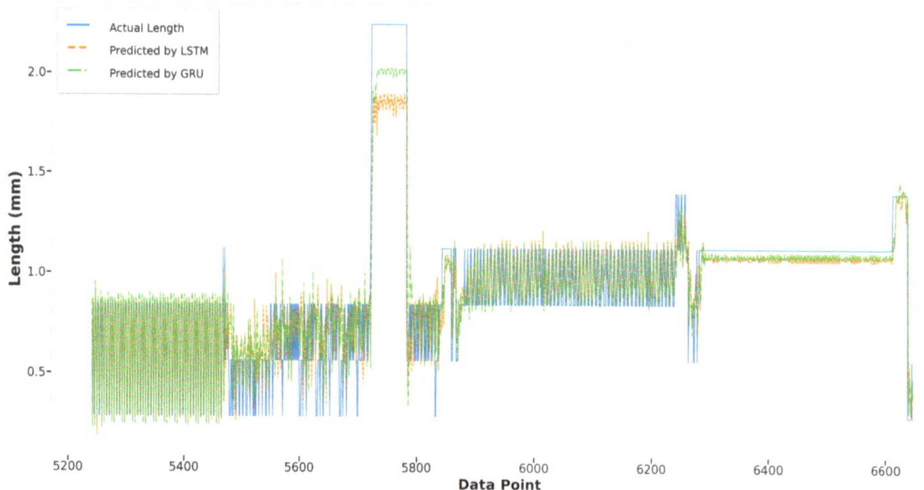

Figure 21. Run14: actual length versus prediction by LSTM and GRU.

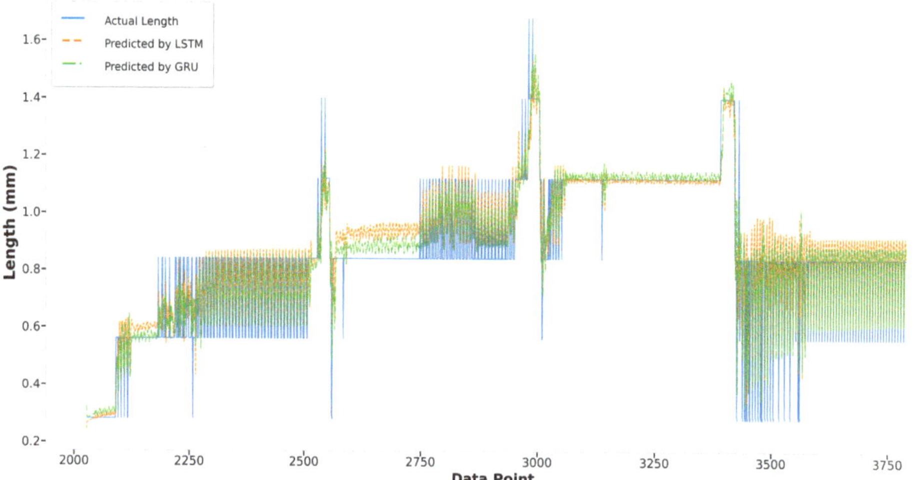

Figure 22. Run23: actual length versus prediction by LSTM and GRU.

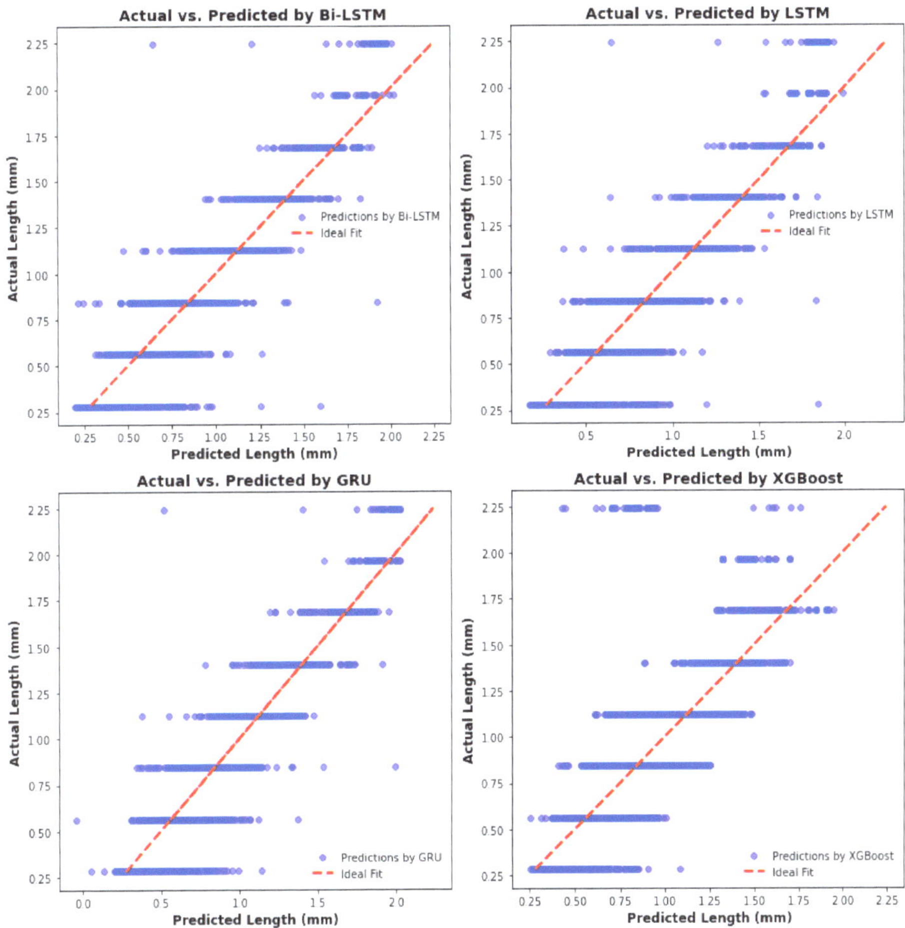

Figure 23. Actual length versus predictions from four algorithms.

Table 5. Evaluation and comparative analysis: melt pool length model.

Algorithms	R-Square	RMSE	MAE	Computation Time (s)	Memory Usage (GB)
XGBoost	0.698	0.1031	0.0629	16.22	0.269
LSTM	0.888	0.0539	0.0412	76.23	1.37
Bi-LSTM	0.902	0.0501	0.0369	120.55	2.65
GRU	0.903	0.0503	0.0381	67.75	1.30

3.3.2. Melt Pool Width Model

In the melt pool width model, Figures 24 and 25 illustrate the predictions made by the Bi-LSTM and XGBoost algorithms for Run14 and Run23, respectively. Those scatter plots show a notable variance in accuracy between the algorithms. For Run14, particularly from data point 5500 to 5900, and in Run23 from data point 3100 to 3400, the predictions by the XGBoost model significantly exceed the actual width, highlighting its lower accuracy compared to the Bi-LSTM model. The Bi-LSTM model more consistently aligns with the actual measurements, particularly in complex segments where the melt pool width fluctuates.

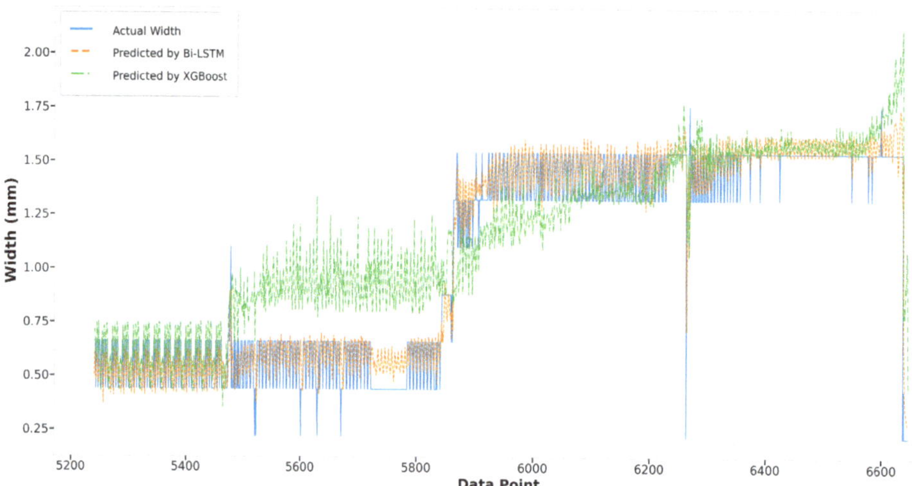

Figure 24. Run14: actual width versus prediction by Bi-LSTM and XGBoost.

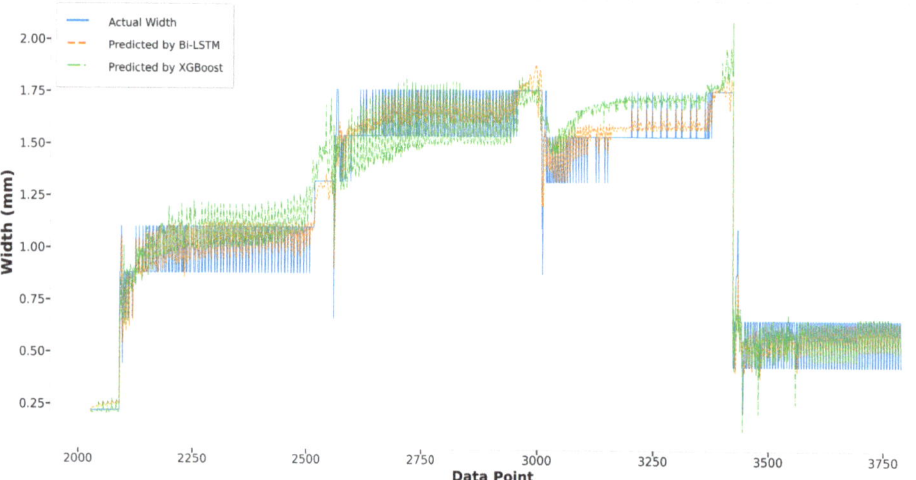

Figure 25. Run23: actual width versus prediction by Bi-LSTM and XGBoost.

Figures 26 and 27 showcase the performance of the LSTM and GRU models for Run14 and Run23, respectively. Both models exhibit similar trends and achieve commendable accuracy in fitting the actual widths in Run14, with the GRU model slightly outperforming the LSTM. Notably, in Run23, while neither model perfectly replicates the fluctuation observed in the actual width measurements, they successfully capture the broader trends. The GRU model consistently demonstrates a slight edge over the LSTM in terms of alignment with the actual data across both runs, indicating its robustness in modeling the melt pool width.

The overall predictive performance of four algorithms is displayed in a scatter plot for comparison, as shown in Figure 28. The RNN algorithms, particularly Bi-LSTM and GRU, exhibit superior performance in predicting sequential data such as melt pool width, evidenced by their close alignment with the ideal fit line. Both models display similar commendable accuracy, effectively capturing the sequential dependencies within the data. In contrast, predictions by the XGBoost algorithm are notably more dispersed, indicating less accuracy. This dispersion becomes especially pronounced when the actual width exceeds 2 mm, where XGBoost predictions significantly deviate from the ideal fit. Table 6 summarizes the comparison among all algorithms, highlighting that the Bi-LSTM model achieves the highest R-square value. However, the GRU model offers comparable accuracy with lower computation time and memory usage—40% and 51% less than the Bi-LSTM model, respectively—demonstrating its greater robustness.

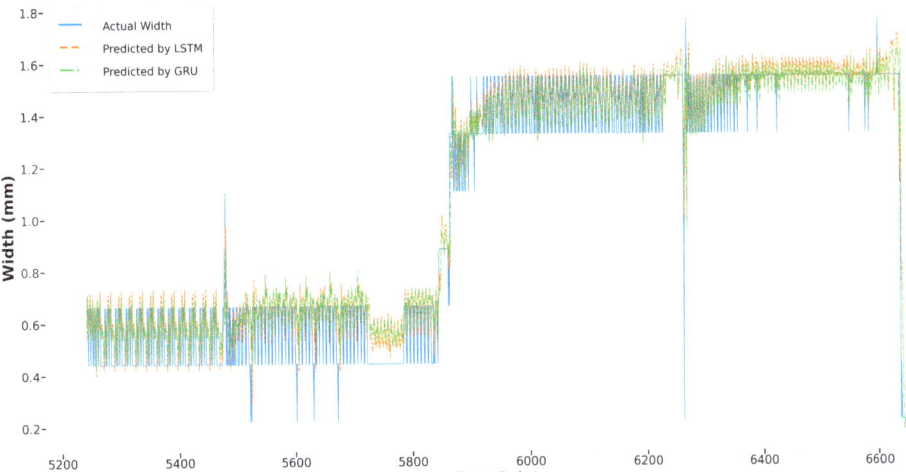

Figure 26. Run14: actual width versus prediction by LSTM and GRU.

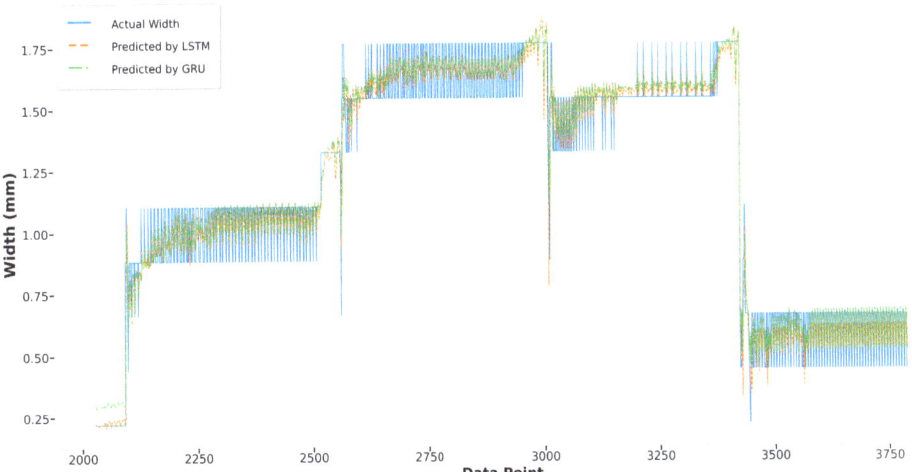

Figure 27. Run23: actual width versus prediction by LSTM and GRU.

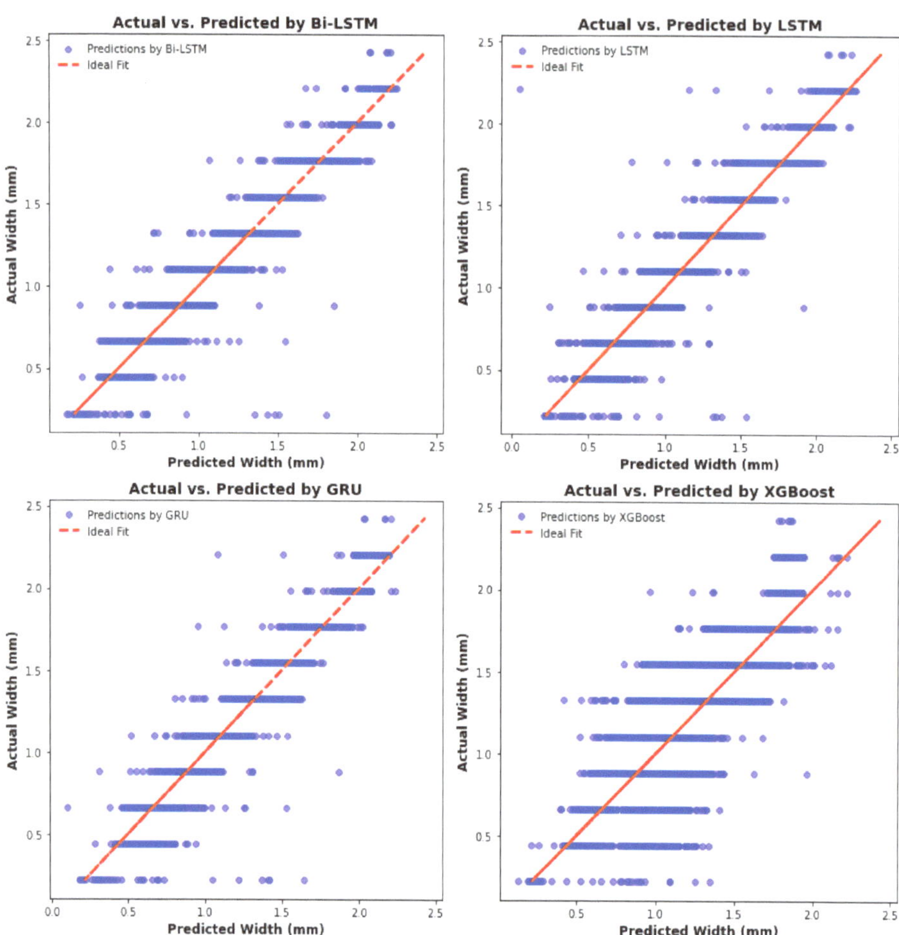

Figure 28. Actual width versus predictions from four algorithms.

Table 6. Evaluation and comparative analysis: melt pool width model.

Algorithms	R-Square	RMSE	MAE	Computation Time (s)	Memory Usage (GB)
XGBoost	0.752	0.0963	0.0762	16.95	0.371
LSTM	0.946	0.0418	0.0313	86.26	1.37
Bi-LSTM	0.952	0.0399	0.0293	128.70	2.65
GRU	0.951	0.04	0.0291	76.73	1.30

3.3.3. Melt Pool Depth Model

In the melt pool depth model, the XGBoost model displays a surprising parity with the Bi-LSTM model in terms of performance in Run14, especially noticeable at the start where XGBoost surpasses Bi-LSTM in accuracy, as shown in Figure 29. In contrast, during Run23 as depicted in Figure 30, although the overall trends of both models align closely with the actual depth measurements, the XGBoost predictions show greater deviations from the actual values, suggesting less consistency compared to the Bi-LSTM model. This indicates that while XGBoost can match the performance of Bi-LSTM in certain scenarios, its performance can be less reliable in others.

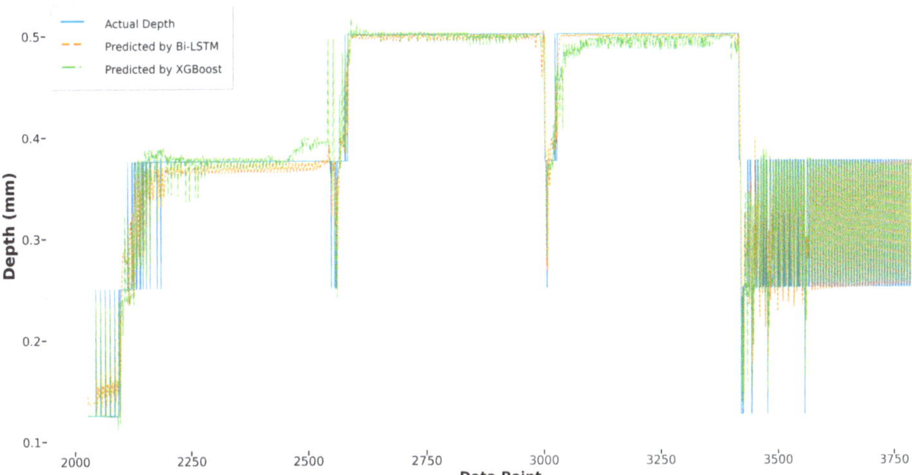

Figure 29. Run14: actual depth versus prediction by Bi-LSTM and XGBoost.

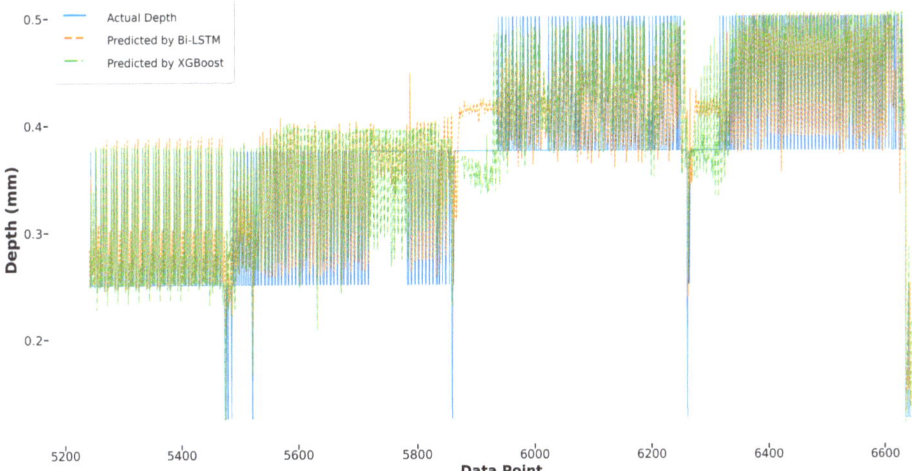

Figure 30. Run23: actual depth versus prediction by Bi-LSTM and XGBoost.

Regarding the LSTM and GRU models, their performance in predicting melt pool depth is commendably consistent, exhibiting similar trends. Both models closely align with the actual values, demonstrating their effectiveness in capturing sequential data characteristics. In Run14, although the predictions start slightly below the actual values, both LSTM and GRU adjust quickly and maintain a good match throughout the data range, as shown in Figure 31. Run23 shows a slight divergence in the predictions from both models, especially in the latter half, where the LSTM model exhibits more deviation than the GRU model, yet both still maintain a general adherence to the trend of actual depth values, as illustrated in Figure 32.

The scatter plots of predictions by all four algorithms are presented in Figure 33. Unlike the melt pool peak temperature and other geometric models, no single model exhibits particularly strong performance. All models deviate from the ideal fit, especially when predicting maximum and minimum melt pool depths. For a more comprehensive comparison and analysis, Table 7 reveals that the XGBoost model has the shortest computation time and lowest memory usage, but relatively lower accuracy. Additionally, the GRU model boasts the highest R-square value and has lower computation time and memory usage—29% and 50% less, respectively, compared to the Bi-LSTM model—highlighting the reliability and robustness of the GRU model.

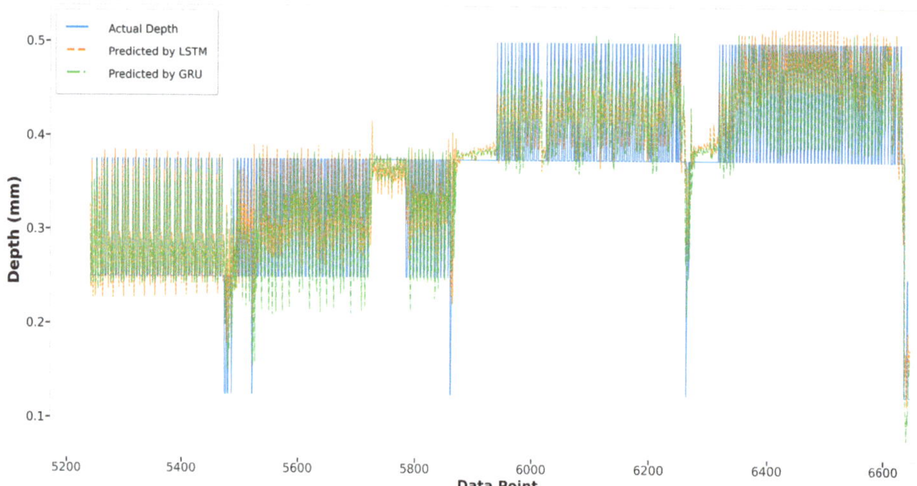

Figure 31. Run14: actual depth versus prediction by LSTM and GRU.

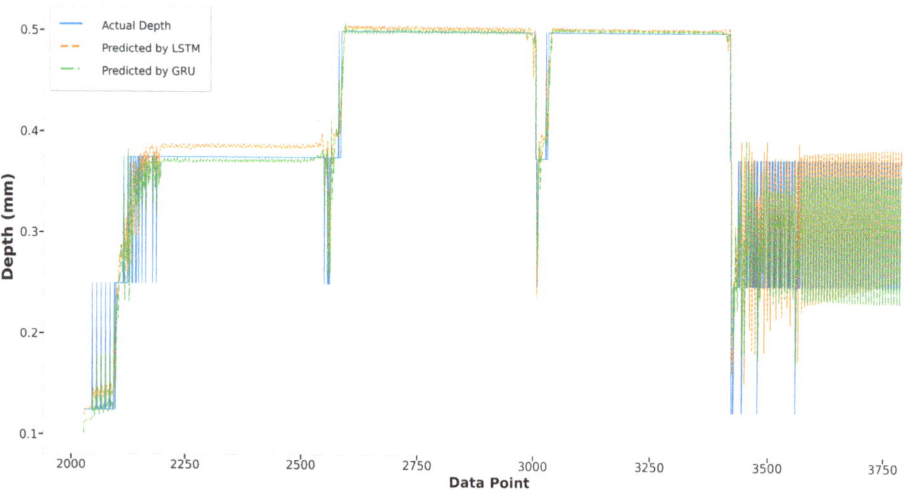

Figure 32. Run23: actual depth versus prediction by LSTM and GRU.

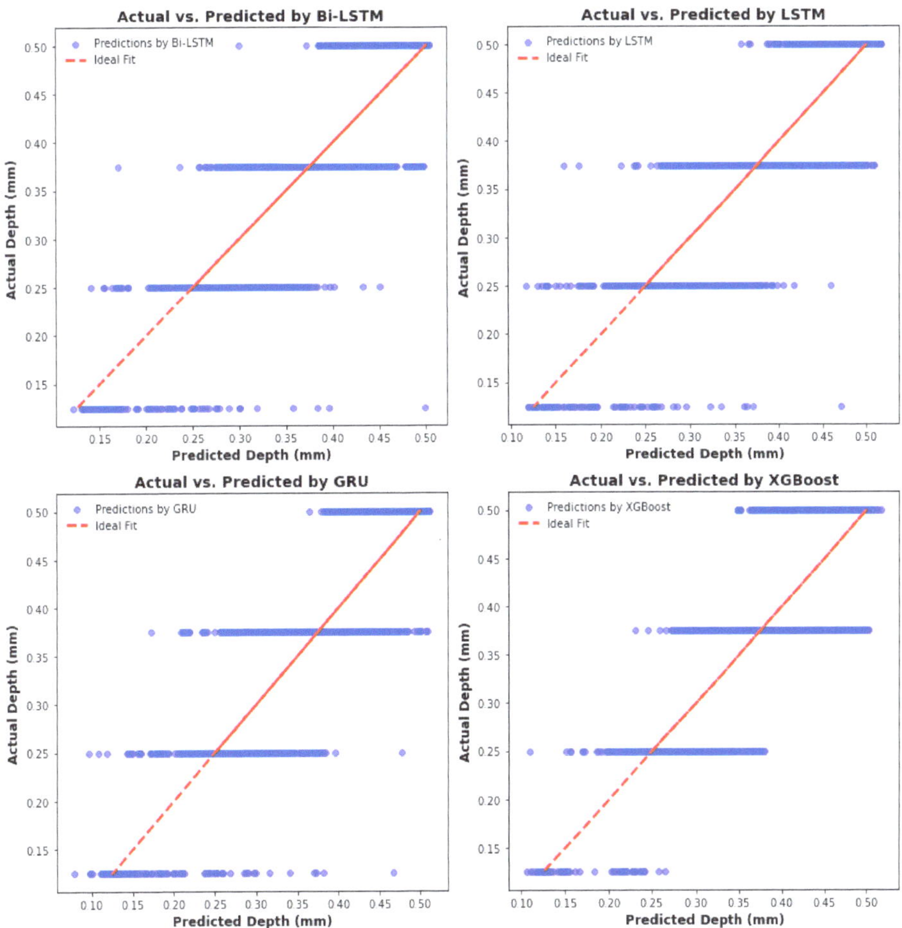

Figure 33. Actual depth versus predictions from four algorithms.

Table 7. Evaluation and comparative analysis: melt pool depth model.

Algorithms	R-Square	RMSE	MAE	Computation Time (s)	Memory Usage (GB)
XGBoost	0.751	0.0892	0.0555	20.20	0.344
LSTM	0.871	0.0479	0.0360	97.69	1.44
Bi-LSTM	0.881	0.0476	0.0359	120.19	2.72
GRU	0.885	0.0420	0.0293	85.43	1.37

4. Conclusions

This study developed a recurrent neural network (RNN)-based surrogate model to predict melt pool characteristics, such as peak temperature, length, width, and depth, in directed energy deposition (DED) processes. By integrating a three-level, three-factor design of experiments and multi-physics simulation data into an LSTM, Bi-LSTM, and GRU framework, the model demonstrates exceptional predictive accuracy for sequential melt pool data under varied processing conditions. The research also presents a comprehensive evaluation and comparative analysis of surrogate models built with different algorithms. Key contributions of this research include:

- **Robust Model Architecture:** Employed advanced RNN architectures—LSTM, Bi-LSTM, and GRU—to effectively capture the sequential and dynamic behavior of melt pools in DED processes.
- **High Predictive Accuracy:** Achieved an R-square of 0.983 for melt pool peak temperature predictions using the Bi-LSTM algorithm. Demonstrated superior performance in melt pool geometry predictions:
  - Melt pool length: R-square of 0.903 with the GRU algorithm.
  - Melt pool width: R-square of 0.952 with the Bi-LSTM algorithm.
  - Melt pool depth: R-square of 0.885 with the GRU algorithm.
- **Efficiency and Robustness:** The GRU-based surrogate model outperformed other algorithms in terms of accuracy, computation time, and memory usage, showing a reduction of at least 29% in computation time and 50% in memory usage, highlighting the model's efficiency and robustness.

Author Contributions: Conceptualization, S.-H.W., U.T., R.J. and M.A.M.; methodology, S.-H.W., U.T. and M.A.M.; software, S.-H.W., U.T. and A.W.M.; validation, S.-H.W. and U.T.; formal analysis, S.-H.W., U.T. and R.J.; investigation, S.-H.W., U.T. and M.A.M.; resources, F.L.; data curation, S.-H.W. and U.T.; writing—original draft preparation, S.-H.W., U.T. and M.A.M.; writing—review and editing, M.A.M. and F.L.; visualization, S.-H.W.; supervision, M.A.M., A.W.M. and F.L.; project administration, F.L.; funding acquisition, F.L. All authors have read and agreed to the published version of the manuscript.

Funding: This study was supported by NSF Grants CMMI 1625736 and NSF EEC 1937128, Product Innovation and Engineering's NAVAIR SBIR Phase II Contract N6833524C0215, and the Center for Aerospace Manufacturing Technologies (CAMT), Intelligent Systems Center (ISC), and Material Research Center (MRC) at Missouri S&T. We greatly appreciate their financial support.

Data Availability Statement: The original contributions presented in the study are included in the article, further inquiries can be directed to the corresponding author.

Conflicts of Interest: The authors declare no conflicts of interest.

References

1. Svetlizky, D.; Das, M.; Zheng, B.; Vyatskikh, A.L.; Bose, S.; Bandyopadhyay, A.; Schoenung, J.M.; Lavernia, E.J.; Eliaz, N. Directed energy deposition (DED) additive manufacturing: Physical characteristics, defects, challenges and applications. *Mater. Today* **2021**, *49*, 271–295. [CrossRef]
2. Xie, J.; Zhou, Y.; Zhou, C.; Li, X.; Chen, Y. Microstructure and mechanical properties of Mg–Li alloys fabricated by wire arc additive manufacturing. *J. Mater. Res. Technol.* **2024**, *29*, 3487–3493. [CrossRef]
3. Madhavadas, V.; Srivastava, D.; Chadha, U.; Raj, S.A.; Sultan, M.T.H.; Shahar, F.S.; Shah, A.U.M. A review on metal additive manufacturing for intricately shaped aerospace components. *CIRP J. Manuf. Sci. Technol.* **2022**, *39*, 18–36. [CrossRef]
4. Saboori, A.; Aversa, A.; Marchese, G.; Biamino, S.; Lombardi, M.; Fino, P. Application of directed energy deposition-based additive manufacturing in repair. *Appl. Sci.* **2019**, *9*, 3316. [CrossRef]
5. Tariq, U.; Wu, S.H.; Mahmood, M.A.; Woodworth, M.M.; Liou, F. Effect of pre-heating on residual stresses and deformation in laser-based directed energy deposition repair: A comparative analysis. *Materials* **2024**, *17*, 2179. [CrossRef]
6. Mohd Yusuf, S.; Cutler, S.; Gao, N. The impact of metal additive manufacturing on the aerospace industry. *Metals* **2019**, *9*, 1286. [CrossRef]
7. Piscopo, G.; Iuliano, L. Current research and industrial application of laser powder directed energy deposition. *Int. J. Adv. Manuf. Technol.* **2022**, *119*, 6893–6917. [CrossRef]
8. Markets, R. Market Opportunities for Directed Energy Deposition Manufacturing. Available online: https://www.researchandmarkets.com/reports/4850372/market-opportunities-for-directed-energy (accessed on 6 June 2024).
9. Brennan, M.; Keist, J.; Palmer, T. Defects in metal additive manufacturing processes. *J. Mater. Eng. Perform.* **2021**, *30*, 4808–4818. [CrossRef]
10. Yuhua, C.; Yuqing, M.; Weiwei, L.; Peng, H. Investigation of welding crack in micro laser welded NiTiNb shape memory alloy and Ti6Al4V alloy dissimilar metals joints. *Opt. Laser Technol.* **2017**, *91*, 197–202. [CrossRef]
11. Chen, Y.; Sun, S.; Zhang, T.; Zhou, X.; Li, S. Effects of post-weld heat treatment on the microstructure and mechanical properties of laser-welded NiTi/304SS joint with Ni filler. *Mater. Sci. Eng. A* **2020**, *771*, 138545. [CrossRef]
12. Ertay, D.S.; Naiel, M.A.; Vlasea, M.; Fieguth, P. Process performance evaluation and classification via in-situ melt pool monitoring in directed energy deposition. *CIRP J. Manuf. Sci. Technol.* **2021**, *35*, 298–314. [CrossRef]

13. Jiang, H.Z.; Li, Z.Y.; Feng, T.; Wu, P.Y.; Chen, Q.S.; Feng, Y.L.; Chen, L.F.; Hou, J.Y.; Xu, H.J. Effect of process parameters on defects, melt pool shape, microstructure, and tensile behavior of 316L stainless steel produced by selective laser melting. *Acta Metall. Sin. (English Lett.)* **2021**, *34*, 495–510. [CrossRef]

14. Liu, M.; Kumar, A.; Bukkapatnam, S.; Kuttolamadom, M. A review of the anomalies in directed energy deposition (DED) processes & potential solutions-part quality & defects. *Procedia Manuf.* **2021**, *53*, 507–518.

15. Zheng, B.; Haley, J.; Yang, N.; Yee, J.; Terrassa, K.; Zhou, Y.; Lavernia, E.; Schoenung, J. On the evolution of microstructure and defect control in 316L SS components fabricated via directed energy deposition. *Mater. Sci. Eng. A* **2019**, *764*, 138243. [CrossRef]

16. Xie, J.; Chen, Y.; Wang, H.; Zhang, T.; Zheng, M.; Wang, S.; Yin, L.; Shen, J.; Oliveira, J. Phase transformation mechanisms of NiTi shape memory alloy during electromagnetic pulse welding of Al/NiTi dissimilar joints. *Mater. Sci. Eng. A* **2024**, *893*, 146119. [CrossRef]

17. Mahmood, M.A.; Popescu, A.C.; Oane, M.; Channa, A.; Mihai, S.; Ristoscu, C.; Mihailescu, I.N. Bridging the analytical and artificial neural network models for keyhole formation with experimental verification in laser melting deposition: A novel approach. *Results Phys.* **2021**, *26*, 104440. [CrossRef]

18. Wu, Y.; Wu, H.; Zhao, Y.; Jiang, G.; Shi, J.; Guo, C.; Liu, P.; Jin, Z. Metastable structures with composition fluctuation in cuprate superconducting films grown by transient liquid-phase assisted ultra-fast heteroepitaxy. *Mater. Today Nano* **2023**, *24*, 100429. [CrossRef]

19. Wu, S.H.; Joy, R.; Tariq, U.; Mahmood, M.A.; Liou, F. *Role of In-Situ Monitoring Technique for Digital Twin Development Using Direct Energy Deposition: Melt Pool Dynamics and Thermal Distribution*; University of Texas at Austin: Austin, TX, USA, 2023.

20. Yeoh, Y. Decoupling Part Geometry from Microstructure in Directed Energy Deposition Technology: Towards Reliable 3D Printing of Metallic Components. Ph.D. Thesis, Nanyang Technological University, Singapore, 2021.

21. Kistler, N.A.; Corbin, D.J.; Nassar, A.R.; Reutzel, E.W.; Beese, A.M. Effect of processing conditions on the microstructure, porosity, and mechanical properties of Ti-6Al-4V repair fabricated by directed energy deposition. *J. Mater. Process. Technol.* **2019**, *264*, 172–181. [CrossRef]

22. Tariq, U.; Joy, R.; Wu, S.H.; Arif Mahmood, M.; Woodworth, M.M.; Liou, F. *Optimization of Computational Time for Digital Twin Database in Directed Energy Deposition for Residual Stresses*; University of Texas at Austin: Austin, TX, USA, 2023.

23. Hooper, P.A. Melt pool temperature and cooling rates in laser powder bed fusion. *Addit. Manuf.* **2018**, *22*, 548–559. [CrossRef]

24. He, W.; Shi, W.; Li, J.; Xie, H. In-situ monitoring and deformation characterization by optical techniques; part I: Laser-aided direct metal deposition for additive manufacturing. *Opt. Lasers Eng.* **2019**, *122*, 74–88. [CrossRef]

25. Nuñez, L., III; Sabharwall, P.; van Rooyen, I.J. In situ embedment of type K sheathed thermocouples with directed energy deposition. *Int. J. Adv. Manuf. Technol.* **2023**, *127*, 3611–3623. [CrossRef]

26. Zhao, M.; Wei, H.; Mao, Y.; Zhang, C.; Liu, T.; Liao, W. Predictions of Additive Manufacturing Process Parameters and Molten Pool Dimensions with a Physics-Informed Deep Learning Model. *Engineering* **2023**, *23*, 181–195. [CrossRef]

27. Wang, Z.; Wang, C.; Zhang, S.; Qiu, L.; Lin, Y.; Tan, J.; Sun, C. Towards high-accuracy axial springback: Mesh-based simulation of metal tube bending via geometry/process-integrated graph neural networks. *Expert Syst. Appl.* **2024**, *255*, 124577. [CrossRef]

28. De Borst, R. Challenges in computational materials science: Multiple scales, multi-physics and evolving discontinuities. *Comput. Mater. Sci.* **2008**, *43*, 1–15. [CrossRef]

29. Darabi, R.; Azinpour, E.; Reis, A.; de Sa, J.C. Multi-scale multi-physics phase-field coupled thermo-mechanical approach for modeling of powder bed fusion process. *Appl. Math. Model.* **2023**, *122*, 572–597. [CrossRef]

30. Zhao, T.; Yan, Z.; Zhang, B.; Zhang, P.; Pan, R.; Yuan, T.; Xiao, J.; Jiang, F.; Wei, H.; Lin, S.; et al. A comprehensive review of process planning and trajectory optimization in arc-based directed energy deposition. *J. Manuf. Process.* **2024**, *119*, 235–254. [CrossRef]

31. Bayat, M.; Dong, W.; Thorborg, J.; To, A.C.; Hattel, J.H. A review of multi-scale and multi-physics simulations of metal additive manufacturing processes with focus on modeling strategies. *Addit. Manuf.* **2021**, *47*, 102278. [CrossRef]

32. Zhu, Q.; Liu, Z.; Yan, J. Machine learning for metal additive manufacturing: Predicting temperature and melt pool fluid dynamics using physics-informed neural networks. *Comput. Mech.* **2021**, *67*, 619–635. [CrossRef]

33. Qi, X.; Chen, G.; Li, Y.; Cheng, X.; Li, C. Applying neural-network-based machine learning to additive manufacturing: Current applications, challenges, and future perspectives. *Engineering* **2019**, *5*, 721–729. [CrossRef]

34. Akbari, P.; Ogoke, F.; Kao, N.Y.; Meidani, K.; Yeh, C.Y.; Lee, W.; Farimani, A.B. MeltpoolNet: Melt pool characteristic prediction in Metal Additive Manufacturing using machine learning. *Addit. Manuf.* **2022**, *55*, 102817. [CrossRef]

35. Zhu, X.; Jiang, F.; Guo, C.; Wang, Z.; Dong, T.; Li, H. Prediction of melt pool shape in additive manufacturing based on machine learning methods. *Opt. Laser Technol.* **2023**, *159*, 108964. [CrossRef]

36. Zhang, Z.; Liu, Z.; Wu, D. Prediction of melt pool temperature in directed energy deposition using machine learning. *Addit. Manuf.* **2021**, *37*, 101692. [CrossRef]

37. Jones, K.; Yang, Z.; Yeung, H.; Witherell, P.; Lu, Y. Hybrid modeling of melt pool geometry in additive manufacturing using neural networks. In *Proceedings of the International Design Engineering Technical Conferences and Computers and Information in Engineering Conference*; American Society of Mechanical Engineers: New York, NY, USA, 2021; Volume 85376, p. V002T02A031.

38. Mahmood, M.A.; Ishfaq, K.; Khraisheh, M. Inconel-718 processing windows by directed energy deposition: A framework combining computational fluid dynamics and machine learning models with experimental validation. *Int. J. Adv. Manuf. Technol.* **2024**, *130*, 3997–4011. [CrossRef]

39. Tariq, U.; Joy, R.; Wu, S.H.; Mahmood, M.A.; Malik, A.W.; Liou, F. A state-of-the-art digital factory integrating digital twin for laser additive and subtractive manufacturing processes. *Rapid Prototyp. J.* **2023**, *29*, 2061–2097. [CrossRef]
40. Lu, X.; Lin, X.; Chiumenti, M.; Cervera, M.; Hu, Y.; Ji, X.; Ma, L.; Yang, H.; Huang, W. Residual stress and distortion of rectangular and S-shaped Ti-6Al-4V parts by Directed Energy Deposition: Modelling and experimental calibration. *Addit. Manuf.* **2019**, *26*, 166–179. [CrossRef]
41. Newkirk, J. Multi-Layer Laser Metal Deposition Process. Ph.D. Thesis, Missouri University of Science and Technology Rolla, Rolla, MO, USA, 2014.
42. Wu, S.H.; Tariq, U.; Joy, R.; Sparks, T.; Flood, A.; Liou, F. Experimental, computational, and machine learning methods for prediction of residual stresses in laser additive manufacturing: A critical review. *Materials* **2024**, *17*, 1498. [CrossRef]
43. Gouge, M.; Michaleris, P.; Denlinger, E.; Irwin, J. The finite element method for the thermo-mechanical modeling of additive manufacturing processes. In *Thermo-Mechanical Modeling of Additive Manufacturing*; Elsevier: Amsterdam, The Netherlands, 2018; pp. 19–38.
44. Dhieb, N.; Ghazzai, H.; Besbes, H.; Massoud, Y. Extreme gradient boosting machine learning algorithm for safe auto insurance operations. In Proceedings of the 2019 IEEE International Conference on Vehicular Electronics and Safety (ICVES), Cairo, Egypt, 4–6 September 2019; pp. 1–5.
45. Yu, Y.; Si, X.; Hu, C.; Zhang, J. A review of recurrent neural networks: LSTM cells and network architectures. *Neural Comput.* **2019**, *31*, 1235–1270. [CrossRef]

 *materials*

Article

Deep-Learning-Based Segmentation of Keyhole in In-Situ X-ray Imaging of Laser Powder Bed Fusion

William Dong [1], Jason Lian [2], Chengpo Yan [2], Yiran Zhong [2], Sumanth Karnati [2], Qilin Guo [1], Lianyi Chen [1,3,*] and Dane Morgan [3,*]

[1] Department of Mechanical Engineering, University of Wisconsin-Madison, Madison, WI 53706, USA; williamdong@berkeley.edu (W.D.); qilin.guo@unl.edu (Q.G.)

[2] Department of Computer Science, University of Wisconsin-Madison, Madison, WI 53706, USA; jlian7@wisc.edu (J.L.); cyan46@wisc.edu (C.Y.); yzhong68@wisc.edu (Y.Z.); karnati2@wisc.edu (S.K.)

[3] Department of Material Science & Engineering, University of Wisconsin-Madison, Madison, WI 53706, USA

* Correspondence: lianyi.chen@wisc.edu (L.C.); ddmorgan@wisc.edu (D.M.)

Abstract: In laser powder bed fusion processes, keyholes are the gaseous cavities formed where laser interacts with metal, and their morphologies play an important role in defect formation and the final product quality. The in-situ X-ray imaging technique can monitor the keyhole dynamics from the side and capture keyhole shapes in the X-ray image stream. Keyhole shapes in X-ray images are then often labeled by humans for analysis, which increasingly involves attempting to correlate keyhole shapes with defects using machine learning. However, such labeling is tedious, time-consuming, error-prone, and cannot be scaled to large data sets. To use keyhole shapes more readily as the input to machine learning methods, an automatic tool to identify keyhole regions is desirable. In this paper, a deep-learning-based computer vision tool that can automatically segment keyhole shapes out of X-ray images is presented. The pipeline contains a filtering method and an implementation of the BASNet deep learning model to semantically segment the keyhole morphologies out of X-ray images. The presented tool shows promising average accuracy of 91.24% for keyhole area, and 92.81% for boundary shape, for a range of test dataset conditions in Al6061 (and one AliSi10Mg) alloys, with 300 training images/labels and 100 testing images for each trial. Prospective users may apply the presently trained tool or a retrained version following the approach used here to automatically label keyhole shapes in large image sets.

Keywords: keyhole; laser powder bed fusion; deep learning; image segmentation

Citation: Dong, W.; Lian, J.; Yan, C.; Zhong, Y.; Karnati, S.; Guo, Q.; Chen, L.; Morgan, D. Deep-Learning-Based Segmentation of Keyhole in In-Situ X-ray Imaging of Laser Powder Bed Fusion. *Materials* **2024**, *17*, 510. https://doi.org/10.3390/ma17020510

Academic Editors: Tuhin Mukherjee and Qianru Wu

Received: 16 October 2023
Revised: 10 January 2024
Accepted: 16 January 2024
Published: 21 January 2024

1. Introduction

Laser powder bed fusion (LPBF), also known as selective laser melting (SLM), is currently one of the most common metal additive manufacturing (AM) techniques [1,2]. During the LPBF process, a focused laser will shoot onto the powder bed selectively and make powders melt, merge, and solidify to build up a part based on the CAD (computer-aided design) model [3]. Using in-situ X-ray imaging to monitor the process, previous studies have found that high intensity of the laser will result in vaporization of the material, which will lead to recoil pressure that pushes the molten metal in the melt pool to create a vapor cavity, named a 'keyhole' [4,5]. In the process and along the printing path, the keyhole experiences inconsistent recoil pressure, surface tension, and Marangoni force, leading to severe and random fluctuations [6,7]. Sometimes, fluctuated keyholes will collapse, and the vapor will be partially trapped inside the melt pool, resulting in undesirable porosity in the final product [4,8].

Many studies have been made to find correlations between the keyhole morphology revealed in the X-ray imaging and the generation of keyhole-induced pores [9,10], with an eventual goal of reducing defects. Due to the complex dynamics of the LPBF process, there is great interest in in-situ data-driven methods to study the defect formation mechanisms [11,12]. As a popular approach to observe the dynamics in the LPBF process, in-situ X-ray imaging can be captured at a frame rate of over one million frames per second, generating enormous keyhole morphology images, and creating a huge potential for data-driven studies with keyholes. However, these studies are greatly constrained by a limited quantity of well-characterized keyhole morphology data, which typically needs to be labeled by humans. More specifically, sufficient data for machine learning-based pore prediction for many types of LPBF systems, conditions, and alloys, will require very large efforts to label keyhole morphologies unless accessible automatic tools that can segment the keyholes are developed. Several automatic tools have recently been explored in previous works: Pyeon et al. developed a non-machine learning-based semi-automatic keyhole region extraction tool [13], and Zhang et al. tested several semantic segmentation and object detection models and compared the performances of extracting keyholes and pores at the same time [14]. However, the filter developed by Pyeon et al. was only tested with clean images without metal powder, and models tested by Zhang et al. segment both keyholes and pores in the same classification. In addition, while the boundary is the most important feature in the keyhole morphology, previously proposed methods are not designed to have high segmentation boundary accuracy and be validated against datasets from different experiments. Therefore, considering the increasing need for keyhole segmentation for large X-ray imaging databases, we here develop an automatic keyhole segmentation tool with high accuracy for both area and boundary and test against datasets with powders from different experiments.

In this paper, a deep-learning-based semantic segmentation tool that is capable of automatic segmentation of keyholes in X-ray images with accurate boundaries was developed. This tool is composed of a filter that standardizes, normalizes, and cleans the X-ray images, and an implementation of BASNet, a Boundary-Aware Segmentation network, that predicts semantic labels [15]. Without any human inputs, this tool only requires users to run algorithms with their X-ray images, which significantly accelerates the keyhole morphology labeling process and enables the possibility of data-driven analysis with a large quantity of morphology data. In the following, the implementation of the method will be illustrated, the performance of the tool will be quantified, and the mechanism and limitation of the method will be discussed at the end. This work was conducted with data derived from different experiments at multiple different times, but the tool provided in this study will likely need to be refit for systems with significant differences from those studied here. However, such refitting can likely be performed quickly through transfer learning, i.e., by starting from the weights in this paper. Development of similar segmentation tasks for X-ray imaging, like segmenting melt pools or spatters in the LPBF process, could also be accelerated by transfer learning from the present model.

2. Methods

The workflow of the whole segmentation process is shown in Figure 1: Raw X-ray images are processed with a designed filter and then fed into the segmentation network, which outputs the semantic labels. The segmentation network needs to be previously trained by filtered images and ground truth labels in the training set.

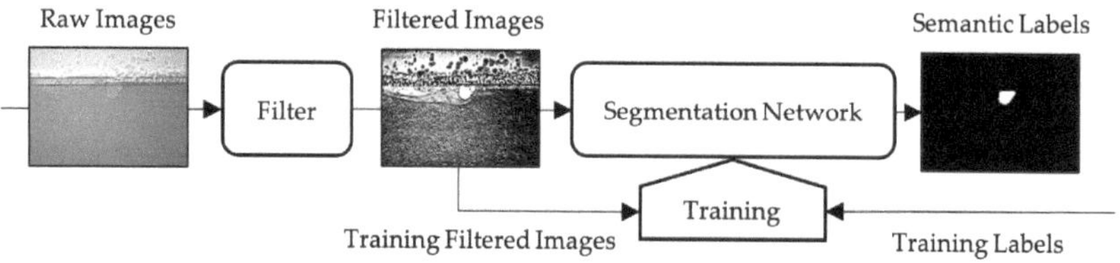

Figure 1. A flowchart illustrating the keyhole segmentation process pipeline.

2.1. Raw Images

Data used in this study were acquired by using in situ X-ray imaging on LPBF processes at the beamline 32-ID-B of the Advanced Photon Source at Argonne National Laboratory. During experiments, the ytterbium fiber laser generated a laser beam and was directed by a galvo scan head toward the metal powder and substrate. While the laser was melting the material, the scanning area was penetrated by X-ray simultaneously, and the shapes of keyholes and pores were projected and converted to visible light by a scintillator, which was recorded by a high-speed camera with a frame rate of 50 kHz [4]. There are 8 X-ray imaging data sets in this work and they were acquired by 8 separate experiments, each with different processing parameters, as shown in Table 1. The experiments were performed with Al6061 in 7 cases and AliSi10Mg in one case. Each data set has 400–500 frames, and 50 of those were labeled in each data set and used in this study. In total, there are 3416 images in the 8 data sets from which 1441 images have a visible keyhole, and 400 images with visible keyholes were labeled for training and testing. The remaining 1975 images show no keyholes since the laser in the LPBF process was not on or in the field of view at the time of imaging, and are not used in the testing and training of this paper.

Table 1. Samples and processing parameters of in situ X-ray imaging experiments.

Experiment	Material	Nanoparticle	Substrate	Power (W)	Scan Speed (m/s)
1	Al6061	10%vol TiC	Printed	385	0.3
2	Al6061	10%vol TiC	Printed	385	0.4
3	Al6061	/	Printed	443	0.4
4	Al6061	/	As-cast	500	0.2
5	Al6061	/	Printed	500	0.4
6	Al6061	/	As-cast	500	0.4
7	Al6061	/	As-cast	500	0.4
8	AlSi10Mg	/	As-cast	500	0.5

2.2. Filtering

There are two major goals when filtering the raw images: (1) standardize images to be acceptable for the segmentation network as inputs, and (2) normalize and clean the image by reducing the differences between datasets and removing stationary obstructions. The filter is built using MATLAB R2023b based on the schematics shown below in Figure 2 and described below.

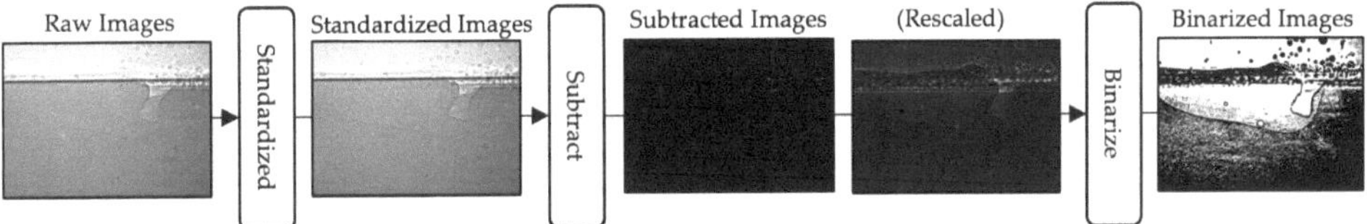

Figure 2. Flowchart illustrating filtering procedures. The subtracted image is rescaled for better representation in the following image, no re-scaling was involved in the "Subtract" and "Binarize" steps.

To meet the requirements of the network, all images need to be at the same resolution, a fixed size, and in the same format. In our case, all images are converted into uint 8, with a size of 700 × 500 pixels, and in PNG format. This step is known as "Standardize".

In different experiment setups, X-ray images will have different brightness, contrast, and sometimes stationary obstructions overlapping with the keyhole area. These factors greatly hinder the segmentation of keyholes. To alleviate these factors, a concurrent subtraction step is designed, and the detailed mechanism will be covered in the Discussion section. For each image, the sum of previous N frames of images is calculated, and the current image is subtracted by the average of previous N frames, where N is a tunable parameter for each dataset, and here 40 previous images are used. This step is termed "Subtract".

To further normalize the images, greyscale images are converted into black and white (binary) images. The images are binarized with a threshold of 0.5. The global average value of all pixels in the whole image is calculated, pixels with a value above the average are marked as 1, and pixels with a value below average are marked as 0. This step is "Binarize".

With these three steps, any raw X-ray images can be transformed into the format used for training and can be input into our training model. Please check the GitHub repository in the Supplementary Material to find a specific implementation of these steps.

2.3. Segmentation Network Training

Before the filtered images are fed into the segmentation network, the network needs to be trained by filtered training images and corresponding ground truth labeling. The ground truths are labeled manually using MATLAB R2023b Image Labeler app, with pixels assigned as 1 in the keyhole region and pixels assigned as 0 in the background, exported with the same resolution, size, and format as training images, creating a binary label for each corresponding image. In this study, experiments/datasets 3 to 8 were picked as training datasets, and their corresponding ground truth labels were picked as the training set, with 50 images and labels for each of the 6 datasets, and 300 images and labels in total.

In this study, we chose the Boundary-Aware Segmentation Network (BASNet) developed by Qin et al. as the semantic segmentation network [15]. The BASNet model was trained at a batch size equal to 1, and 70 epochs, leading to 21,000 iterations in total, and the trained model can be found in the GitHub repository in the Supplementary Material. The model appeared to be well-converged by this number of epochs and the loss curve can be found in the Results and Discussion section. The source code for training was modified to add a testing step with datasets 1 and 2 after each epoch, where the model was tested to generate the labels for images in datasets 1 and 2. The labels were compared with the ground truths and the testing losses for these two datasets were also calculated after each epoch of training. Convergence on testing loss was also observed, with more details in the Results and Discussion section.

2.4. Deep Learning Segmentation on Test Data

Experiments/datasets 1 and 2 were picked as the testing datasets for the segmentation model, and the ground truth was labeled in advance as a comparison to prediction. Filtered X-ray images for datasets 1 and 2 were fed into the trained BASNet model to generate predicted labels, which need to be binarized to convert the label from gradient to binary images. The sample X-ray image, its predicted label, and comparisons with ground truth are shown in Figure 3.

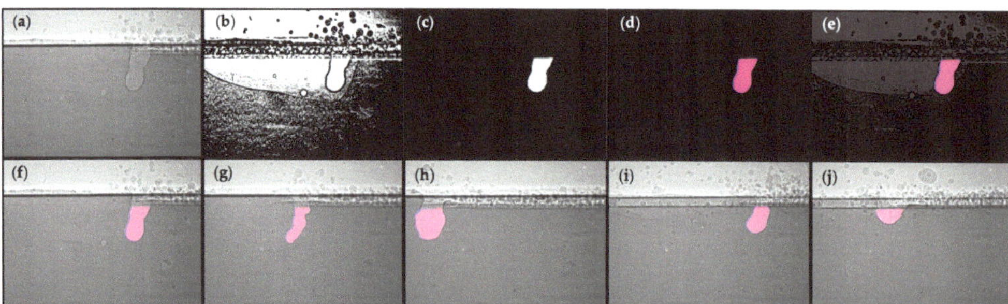

Figure 3. Sample X-ray image, its predicted label, and comparisons: (**a**) raw image; (**b**) filtered image; (**c**) predicted label; (**d**) overlay of predicted label and ground truth, where predicted label is in red, ground truth is in blue, intersection is overlapped as purple (Since there are not much unmatched predicted label and ground truth, red and blue area are barely visible for the given images); (**e**) overlay on filtered image; (**f**) overlay on raw image; (**g,h**) overlay on raw images for other two frames in testing dataset 1; (**i,j**) overlay on raw images for two frames in testing dataset 2.

Testing on data from different experiments can allow for evaluation of the actual performance of this tool when prospective users implement this tool to acquire keyhole morphology on their own X-ray images. The performance of this model on testing datasets will be quantified in the Results section.

3. Results

3.1. Training and Loss Function

The BASNet model adopts a predict-refinement structure, where the input image first passes through a predict module (encoder–decoder) and then a residual refinement module to finally generate the segmentation. The segmentation generated by the refinement module is the final output of the model, and there are 7 "side outputs", or intermediate outputs, which are the outputs of every stage of the decoder in the predict module, and are also the inputs of their upcoming stage. The loss function for optimization of BASNet takes the summation of loss values of all final outputs and 7 side outputs generated along the network, which is named "summation loss". While this is useful for training, the performance a user cares about is the loss from the final output, which is named "final output loss", as that is what will be used in applications. Both training summation loss and the final output loss for the training and test data were recorded at the end of each epoch, shown in Figure 4 below. As the training loss curve shows in Figure 4a, the model gradually converged to a low and consistent summation loss value as training proceeded. The model also performed nearly as well on test datasets, which also gradually converged to a low loss value along with the training loss. During training, the final output loss for the training datasets and test datasets was also calculated and is shown in Figure 4b below. The convergence of the training and test datasets onto low loss values indicates the high accuracy of segmentation of the trained model on both training and test datasets.

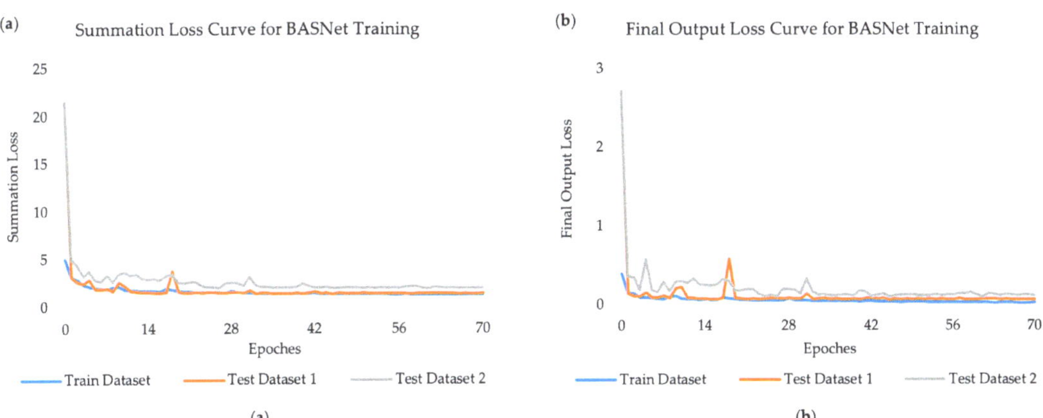

Figure 4. Recorded loss curve for BASNet training process for training datasets (experiment 3–8), test dataset 1 and 2, from 0 epoch to 70 epochs: (**a**) training and testing data loss curve for summation loss; (**b**) training and testing data loss curve for final output loss.

3.2. Testing and Performance Matrices

To evaluate the performance of the tool, by comparing with the predicted label and ground truth, two matrices are calculated to quantify the segmentation accuracy of the pipeline: intersection over union (IoU) and boundary F-score (BF-score) (both defined below). Both IoU and BF-score are in the range [0, 1], an IoU closer to 1 means a better match in area, and a BF-score close to 1 means a better match on the boundary [16]. These two metrics are calculated based on the predicted label and ground truth for all images in both testing datasets 1 and 2, based on Equations (1) to (4) below, where PL and PLB stand for the model-predicted label and its boundary, and GT and GTB stand for the ground truth label and its boundary, which represents the actual keyhole region. In the IoU calculation, the intersection area and union area of the predicted label and ground truth are calculated, and a ratio of intersection over union that is close to 1 shows both successful coverage of the actual keyhole area by the model predicted labeling and little overestimation of the actual keyhole area by the predicted labeling. In the BF-score calculation, precision represents the percentage of the model-predicted boundary that matches the actual keyhole boundary, and high precision means the model predicted a more correct boundary. Recall represents the percentage of the actual keyhole boundary that is predicted correctly by the model-predicted boundary, and high recall indicates more actual keyhole boundary is successfully predicted by the model. The BF-score is calculated by the multiplication over the sum of the precision and recall times by 2, and a score close to 1 indicates both high precision and recall by the model prediction. In Equations (2) and (3), the threshold is set for the maximum distance between two boundaries is 1 pixel, meaning that any portion of the boundary that exceeds 1 pixel distance from the other will not count in the numerator. This is a demanding criterion, representing just a fraction of a percent of the dimensions of the filtered images, which in this study are 700 by 300 pixels.

$$\text{IoU} = (\text{Area Intersection of PL and GT})/(\text{Area Union of PL and GT}) \tag{1}$$

$$\text{Precision} = (\text{Portion of PLB with distance to GTB within the threshold})/(\text{Full PLB}) \tag{2}$$

$$\text{Recall} = (\text{Portion of GTB with distance to PLB within the threshold})/(\text{Full GTB}) \tag{3}$$

$$\text{BF-score} = 2 \times (\text{Precision} \times \text{Recall})/(\text{Precision} + \text{Recall}) \tag{4}$$

As shown in Table 2 below, IoU and BF-score are high and close to 1 for both datasets 1 and 2 for run 1, which means that this method successfully segments out the keyhole region in test datasets. Considering IoU, this method is on the same level as other segmentation tools proposed in previous works, and a high BF-score further validates the segmentation accuracy on the boundary [14]. The same tool is later tested using cross-validation, with three more runs trained and tested as Table 3 below. The IoU and BF-score for cross-validation are also shown in Table 2, with most testing datasets showing similar values for both matrices, with an average IoU of 0.9124 and average BF-score of 0.9281, suggesting that this tool is very accurate for random 75% training/25% test splits. However, the proposed tool might also encounter segmentation errors, as testing dataset 3 for run 2 shows relatively lower IoU and BF-score values, which will be further covered in the Discussion section.

Table 2. IoU and BF-score for cross validation with 6 training sets and 2 testing sets, 4 runs in total.

Run	Testing Dataset	IoU	BF-Score
1	1	0.9381	0.9514
1	2	0.8923	0.9098
2	3	0.8333	0.8603
2	4	0.9321	0.9421
3	5	0.9195	0.9352
3	6	0.9190	0.9351
4	7	0.9130	0.9313
4	8	0.9518	0.9595
Average		**0.9124**	**0.9281**

Table 3. Cross validation training and testing datasets assignments.

Run	Dataset 1 and 2	Dataset 3 and 4	Dataset 5 and 6	Dataset 7 and 8
1	Test	Train	Train	Train
2	Train	Test	Train	Train
3	Train	Train	Test	Train
4	Train	Train	Train	Test

4. Discussion

We constructed a pipeline for keyhole region semantic segmentation in in-situ LPBF X-ray Imaging. The whole pipeline consists of two main components, a filter that standardizes, normalizes, and cleans raw datasets, and deep learning segmentation labels the keyhole region from the filtered images. In different experiment setups, X-ray images have different brightness, contrast, and sometimes stationary obstructions overlapping with the keyhole area. These factors greatly hinder the segmentation of keyholes.

Therefore, the intuition of the design of the filter part in the pipeline is to manage all training data in a consistent fashion, so as to reduce the effects of attributes of the datasets on the prediction of keyhole segmentation. From the eight raw datasets used in this experiment, images from four datasets have a dimension of 712×512 pixels, and images from four other datasets have a dimension of 896×448 pixels, all in TIF format. Theoretically, having images with different sizes should still be applicable for training, as there is a rescaling step in the training algorithm, but a consistent size and file format will make image labeling and manipulation much easier. Hence, the designed Standardize step converts all images into 700×500 pixels and PNG format.

As shown in Table 4, raw images from different datasets have vastly different brightness and contrast values, and the differences can be mitigated with the Subtract and the Binarize step to normalize images from all datasets. Firstly, the Subtract step greatly reduces the differences by subtracting the average image of the whole dataset, leaving only the features of each image relative to the dataset. Then, the differences are further alleviated by the Binarize step, which reduces the greyscale difference across different datasets by turning the image into black and white, which will have contrast of 1.

Table 4. Average brightness and average contrast value for raw, subtracted, and binarized images for all images in 8 datasets. For each image, brightness is calculated by the mean of all pixels' value over the white value (255 for uint 8), and contrast is calculated by the range of pixels' value (maximum − minimum) over the white value. Values from all 50 images for each dataset are averaged.

Dataset	Average Brightness			Average Contrast		
	Raw	Subtracted	Binarized	Raw	Subtracted	Binarized
1	0.5852	0.0079	0.0363	0.8034	0.4820	1.0000
2	0.5802	0.0091	0.0374	0.7496	0.4958	1.0000
3	0.0131	0.0001	0.0314	0.0278	0.0155	1.0000
4	0.0113	0.0001	0.0275	0.0255	0.0129	1.0000
5	0.0131	0.0001	0.0328	0.0288	0.0156	1.0000
6	0.5841	0.0097	0.0350	0.7833	0.5260	1.0000
7	0.0115	0.0001	0.0215	0.0264	0.0119	1.0000
8	0.5876	0.0083	0.0389	0.7503	0.4261	1.0000

Despite brightness and contrast, stationary obstruction also greatly hinders the segmentation of keyholes, especially when they are spatially overlapped with the keyhole area. These obstructions are stationary and are affiliated with a particular dataset, which yields inconsistent X-ray images across different datasets and influences the prediction of the keyhole area. As shown in Figure 5, the obstruction can also be resolved with the Subtraction step, as the subtraction value shown in Figure 5b contains the stationary patterns, and stationary obstructions can be removed for subtracted and binarized images, as shown in Figure 5c–e. Note that the deep learning model segmentations described below with just the Standardize step and just the Subtract step were also tried, but the model did not perform well and failed to identify the keyhole region in most frames.

For the deep learning segmentation, BASNet utilizes a hybrid fusion loss function to achieve training supervision of prediction on multiple levels: pixel level, patch level, and map level, instead of barely relying on IoU, which could lead to insufficient prediction of structural properties on the patch level. Along with the prediction-refinement structure, BASNet is capable of semantic segmentation for boundary-sensitive cases like the detailed morphology in keyhole segmentation. The boundary prediction performance is further validated with the BF-score in the Results section. Other popular models like UNet and Deep LabV3+ were also tried in the pipeline, but they all failed to map the boundary in most cases even when with a high IoU score [17,18].

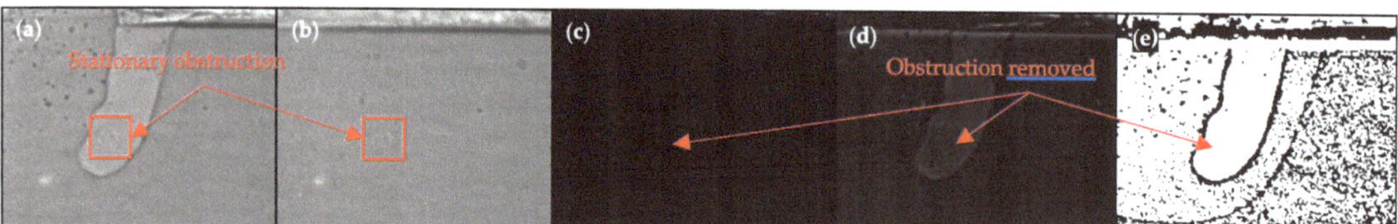

Figure 5. The Subtract step in the filter removes a stationary obstruction affiliated to the dataset: (**a**) a standardized image with the obstruction inside the keyhole; (**b**) the subtraction value as an image, average of 40 previous images in the same dataset, containing the obstruction; (**c**) the subtracted image, where the obstruction is removed by the Subtract step; (**d**) the rescaled subtracted image for visualization; (**e**) binarized image, the output of the filter, with no obstruction.

However, the proposed tool may also encounter segmentation errors, like run 2 testing dataset 3 mentioned in the Results section. Two failed segmentation examples are shown in Figure 6, where in the first example (Figure 6a,b), BASNet predicts no keyhole region, resulting in an IoU and BF-score of less than 0.1. In the second example, shown in Figure 6c,d, BASNet failed to predict the "tail" of the keyhole region. As shown in Figure 6e, a small number of similar errors contribute to the relatively low average IoU and BF-score, while in most cases, BASNet can accurately label the keyhole region. These errors could be attributed to the small keyhole size of dataset 3, which leads to inconsistency in the prediction area with other datasets. Consistency can potentially be achieved by adding a cropping module in the filter or introducing more training data with similar sizes to reduce the effect of keyhole size on the prediction of BASNet. In addition, another factor behind segmentation errors is the fuzzy imaging background for dataset 3, which results in unclear keyhole region contour of the filtered images. Further fine-tuning of the filter parameter N could potentially improve the clarity of the image with a distinct keyhole boundary.

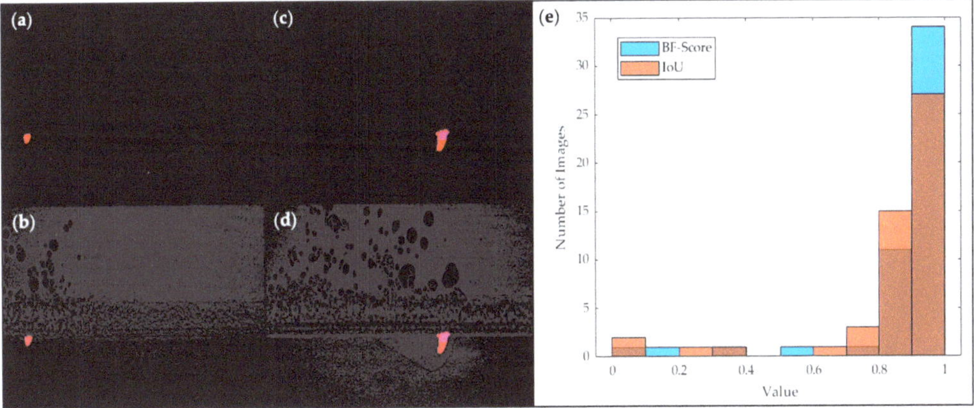

Figure 6. Segmentation errors in run 2, testing dataset 3: (**a**) overlay of the predicted label and ground truth of a failed segmentation example, no label is predicted by BASNet (predicted label in blue, ground truth in red, intersection in purple); (**b**) overlay on the input filtered image; (**c**) another failed segmentation example, where the "tail" of the keyhole region is missed by prediction; (**d**) overlay on the input filtered image; (**e**) histogram of the distribution of tested images in testing dataset 3 in run 2 regarding two evaluation matrices (Number of images with respect to BF-Score in blue, IoU in red, and overlapped in darker orange). Blue area in (**a–d**) is barely visible because there is little predicted area that is not ground truth.

So far, the current algorithm cannot process X-ray images with significant differences in experiment setups, and re-training might be needed for prospective users to implement the algorithm. In the future, more training images will be labeled to further optimize this algorithm and to enhance the versatility and robustness of the model.

5. Conclusions

In this study, a deep-learning-based segmentation tool that is capable of automatic segmentation of keyhole morphology in X-ray images with a filter and a trained network pipeline was developed. This tool is validated, with an average IoU of 0.9124 and an average BF-score of 0.9281 on X-ray images from different experiments, proving its high accuracy both in area and boundary, with cross-validation of 300 training and 100 testing images/labeling for each trail.

This work illustrates a repeatable approach for prospective users to automatically generate massive keyhole morphology data with high accuracy on area and boundary from X-ray images. Sufficient morphology data will support developing data-driven analysis of LPBF processes to further improve the quality of additively manufactured products.

Supplementary Materials: The following supporting information can be downloaded at: https://github.com/WilliamDongSH/KeyholeSeg, Computer Source Code S1: MATLab filter for keyhole segmentation; Computer Source Code S2: MATLab checker for keyhole segmentation accuracy performance; Computer Source Code S3: Modified BASNet code with loss visualization and logging; Computer Source Code S4: Trained BASNet model.

Author Contributions: Conceptualization, L.C. and D.M.; methodology, W.D., J.L., C.Y., Y.Z., S.K. and D.M.; software, W.D., J.L., C.Y., Y.Z. and S.K.; validation, W.D.; formal analysis, W.D., J.L., C.Y., Y.Z., S.K. and D.M.; investigation, W.D. and D.M.; resources, W.D., L.C. and D.M.; data curation, W.D., Q.G.; writing—original draft preparation, W.D., L.C. and D.M.; writing—review and editing, W.D., L.C. and D.M.; visualization, W.D.; supervision, L.C. and D.M.; project administration, Q.G., D.M., L.C.; funding acquisition, W.D. and L.C. All authors have read and agreed to the published version of the manuscript.

Funding: This research is funded by the U.S. National Science Foundation grant numbers [2002840 and 2011354] except for D.M., who was supported by U.S. National Science Foundation Training-based Workforce Development for Advanced Cyberinfrastructure (CyberTraining) award No. 2017072.

Institutional Review Board Statement: Not applicable.

Informed Consent Statement: Not applicable.

Data Availability Statement: The computer source code is publicly available at: https://github.com/WilliamDongSH/KeyholeSeg. Data used in this study is available from the corresponding author upon reasonable request.

Acknowledgments: The authors would like to thank Ryan Jacobs and Yingwei Son for their technical help in this study. This research used resources of the Advanced Photon Source, a US Department of Energy (DOE) Office of Science user facility operated for the DOE Office of Science by Argonne National Laboratory under contract DE-AC02-06CH11357.

Conflicts of Interest: The authors declare no conflicts of interest.

References

1. Chowdhury, S.; Yadaiah, N.; Prakash, C.; Ramakrishna, S.; Dixit, S.; Gupta, L.R.; Buddhi, D. Laser Powder Bed Fusion: A State-of-the-Art Review of the Technology, Materials, Properties & Defects, and Numerical Modelling. *J. Mater. Res. Technol.* **2022**, *20*, 2109–2172. [CrossRef]
2. Abd-Elaziem, W.; Elkatatny, S.; Abd-Elaziem, A.-E.; Khedr, M.; Abd El-baky, M.A.; Hassan, M.A.; Abu-Okail, M.; Mohammed, M.; Järvenpää, A.; Allam, T.; et al. On the Current Research Progress of Metallic Materials Fabricated by Laser Powder Bed Fusion Process: A Review. *J. Mater. Res. Technol.* **2022**, *20*, 681–707. [CrossRef]
3. Dev Singh, D.; Mahender, T.; Raji Reddy, A. Powder Bed Fusion Process: A Brief Review. *Mater Today Proc.* **2021**, *46*, 350–355. [CrossRef]
4. Zhao, C.; Parab, N.D.; Li, X.; Fezzaa, K.; Tan, W.; Rollett, A.D.; Sun, T. Critical Instability at Moving Keyhole Tip Generates Porosity in Laser Melting. *Science 1979* **2020**, *370*, 1080–1086. [CrossRef] [PubMed]
5. Martin, A.A.; Calta, N.P.; Hammons, J.A.; Khairallah, S.A.; Nielsen, M.H.; Shuttlesworth, R.M.; Sinclair, N.; Matthews, M.J.; Jeffries, J.R.; Willey, T.M.; et al. Ultrafast Dynamics of Laser-Metal Interactions in Additive Manufacturing Alloys Captured by in Situ X-Ray Imaging. *Mater. Today Adv.* **2019**, *1*, 100002. [CrossRef]
6. Zhang, Y.; Zhang, J. Modeling of Solidification Microstructure Evolution in Laser Powder Bed Fusion Fabricated 316L Stainless Steel Using Combined Computational Fluid Dynamics and Cellular Automata. *Addit. Manuf.* **2019**, *28*, 750–765. [CrossRef]
7. Alphonso, W.E.; Baier, M.; Carmignato, S.; Hattel, J.H.; Bayat, M. On the Possibility of Doing Reduced Order, Thermo-Fluid Modelling of Laser Powder Bed Fusion (L-PBF)—Assessment of the Importance of Recoil Pressure and Surface Tension. *J. Manuf. Process.* **2023**, *94*, 564–577. [CrossRef]
8. Qu, M.; Guo, Q.; Escano, L.I.; Clark, S.J.; Fezzaa, K.; Chen, L. Mitigating Keyhole Pore Formation by Nanoparticles during Laser Powder Bed Fusion Additive Manufacturing. *Addit. Manuf. Lett.* **2022**, *3*, 100068. [CrossRef]
9. Wang, L.; Zhang, Y.; Chia, H.Y.; Yan, W. Mechanism of Keyhole Pore Formation in Metal Additive Manufacturing. *NPJ Comput. Mater.* **2022**, *8*, 22. [CrossRef]
10. Bayat, M.; Thanki, A.; Mohanty, S.; Witvrouw, A.; Yang, S.; Thorborg, J.; Tiedje, N.S.; Hattel, J.H. Keyhole-Induced Porosities in Laser-Based Powder Bed Fusion (L-PBF) of Ti6Al4V: High-Fidelity Modelling and Experimental Validation. *Addit. Manuf.* **2019**, *30*, 100835. [CrossRef]
11. Chen, L.; Yao, X.; Tan, C.; He, W.; Su, J.; Weng, F.; Chew, Y.; Ng, N.P.H.; Moon, S.K. In-Situ Crack and Keyhole Pore Detection in Laser Directed Energy Deposition through Acoustic Signal and Deep Learning. *Addit. Manuf.* **2023**, *69*, 103547. [CrossRef]
12. Ren, Z.; Gao, L.; Clark, S.J.; Fezzaa, K.; Shevchenko, P.; Choi, A.; Everhart, W.; Rollett, A.D.; Chen, L.; Sun, T. Machine Learning–Aided Real-Time Detection of Keyhole Pore Generation in Laser Powder Bed Fusion. *Science 1979* **2023**, *379*, 89–94. [CrossRef] [PubMed]
13. Pyeon, J.; Aroh, J.; Jiang, R.; Verma, A.K.; Gould, B.; Ramlatchan, A.; Fezzaa, K.; Parab, N.; Zhao, C.; Sun, T.; et al. Time-Resolved Geometric Feature Tracking Elucidates Laser-Induced Keyhole Dynamics. *Integr. Mater. Manuf. Innov.* **2021**, *10*, 677–688. [CrossRef]
14. Zhang, J.; Lyu, T.; Hua, Y.; Shen, Z.; Sun, Q.; Rong, Y.; Zou, Y. Image Segmentation for Defect Analysis in Laser Powder Bed Fusion: Deep Data Mining of X-Ray Photography from Recent Literature. *Integr. Mater. Manuf. Innov.* **2022**, *11*, 418–432. [CrossRef]
15. Qin, X.; Fan, D.-P.; Huang, C.; Diagne, C.; Zhang, Z.; Sant'Anna, A.C.; Suàrez, A.; Jägersand, M.; Shao, L. Boundary-Aware Segmentation Network for Mobile and Web Applications. *arXiv* **2021**, arXiv:2101.04704.
16. Csurka, G.; Larlus, D.; Perronnin, F. What Is a Good Evaluation Measure for Semantic Segmentation? In Proceedings of the British Machine Vision Conference, Bristol, UK, 9–13 September 2013; Burghardt, T., Damen, D., Mayol-Cuevas, W., Mirmehdi, M., Eds.; Visual Information Laboratory (VIL): Bristol, UK, 2013; pp. 32.1–32.11.

17. Chen, L.-C.; Zhu, Y.; Papandreou, G.; Schroff, F.; Adam, H. Encoder-Decoder with Atrous Separable Convolution for Semantic Image Segmentation. In Proceedings of the Computer Vision—ECCV 2018, Munich, Germany, 8–14 September 2018; Ferrari, V., Hebert, M., Sminchisescu, C., Weiss, Y., Eds.; Springer International Publishing: Cham, Switzerland, 2018; pp. 833–851.
18. Ronneberger, O.; Fischer, P.; Brox, T. U-Net: Convolutional Networks for Biomedical Image Segmentation. In Proceedings of the Medical Image Computing and Computer-Assisted Intervention—MICCAI 2015, Munich, Germany, 5–9 October 2015; Navab, N., Hornegger, J., Wells, W.M., Frangi, A.F., Eds.; Springer International Publishing: Cham, Switzerland, 2015; pp. 234–241.

materials

Article

Data-Driven Prediction and Uncertainty Quantification of Process Parameters for Directed Energy Deposition

Florian Hermann [1,2,*], Andreas Michalowski [1,3], Tim Brünnette [4], Peter Reimann [1], Sabrina Vogt [2] and Thomas Graf [1,3]

1   Graduate School of Excellence Advanced Manufacturing Engineering (GSaME), University of Stuttgart, Nobelstraße 12, 70569 Stuttgart, Germany; andreas.michalowski@ifsw.uni-stuttgart.de (A.M.); peter.reimann@gsame.uni-stuttgart.de (P.R.); thomas.graf@ifsw.uni-stuttgart.de (T.G.)
2   TRUMPF Laser- und Systemtechnik GmbH, Johann-Maus-Straße 2, 71254 Ditzingen, Germany
3   Institut für Strahlwerkzeuge (IFSW), University of Stuttgart, Pfaffenwaldring 43, 70569 Stuttgart, Germany
4   Institut für Wasser- und Umweltsystemmodellierung (IWS), University of Stuttgart, Pfaffenwaldring 5a, 70569 Stuttgart, Germany; tim.bruennette@iws.uni-stuttgart.de
*   Correspondence: florian.hermann@gsame.uni-stuttgart.de

Abstract: Laser-based directed energy deposition using metal powder (DED-LB/M) offers great potential for a flexible production mainly defined by software. To exploit this potential, knowledge of the process parameters required to achieve a specific track geometry is essential. Existing analytical, numerical, and machine-learning approaches, however, are not yet able to predict the process parameters in a satisfactory way. A trial-&-error approach is therefore usually applied to find the best process parameters. This paper presents a novel user-centric decision-making workflow, in which several combinations of process parameters that are most likely to yield the desired track geometry are proposed to the user. For this purpose, a Gaussian Process Regression (GPR) model, which has the advantage of including uncertainty quantification (UQ), was trained with experimental data to predict the geometry of single DED tracks based on the process parameters. The inherent UQ of the GPR together with the expert knowledge of the user can subsequently be leveraged for the inverse question of finding the best sets of process parameters by minimizing the expected squared deviation between target and actual track geometry. The GPR was trained and validated with a total of 379 cross sections of single tracks and the benefit of the workflow is demonstrated by two exemplary use cases.

Keywords: machine learning; Gaussian Process Regression; directed energy deposition; single track geometry; uncertainty quantification; user-centric decision making; expert knowledge

Citation: Hermann, F.; Michalowski, A.; Brünnette, T.; Reimann, P.; Vogt, S.; Graf, T. Data-Driven Prediction and Uncertainty Quantification of Process Parameters for Directed Energy Deposition. *Materials* **2023**, *16*, 7308. https://doi.org/10.3390/ma16237308

Academic Editors: Tuhin Mukherjee and Qianru Wu

Received: 24 October 2023
Revised: 20 November 2023
Accepted: 22 November 2023
Published: 24 November 2023

1. Introduction

Manufacturing companies face the challenge of ever shorter development and product life cycles and individualized products [1–3]. Software-defined manufacturing is an approach that enables flexible and reconfigurable systems and is therefore able to handle these challenges [4]. The successful implementation of software-defined manufacturing requires production systems that are as flexible and universal as possible [5] and that are sufficiently defined via software so that they can flexibly adapt to changing specifications [6,7]. Laser-based directed energy deposition with metal powder (DED-LB/M) offers such a flexible process, as it can be used for coating, welding, repairing and additive manufacturing without major change in hardware [8–10]. To weld single DED tracks, which are the basis of all mentioned applications, powder is transported to the process zone and the laser beam melts both powder and workpiece leading to a metallurgic bonding [11–13]. The geometry of the DED tracks and the corresponding height of the produced layers are influenced by the process parameters such as velocity v, laser power P, powder flow rate $\dot{m}$ and the diameter d_{L} of the laser beam on the surface of the workpiece. These parameters are all specified and adjusted via software. However, finding suitable process parameters to achieve the required track geometry for each application is currently a highly manual process relying on the individual process knowledge of the operator. Defining the geometry and process parameters in the software layer without performing prior experiments may yet offer a promising step towards software-defined manufacturing. Therefore, models that enable the prediction of the process parameters that yield the desired track geometry are essential for implementing a software-defined workflow.

Physics-based models provide valuable information about the formation of single tracks in DED. Ahsan and Pinkerton [14], for example, propose an analytical-numerical model to predict the geometry of single tracks, El Cheikh et al. [15] analytically describe the geometry of single DED tracks, Gao et al. [16] established a three-dimensional numerical model to predict the single track geometry and temperature distribution for single-tracks, Huang et al. [17] developed a physics-based process model for the prediction of the geometry of single tracks and multi-layer deposition and Zhang et al. [18] developed a three dimensional transient model for evolving temperature fields of thin walls. Despite their undeniable added value, none of the aforementioned models can represent the full complexity of the process. For instance, thermophysical properties are assumed to be constant in Ahsan and Pinkerton [14] and Huang et al. [17], heat convection is neglected in Huang et al. [17], the influence of molten pool fluid and the heat loss caused by vaporization of powder is ignored in Gao et al. [16], the heat that is incorporated into the melt-pool by the powder is neglected and assumptions about the catch efficiency of the powder are made in Zhang et al. [18], the absorption coefficient is determined experimentally in El Cheikh et al. [15] and some input values for the simulation such as the intensity profile of the laser in Gao et al. [16] are prone to some uncertainty. That is why physics-based models lose predictive accuracy in consideration of the process variability [19].

In recent years, data-driven models are becoming increasingly popular for performing such tasks as they are less computationally expensive and do not require that assumptions be made about the underlying physical process [20]. Hereby, deterministic models such as artificial neural networks [21–28] or Regression trees (RT) [24,29] are for example applied to predict the track geometry as a function of the process parameters in DED. However, deterministic models cannot provide uncertainty quantification (UQ), which is crucial for reliable additive manufacturing due to the various sources of uncertainties in additive manufacturing [30–33]. Probabilistic machine learning models such as Gaussian Process Regression (GPR) [34,35] can account for this UQ and have been applied in laser powder bed fusion (LPBF) to predict the melt pool geometry [36–41] or in DED to predict the mechanical properties [42], the component height [43], the geometry of single tracks [44,45], or melt pool geometry [46,47] based on the process parameters. The inverse problem, i.e., the determination of a suitable process to produce the desired track geometry, can principally be solved by combining the regression model with an optimization algorithm. In this context, GPR may be combined with a global optimization algorithm, for example, to minimize distortion in fused filament fabrication (FFF) [48], to optimize the microstructure in electron beam melting (EBM) [32], to reduce the surface roughness and the geometric deviation in LPBF, or to optimize the parameters with respect to the mechanical properties in DED [49]. Mondal et al. [50] trained a GPR model with simulation data for predicting the melt pool geometry as a function of laser power P and velocity v, and performed a Bayesian optimization to determine the optimal parameter combination to keep the geometry of the melt pool at a suitable value.

However, with an increasing number of considered process parameters, different sets of parameters may lead to the same processing results. Therefore, we include the prediction uncertainty of each combination of process parameters as well as expert knowledge when selecting the best parameters to manufacture a desired track geometry. Thus, this paper presents a novel workflow to select multiple parameter combinations that are most likely to yield the desired track geometry in DED. This is achieved by combining a GPR-model with an optimization algorithm that identifies multiple suitable sets of process parameters based both on the deviation from the targeted geometry of the single tracks as well as on the uncertainty of the prediction. The consideration of several possible parameter combinations allows the user to select the process parameters that best suit the application in question and make an informed decision on how to manufacture the component. Section 2 introduces the workflow on a general level, highlighting the interaction between the building blocks. Section 3.1 describes how the experimental data are obtained, while Section 3.2 introduces the GPR model. The data are subsequently used in Section 4 to validate the workflow, give exemplary applications, and discuss quantitative results apparent in the given DED process.

2. Prediction Workflow

As schematically shown in Figure 1, the workflow to determine suitable processing parameters consists of three main elements:

- Regression models
- Identification of optimal process parameters
- Application

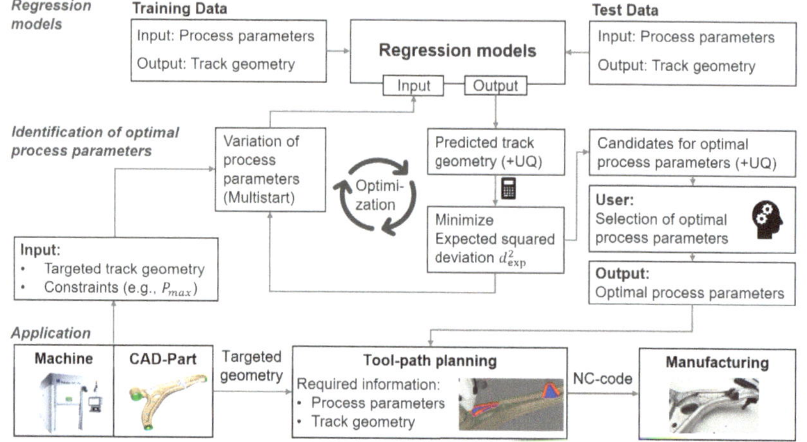

Figure 1. Workflow for finding optimal process parameters.

The manufacturing of the component takes place in the application layer that is displayed at the bottom of Figure 1. Based on the requirements from the application, the user defines the targeted geometry of a single track and the constraints on the process parameters. The targeted track geometry results from the geometry of the component (usually a CAD-part), and the constraints on the process parameters are mostly given by the limits of the given machine. In return, the application layer requires information about the optimal process parameters that lead to the targeted track geometry in order to be able to perform the toolpath planning. The toolpaths and process parameters are stored in a numerical control (NC) code that is readable by the machine and that enables manufacturing of the component. The probabilistic regression models that are displayed at the top of Figure 1 are essential to identify the required process parameters. These models predict the geometry of single tracks based on the process parameters and offer an uncertainty quantification (UQ) of the prediction. Parts of the available data, which are described in Section 3.1, were used to train the models and the rest of the data were used to test the performance of the models.

To answer the inverse question of finding optimal process parameters for a given targeted geometry, we implemented the optimization workflow that is displayed in the middle of Figure 1. Finding the optimal set of process parameters to achieve a specific track geometry may be a trade-off between accuracy and uncertainty. This choice depends on whether one prefers a precise prediction with tight tolerances but a high uncertainty or a less accurate prediction that is fulfilled with a higher probability. While both accuracy and uncertainty could be kept as separate optimization goals, we incorporate them into a single measure of optimality. The expected squared deviation $d^2_{\exp}$ between the predicted values $y(\mathbf{x})$ of a geometric property (such as height, width or melting depth) of the track obtained with a set $\mathbf{x}$ of processing parameters (e.g. laser power, beam diameter, velocity and powder flow rate) and the targeted value z of each geometric property can be expressed by

$$d^2_{\exp}(\mathbf{x}) = \mathbb{E}\left[||y(\mathbf{x}) - z||^2\right] = \mathrm{Var}[y(\mathbf{x})] + ||\bar{y}(\mathbf{x}) - z||^2. \tag{1}$$

Using the approach of GPR, the predicted values $y(\mathbf{x})$ are subject to a Gaussian probability distribution with the expected value $\bar{y}(\mathbf{x})$. The value of $d^2_{\exp}(\mathbf{x})$ is then minimized in order to identify the optimal process parameters. As we want to identify multiple sets of process parameters at different parameter ranges, we use Newton optimization in combination with a multistart strategy. The step size $\Delta\mathbf{x}$ of the Newton algorithm is limited to ensure that the algorithm converges to the closest local minimum and the multistart strategy augments the probability that all relevant local minima are identified during the optimization. The process parameters $\mathbf{x}$ are varied within the constraints on the search space as given by the user (depending on the used machine) with a step size that is varied based on the hyperparameters of the GPR models. The process parameters corresponding to the identified local minima are subsequently delivered to the application layer, which allows the parameters that best suit the application to be selected.

The principle of the identification of the most promising processing parameters and the involved quantities are illustrated by Figure 2.

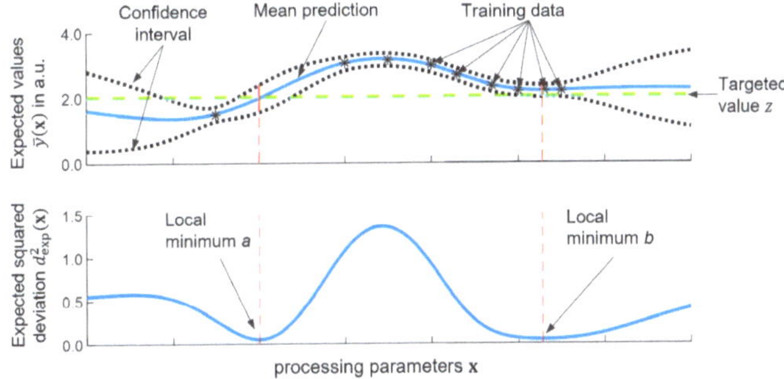

Figure 2. Fictional illustration of the approach. (**Top**): expected values $\bar{y}(\mathbf{x})$ (blue), variation of $y(\mathbf{x})$ as given by the 95% confidence interval (black dotted) and targeted value z (green dashed). (**Bottom**): expected squared deviation $d^2_{\exp}(\mathbf{x})$ in relation to the targeted value z.

The stars in the upper diagram represent the training data, which are obtained from experiments. The GPR then yields the expected values $\bar{y}(\mathbf{x})$ (blue, top graph) and the variation of $y(\mathbf{x})$, as represented by the 95% confidence interval (black dotted). The green dashed line represents the targeted value z. The red dashed lines mark the local minima of the expected squared deviation $d^2_{\exp}(\mathbf{x})$ defining the most promising sets of processing parameters $\mathbf{x}_{a,b}$. At the local minimum a, the expected value $\bar{y}(\mathbf{x}_a)$ equals the targeted value z, but the variance of $y(\mathbf{x}_a)$ is larger than at the local minimum b, where the expected value $\bar{y}(\mathbf{x}_b)$, however, does not correspond exactly to the desired value z. Hence, although the certainty of the predicted value $\bar{y}(\mathbf{x}_b)$ is higher at the local minimum b, the application of the corresponding processing parameters $\mathbf{x}_b$ is expected to result in a value $y(\mathbf{x}_b)$ slightly larger than the target z. Conversely, this means that the expected value $\bar{y}(\mathbf{x})$ at the local minimum a is more accurate but has a higher uncertainty compared to the one obtained at the minimum b. Both minima are, however, of similar quality with respect to the expected squared deviation $d^2_{\exp}(\mathbf{x})$. Depending on whether uncertainty or accuracy is more important and depending on which of the processing parameters better suit the application in question, the user may either select local minimum *a* or *b*. The choice of a local optimization with multistart instead of a global optimization such as Bayesian optimization or Upper Confidence Bound (UCB) enables the identification of multiple sets of suitable process parameters at different parameter ranges and allows the user to select the set of parameters that best suits the application in question.

3. Materials and Methods

To apply the described workflow, the regression models were trained with and tested on experimental results. Section 3.1 describes the experimental set-up and Section 3.2 describes the details of the regression models and how they were trained.

3.1. Experimental Data

To collect the necessary data for the training of the regression models for the different geometrical features of the DED welding tracks, a total of 379 single tracks were produced on a 10 mm thick plate of AlMg3 using different process parameters. Some of the tracks are exemplarily shown on the left in Figure 3. The DED-LB/M was performed on the five-axis laser machine TruLaser Cell 3000 using the 4 kW disk laser TruDisk4001 with a wavelength of 1030 nm and a beam parameter product of 4 mm × mrad, a laser light cable with a diameter of 100 μm and the optics focusLine Professional all from TRUMPF Laser- und Systemtechnik GmbH, Ditzingen, Germany. The AlSi10Mg powder from Carpenter Additive (CA), Widnes, UK, with particle diameters ranging from 45 to 107 μm, was fed using a vibratory feeder from Medicoat AG, Mägenwil, Switzerland and a helium gas flow with a flow rate of 10 L/ min. The multijet nozzle from TRUMPF Laser- und Systemtechnik GmbH, with seven jets arranged coaxially around the laser beam was used as a powder nozzle and Argon with a flow rate of 12 L/ min used to shield the process zone from the atmosphere. The powder was melted by a defocused laser beam with a focus diameter of 200 μm and a variable diameter d_L on the surface of the substrat. Cross-sections of the DED tracks were prepared by cutting, grinding, polishing and etching with a water solution containing 10% sodium hydroxide.

The depth d_w, the width w and the height h of the tracks were measured from the resulting cross-sections, as displayed on the right in Figure 3, by means of an optical microscope. The laser power P, the mass supply rate $\dot{m}$ of the powder, the diameter d_L of the laser beam on the surface, and the velocity v have a significant influence on the geometry of the resulting track and were therefore varied over a wide range and in variable steps: P between 1 and 4 kW in 16 steps, $\dot{m}$ between 0 and 42 g/min in 21 steps, d_L between 1 and 2 mm in 4 steps and v between 0.75 and 20 m/min in 14 steps. Each combination of parameters was repeated at least three times, resulting in a total of 379 single tracks.

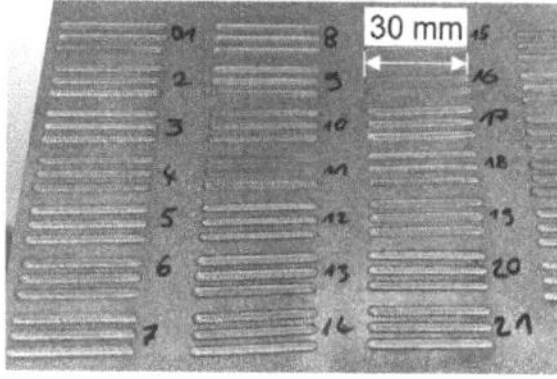

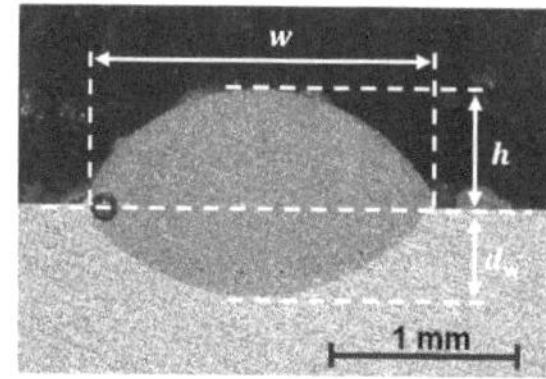

Figure 3. Picture of multiple separate DED tracks (**left**) and microscopical image of a cross section of a single DED track (**right**).

3.2. Training of Regression Models

Since the identification of optimal process parameters described in Section 2 is based on the probability of producing the targeted track geometry, a prediction model with built-in uncertainty quantification is required. The specific mathematical tool employed in this work is Gaussian Process Regression (GPR), which is capable of quantifying the uncertainty. We trained one regression model to predict the width w, one for predicting the height h and one for the prediction of the depth d_w of a single track. The models use the laser power P, the mass supply rate $\dot{m}$, the velocity v and the beam diameter d_L as the relevant processing parameters. A fraction of 80% of the acquired experimental data was randomly selected to train the models using k-fold cross-validation with five folds. The use of k-fold cross-validation enables a more robust model, which is less sensitive to the sampling of the data and the specification of the prior [51]. The remaining 20% of the data were used to test the performance of the data-driven models. All the repetitions of the experiments with the same set of processing parameters were kept in the same control group to make sure that the test data set only contained parameter combinations that the models were not trained on before. We integrated linear basis functions into our models because we assume that a linear trend of the mean-function can be continued to some extent when extrapolated at the boundaries of our experimental domain. For this, we used the same approach as described in ([34], Section 2.7) where the dependant variable $y(\mathbf{x})$ of the regression model, which is the predicted track geometry in our application, is modelled by

$$y(\mathbf{x}) = f(\mathbf{x}) + \mathbf{q}(\mathbf{x})^T \beta. \tag{2}$$

Here, $\mathbf{q}(\mathbf{x})$ contains the linear basis functions and β the corresponding coefficients, which are determined from the data. The function $f(\mathbf{x})$ denotes the prediction of a Gaussian process at the parameter vector $\mathbf{x}$ with a zero mean-function and a squared exponential kernel k, which is defined by

$$k(\mathbf{x}_i, \mathbf{x}_j) = \sigma_f^2 \exp\left[-\frac{1}{2} \sum_{m=1}^{d} \frac{(x_{j,m} - x_{i,m})^2}{(l_m)^2} \right] \tag{3}$$

for two input vectors $\mathbf{x}_i$ and $\mathbf{x}_j$ and their elements $x_{i,m}$ and $x_{j,m}$ in the dimension m. For our application the input vector $\mathbf{x}$ contains the processing parameters and has $d = 4$ dimensions (P, $\dot{m}$, d_L, v). The hyperparameters of the kernel k, i.e., the length scales l_m and the signal variance σ_f^2 are determined from the data. Homoscedastic noise is assumed. Therefore, the corresponding covariance matrix C is defined by

$$C_{ij} = k(\mathbf{x}_i, \mathbf{x}_j) + \sigma^2 \delta_{ij}, \tag{4}$$

where δ_{ij} is the Kronecker delta and σ^2 is the noise variance. The predicted mean $\bar{f}(\mathbf{x})$ and the variance $\mathrm{Var}[f(\mathbf{x})]$ of the prediction are calculated by

$$\bar{f}(\mathbf{x}) = \mathbf{k}^T \mathbf{C}^{-1} \mathbf{t} \tag{5}$$

$$\mathrm{Var}[f(\mathbf{x})] = c - \mathbf{k}^T \mathbf{C}^{-1} \mathbf{k}, \tag{6}$$

where the scalar $c = k(\mathbf{x}, \mathbf{x})$, the vector $\mathbf{k}$ has the elements $k(\mathbf{x}_n, \mathbf{x})$ for $n = 1, \ldots, N$ and the vector $\mathbf{t}$ contains the measured target values at the input points $\mathbf{x}_n$. N represents the number of data points used to train the models. The data are normalized in the input space and in the output space before training the models.

4. Results and Discussion

To show that the workflow described above can be successfully applied to predict the processing parameters required to produce the desired geometry of single tracks in DED-LB/M, we applied our workflow and regression models to the previously described test data. The predictive quality of the models as a function of the process parameters is discussed in Section 4.1. Section 4.2 is then devoted to the inverse problem of finding optimal process parameters. The workflow is found to yield plausible results and we show how the user can interact and profit in realistic scenarios.

4.1. Analysis of Regression Models

For each quantity of interest, i.e., track-width, track-height, and track-depth, a separate Gaussian process was trained with the input parameters laser power P, mass supply rate $\dot{m}$ of the powder, laser beam diameter d_L and velocity v. The resulting length scales l_m, which are determined via automated relevance determination (ARD), show the influence of the input parameters on the respective output quantity, whereby a small length scale indicates possible changes in output even for small changes in input. The coefficients β of the linear model in Equation (7) provide information about the offset and the linear trends regarding the influence of the input parameters on the respective output quantity. The mean-absolute-error (MAE) between the mean value $\bar{y}(\mathbf{x})$ of the prediction and the corresponding value $y_\mathrm{test}(\mathbf{x})$ of the test data set as well as the coefficient of determination R^2, which is calculated by

$$R^2 = 1 - \frac{\sum(y_\mathrm{test}(\mathbf{x}) - \bar{y}(\mathbf{x}))^2}{\sum(y_\mathrm{test}(\mathbf{x}) - \bar{y}_\mathrm{test})^2} \, , \tag{7}$$

were used as a measure to evaluate the accuracy. Hereby, $\bar{y}_\mathrm{test}$ is the average value of the test data set. The resulting numerical values are summarized in Table 1. The expected prediction errors that arise for track-width w, track-height h and track-depth d_w are tolerable for most applications. The two values R^2 and MAE only consider the mean prediction of the models and therefore enable comparison to other deterministic models. However, in the following, we present capabilities that are exclusive to the probabilistic paradigm.

Table 1. Optimal hyperparameters and accuracy of the GPR models.

	σ	l_m **per Predictor** $(P/\dot{m}/d_\mathrm{L}/v)$	β**: Coefficients of Linear** **Basis** $(1/P/\dot{m}/d_\mathrm{L}/v)$	R^2	**MAE**
w	0.08	1.05/0.03/1.05/0.15	[2.06/0.31/0.08/0.09/−0.65]	0.89	0.11 mm
h	0.04	1.62/0.25/1.95/0.05	[0.53/0.03/0.45/−0.01/−0.48]	0.88	0.04 mm
d_w	0.02	1.17/0.04/1.90/0.17	[0.52/0.13/−0.05/0.04/−0.13]	0.91	0.04 mm

For the discussion of the influence of the process parameters on the processing result, Figure 4 exemplarily shows the predicted dependence of the width (in red), the height (in blue), and the depth (in green) of the tracks on the laser power. All other parameters are kept constant: $v = 2$ m/min, $d_\mathrm{L} = 2$ mm and $\dot{m} = 2$ g/min. The pale-colored areas represent the 95 % confidence intervals around the predicted values (solid lines).

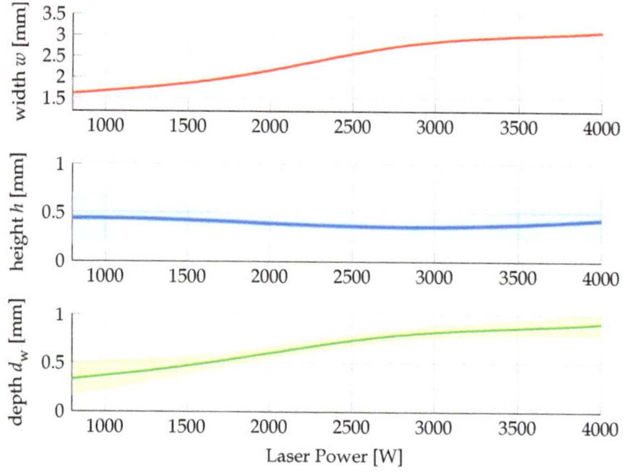

Figure 4. Dependence of the predicted track geometry on the applied laser power for $v = 2$ m/min, $d_\mathrm{L} = 2$ mm and $\dot{m} = 2$ g/min.

The track width and track depth both increase steadily with increasing laser power. This can be explained by an increase of the volume of the melt pool. The height of the track does not change significantly. The fact that the product of the track width and the height increases despite the constant powder flow rate indicates that increasing the laser power yields an increased powder efficiency. The laser power that was applied in the training data mostly ranged between 2000 W and 3400 W. This is why the uncertainties of the predictions are lower in this range of laser power and increase significantly when the predictions are made for laser powers above or below this range.

4.2. Identification of Optimal Process Parameters

4.2.1. Multiple Local Minima

The proposed optimization approach is able to find multiple sets of optimal process parameters corresponding to the local minima of the expected squared deviation. This allows the user to select the parameter combination that best fits a given application.

To illustrate how our optimization routine identifies these local minima, we exemplarily defined a targeted geometry with $w = 2.5\,\text{mm}$, $h = 0.45\,\text{mm}$ and $d_\text{w} = 0.6\,\text{mm}$. In the optimization procedure the process parameters were varied with equidistant steps that correspond to one non-standardized length scale of the GPR process, i.e., $\Delta P = 936\,\text{W}$, $\Delta d_\text{L} = 0.33\,\text{mm}$ and $\Delta v = 0.35\,\text{m/min}$. The mass supply rate $\dot{m}$ of the powder was adapted in a way that the mass per distance remains constant at $2.1\,\text{g/m}$. These initial parameters yield the different local optima listed in Table 2. They are sorted from the lowest to highest expected squared deviation and are also provided to the user in this way.

Table 2. Identified process parameters when searching for local minima of the expected squared deviation.

No.	P [W]	$\dot{m}$ [g/min]	d_L [mm]	v [m/min]	d_exp^2 [mm^2]
a	2836	3.0	1.3	1.5	0.014
b	2815	3.3	1.8	2.0	0.027
c	2669	1.6	1.6	1.0	0.060
d	3399	3.1	1.0	2.3	0.064
e	2173	2.1	2.3	1.0	0.071
f	2173	2.0	1.5	1.0	0.076
g	3054	7.3	2.1	4.0	0.082
h	2965	7.3	2.3	3.9	0.083

The corresponding predicted track geometry is shown in Figure 5 together with the 95% confidence interval. The black dashed lines represent the targeted values. It is evident that there is more than one set of process parameters leading to the targeted geometry and that the combination of the multistart with our optimization algorithm is able to identify these suitable sets of process parameters. A comparison with a gridsearch optimization revealed that our optimization identifies all interesting local minima.

By proposing several sets of suitable processing parameters, we provide the user with sufficient information to asses which process parameters best suit a given application. For our exemplary target geometry, local optimum a, cf. Figure 5, exhibits the smallest expected squared deviation and the expected depth and height are significantly closer to the targeted value as compared to local optimum b. When the height and the depth of a track are critical for the application, the user will most likely opt for the parameter set a. If multiple local minima are of similar quality the user may also consider further criteria for the selection of process parameters based on his or her expert knowledge: higher velocities may be preferred for economic reasons or lower laser power may be preferred when dealing with heat sensitive parts.

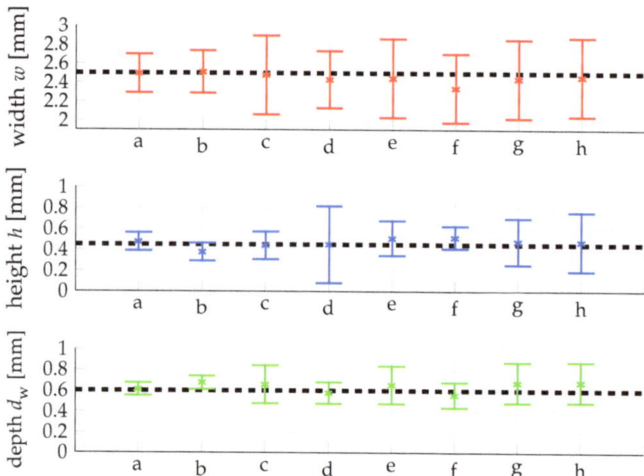

Figure 5. Predicted track geometry for the different local minima listed in Table 2.

4.2.2. Optimal Process Parameters at Different Velocities

In many industrial applications, certain cycle times and thus velocities have to be achieved in order to be economical [52]. In the following, we therefore demonstrate how our workflow can be used to find the optimal processing parameters at a given velocity using an exemplary targeted track geometry of $w = 2$ mm, $h = 0.45$ mm and $d_{\mathrm{w}} = 0.5$ mm and perform the optimization for ten different fixed velocities from 2 to 20 m/min. The parameters that are most likely to yield the targeted geometry at the given velocity are listed in Table 3.

Table 3. Optimal process parameters with lowest expected squared deviation at different velocities.

v [m/min]	P [W]	$\dot{m}$ [g/min]	d_{L} [mm]	$d^2_{\exp}$ [mm²]
2	1856	3.2	2.0	0.018
4	2188	8.4	2.0	0.014
6	2684	12.6	2.0	0.028
8	3329	16.8	1.1	0.020
10	3867	21.0	2.0	0.025
12	3647	25.1	2.8	0.087
14	4000	29.2	2.7	0.095
16	4000	33.6	3.0	0.130
18	4000	37.7	3.0	0.251
20	4000	42	3.0	0.457

The mean of the predicted geometries (coloured circles connected by dashed lines), including their 95% confidence interval (pale-couloured areas) for these parameters, are displayed in Figure 6. The targeted geometry is indicated by the black dashed lines. The predictions match the target reasonably well with an expected squared deviation of less than $0.03\,\mathrm{mm}^2$ up to a velocitiy of 10 m/min. For higher velocities, the prediction quality deteriorates both in closeness to the targeted value and certainty for all three geometrical features, cf. Figure 6. The increased uncertainty observed for all three geometric characteristics at feedrates above 10 m/min is due to the low number of training data in the range of these identified parameter combinations. The mean of the predicted track width deviates downwards from the targeted value for feedrates above 12 m/min and the mean of the predicted track depth for velocities above 16 m/min. These observations are consistent with those of [53], where it was observed that a maximum laser power of 4 kW is insufficient to weld tracks with a width of 2 mm up to velocities of 20 m/min and that both the maximum track width and track depth decrease for velocities above 10 m/min due to the limited laser power. The information about the processing parameters, the uncertainty of the prediction, and the deviation from the targeted geometry enables a user to make an informed decision regarding the economic stipulations and quality requirements.

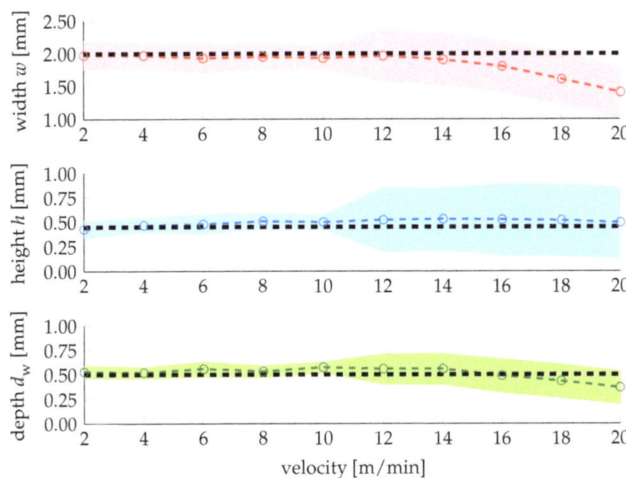

Figure 6. Predicted track geometry for the optimal process parameters listed in Table 3.

5. Conclusions

In conclusion, it was shown that the combination of an optimization workflow and expert knowledge with probabilistic regression models such as GPR enables us to predict the process parameters needed to achieve a specific track geometry for laser-based directed energy deposition using metal powder (DED-LB/M). The validation with a large number of individual tracks revealed a good agreement between the test data and the predictions of the regression models. The usefulness and applicability of the proposed workflow for a user to make an informed decision on optimal process parameters as well as for receiving optimal process parameters at different velocities has been demonstrated with two exemplary targeted geometries of a single track. The proposed workflow thus provides a promising step towards software-defined manufacturing.

Even though the potential of that workflow has been shown, further investigation may be undertaken. The models are so far only applied to isolated welding tracks on a plane sample. A generalization of the models towards more complex geometric scenarios may enable a much wider range of uses. This incorporates many new challenges, such as the detection and the addition of other relevant parameters and effects such as heat accumulation. The latter challenge may be tackled by the integration of uncertainty-aware temperature predictions, as proposed by Sideris et al. [20]. Furthermore, to get closer to an efficient industrial application, the generation of the data for this data-based model can be automated to make it less time-consuming. The combination of automated and optimized selection of parameters with automated data acquisition allows for quick training of the model and a rapid adaption of the workflow to new circumstances such as the use of different alloys. Bayesian optimization or upper-confidence bound (UCB), both of which merge the mean and variance of a prediction, may be used in order to design a data-efficient adaptive experimental design. One possibility to adapt the model to new environments is to train only the deviation from the old model instead of training the model from scratch.

Author Contributions: Conceptualization, F.H. and T.B.; methodology, F.H., T.B. and A.M.; software, F.H.; validation, F.H., T.B. and A.M.; formal analysis, F.H., T.B., A.M. and P.R.; investigation, F.H.; resources, S.V. and F.H.; data curation, F.H.; writing—original draft preparation, F.H. and T.B.; writing—review and editing, P.R., A.M., S.V. and T.G.; visualization, F.H.; supervision, S.V. and T.G.; project administration, S.V. and T.G.; funding acquisition, S.V. and T.G. All authors have read and agreed to the published version of the manuscript.

Funding: This work was supported by the Landesministerium für Wissenschaft, Forschung und Kunst Baden-Württemberg (Ministry of Science, Research and the Arts of the State of Baden-Wurttemberg) within the Nachhaltigkeitsförderung (sustainability support) of the projects of the Exzellenzinitiative II and the German Federal Ministry of Education and Research (BMBF) within the framework concept "Forschungscampus" ARENA2036 (funding number: 02P18Q643). We also thank the Deutsche Forschungsgemeinschaft (DFG, German Research Foundation) for supporting this work by funding SFB 1313, Project Number 327154368.

Institutional Review Board Statement: Not applicable.

Informed Consent Statement: Not applicable.

Data Availability Statement: The data presented in this study are available on request from the corresponding author.

Conflicts of Interest: Author Sabrina Vogt was employed by the company TRUMPF Laser- und Systemtechnik GmbH. The remaining authors declare that the research was conducted in the absence of any commercial or financial relationships that could be construed as a potential conflict of interest. The funders had no role in the design of the study; in the collection, analyses, or interpretation of data; in the writing of the manuscript; or in the decision to publish the results.

Abbreviations

The following abbreviations are used in this manuscript:

GPR	Gaussian Process Regression
RT	Regression tree
UQ	Uncertainty quantification
DED	Directed energy deposition
DED-LB/M	Laser-based directed energy deposition using metal powder
LPBF	Laser powder bed fusion
FFF	Fused filament fabrication
EBM	Electron beam melting
CAD	Computer aided design
ARD	Automatic relevance determination
MAE	Mean absolute error
v	Velocity
P	Laser power
$\dot{m}$	Powder flow rate
d_{L}	Laser beam diameter on the surface of the substrate
$d_{\exp}^2$	Expected squared deviation between prediction and target value
z	Target value of geometry characteristic
$y(\mathbf{x})$	Prediction performed with the GPR model
d_{w}	Depth of of a single DED track
w	Depth of a single DED track
h	Height of a single DED track
R^2	Coefficient of determination
$\mathbf{q}(\mathbf{x})$	Vector with linear basis functions
β	Coefficients of linear basis
k	Kernel of GPR model
l_m	Length scale
σ_f^2	Signal variance
δ_{ij}	Kronecker delta
y_{test}	Measured values in the test data set
$\bar{y}_{\mathrm{test}}$	Mean value of y_{test}

References

1. Schuh, G.; Rudolf, S.; Riesener, M. Design for industrie 4.0. In Proceedings of the 14th International Design Conference, Cavtat, Dubrovnik, 16–19 May 2016.
2. Maalouf, E.; Daaboul, J.; Le Duigou, J.; Hussein, B. Production management for mass customization and smart cellular manufacturing system: NSGAII and SMPSO for factory-level planning. *Int. J. Adv. Manuf. Technol.* **2022**, *120*, 6833–6854. [CrossRef]
3. Mourtzis, D.; Doukas, M.; Vandera, C. Smart mobile apps for supporting product design and decision-making in the era of mass customisation. *Int. J. Comput. Integr. Manuf.* **2017**, *30*, 690–707. [CrossRef]
4. Thames, L.; Schaefer, D. Software-defined Cloud Manufacturing for Industry 4.0. *Procedia CIRP* **2016**, *52*, 12–17. [CrossRef]
5. Xu, L.; Chen, L.; Gao, Z.; Moya, H.; Shi, W. Reshaping the Landscape of the Future: Software-Defined Manufacturing. *Computer* **2021**, *54*, 27–36. [CrossRef]
6. Lechler, A.; Riedel, O.; Coupek, D. Virtual representation of physical objects for software defined manufacturing. In Proceedings of the 24th International Conference on Production Research (ICPR 2017), Posnan, Poland, 30 July–3August 2017. [CrossRef]
7. Barwasser, A.; Lentes, J.; Riedel, O.; Zimmermann, N.; Dangelmaier, M.; Zhang, J. Method for the development of Software-Defined Manufacturing equipment. *Int. J. Prod. Res.* **2023**, *61*, 6467–6484. [CrossRef]
8. Poprawe, R. *Lasertechnik für die Fertigung*; VDI-Buch, Springer: Berlin/Heidelberg, Germany, 2005.
9. Cavaliere, P. *Laser Cladding of Metals*; Springer International Publishing: Cham, Switzerland, 2021. [CrossRef]
10. Mahamood, R.M. *Laser Metal Deposition Process of Metals, Alloys, and Composite Materials*; Springer International Publishing: Cham, Switzerland, 2018. [CrossRef]
11. Toyserkani, E.; Khajepour, A.; Corbin, S. *Laser Cladding*; CRC Press: Boca Raton, FL, USA, 2005.
12. Moeller, M. *Prozessmanagement fuer das Laser-Pulver-Auftragschweissen*; Springer: Berlin/Heidelberg, Germany, 2021. [CrossRef]
13. Huegel, H.; Graf, T. (Eds.) Additive Verfahren. In *Materialbearbeitung mit Laser*; Springer Fachmedien Wiesbaden: Wiesbaden, Germany, 2022; pp. 415–454.
14. Ahsan, M.N.; Pinkerton, A.J. An analytical–numerical model of laser direct metal deposition track and microstructure formation. *Model. Simul. Mater. Sci. Eng.* **2011**, *19*, 055003. [CrossRef]
15. El Cheikh, H.; Courant, B.; Hascoët, J.Y.; Guillén, R. Prediction and analytical description of the single laser track geometry in direct laser fabrication from process parameters and energy balance reasoning. *J. Mater. Process. Technol.* **2012**, *212*, 1832–1839. [CrossRef]
16. Gao, J.; Wu, C.; Hao, Y.; Xu, X.; Guo, L. Numerical simulation and experimental investigation on three-dimensional modelling of single-track geometry and temperature evolution by laser cladding. *Opt. Laser Technol.* **2020**, *129*, 106287. [CrossRef]
17. Huang, Y.; Khamesee, M.B.; Toyserkani, E. A new physics-based model for laser directed energy deposition (powder-fed additive manufacturing): From single-track to multi-track and multi-layer. *Opt. Laser Technol.* **2019**, *109*, 584–599. [CrossRef]
18. Zhang, D.; Feng, Z.; Wang, C.; Liu, Z.; Dong, D.; Zhou, Y.; Wu, R. Modeling of Temperature Field Evolution During Multilayered Direct Laser Metal Deposition. *J. Therm. Spray Technol.* **2017**, *26*, 831–845. [CrossRef]
19. Chadha, U.; Selvaraj, S.K.; Lamsal, A.S.; Maddini, Y.; Ravinuthala, A.K.; Choudhary, B.; Mishra, A.; Padala, D.; M, S.; Lahoti, V.; et al. Directed Energy Deposition via Artificial Intelligence-Enabled Approaches. *Complexity* **2022**, *2022*, 2767371. [CrossRef]
20. Sideris, I.; Crivelli, F.; Bambach, M. GPyro: Uncertainty-aware temperature predictions for additive manufacturing. *J. Intell. Manuf.* **2023**, *34*, 243–259. [CrossRef]
21. Caiazzo, F.; Caggiano, A. Laser Direct Metal Deposition of 2024 Al Alloy: Trace Geometry Prediction via Machine Learning. *Materials* **2018**, *11*, 444. [CrossRef] [PubMed]
22. Pant, P.; Chatterjee, D. Prediction of clad characteristics using ANN and combined PSO-ANN algorithms in laser metal deposition process. *Surfaces Interfaces* **2020**, *21*, 100699. [CrossRef]
23. Feenstra, D.R.; Molotnikov, A.; Birbilis, N. Utilisation of artificial neural networks to rationalise processing windows in directed energy deposition applications. *Mater. Des.* **2021**, *198*, 109342. [CrossRef]
24. Gao, J.; Wang, C.; Hao, Y.; Liang, X.; Zhao, K. Prediction of TC11 single-track geometry in laser metal deposition based on back propagation neural network and random forest. *J. Mech. Sci. Technol.* **2022**, *36*, 1417–1425. [CrossRef]
25. Bhardwaj, T.; Shukla, M. Laser Additive Manufacturing- Direct Energy Deposition of Ti-15Mo Biomedical Alloy: Artificial Neural Network Based Modeling of Track Dilution. *Lasers Manuf. Mater. Process.* **2020**, *7*, 245–258. [CrossRef]
26. Liu, H.; Qin, X.; Huang, S.; Jin, L.; Wang, Y.; Lei, K. Geometry Characteristics Prediction of Single Track Cladding Deposited by High Power Diode Laser Based on Genetic Algorithm and Neural Network. *Int. J. Precis. Eng. Manuf.* **2018**, *19*, 1061–1070. [CrossRef]
27. Saqib, S.; Urbanic, R.J.; Aggarwal, K. Analysis of Laser Cladding Bead Morphology for Developing Additive Manufacturing Travel Paths. *Procedia CIRP* **2014**, *17*, 824–829. [CrossRef]
28. Narayana, P.L.; Kim, J.H.; Lee, J.; Choi, S.W.; Lee, S.; Park, C.H.; Yeom, J.T.; Reddy, N.G.S.; Hong, J.K. Optimization of process parameters for direct energy deposited Ti-6Al-4V alloy using neural networks. *Int. J. Adv. Manuf. Technol.* **2021**, *114*, 3269–3283. [CrossRef]
29. Lee, S.; Peng, J.; Shin, D.; Choi, Y.S. Data analytics approach for melt-pool geometries in metal additive manufacturing. *Sci. Technol. Adv. Mater.* **2019**, *20*, 972–978. [CrossRef] [PubMed]
30. Pham, T.Q.D.; Hoang, T.V.; van Tran, X.; Fetni, S.; Duchêne, L.; Tran, H.S.; Habraken, A.M. Uncertainty Quantification in the Directed Energy Deposition Process Using Deep Learning-Based Probabilistic Approach. *Key Eng. Mater.* **2022**, *926*, 323–330. [CrossRef]

31. Hu, Z.; Mahadevan, S. Uncertainty quantification and management in additive manufacturing: Current status, needs, and opportunities. *Int. J. Adv. Manuf. Technol.* **2017**, *93*, 2855–2874. [CrossRef]
32. Wang, Z.; Liu, P.; Ji, Y.; Mahadevan, S.; Horstemeyer, M.F.; Hu, Z.; Chen, L.; Chen, L.Q. Uncertainty Quantification in Metallic Additive Manufacturing through Physics-Informed Data-Driven Modeling. *JOM* **2019**, *71*, 2625–2634. [CrossRef]
33. Gholaminezhad, I.; Assimi, H.; Jamali, A.; Vajari, D.A. Uncertainty quantification and robust modeling of selective laser melting process using stochastic multi-objective approach. *Int. J. Adv. Manuf. Technol.* **2016**, *86*, 1425–1441. [CrossRef]
34. Rasmussen, C.E.; Williams, C.K.I. *Gaussian Processes for Machine Learning*; Adaptive computation and machine learning; MIT Press: Cambridge, MA, USA, 2006.
35. Bishop, C.M. *Pattern Recognition and Machine Learning*; Information science and statistics; Springer: New York, NY, USA, 2006.
36. Meng, L.; Zhang, J. Process Design of Laser Powder Bed Fusion of Stainless Steel Using a Gaussian Process-Based Machine Learning Model. *JOM* **2020**, *72*, 420–428. [CrossRef]
37. Saunders, R.; Rawlings, A.; Birnbaum, A.; Iliopoulos, A.; Michopoulos, J.; Lagoudas, D.; Elwany, A. Additive Manufacturing Melt Pool Prediction and Classification via Multifidelity Gaussian Process Surrogates. *Integr. Mater. Manuf. Innov.* **2022**, *11*, 497–515. [CrossRef]
38. Tapia, G.; Khairallah, S.; Matthews, M.; King, W.E.; Elwany, A. Gaussian process-based surrogate modeling framework for process planning in laser powder-bed fusion additive manufacturing of 316L stainless steel. *Int. J. Adv. Manuf. Technol.* **2018**, *94*, 3591–3603. [CrossRef]
39. Olleak, A.; Xi, Z. Calibration and Validation Framework for Selective Laser Melting Process Based on Multi-Fidelity Models and Limited Experiment Data. *J. Mech. Des.* **2020**, *142*, 081701. [CrossRef]
40. Moges, T.; Yang, Z.; Jones, K.; Feng, S.; Witherell, P.; Lu, Y. Hybrid Modeling Approach for Melt-Pool Prediction in Laser Powder Bed Fusion Additive Manufacturing. *J. Comput. Inf. Sci. Eng.* **2021**, *21*, 050902. [CrossRef]
41. Ren, Y.; Wang, Q.; Michaleris, P. A Physics-Informed Two-Level Machine-Learning Model for Predicting Melt-Pool Size in Laser Powder Bed Fusion. *J. Dyn. Syst. Meas. Control* **2021**, *143*, 121006. [CrossRef]
42. Yan, F.; Chan, Y.C.; Saboo, A.; Shah, J.; Olson, G.B.; Chen, W. Data-Driven Prediction of Mechanical Properties in Support of Rapid Certification of Additively Manufactured Alloys. *Comput. Model. Eng. Sci.* **2018**, *117*, 343–366. [CrossRef]
43. Lee, J.A.; Sagong, M.J.; Jung, J.; Kim, E.S.; Kim, H.S. Explainable machine learning for understanding and predicting geometry and defect types in Fe-Ni alloys fabricated by laser metal deposition additive manufacturing. *J. Mater. Res. Technol.* **2023**, *22*, 413–423. [CrossRef]
44. Wang, S.; Zhu, L.; Fuh, J.Y.H.; Zhang, H.; Yan, W. Multi-physics modeling and Gaussian process regression analysis of cladding track geometry for direct energy deposition. *Opt. Lasers Eng.* **2020**, *127*, 105950. [CrossRef]
45. Hermann, F.; Chen, B.; Ghasemi, G.; Stegmaier, V.; Ackermann, T.; Reimann, P.; Vogt, S.; Graf, T.; Weyrich, M. A Digital Twin Approach for the Prediction of the Geometry of Single Tracks Produced by Laser Metal Deposition. *Procedia CIRP* **2022**, *107*, 83–88. [CrossRef]
46. Menon, N.; Mondal, S.; Basak, A. Multi-Fidelity Surrogate-Based Process Mapping with Uncertainty Quantification in Laser Directed Energy Deposition. *Materials* **2022**, *15*, 2902. [CrossRef] [PubMed]
47. Menon, N.; Mondal, S.; Basak, A. Linking processing parameters with melt pool properties of multiple nickel-based superalloys via high-dimensional Gaussian process regression. *J. Mater. Inform.* **2023**, *3*, 7. [CrossRef]
48. Nath, P.; Olson, J.D.; Mahadevan, S.; Lee, Y.T.T. Optimization of fused filament fabrication process parameters under uncertainty to maximize part geometry accuracy. *Addit. Manuf.* **2020**, *35*, 101331. [CrossRef] [PubMed]
49. Zhang, Y.; Karnati, S.; Nag, S.; Johnson, N.; Khan, G.; Ribic, B. Accelerating Additive Design with Probabilistic Machine Learning. *ASME J. Risk Uncertain. Eng. Syst. Part B Mech. Eng.* **2022**, *8*, 011109. [CrossRef]
50. Mondal, S.; Gwynn, D.; Ray, A.; Basak, A. Investigation of Melt Pool Geometry Control in Additive Manufacturing Using Hybrid Modeling. *Metals* **2020**, *10*, 683. [CrossRef]
51. Cawley, G.C.; Talbot, N.L. On over-fitting in model selection and subsequent selection bias in performance evaluation. *J. Mach. Learn. Res.* **2010**, *11*, 2079–2107.
52. Hassen, A.A.; Noakes, M.; Nandwana, P.; Kim, S.; Kunc, V.; Vaidya, U.; Love, L.; Nycz, A. Scaling Up metal additive manufacturing process to fabricate molds for composite manufacturing. *Addit. Manuf.* **2020**, *32*, 101093. [CrossRef]
53. Hermann, F.; Vogt, S.; Göbel, M.; Möller, M.; Frey, K. Laser Metal Deposition of AlSi10Mg with high build rates. *Procedia CIRP* **2022**, *111*, 210–213. [CrossRef]

Article

Multiscale Simulation of Laser-Based Direct Energy Deposition (DED-LB/M) Using Powder Feedstock for Surface Repair of Aluminum Alloy

Xiaosong Zhou [1], Zhenchao Pei [1], Zhongkui Liu [2,3], Lihang Yang [2,3], Yubo Yin [2,3], Yinfeng He [2,4], Quan Wu [1] and Yi Nie [2,3,*]

1   School of Mechanical and Electrical Engineering, Guizhou Normal University, Guiyang 550001, China; ivzhouxiaosong@hotmail.com (X.Z.); pzc330230313@163.com (Z.P.); wu_quan@gznu.edu.cn (Q.W.)
2   Nottingham Ningbo China Beacons of Excellence Research and Innovation Institute, University of Nottingham Ningbo China, Ningbo 315100, China; zhongkui.liu@nottingham.edu.cn (Z.L.); lihang.yang2@nottingham.edu.cn (L.Y.); yinfeng.he@nottingham.edu.cn (Y.H.)
3   Faculty of Science and Engineering, University of Nottingham Ningbo China, Ningbo 315100, China
4   Faculty of Engineering, University of Nottingham, Nottingham NG7 2RD, UK
*   Correspondence: yi-nie@nottingham.edu.cn; Tel.: +86-13566358684

Abstract: Laser-based direct energy deposition (DED-LB/M) has been a promising option for the surface repair of structural aluminum alloys due to the advantages it offers, including a small heat-affected zone, high forming accuracy, and adjustable deposition materials. However, the unequal powder particle size during powder-based DED-LB/M can cause unstable flow and an uneven material flow rate per unit of time, resulting in defects such as pores, uneven deposition layers, and cracks. This paper presents a multiscale, multiphysics numerical model to investigate the underlying mechanism during the powder-based DED-LB/M surface repair process. First, the worn surfaces of aluminum alloy components with different flaw shapes and sizes were characterized and modeled. The fluid flow of the molten pool during material deposition on the worn surfaces was then investigated using a model that coupled the mesoscale discrete element method (DEM) and the finite volume method (FVM). The effect of flaw size and powder supply quantity on the evolution of the molten pool temperature, morphology, and dynamics was evaluated. The rapid heat transfer and variation in thermal stress during the multilayer DED-LB/M process were further illustrated using a macroscale thermomechanical model. The maximum stress was observed and compared with the yield stress of the adopted material, and no relative sliding was observed between deposited layers and substrate components.

Keywords: laser direct energy deposition; surface repair; aluminum alloy; multiscale simulation; molten pool; thermal stress

Citation: Zhou, X.; Pei, Z.; Liu, Z.; Yang, L.; Yin, Y.; He, Y.; Wu, Q.; Nie, Y. Multiscale Simulation of Laser-Based Direct Energy Deposition (DED-LB/M) Using Powder Feedstock for Surface Repair of Aluminum Alloy. *Materials* **2024**, *17*, 3559. https://doi.org/10.3390/ma17143559

Academic Editor: Joan-Josep Suñol

Received: 8 June 2024
Revised: 6 July 2024
Accepted: 12 July 2024
Published: 18 July 2024

1. Introduction

Aluminum alloys are a crucial group of materials in the aerospace and automotive fields owing to their outstanding specific stiffness and strength [1,2]. However, flaws such as wear, cracks, and holes are unavoidable after a continuous external load, which undermines their reliability in servicing conditions and can sometimes result in fatal fractures. Compared with the cost of direct disposal or replacement of whole parts, surface repairs of the compromised parts has always been an attractive alternative choice. Traditional surface repair methods, including casting and forging, rely on producing an entire replacement body at local places, which is time-consuming and expensive [3]. The alternative solution, e.g., metal patching on the damaged surface, can only be a temporary backup in an emergency with the sacrifice of property consistency and the introduction of extra weight [4].

Laser-based direct energy deposition (DED-LB/M), often also referred to as laser metal deposition (LMD), laser cladding (LC), or laser welding (LW), utilizes a laser source to heat the powder or wire feedstock material to form a molten pool deposited and cooled down on the substrate, and is a promising choice for the surface repair of compromised parts [3]. However, defects such as pores, unevenly deposited layers, and cracks are always present during the powder-based DED-LB/M process due to the rapid heating and cooling and the non-uniform powder particle stream. The complex physical phenomenon during the powder-based DED-LB/M process can be observed by an in situ high-speed camera [5] or via X-ray imaging [6] for molten dimensions. The heat or temperature field has also been directly [7] or indirectly [8] captured by a suitable monitoring setup. Nevertheless, such expensive testing facilities only provide researchers with data regarding the ongoing phenomenon and with limited information on the underlying physics. Computational simulation provides an insightful method to reveal the complicated physics inside the molten pool and resolve the mechanism that produces defects during the powder-based DED-LB/M process [9].

To investigate the fluid flow inside the molten pool during the powder-based DED-LB/M process, the challenge is to suitably capture the interaction between the discrete-based non-uniform powder feeding stream and the continuum-based molten pool. Wang et al. [10,11] and Bayat et al. [12] applied the discrete phase method (DPM) to simplify the supplied metal powder as Lagrangian particles and restored the physical mass, thermal energy, and kinetic energy exchange during the powder-based DED-LB/M process by using FLOW-3D software (https://www.flow3d.com/). The physical shape of powder particles and the interaction between laser and metal powder are neglected by setting the temperature of the deposited Lagrangian particles before entering the molten pool as their liquidus temperature. Sun et al. [13] coupled DPM with the volume of fluid (VOF) to capture the interactions between the powder stream and molten pool and adopted the enthalpy–porosity method to account for phase change. By assigning the powder particles to their liquidus temperature, as granted by the nature of the coupled method, the powder particles can transform from a discrete phase to a continuous phase when they are injected at the local molten pool surface. The fluid in the regions where the powder was injected was captured by a downward flow and a reduction in temperature. Good agreements between the simulation model and experiments were obtained.

Considering the dynamic behaviors, physical size, and shape of the delivered powder, it is appropriate and more convincing to employ the discrete element method (DEM) to model the powder-based DED-LB/M process [14]. Aggarwal et al. [15] developed the coupled DEM and finite volume method (FVM) to investigate the interactions between the molten pool and the impacting powder particles. It was observed that the momentum introduced by the impacting powder particles outweighs the Marangoni effect, thereby stimulating the melting of the metal in the molten pool, which is in contrast to past findings. The obtained morphology of the molten pool and temperature fields were verified by the conducted experimental measurements. Khairallah et al. [16] employed the ALE3d multiphysics code to develop a high-fidelity mesoscale numerical model for the powder-based DED-LB/M process. It was predicted that the laser absorptivity would stay around 0.4 regardless of variations in process parameters. Although the high-fidelity nature of mesoscale molten pool simulation contributes to its accurate spatial and temporal results, the huge computational burden limits the situations in which it can be applied.

In addition to the above models of fluid flow, the rapid heat transfer and variation in thermal stress accompanied by the molten pool have also been studied. Srivastava et al. [17] developed a thermomechanical model for the arc-based DED process to quantify the residual stresses and deformations of the produced components. Stender et al. [18] employed a heat source to heat the activated elements of composite materials at each time step. The temperature and topology of the materials are then transferred to a solid mechanics analysis, allowing for the computation of displacement and stress fields. Li et al. [19,20] focused on calculating the thermal history at the macroscopic scale and directly used it for thermomechanical resolution without considering local thermal histories. Bresson et al. [21] systematically defined the spatiotemporal boundaries at each layer and the modeling strategies of the heat source's initial and boundary conditions. This approach was used to determine the position of porosity and calculate the thermal history, which is crucial for understanding the formation of residual stresses and deformation generated during the DED-LB/M process.

Existing computational studies on DED-LB/M have been primarily focused on the additive manufacturing (AM) aspect to layer material up, with limited attention on the surface repair aspect to restore flaws such as holes and cracks. This paper aims to fill the research gap in the multiscale and multiphysics simulation of the coaxial-powder-based DED-LB/M technology for surface repair or remanufacturing. A mesoscale coupled DEM-FVM model and a macroscale thermomechanical coupled model are established to illustrate the temperature field, velocity field, and pressure field variations in the molten pool, analyze the morphology evolution of the molten pool, evaluate the repair effectiveness of different-sized surface defects, and reveal the formation mechanisms of pores or uneven deposited layers and of residual stress and deformation during the DED-LB/M surface repair process.

2. Physics during the Powder-Based DED-LB/M Process

The physical phenomenon involved in the powder-based DED-LB/M process starts with the laser heat source (e.g., a Gaussian laser beam)-induced thermal radiation on the delivered metal powder and the rapid melting and formation of a molten pool, shown in in Figure 1. Within the molten pool, thermal radiation, heat diffusion, and evaporative heat dissipation occur simultaneously. The molten pool and the surrounding air generate a dynamic flow of multiphase fluid. The flow of the molten pool is influenced by surface tension, mushy zone drag forces, and the Marangoni effect. As the laser scans over the region, the molten pool rapidly cools and solidifies, forming a deposit track. The stability of the molten pool during the powder-based DED-LB/M process affects the surface quality and internal structure of the deposited layer. Multiple deposit tracks form a deposit layer, and the deposit layers stack on top of each other to create the final deposit surface. During the powder-based DED-LB/M process, the material is built up layer by layer, with each layer being radiatively heated by the laser heat source. Heat conduction occurs within the material, resulting in rapid temperature increases and decreases. The material also undergoes repeated heating and cooling, leading to the formation of significant temperature gradients, thermal stresses, and thermal deformations within the material.

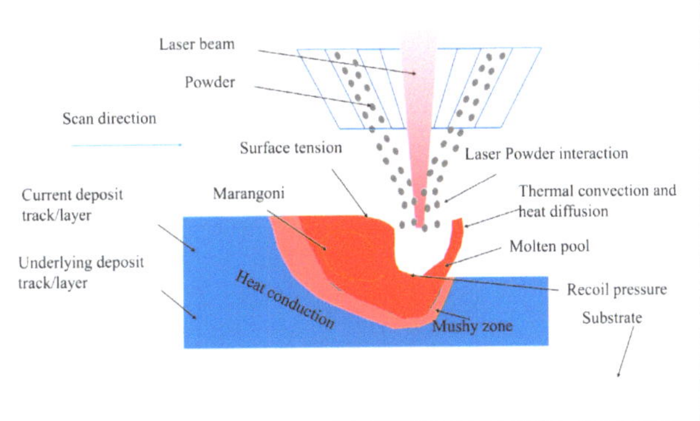

Figure 1. Physical processes of the laser-based direct energy deposition (DED-LB/M) process.

The physical processes involved in the interaction between laser and metal powder can be divided into four scales based on the desired research objectives. At the nanoscale or grain scale (<0.001–0.01 mm), the growth process of metal grains and the resulting internal grain structure can be studied based on the computed temperature field. At the mesoscale or powder scale (0.01–1.0 mm), a model considering particle morphology and size distribution can calculate the transportation and deposition processes of particles. By combining computational fluid dynamics with multiphysics analysis, the morphology and flow conditions of the molten pool can be studied, and the detailed morphology and internal porosity of the molten track can be predicted. The macroscale can be further divided into two levels: the scanning spacing scale (0.1–10 mm) and the part scale (1–250 mm). In terms of the scanning spacing scale, the deposition processes of individual scan tracks can be studied. Through thermomechanical coupling analysis, the evolution conditions, including temperature fields, stress fields, and deformation fields, during the deposition process can be obtained, allowing for the investigation of thermal stresses and deformations on the surface of the substrate part and between deposit layers. At the part scale (1–250 mm), the surface deposition process of large-sized components is considered. Similarly, with the aid of thermomechanical coupling analysis, the stress and deformation distribution of the parts, molten tracks, as well as between deposit layers during the powder-based DED-LB/M process can be studied.

This work will focus on the numerical simulation of the molten pool multiphysics fields at the mesoscale and a thermomechanical analysis at the macroscale during the surface repair of substrate components using the powder-based DED-LB/M technique. By accessing the numerical results, an evaluation based on the adopted process parameters will be presented.

3. Coupled DEM-FVM Modeling of the Molten Pool

Before establishing the numerical model for the molten pool flow process, it is necessary to reasonably simplify the complex physical process to ensure efficiency. This model will adopt incompressible Newtonian fluid laminar flow [22], neglect mass loss caused by metal vaporization [23], and disregard the influence of volume changes due to metal density variations [23]. The specific implementation process of the simulation is as follows.

3.1. Mathematical Modeling

The process of forming, flowing, and cooling a molten pool involves multiphysics fluid flow. The entire process is governed by the continuity equation, momentum conservation equation, and energy conservation equation. To fully account for the various physical factors affecting the molten pool during the powder-based DED-LB/M process, a two-phase fluid flow model is employed. Since the studied fluid is assumed to be incompressible, and the mass loss induced by gasification is ignored, the continuity equation in the computational domain becomes Equation (1):

$$\nabla \cdot u = 0 \tag{1}$$

The finite volume method (FVM) is applied to track the metal–gas interface in the two-phase fluid flow model. The equations governing the law of element volume ratios for the metal and gas phases are Equations (2) and (3):

$$\frac{\partial \alpha_1}{\partial t} + \nabla \cdot (\alpha_1 u) = 0 \tag{2}$$

$$\alpha_1 + \alpha_2 = 1 \tag{3}$$

where t represents time, and α_1 and α_2 represent the elemental volume ratio of the gas phase and the metal phase, respectively. When $\alpha_2 = 0$, the region is fully occupied by the gas phase, while when $\alpha_2 = 1$, it means that the region is fully occupied by the metal phase.

The α_2 value ranges from 0 to 1 and accounts for the metal–gas interfacial region. According to the value of α_2, the unit normal vector $\boldsymbol{n}$ at the gas–metal interface is derived as follows:

$$n = \frac{\nabla \alpha_2}{|\nabla \alpha_2|} \tag{4}$$

The curvature κ at the metal–gas interface is calculated as:

$$\kappa = -\nabla \cdot \boldsymbol{n} \tag{5}$$

The momentum conservation equation for the two-phase fluid flow domain is expressed as follows:

$$\bar{\rho}\frac{\partial u}{\partial t} + \bar{\rho}(u \cdot \nabla)u = -\nabla P + \bar{\mu}\nabla^2 u + \bar{\rho}g + S_{mom} \tag{6}$$

among which:

$$\bar{\rho} = \alpha_1 \rho_1 + \alpha_2 \rho_2 \tag{7}$$

$$\bar{\mu} = \alpha_1 \mu_1 + \alpha_2 \mu_2 \tag{8}$$

where $\bar{\rho}$, ρ_1, and ρ_2 are the density of the mixed gas and metal phase, the density of the gas phase, and the density of the metal phase, respectively. $\bar{\mu}$, μ_1, and μ_2 represent the dynamic viscosity of the mixed gas and metal phase, dynamic viscosity of the gas phase, and dynamic viscosity of the metal phase; P is the pressure; and S_{mom} indicates any remaining momentum source terms, which include three components, as shown in Equation (9):

$$S_{mom} = S_b + S_m + \left(f_{sn} + f_{st} + P_r\right)|\nabla \alpha_1| \tag{9}$$

where S_b is the buoyancy force, S_m is the mushy zone drag force used to characterize the fluidity discrepancy induced by liquid–solid phase transition, P_r is the recoil pressure due to metal evaporation, and f_{sn} and f_{st} are the surface tension and Marangoni effect at the interface between the liquid metal and gas, respectively. The term $|\nabla \alpha_1|$ is used to incorporate the surface forces into the volume forces.

The buoyancy force S_b is considered using Boussinesq approximation [24], expressed as given in Equation (10):

$$S_b = \bar{\rho}g\beta\left(T - T_{ref}\right) \tag{10}$$

where g is the acceleration due to gravity, β refers to the thermal expansion related to the buoyancy force, T is the temperature, and T_ref is the reference temperature, typically set as the liquidus temperature. The drag force in the mushy zone $S_m(mushy)$ is calculated as follows [20]:

$$S_m(mushy) = -C\left[\frac{(1-f_l)^2}{f_l^3 + C_m}\right]u \tag{11}$$

$$f_l = \begin{cases} 0 & if\ T < T_s \\ \frac{T-T_S}{T_L-T_S} & if\ T_S \le T \le T_L \\ 1 & if\ T > T_L \end{cases} \tag{12}$$

where C is a constant; its value is set to be large enough to ensure that the velocity decreases to zero when the local region fully solidifies. Typically, it is set to 105 or larger. f_l is the liquid fraction of the metal phase, T is the temperature, and T_L and T_S represent the liquidus temperature and solidus temperature of the metal phase, respectively. C_m is a custom small value used to avoid singularity in the mushy region during the calculation of the drag force.

The recoil pressure P_r acts normal to the local free surface, which is calculated as a function of the liquid surface temperature, defined as follows [21,22]:

$$S_{recoil} = 0.54 P_a exp\left(\frac{L_v M(T - T_V)}{RTT_V}\right) \tag{13}$$

where P_a is the ambient pressure, T_V is the vaporization temperature of the metal phase, L_v is the latent heat of vaporization, M is the molar mass, and R is the universal gas constant. The surface tension $S_m(tension)$ can be obtained using a continuous surface force (CSF) model [23]:

$$f_{sn} = \sigma \kappa n \tag{14}$$

where σ is the surface tension coefficient. The Marangoni effect f_{st} can be expressed as follows [23]:

$$S_m(Marangoni) = \frac{d\sigma}{dT}[\nabla T - (n \cdot \nabla T)n] \tag{15}$$

where the coefficient $\frac{d\sigma}{dT}$ represents the rate of variation in the surface tension with respect to temperature.

The energy conservation equation can be expressed as:

$$\frac{\partial \overline{\rho c_e} T}{\partial t} + \nabla \cdot (\overline{\rho u c_e} T) = \nabla \cdot \left(\overline{k} \nabla T\right) + S_h \tag{16}$$

where $\overline{c_e}$ is the equivalent specific heat capacity, and the calculation formula can be expressed as:

$$\overline{c_e} = \begin{cases} \alpha_2\left(c_2 + \frac{L_f}{T_L - T_S}\right) + \alpha_1 c_1 & T_L < T < T_S \\ \alpha_1 c_1 + \alpha_2 c_2 & T \geq T_L \text{ or } T \leq T_S \end{cases} \tag{17}$$

where c_1 and c_2 represent the specific heat of the gas phase and the metal phase, respectively; L_f is the latent heat of fusion; and $\overline{k}$ is the thermal conductivity of the mixed gas and metal phase. The expression for $\overline{k}$ is:

$$\overline{k} = \alpha_1 k_1 + \alpha_2 k_2 \tag{18}$$

where k_1 and k_2 represent the thermal conductivities of the air phase and metal phase, respectively. The last term, S_h, represents the additional heat source terms applied to the surface of the molten film. It can be expressed as a combination of convective heat dissipation S_c, radiative heat dissipation S_r, vaporization heat dissipation S_v, and Gaussian beam heating S_l [23]:

$$S_h = (S_c + S_r + S_v + S_l)|\nabla \alpha_1| \tag{19}$$

where the term $|\nabla \alpha_1|$ is used to incorporate the surface heat dissipation terms into the volume heat dissipation:

$$S_c = h_c(T - T_a) \tag{20}$$

$$S_r = k_B \varepsilon \left(T^4 - T_a^4\right) \tag{21}$$

$$S_v = -\varphi \frac{L_v M}{\sqrt{2\pi MRT}} P_a exp\left[\frac{L_v M(T - T_v)}{RTT_v}\right] \tag{22}$$

$$S_l = \frac{2\eta P_{laser}}{\pi r^2} exp\left(-2\frac{(z - z_0 - vt)^2 + (x - x_0)^2}{R^2}\right) \tag{23}$$

Here, h_c is the convective heat transfer coefficient; T_a denotes the ambient temperature; P_a is the atmospheric pressure; k_B is the Stefan–Boltzmann constant; ε is the surface emissivity; φ is the evaporation coefficient, typically 0.82 [25]; L_v is the latent heat of vaporization; T_v is the vaporization temperature; η is the metal laser absorption coefficient; P_{laser} is the laser power; r is the laser spot radius; x and z are the horizontal coordinates of the beam center during laser movement; x_0 and z_0 are the initial horizontal plane coordinates of the beam center; and v is the speed at which the laser moves along the z-axis.

3.2. Numerical Implementation

ANSYS 2022 R1 Fluent was used to simulate the molten pool during the powder-based DED-LB/M process in this work, which provides a suitable way to program corresponding user-defined functions for the temperature-dependent materials' properties and source terms in the governing equations. The CT scan results, which indicated the flaw size distribution on an aluminum alloy component, are shown in Figure 2a. To facilitate the numerical computation, a specific location on the component was selected and simplified as a flat aluminum alloy plate with dimension of $14 \times 7 \times 2$ mm, shown in Figure 2b. The air phase was also included in the numerical model and placed at the top of the metal plate to capture the interaction between the molten pool and the surrounding air. According to the CT-scanned flaw size, equivalent defects were created on the metal plate surface, with spherical diameters of 0.1, 0.5, and 0.9 mm and uniformly distributed along the deposit track. It is assumed that the required powder for deposition had already been delivered to the part surface before the high-energy laser beam was applied. The model considered a gravitational acceleration of 9.81 m/s^2 (in the negative y-axis direction). The laser beam was assumed to irradiate the particle surface along the vertical direction (in the negative y-axis direction) and scan along the positive z-axis from the origin. The direct coupled DEM/FVM was employed to simulate the interaction between laser beam and powder particles. The DEM was used to calculate the initial positions and particle size information after particle delivery, shown in Figure 2c. This information was then transferred to the fluid computational domain as initialization data for the particle material, depicted in Figure 2d. The FVM was employed in the subsequent fluid dynamics calculations to capture the metal–air interface position, shown in Figure 2e.

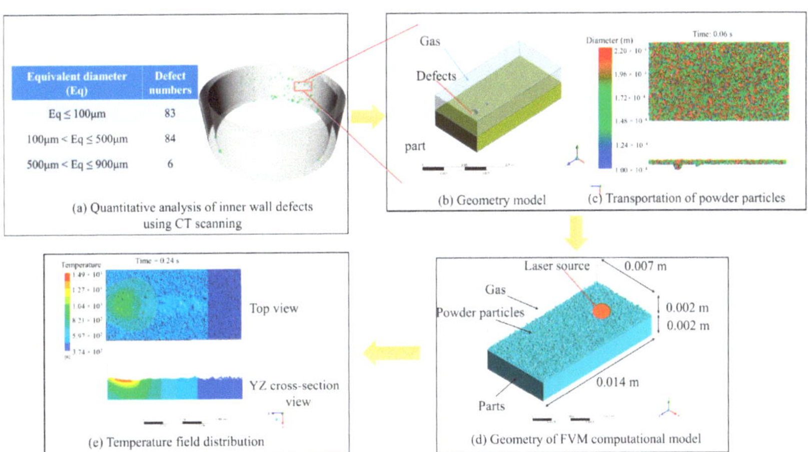

Figure 2. Roadmap for the coupled DEM-FVM modeling of the molten pool.

The simulation of the powder delivery process was performed using the EDEM 2020 software. The physical shape of the delivered metal powder is defined as spherical, and their size distribution follows a normal distribution. The average particle size is 80 μm. The minimum particle size is 50 μm, and the maximum particle size is 110 μm. The material properties of both the metal particles and the metal substrate in the model include the Poisson ratio set to 0.334, the density set to 2700 kg/m^3, and Young's modulus set to 6.67×10^{10} Pa. The interaction coefficients between particle–particle and particle–substrate were defined as well. The restitution coefficient was set to 0.75, the static friction coefficient to 0.3, and the dynamic friction coefficient to 0.01. To simulate the powder delivery process, 6000 particles were first generated above the metal plate and then fell freely under gravity. The calculations revealed that the powder particles uniformly fell onto the metal plate surface, resulting in a powder thickness of 0.2 mm after 0.06 s, shown in Figure 3a. All spherical flaws on the surface were fully filled with the delivered powder particles. The majority of particles had velocities less than 6.53×10^{-3} m/s, indicating that the powder transport had been completed, depicted in Figure 3b.

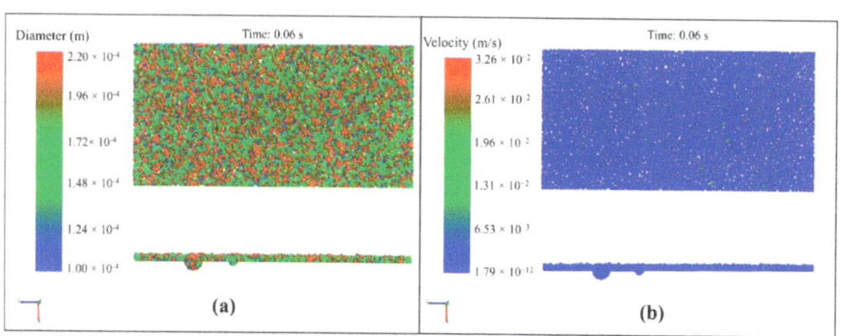

Figure 3. Simulation results of particle dynamics. (**a**) Particle positions and size distribution; (**b**) particle velocity distribution.

The FVM simulation of the molten pool involves various aspects such as meshing of the computational domain, material selection, boundary condition and processing parameter settings, control of the numerical solver, etc. Regarding meshing, a fine mesh was adopted at the metal–air interface, where the molten pool is generated, shown in Figure 4. To ensure the feasibility of capturing essential physics in the laminar flow of the molten pool, the smallest mesh size was set to 0.02 mm while the largest mesh size was set to 0.8 mm. This took the minimal flaw size of 0.1 mm into consideration, as well as the largest geometry size of 14 mm. The tetrahedral element was adopted for the meshing, resulting in a total 14,549,170 elements in the model. The material parameters of the air and adopted aluminum alloy (AlSi10Mg) in the simulation are listed in Tables 1 and 2. The processing parameters for the FVM model included a laser spot diameter of 3.5 mm, laser power P_{laser} of 1600 W, scanning speed of 2160 mm/min, and scanning spacing of 1.2 mm. For the boundary conditions, the bottom and side surfaces of the metal plate were defined as adiabatic and non-slip walls. The top and side surfaces of the air were set as pressure outlets with a static pressure of 0. The loading of the heat and momentum source terms was implemented using Fluent's User-Defined Function (UDF). The calculations employ a dynamic time step with an initial time step of 1×10^{-8} s, and the total physical time considered in the simulation is 0.24 s.

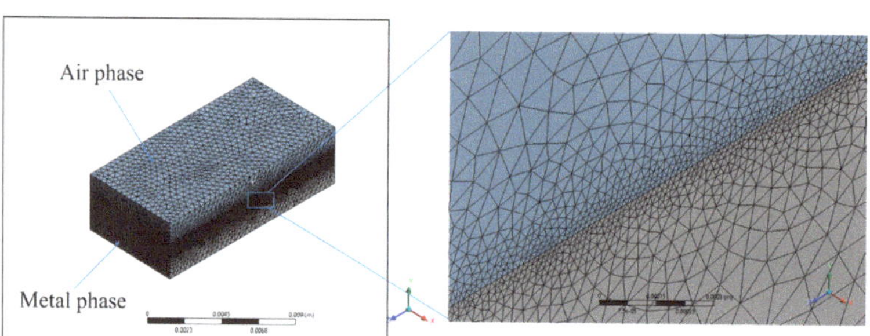

Figure 4. Meshing condition in the computational model.

Table 1. Material properties of air.

Symbol	Definition	Value
ρ_1	Density of gas phase (kg·m^{-3})	1.225
c_1	Specific heat of gas phase (J·kg^{-1}·K^{-1})	1006.43
k_1	Thermal conductivity of gas phase (W·m^{-1}·K^{-1})	0.0242
μ_1	Dynamic viscosity of gas phase (kg·m^{-1}·s^{-1})	1.7894×10^{-5}

Table 2. Material properties of aluminum alloy.

Symbol	Definition	Value
T_s	Solidus temperature (K)	890 [26]
T_L	Liquidus temperature (K)	929 [26]
T_v	Vaporization temperature (K)	2743 [26]
μ_2	Liquidus dynamic viscosity (kg·m^{-1}·s^{-1})	$0.00223exp(12200/8.3144T)$ [27]
ρ_2	Solidus density (kg·m^{-3})	2719
	Liquidus density (kg·m^{-3})	$2828 - 0.3636T$
c_2	Solidus specific heat (J·kg^{-1}·K^{-1})	$798.85 + 0.3324T + 9 \times 10^{-5}T^2$ [26]
	Liquidus specific heat (J·kg^{-1}·K^{-1})	1220 [26]
k_2	Solidus thermal conductivity (W·m^{-1}·K^{-1})	$124.66 + 0.0561T + 1 \times 10^{-5}T^2$ [26]
	Liquidus thermal conductivity (W·m^{-1}·K^{-1})	61 [26]
L_f	Latent heat of fusion (J·kg^{-1})	3.83×10^5 [26]
L_v	Latent heat of vaporization (J·kg^{-1})	1.087×10^7 [26]
h_c	Convective heat transfer coefficient (W·m^2·K)	10
σ	Surface tension coefficient (kg·s^{-2})	$0.914 - 0.00035(T - 890)$ [28]
R	Universal gas constant (J·mol^{-1}·K^{-1})	8.314
k_B	Stefan–Boltzmann constant (W·m^{-2}·K^{-4})	5.67×10^{-8}
η	Laser beam absorptivity	0.35 [29]
M	Molar mass (kg·mol^{-1})	0.026982
ε	Surface emissivity	0.3

3.3. Result Analysis

The morphology of the molten pool and the evolution of the temperature field at different time points are illustrated in Figure 5. The computation reveals that surface flaws of 100 μm and 500 μm are successfully repaired. However, the surface flaw of 900 μm is not successfully repaired, resulting in a depression. The failure at the larger flaw size position is mainly due to an insufficient powder feed rate. A significant amount of unmelted particles and a rough surface are observed at the starting position of the laser scan. This can be attributed to inadequate energy absorption at the starting position, making it difficult to form a fully developed molten pool. The formation of the molten pool occurs around 0.01 s and stabilizes at around 0.2 s, forming a semi-ellipsoidal shape. Certain localized regions within the molten pool experience an excessive temperature gradient and fluid flow, which affects the deposition stability and leads to residual porosity within the deposit layer.

Figure 6 demonstrates the effects of increased powder layer thickness (or powder feed rate) to 0.4 mm on the deposition and repair outcomes. The results indicate that with the increased powder layer thickness, the surface flaws of various sizes disappear. However, this improvement comes at the cost of material loss on the substrate surface, leading to irreversible damage to the part dimensions. The thicker powder layer carries a larger amount of molten material along the scanning direction and eventually forms a hump, making it challenging to successfully repair surface flaws on the plate and even causing damage to the part. Therefore, it is crucial to use an appropriate powder feed rate and laser processing parameters to achieve a stable molten pool, smooth deposited layer, and effective repair of surface flaws.

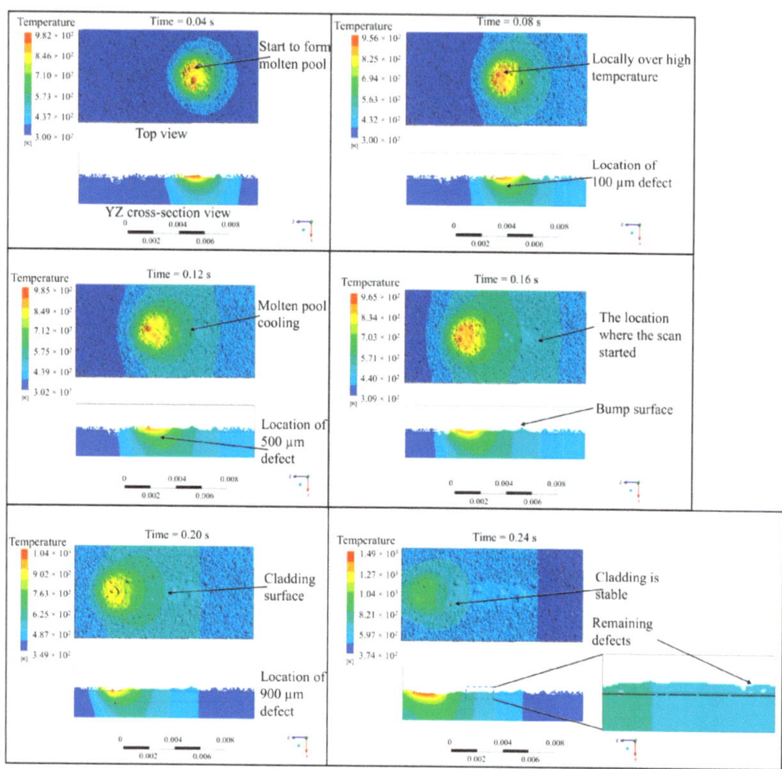

Figure 5. Morphology and temperature field evolution of the molten pool during the powder-based DED-LB/M process.

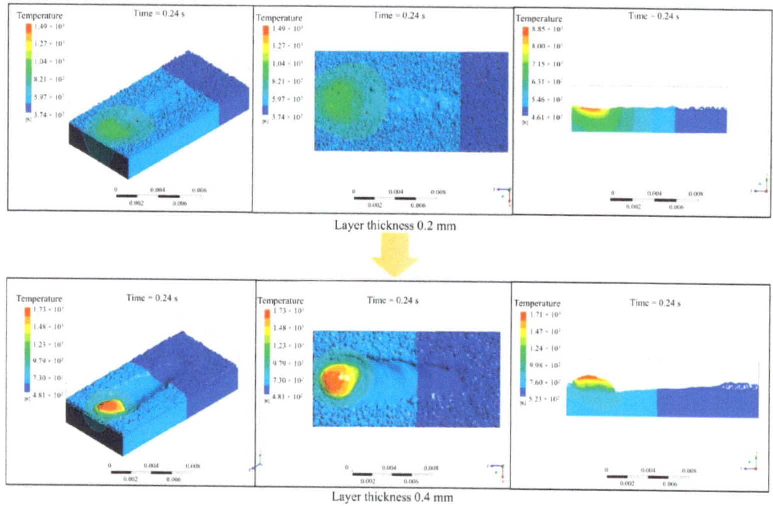

Figure 6. Comparison of repair qualities with different layer thicknesses or powder feed rates.

4. Coupled Thermostructural Modeling of the Deposit Layers

4.1. Mathematical Modeling

The surface repair process using powder-based DED-LB/M refers to the first or second deposit layers. To compute thermal stress and deformation of the deposit track, a macroscale coupled thermostructural simulation is employed. The general form of the governing equation for the thermal conduction during the powder-based DED-LB/M process is presented as follows:

$$K(T) \cdot \left(\frac{\partial^2 T}{\partial x^2} + \frac{\partial^2 T}{\partial y^2} + \frac{\partial^2 T}{\partial z^2} \right) + F = \frac{\partial H(T)}{\partial t} \tag{24}$$

where $K(T)$ is the temperature-dependent material's thermal conductivity, T is the current temperature, F is the heat flux, t is time, and x, y, and z are the spatial directions. $H(T)$ is the temperature-dependent enthalpy, representing the latent heat evolution by phase transformation effect.

When $T \leq T_s$,

$$H(T) = \rho \int_{T_0}^{T} C_p dT \tag{25}$$

When $T_s \leq T \leq T_l$,

$$H(T) = \rho \int_{T_0}^{T_s} C_p dT + \rho L_f \left(\frac{T - T_s}{T_l - T_s} \right) \tag{26}$$

When $T > T_l$,

$$H(T) = \rho \int_{T_0}^{T_s} C_p dT + \rho L_f + \rho \int_{T_l}^{T} C_p dT \tag{27}$$

where T_l is the liquidus temperature, T_s is the solidus temperature, ρ is the density, C_p is the specific heat, L_f is the latent heat of melting, and T_0 is the room temperature, assumed to be 22 °C.

Structural analysis was used to compute residual stresses, which is generated by the strains corresponding to thermal expansion, contraction caused by temperature variation, and the nonelastic strains resulting from plastic deformation. The total strain increment vector can be expressed via the superposition of elastic, plastic, and thermal components, in the following form:

$$\Delta \varepsilon_{ij}^{tol} = \Delta \varepsilon_{ij}^{e} + \Delta \varepsilon_{ij}^{th} + \Delta \varepsilon_{ij}^{p} \tag{28}$$

where $\Delta \varepsilon_{ij}^{e}$ is the elastic strain increment, $\Delta \varepsilon_{ij}^{th}$ is the thermal strain increment, and $\Delta \varepsilon_{ij}^{p}$ is the plastic strain increment. The resulting stress increments are calculated via increments in elastic strain as follows:

$$\Delta \sigma_{ij} = D_{ijlm} \Delta \varepsilon_{ij}^{e} \tag{29}$$

where E is Young's modulus, v is Poisson's ratio, and D_{ijlm} is the elastic stiffness tensor provided by Hook's law:

$$D_{ijlm} = \frac{E}{1+v} \left[\frac{1}{2} \left(\delta_{il} \delta_{jm} + \delta_{lm} \delta_{ij} \right) + \frac{v}{1 - 2v} \delta_{ij} \delta_{lm} \right] \tag{30}$$

where δ is the Dirac function.

Combining Equation (27) with Equation (28) yields:

$$\Delta \sigma_{ij} = D_{ijlm} \left(\Delta \varepsilon_{ij}^{tol} - \Delta \varepsilon_{ij}^{th} - \Delta \varepsilon_{ij}^{p} \right) \tag{31}$$

where $\Delta \varepsilon_{ij}^{th} = \alpha \delta_{lm} \Delta T$, and α and ΔT are thermal expansion coefficient and temperature increment, respectively.

For the purpose of data mapping, the structural analysis utilizes the same mesh as the thermal analysis but with modified element and material properties. Material properties that vary with temperature are employed.

4.2. Numerical Implementation

ANSYS 2022 R1 Workbench was used to simulate the deposit layer during the powder-based DED-LB/M process in this work, which offers a suitable platform to perform both thermal and mechanical analysis and the data transfer between them. As depicted in Figure 7, the macroscale coupled thermostructural simulation employed the same metal plate, with a dimension of $14 \times 7 \times 2$ mm, for consideration. The flaws with sizes of 0.1, 0.5, and 0.9 mm were incorporated on the plate's surface as well. The actual metal deposit layer is assumed to be a finite element layer to ensure computational efficiency. The deposition process involved two layers, where deposit layers 1 and 2 were represented as finite element layers. The layers were incrementally added using the element birth and death method. A moving Gaussian heat source was applied for heat conduction in each finite element layer. The simulation incorporated direct coupling between thermal and structural mechanics, with the temperature field acting as a thermal load applied to subsequent structural analysis.

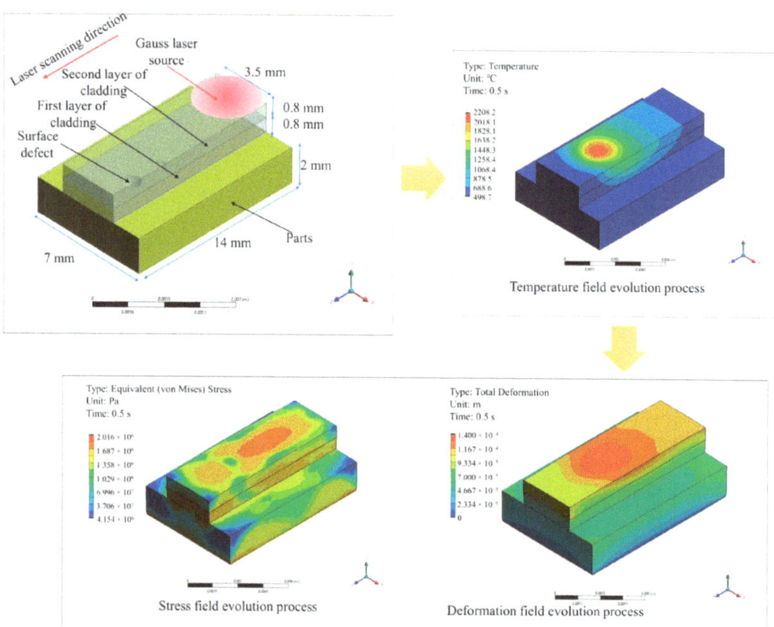

Figure 7. Roadmap of coupled thermostructural simulation of the powder-based DED-LB/M process.

The establishment of this numerical model involves meshing, material property settings, boundary conditions, process parameters, and control of the computation process. The model employed mesh refinement at the locations of the surface defects on the metal plate to capture significant temperature and stress gradients. The minimum mesh size was set to 0.01 mm, while the maximum grid size was set to 0.4 mm. The computational domain was meshed using the tetrahedral elements, resulting in a total of 281,861 grid cells. Both the deposit material and the metal plate material are AlSi10Mg, and their thermodynamic material properties, varying with temperature, were obtained from the material library of ANSYS, shown in Figure 8.

The boundary conditions of the coupled thermostructural model were set so that the thermal convection with air was applied on the surface of the metal plate, and a fixed support was imposed on the bottom surface of the component. On the surfaces of each deposit layer, a Gaussian heat source moving in the positive Z-axis direction was defined using the Ansys Parametric Design Language (APDL) language. The processing parameters adopted were exactly the same as those of the coupled DEM-FVM model, where the diameter of the laser spot is 3.5 mm, the laser power is 1600 W, the scanning speed is 2160 mm/min, and the absorption coefficient of the metal material to laser energy is 0.35. For the thermal calculation, the initial temperature is set to 22 °C. The calculations are performed using a dynamic time step approach, with an initial time step of 3×10^{-3} s, a minimum time step of 3×10^{-4} s, and a maximum time step of 3×10^{-2} s.

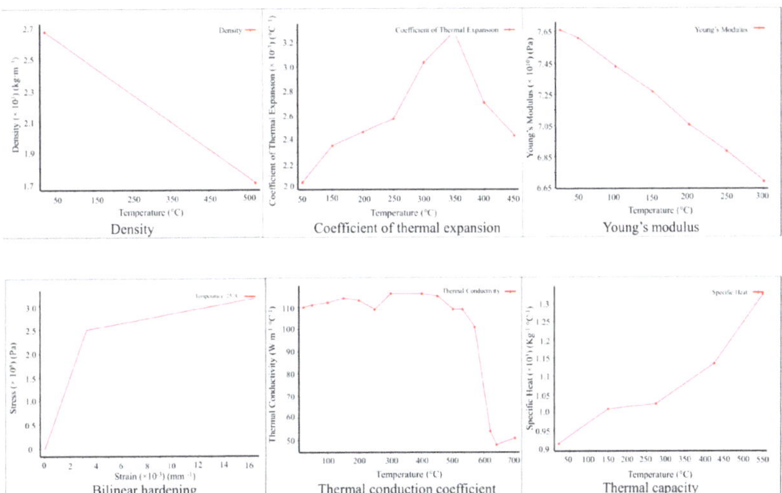

Figure 8. Temperature-dependent parameters of the aluminum alloy.

4.3. Result Analysis

Figure 9 illustrates the temperature field variation during the powder-based DED-LB/M process. The analysis reveals that the maximum temperature gradually accumulated as each layer is deposited. The maximum temperature was concentrated at the scanning position of the laser beam, conforming to the Gaussian distribution characteristic of the energy. During the first deposit layer, the entire surface and interior of the deposit layer reached a temperature above the melting point, allowing for the formation of a sufficiently molten pool for repairing the component. The depth of the molten pool was sufficient to cover surface flaws of 0.1 mm and 0.5 mm, but it was insufficient to cover surface flaws with a depth of 0.9 mm. During the second deposit layer, the entire surface and interior of the deposit layer reached a temperature above the melting point, ensuring interlayer bonding.

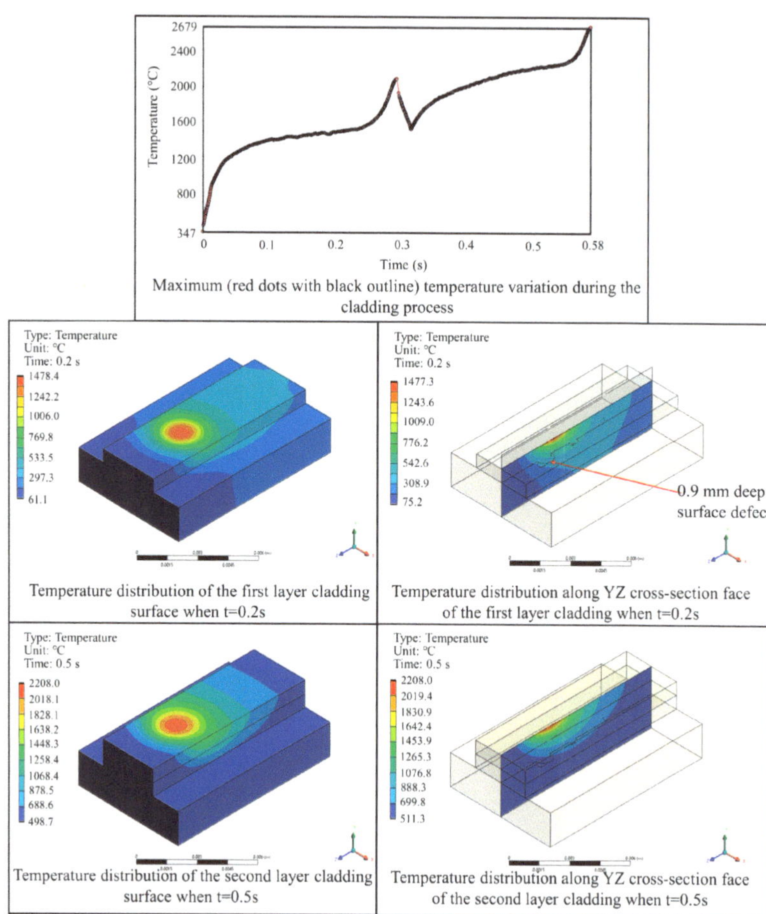

Figure 9. Evolution of temperature distribution during the powder-based DED-LB/M process.

Figure 10 illustrates the variation in the von Mises stress field during the powder-based DED-LB/M process. The analysis reveals that the maximum von Mises stress occurred at the bottom surface of the metal plate in the first deposit layer, measuring 236.86 MPa. As the second layer was deposited, the maximum von Mises stress decreased to 184.76 MPa, below the yield stress of the material, indicating that issues such as cracking are unlikely to occur during the deposition process. During the second deposit layer, the maximum von Mises stress appeared at the interfacial region between layers. No significant stress concentration was observed at the locations of the component's flaws, demonstrating the feasibility of utilizing the powder-based DED-LB/M for surface repair of an aluminum alloy component.

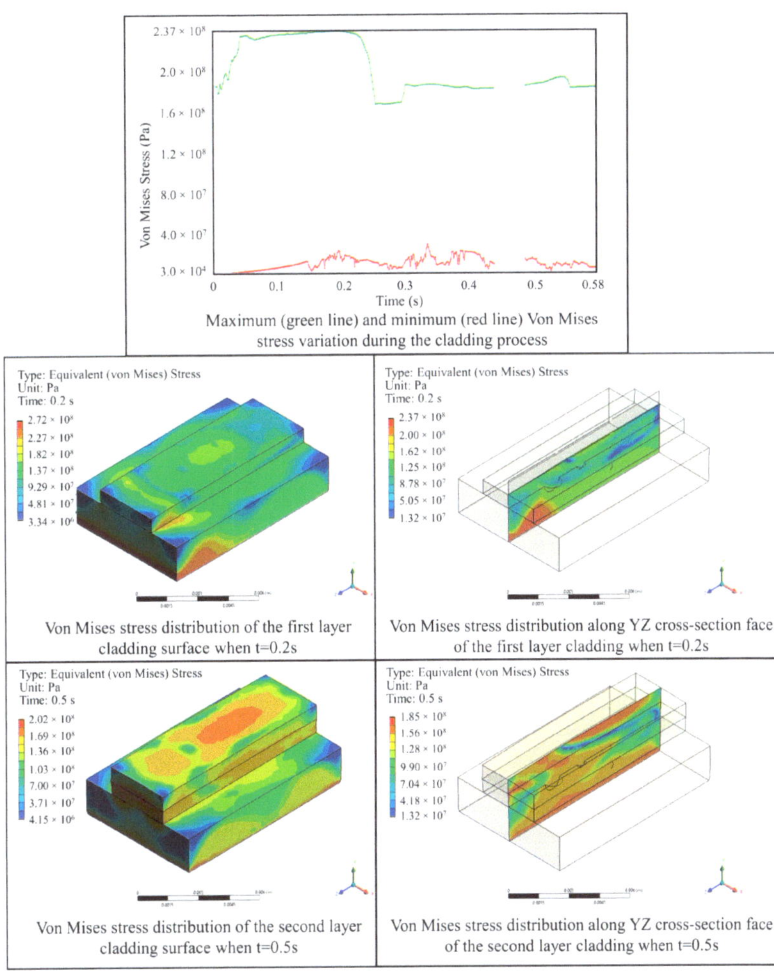

Figure 10. Evolution of von Mises stress distribution during the powder-based DED-LB/M process.

The deformation field evolution during the powder-based DED-LB/M process is shown in Figure 11. The simulation results show that the maximum deformation gradually accumulated throughout the deposition process, and it reached its maximum value at the end of the second deposit layer (0.179 mm). The maximum deformation always occurred at the surface of the deposit layer and within the spot radius of the laser beam. This was attributed to the rapid temperature increase and thermal expansion of the material. No relative sliding resulting from deformation was observed at the surface flaw locations of the metal plate, indicating the reliability of the powder-based DED-LB/M process.

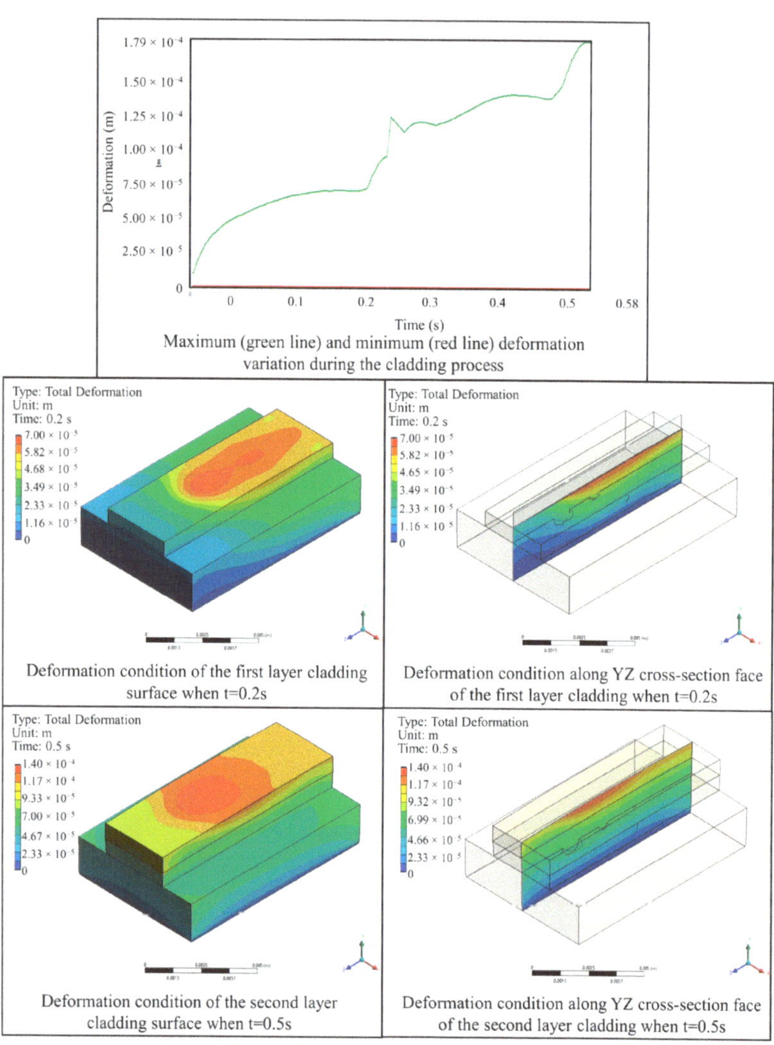

Figure 11. Evolution of deformation condition during the deposition process.

5. Conclusions

Both a mesoscale coupled DEM-FVM model of the molten pool and a macroscale coupled thermostructural model of the deposit layer for the powder-based DED-LB/M process were developed. The software ANSYS 2022 R1 Fluent and Workbench were employed to perform the mesoscale and macroscale simulations, respectively. The gradient grid size was applied to resolve tiny features for both the mesoscale (0.02 to 0.8 mm) and the macroscale models (0.01 to 0.4 mm). The computational time for the mesoscale coupled DEM-FVM model was 15 h, while that for the macroscale coupled thermostructural model was 12 h. The feasibility of repairing surface flaws using powder-based DED-LB/M technology was demonstrated, and the detailed conclusions are listed below:

(1) Micropores and bumping accompany the powder-based DED-LB/M process due to the extensive flow of the molten pool, and larger surface flaw sizes tend to result in an uneven deposit layer due to insufficient material supply. However, too much

powder feed on the surface will lead to agglomeration of the molten materials along the scanning direction and severe damage of the metal base plate.

(2) The maximum von Mises stress is far less than the yield stress of the adopted material, and no stress concentration exists during the powder-based DED-LB/M process. The total deformation will accumulate during the powder-based DED-LB/M process, and maximum deformation always occurs within the laser beam spot. No relative sliding phenomenon is observed between deposit layers.

The proposed multiscale model on the powder-based DED-LB/M for the surface repair of aluminum alloys would be necessary for a CAM engineer and machine operator to virtually validate and optimize the processing parameters before the actual manufacturing. Future work is expected to consider the mass loss caused by metal vaporization and the volume changes due to metal density variations to further increase the validity of the proposed mesoscale coupled DEM-FVM model of a molten pool. The thermal stresses during the surface repair of typical curved surfaces should be investigated using the established macroscale coupled thermostructural model. The effects of processing parameters such as laser absorptivity, laser scanning speed, and spot radius on the surface repair quality of the aluminum alloy component should also be studied. Simultaneously, a systematic experiment to verify the numerical model and reveal some new phenomena during the powder-based DED-LB/M process is expected. This experiment will likely use an in situ high-speed camera or X-ray imaging.

Author Contributions: Conceptualization and methodology, X.Z., Z.P., Y.N. and Q.W.; software and calculation, Z.P. and Y.N.; validation, Q.W.; formal analysis, investigation, and resources, X.Z., Z.L., L.Y., Y.Y. and Y.H.; writing—original draft preparation, X.Z.; writing—review and editing, Z.L., L.Y. and Q.W.; supervision, Y.H. and Y.N.; funding acquisition, X.Z. All authors have read and agreed to the published version of the manuscript.

Funding: This research was funded by the Science and Technology Project of Guizhou Province, Grant No. ZK [2022]-326, [2024]-181; Youth Science and Technology Talent Growth Project of Guizhou Province, Grant No. KY [2022]-169; and Serve the Four New Modernizations Science and Technology Project of Guizhou Province, Grant No. [2022]-005. This work was also supported by the Nottingham Ningbo China Beacons of Excellence Research and Innovation Institute (budget code I01220400001).

Institutional Review Board Statement: Not applicable.

Informed Consent Statement: Not applicable.

Data Availability Statement: Due to privacy concerns, the ANSYS 2022 R1 Fluent UDF code presented in this study is only available on request from the corresponding author.

Conflicts of Interest: The authors declare no conflicts of interest.

References

1. Li, S.S.; Yue, X.; Li, Q.Y.; Peng, H.L.; Dong, B.X.; Liu, T.S.; Yang, H.Y.; Fan, J.; Shu, S.L.; Qiu, F. Development and applications of aluminum alloys for aerospace industry. *J. Mater. Res. Technol.* **2023**, *27*, 944–983. [CrossRef]
2. Kusoglu, I.M.; Gökce, B.; Barcikowski, S. Research trends in laser powder bed fusion of Al alloys within the last decade. *Addit. Manuf.* **2020**, *36*, 101489. [CrossRef]
3. Gibson, I.; Rosen, D.W.; Stucker, B.; Khorasani, M.; Rosen, D.; Stucker, B.; Khorasani, M. *Additive Manufacturing Technologies*; Springer: Berlin/Heidelberg, Germany, 2021; Volume 17.
4. Masroor, Z.; Rauf, A.A.; Mustafa, F.; Hussain, S.W. Crack repairing of aluminum alloy 2024 by reinforcement of Al_2O_3 and B_4C particles using friction stir processing. In Proceedings of the 2019 Sixth International Conference on Aerospace Science and Engineering (ICASE), Islamabad, Pakistan, 12–14 November 2019; pp. 1–6.
5. Da Silva, A.; Frostevarg, J.; Kaplan, A.F. Melt pool monitoring and process optimisation of directed energy deposition via coaxial thermal imaging. *J. Manuf. Process.* **2023**, *107*, 126–133. [CrossRef]
6. Stavropoulos, P.; Pastras, G.; Tzimanis, K.; Bourlesas, N. Addressing the challenge of process stability control in wire DED-LB/M process. *CIRP Ann.* **2024**, *in press*. [CrossRef]
7. Akbari, M.; Kovacevic, R. Closed loop control of melt pool width in robotized laser powder—Directed energy deposition process. *Int. J. Adv. Manuf. Technol.* **2019**, *104*, 2887–2898. [CrossRef]

8. Wang, H.; Gould, B.; Parab, N.; Zhao, C.; Greco, A.; Sun, T.; Wolff, S.J. High-speed synchrotron X-ray imaging of directed energy deposition of titanium: Effects of processing parameters on the formation of entrapped-gas pores. *Procedia Manuf.* **2021**, *53*, 148–154. [CrossRef]

9. Stavropoulos, P.; Foteinopoulos, P. Modelling of additive manufacturing processes: A review and classification. *Manuf. Rev.* **2018**, *5*, 2. [CrossRef]

10. Wang, S.; Zhu, L.; Fuh, J.Y.H.; Zhang, H.; Yan, W. Multi-physics modeling and Gaussian process regression analysis of cladding track geometry for direct energy deposition. *Opt. Lasers Eng.* **2020**, *127*, 105950. [CrossRef]

11. Wang, S.; Zhu, L.; Dun, Y.; Yang, Z.; Fuh, J.Y.H.; Yan, W. Multi-physics modeling of direct energy deposition process of thin-walled structures: Defect analysis. *Comput. Mech.* **2021**, *67*, 1229–1242. [CrossRef]

12. Bayat, M.; Nadimpalli, V.K.; Biondani, F.G.; Jafarzadeh, S.; Thorborg, J.; Tiedje, N.S.; Bissacco, G.; Pedersen, D.B.; Hattel, J.H. On the role of the powder stream on the heat and fluid flow conditions during directed energy deposition of maraging steel—Multiphysics modeling and experimental validation. *Addit. Manuf.* **2021**, *43*, 102021. [CrossRef]

13. Sun, Z.; Guo, W.; Li, L. Numerical modelling of heat transfer, mass transport and microstructure formation in a high deposition rate laser directed energy deposition process. *Addit. Manuf.* **2020**, *33*, 101175. [CrossRef]

14. Chen, H.; Sun, Y.; Yuan, W.; Pang, S.; Yan, W.; Shi, Y. A review on discrete element method simulation in laser powder bed fusion additive manufacturing. *Chin. J. Mech. Eng. Addit. Manuf. Front.* **2022**, *1*, 100017. [CrossRef]

15. Aggarwal, A.; Chouhan, A.; Patel, S.; Yadav, D.K.; Kumar, A.; Vinod, A.R.; Prashanth, K.G.; Gurao, N.P. Role of impinging powder particles on melt pool hydrodynamics, thermal behaviour and microstructure in laser-assisted DED process: A particle-scale DEM–CFD–CA approach. *Int. J. Heat Mass Transf.* **2020**, *158*, 119989. [CrossRef]

16. Khairallah, S.A.; Chin, E.B.; Juhasz, M.J.; Dayton, A.L.; Capps, A.; Tsuji, P.H.; Bertsch, K.M.; Perron, A.; McCall, S.K.; McKeown, J.T. High fidelity model of directed energy deposition: Laser-powder-melt pool interaction and effect of laser beam profile on solidification microstructure. *Addit. Manuf.* **2023**, *73*, 103684. [CrossRef]

17. Srivastava, S.; Garg, R.K.; Sharma, V.S.; Sachdeva, A. Measurement and Mitigation of Residual Stress in Wire-Arc Additive Manufacturing: A Review of Macro-Scale Continuum Modelling Approach. *Arch. Comput. Methods Eng.* **2020**, *28*, 3491–3515. [CrossRef]

18. Stender, M.E.; Beghini, L.L.; Sugar, J.D.; Veilleux, M.G.; Subia, S.R.; Smith, T.R.; San Marchi, C.W.; Brown, A.A.; Dagel, D.J. A thermal-mechanical finite element workflow for directed energy deposition additive manufacturing process modeling. *Addit. Manuf.* **2018**, *21*, 556–566. [CrossRef]

19. Li, C.; Liu, J.; Fang, X.; Guo, Y. Efficient predictive model of part distortion and residual stress in selective laser melting. *Addit. Manuf.* **2017**, *17*, 157–168. [CrossRef]

20. Li, C.; Liu, Z.; Fang, X.; Guo, Y. On the simulation scalability of predicting residual stress and distortion in selective laser melting. *J. Manuf. Sci. Eng.* **2018**, *140*, 041013. [CrossRef]

21. Bresson, Y.; Tongne, A.; Baili, M.; Arnaud, L. Global-to-local simulation of the thermal history in the laser powder bed fusion process based on a multiscale finite element approach. *Int. J. Adv. Manuf. Technol.* **2023**, *127*, 4727–4744. [CrossRef]

22. Gu, H.; Wei, C.; Li, L.; Han, Q.; Setchi, R.; Ryan, M.; Li, Q. Multi-physics modelling of molten pool development and track formation in multi-track, multi-layer and multi-material selective laser melting. *Int. J. Heat Mass Transf.* **2020**, *151*, 119458. [CrossRef]

23. Cao, L. Numerical simulation of the impact of laying powder on selective laser melting single-pass formation. *Int. J. Heat Mass Transf.* **2019**, *141*, 1036–1048. [CrossRef]

24. Barletta, A. The Boussinesq approximation for buoyant flows. *Mech. Res. Commun.* **2022**, *124*, 103939. [CrossRef]

25. Lee, Y.; Zhang, W. Modeling of heat transfer, fluid flow and solidification microstructure of nickel-base superalloy fabricated by laser powder bed fusion. *Addit. Manuf.* **2016**, *12*, 178–188. [CrossRef]

26. Valencia, J.J.; Quested, P.N. Thermophysical properties. In *Metals Process Simulation*; ASM International: Almere, The Netherlands, 2010; pp. 18–32.

27. Battezzati, L.; Greer, A. The viscosity of liquid metals and alloys. *Acta Metall.* **1989**, *37*, 1791–1802. [CrossRef]

28. Mills, K.C. *Recommended Values of Thermophysical Properties for Selected Commercial Alloys*; Woodhead Publishing: Cambridge, UK, 2002.

29. Pierron, N.; Sallamand, P.; Jouvard, J.-M.; Cicala, E.; Mattei, S. Determination of an empirical law of aluminium and magnesium alloys absorption coefficient during Nd: YAG laser interaction. *J. Phys. D Appl. Phys.* **2007**, *40*, 2096. [CrossRef]

materials

Article

Machine Learning-Enabled Quantitative Analysis of Optically Obscure Scratches on Nickel-Plated Additively Manufactured (AM) Samples

Betelhiem N. Mengesha [1], Andrew C. Grizzle [1], Wondwosen Demisse [1], Kate L. Klein [1], Amy Elliott [2] and Pawan Tyagi [1,*]

[1] Mechanical Engineering, University of the District of Columbia, Washington, DC 20008, USA; andrew.grizzle@udc.edu (A.C.G.); wondwosen.demisse@udc.edu (W.D.); kate.klein@udc.edu (K.L.K.)
[2] Manufacturing Demonstration Facility, 2350 Cherahala Boulevard, Knoxville, TN 37932, USA; elliottam@ornl.gov
* Correspondence: ptyagi@udc.edu

Abstract: Additively manufactured metal components often have rough and uneven surfaces, necessitating post-processing and surface polishing. Hardness is a critical characteristic that affects overall component properties, including wear. This study employed K-means unsupervised machine learning to explore the relationship between the relative surface hardness and scratch width of electroless nickel plating on additively manufactured composite components. The Taguchi design of experiment (TDOE) L9 orthogonal array facilitated experimentation with various factors and levels. Initially, a digital light microscope was used for 3D surface mapping and scratch width quantification. However, the microscope struggled with the reflections from the shiny Ni-plating and scatter from small scratches. To overcome this, a scanning electron microscope (SEM) generated grayscale images and 3D height maps of the scratched Ni-plating, thus enabling the precise characterization of scratch widths. Optical identification of the scratch regions and quantification were accomplished using Python code with a K-means machine-learning clustering algorithm. The TDOE yielded distinct Ni-plating hardness levels for the nine samples, while an increased scratch force showed a non-linear impact on scratch widths. The enhanced surface quality resulting from Ni coatings will have significant implications in various industrial applications, and it will play a pivotal role in future metal and alloy surface engineering.

Keywords: unsupervised machine learning; K-means clustering; additive manufacturing; nickel plating; hardness; scratch test

Citation: Mengesha, B.N.; Grizzle, A.C.; Demisse, W.; Klein, K.L.; Elliott, A.; Tyagi, P. Machine Learning-Enabled Quantitative Analysis of Optically Obscure Scratches on Nickel-Plated Additively Manufactured (AM) Samples. *Materials* **2023**, *16*, 6301. https://doi.org/10.3390/ma16186301

Academic Editors: Tuhin Mukherjee and Qianru Wu

Received: 20 August 2023
Revised: 8 September 2023
Accepted: 14 September 2023
Published: 20 September 2023

1. Introduction

Machine learning (ML) has received a great deal of attention recently, particularly as a result of recent developments in the field of deep learning [1,2]. Artificial intelligence has become a central focus across various research fields and in additive manufacturing, like engineering disciplines [3–6], as it offers a unified framework through which to integrate intelligent decision making into numerous fields [7,8]. There are several forms of additive manufacturing, such as binder-jetting-based metal additive manufacturing (BJAM). In this study, we focused on stainless-steel and bronze composite samples that were manufactured using the binder-jetted method [9]. Binder jetting is a 3D printing process that involves the deposition of an adhesive binding agent onto thin layers of powdered material. The printer head moves over the build platform depositing binder droplets, and it then prints each layer in a way that is not dissimilar to 2D printers that print ink on paper. After each layer is complete, the powder bed moves downward, and the printer spreads a new layer of powder onto the build area. The process goes on layer by layer until all parts are complete. After printing, the parts are in a green, or unfinished, state, and they require additional post-processing before they are ready to use. Often, the operator adds an infiltrating substance to improve the mechanical properties of the parts. The infiltrate substance is usually bronze in the case of metal 3D prints [9,10]. The BJAM process has a significant scope of improvement with the application of ML and deep learning [11,12]. The further advancement of AM is critically dependent on the post-processing of completed parts and the use of ML in solving problems.

While BJAM and other additive manufacturing processes have been widely used for rapid prototyping [13,14], some of its constraints revolve around reliability and control [15,16]. Compared to subtractive manufacturing procedures, AM creates objects with poor surface smoothness [9,17]. As produced, surface quality has a negative effect on the tribological behavior of printed parts. It is well understood that rough surfaces tend to experience faster wear compared to smooth surfaces [18,19]. Therefore, it becomes crucial to thoroughly investigate and regulate the surface roughness of AM parts via different approaches to deal with interior and exterior surface quality [20]. The study and control of surface roughness in AM components are essential in order to enhance their durability, reduce wear, and improve overall performance. By understanding and managing the surface roughness, it is possible to optimize the tribological characteristics and extend the lifespan of AM parts [21]. In the context of BJAM, coatings are indispensable for maintaining integrity in a reactive environment. Due to the use of soft and hard materials, a binder-jetted part may be susceptible to corrosion and may display non-uniform mechanical properties. Hence, a protective coating, depending upon the end use, may be a necessity when utilizing BJAM parts.

Among various coatings, electroless nickel coating has been widely applied and studied for conventional engineering components. Electroless nickel plating has been extensively studied with various plating baths so as to identify the optimal conditions for achieving desired qualities such as corrosion resistance, wear resistance, and hardness. Through systematic experimentation and analysis, researchers have aimed to identify the ideal parameters and bath compositions that can lead to electroless nickel coatings with excellent performance in terms of corrosion resistance, wear resistance, and hardness. These efforts contribute to the development and application of electroless nickel plating as a reliable surface treatment method for enhancing the functional properties of various materials [22,23].

In this research, we explored electroless nickel coatings for BJAM parts. The major challenge was experienced in analyzing the hardness of nickel coating via the standard scratch test process. A standard method for determining surface hardness using scratch testing involves running a diamond stylus across the coated surface while applying increasing force until adhesion failure is observed [24]. In the study, it was observed that the surface of the Ni-plated samples exhibited minor scratches, which posed challenges when aiming to accurately capture 3D images with a light microscope. The highly reflective surface resulted in an oversaturation and lens reflection artifacts in the images, making it difficult to quantify scratch widths effectively. To overcome this limitation, scanning electron microscopy (SEM) was employed to generate a 3D height map of the area, thus providing a higher resolution for measuring scratch widths. To address this challenge in the postprocessing of BJAM parts, we applied ML. ML algorithms have proven valuable in addressing various problem-solving tasks such as regression, classification, and forecasting [25]. ML can be broadly categorized into four types based on the learning approach used: supervised, unsupervised, semi-supervised, and reinforcement learning. In unsupervised ML, the algorithm predicts outputs without any explicit supervision, and it relies on unlabeled datasets. One prominent approach in unsupervised ML is clustering, which involves extracting natural groups from data based on their similarities [26,27]. The K-means algorithm is the most well-known and often-used unsupervised clustering method [28]. The K-means cluster seeks to determine the centroid of each cluster and assign the data points to the nearest centroid. The centroid is the arithmetic mean of all the points belonging to the cluster [29]. It iteratively calculates the cluster centroids repeatedly, and adjusts the parameters until a negligible change is observed [30]. There is no need for a training dataset since it is a type of unsupervised ML, and computation is conducted on the real dataset [31].

To the best of our knowledge, for the first time, we explored the application of the K-means ML approach to successfully analyze the scratch on electroless nickel films coated on BJAM. To automate the analysis process and extract scratch data from the images and height maps, a Python script was originally developed. The script utilized the K-means algorithm, an unsupervised machine learning method, to segment and identify the scratches. Applying the K-means algorithm meant that the scratch data could be effectively extracted, thus allowing for a quantitative analysis and characterization of the scratches. The outcomes of this study demonstrated the utility of unsupervised machine learning techniques, such as the K-means algorithm, in addressing challenges encountered in materials science. By leveraging these methods, researchers can overcome limitations in traditional image analysis approaches and obtain valuable insights from complex surface data, such as scratch measurements.

2. Materials and Methods

The focus of this paper is on the application of ML-based image analysis approaches for successfully studying scratches that are created on nickel-coated BJAM samples. The BJAM samples used in this study were manufactured by the ExOne® (Huntington, PA, USA). The stainless-steel 420 powder was shaped with a binder jet 3D printer that was made by ExOne®. Binder jetting works by spreading powder into a layer, and an inkjet printhead is used to selectively deposit a binder into the layer of powder. As the process proceeds, the powder and binder are layered to form a 3D shape in the powder bed. The average particle size is between 15 and 30 microns, and the binding agent is a polymer. The print is then heated to 200 °C to evaporate the solvent from the binder. Once dried, the parts are removed from the powder bed and set up for post-processing. The post-process consists of adding the part to a crucible filled with a measured amount of bronze alloy. The crucible with the part and bronze is heated to around 1100 °C for 1–2 h, which allows the bronze to melt and infiltrate the porous stainless-steel print. Infiltration is driven by the wetting of the molten bronze and the steel, the surface tension of the molten bronze, and the resulting capillary forces between the stainless-steel particles. Once solidified, the resulting part is a stainless-steel and bronze metal-matrix composite, which is approximately 60% stainless steel and 40% bronze by volume. In the follow-up SEM and EDS analyses with Phenom XL SEM purchased from Nanoscience®, (Phoenix, AZ, USA), we observed elemental analysis results that were specific to how many 420 stainless-steel powder particles were present in the imaging area, as well as the variation in the shape and size of each particle. Hence, due to the limitation of EDS in providing consistent results, we report on the percentage of stainless steel and bronze based on the manufacturing process.

Importantly, this paper is mainly about post-manufacturing surface property improvement where the surface properties of BJAM is a critical factor. As a part of post-processing, we developed an empirical model targeting the smooth surface morphology of several micron-thick nickel depositions on nine binder-jetted 420 stainless steel/bronze components. The electroless plating solution was acquired from the Surface Technology Incorporated® company. The experiment plan for nine samples was based on the Taguchi design of experiment, which enables the study of multiple variables and their levels in fewer experiments when compared to the experimental plan where one variable is varied at a time [32]. In this investigation, there were four factors with three levels each. The plating bath solution's phosphorus levels consisted of low (1–4%), medium (6–9%), and high (10–13%). The temperature levels included low (recommended −10 °C), medium (recommended), and high (recommended +10 °C). For low and medium phosphorus, the recommended temperature was set at 90 °C, while for high phosphorus, it was 85 °C. The surface cleaning preparation factor encompassed three levels: organic solution cleaning, plasma cleaning, and chempolishing. Chempolishing-based surface finishing details are published elsewhere [17]. The plasma was produced by 100 W of RF power, at a 30 SCCM Ar flow rate, and at 320 mTorr pressure to etch the binder-jetted samples isotropically. Plasma cleaning was done with SPI Plasma Prep II (West Chester, PA, USA). The fourth factor, plating thickness, also comprised three levels. We targeted depositing at 20, 30, and 40 μm thicknesses, which were determined from the manufacturer-provided data sheet for the three plating solutions. Table 1 depicts the L9 orthogonal array and each sample's name ID, which are utilized in the discussion section when referring to each sample.

Table 1. The L9 orthogonal array of the nine experiments for investigating nickel plating.

Exp. Run	Phosph. Level	Temp. ($^{\circ}$C)	Surface Prep.	Thickness (μm)	ID
1	Low	85	Organic	20	OC1
2	Low	95	Plasma	30	PC1
3	Low	105	Chempolish	40	CP1
4	Medium	85	Plasma	40	PC2
5	Medium	95	Chempolish	20	CP2
6	Medium	105	Organic	30	OC2
7	High	75	Chempolish	20	CP3
8	High	85	Organic	30	OC3
9	High	95	Plasma	40	PC3

After completing the nickel-plating process as per the plan mentioned in Table 1, scratch testing was performed with a Taber Scratch tester®. Scratch geometry analysis is a critical step in determining the toughness of films, and the surface hardness of the composite samples was evaluated by varying the scratch load gradually from 8 N to 15 N. The samples were divided into three groups based on their surface preparation. In general, the trend observed in the graphs indicates that, as the scratch load remains constant, the hardness tends to decrease as the scratch width becomes deeper and wider. This implies that a deeper and wider scratch shows a low surface hardness. Moreover, the relationship between scratch width and applied scratch load is directly proportional, meaning that, as the applied load increases, the scratch width also increases. However, it is important to note that the increase in scratch width is non-linear, suggesting that the relationship between load and width may not be strictly linear. However, since nickel plating makes the surface quite shiny, it became difficult to determine the scratch depth and width profile accurately from an analysis of the optical images. We developed a solution to this problem by relying on the SEM images of the scratches, which produced better depth contrast. The SEM images were visually marked for the location of the scratch, and the K-means machine learning algorithm was applied. The following section describes the K-means algorithm adopted in this study.

K-means clustering is an iterative method that aims to divide a dataset into a predetermined number, K, of distinct clusters or subgroups based on their attributes. The goal is to create clusters that are as dissimilar from each other as possible while making the data points within each cluster as similar as possible. The process begins by randomly assigning K centroids, which serve as the initial center points for the clusters, as shown in Figure 1b. Each data point in the dataset is then assigned to the cluster with the nearest centroid based on a chosen distance metric, typically the Euclidean distance, as shown in Figure 1c. This assignment step ensures that data points are allocated to the group that is closest to them in terms of attribute similarity. After assigning all the data points to clusters, the algorithm recalculates the centroids of each cluster by computing the mean (arithmetic average) of all the data points within the cluster. This updating step adjusts the centroids' positions to reflect the clusters' new center points based on the reassigned data points, as shown in Figure 1d. The algorithm iterates between the assignment and update steps until convergence is reached. Convergence is determined by assessing whether there has been a substantial change in the centroids compared to the previous iteration. If the centroids remain largely unchanged, or if the maximum number of iterations is reached, the algorithm terminates. To determine whether a data point belongs to a particular cluster, the algorithm compares the distance between the data point and the centroid of that cluster. Centroid of the cluster is shown by the red and black "x" for two groups in Figure 1b–d. If the distance is less than a certain threshold, which is often represented by the within-cluster sum of squares or a cost function, the data point is assigned to that cluster. Throughout the iterations, the algorithm strives to minimize the cost function by adjusting the positions of the centroids. This process leads to the formation of well-defined clusters that are distinct from each other, with reduced variability within each cluster. The data points within each cluster become more homogeneous or similar to each other in terms of their attributes.

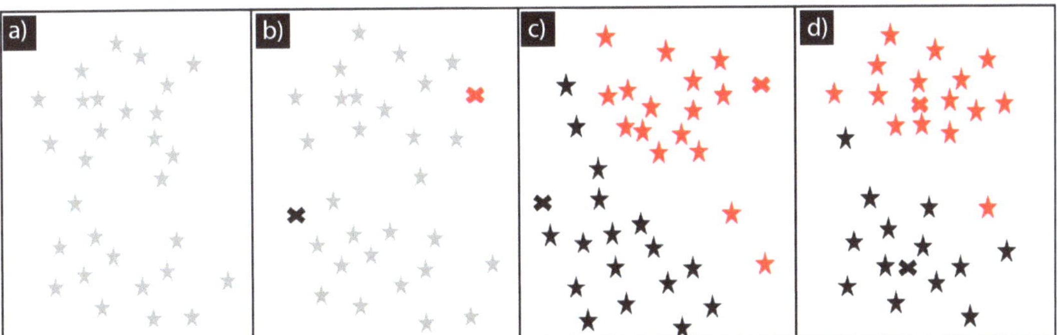

Figure 1. K-means clustering process illustration: (**a**) row data, (**b**) random initializing centroids to represent the center of a cluster, (**c**) for each data point, calculate its distance to each centroid and assign data points to the nearest centroid, and (**d**) update the centroids by computing the mean of all data points assigned to each cluster and repeat until convergence.

In K-means clustering, the hyper-parameter K is predetermined before the training process begins. The letter "K" represents the number of clusters that the algorithm aims to create. This value is typically determined based on prior knowledge or domain expertise. The objective function in K-means clustering involves minimizing the total within-cluster sum of squares, also known as inertia or distortion. The objective function can be mathematically expressed as follows:

$$J = \sum_{j=1}^{k} \sum_{i=1}^{n} \left\| X_i^j - C_j \right\|^2 \tag{1}$$

where J is the objective function, and k and n are the numbers of clusters and cases, respectively. X is the case i, and C is the centroid for cluster j. The term in absolute value is known as the distance function.

The number of clusters in the K-means method represent the moving centroids within the data. The elbow method helps determine the optimal number of clusters by evaluating the distortion or inertia for the different values of "K". The elbow point shown in Figure 2, where the distortion begins to reduce linearly, is chosen as the ideal number of clusters. This method ensures a balance between capturing the right data structure and avoiding overfitting.

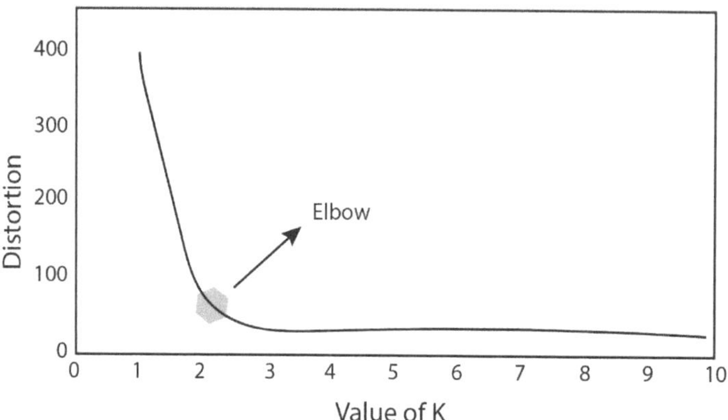

Figure 2. The elbow method being used to find the best K value for the K-means clustering in this paper.

3. Results and Discussion

The surface hardness of the composite samples was evaluated by conducting a scratch test, where the scratch load was gradually increased from 8 N to 15 N. The continuous and highly reflective films did not exhibit noticeable micropores. The samples were divided into three groups based on their surface preparation. In general, the trend observed in the graphs indicated that, as the scratch load remains constant, the hardness tends to decrease as the scratch width becomes wider. This implies that a wider scratch shows a low surface hardness. Moreover, the relationship between scratch width and applied scratch load is directly proportional, meaning that, as the applied load increases, the scratch width also increases. However, it is important to note that the increase in scratch width is non-linear, suggesting that the relationship between load and width may not be strictly linear.

The process of quantifying the scratch width is depicted in Figure 3. In Figure 3b, the shaded area represents the region identified as the scratch. To accomplish this, an individual performed the shading manually using basic image editing software, such as Microsoft Paint. Since the image is in grayscale, consisting of shades of black and white, a K-means clustering algorithm was employed to separate the darker scratched area from the rest of the image. The K-means clustering algorithm is a technique used to partition data into distinct clusters based on their similarity. In this case, it was applied to the grayscale image to create two clusters: one representing black and one representing white. By analyzing the intensity values of the pixels in the image, the algorithm assigned each pixel to one of the two clusters based on its similarity to either black or white.

After the K-means clustering was performed, the resulting clusters provided the coordinates of the pixels within the image that belonged to the black cluster, which represented the scratched area. These coordinates were then utilized to identify the corresponding region on the SEM height map, which provides three-dimensional information about the sample's surface. Figure 3c illustrates the scratch area on the SEM height map. By using the coordinates obtained from the K-means clustering, the scratched region was precisely located and delineated by a boundary line. This enabled a visual representation of the boundaries of the scratch. Finally, in Figure 3d, a more detailed view of the scratch limits is depicted on the contour plot produced by the Phenom XL SEM 3D reconstruction software. The boundary line clearly indicates the extent and shape of the scratch, thereby providing a comprehensive understanding of the scratch width and its specific location on the sample's surface.

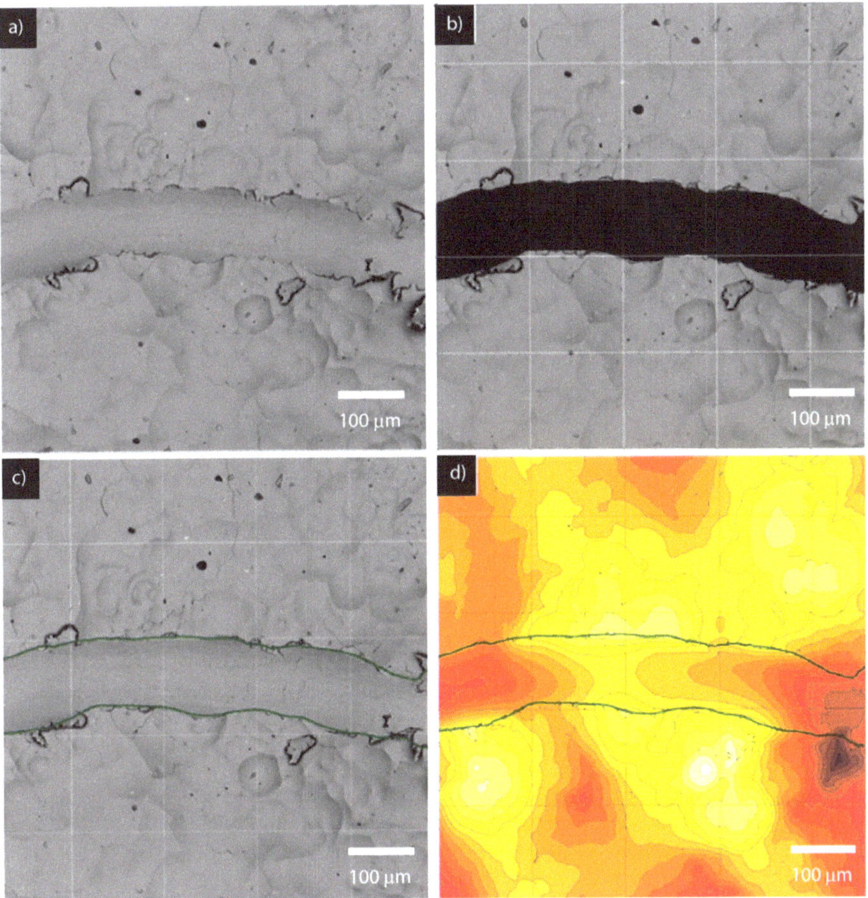

Figure 3. SEM images of the Ni-plated composite surface. (**a**) Raw image obtained using a full backscatter detector (BSD), (**b**) shaded section used to differentiate between scratches and Ni-plating, (**c**) scratch mask coordinates obtained using K-means clustering of the shaded section in (**b**), and (**d**) a contour map of the height map obtained from SEM with a scratch mask for quantifying scratch width.

Once the scratch width data were obtained, they were grouped according to the surface cleaning preparations, as shown in Figure 4. The data were divided into three groups: the first group consisted of samples that underwent chempolishing surface-cleaning preparation, the second group comprised samples that were prepared with organic cleaning, and the third group included samples that were prepared with plasma cleaning. In Figure 4a, the first group is depicted, which contains three samples that underwent chempolishing surface preparation. The CP1 sample shows a scratch width in the ~80–~100 μm range as the load increased from 8 to 12 N. Around 13 N, the scratch width varied, indicating the appearance of more burrs along the scratch contour, thus causing significant jaggedness. Further increase in the load brought the scratch width into the short range (Figure 4a). ML scratch analysis was effective in observing an increase in the average scratch width for the CP2 samples that were subjected to an increasing load (Figure 4a). The average scratch width increased marginally from 60 to 80 μm as load increased 8 to 11N; after that, the load scratch width fluctuated between ~60 to ~100 μm with a large standard deviation that showed the change in material response from a smooth plastic transformation to more burs along the scratch profile. Interestingly, for the 15 N load, a non-uniformity in scratch width was observed, similar to CP1 (Figure 4a). A similar trend was also observed for CP3 as the scratch load increased from 8–15 N. However, for the CP3 sample, the starting average scratch width was around 120 μm for 8 N. This study suggests that the nickel plating hardness on the CP2 sample was around two times more than the plating hardness of the CP3 sample (Figure 1a). By comparing CP1, CP2, and CP3 data, it becomes clear that a significant and clear transition in failure mode occurs between the 12–14 N load range.

The effect of different plating parameters was also studied on the organically cleaned sample (OC group) in Figure 4b. The OC1 sample showed a rather quick jump in average scratch width from the average ~85 μm to the ~120 μm range; the OC1 scratch width remained rather consistent for most of the load range. For the 15 N load, the scratch width was quite non-uniform and appeared with a large variation (Figure 4b). Similarly, the OC2 sample followed the trend observed with OC1. However, the starting scratch width was significantly lower than that observed on OC1. Interestingly, for OC3, the scratch width increased gradually up to 11 N from the ~90 μm to ~120 μm range; after that, the scratch width kept increasing. It appears that for the OC samples, the scratching mechanism was altered in the early stage when compared to the CP samples.

In the case of plasma-treated samples (PC1–3), scratch widths were analyzed. The PC1 sample showed an average scratch width in the ~70 to ~100 μm range as the load increased from 8 to 14 N (Figure 4c). Interestingly, the scratch widths for the PC2 samples increased linearly as the load increased from 8 to 15 N, and the smallest variation was observed in this sample. For the PC3 sample, the scratch width roughly increased with the load. This large variation was attributed to the chempolishing impact on surface morphology because chempolishing can selectively etch one of the components of the BJAM part, thus resulting in a rougher surface. The CP2 samples exhibited a lower average scratch width when compared to CP1 and CP3. This meant that the CP2 samples that had a medium phosphorous (P) nickel coating applied were harder. The CP3 samples showed a significantly larger scratch width with high scattering. It is possible that the nickel coating quality varied significantly when a high-P nickel coating was attempted. Moving on to Figure 4b, the second group represents the three samples that underwent organic cleaning preparation. It is noteworthy that, unlike chempolishing, the organic cleaning process did not impact the BJAM sample surface. Due to better surface smoothness, there was, in general, less scattering. Mid-P nickel coating produced a ~70 μm scratch width, which was nearly 30% lower than the low- and high-phosphorous nickel coatings (Figure 4b). On average, this group was the second hardest, with OC2 (the second organically cleaned sample) showing a high surface hardness that was quite close to the PC2 (the second plasma cleaning) sample. Figure 4c displays the third group, which comprises the samples that underwent plasma cleaning preparation. Plasma cleaning isotopically cleaned the BJAM sample to render a smoother surface. As a result, in general, there was less scattering in the scratch width data. The PC2 samples showed a ~60 μm scratch width, which was clearly more severe than the PC1 samples where low-P solutions were used for Ni coating. Notably, PC2—which represents a combination of a mid-phosphorus level, a temperature 10 degrees lower than the recommended value, and optimal time parameters—demonstrated the highest hardness among all of the samples and had the smallest scratch width. Based on these ML-enabled findings, it is recommended to utilize plasma and organic cleaning methods when aiming for a harder surface. The plasma cleaning method, particularly represented by PC2, resulted in the hardest surface, while the organic cleaning method, particularly represented by OC2, showed a relatively high surface hardness comparable to PC2. Therefore, for applications where a harder surface is desired, the utilization of plasma and organic cleaning methods is recommended based on an analysis of the scratch width and surface hardness data.

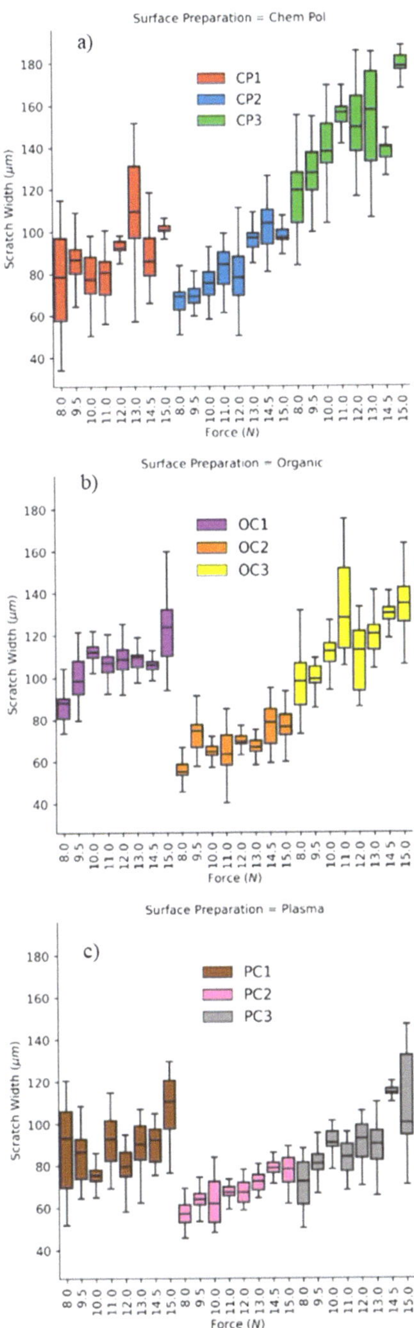

Figure 4. Scratch width measured after a K-means clustering of the samples with different scratch loads: (**a**) samples prepared by chempolishing, (**b**) samples prepared with organic cleaning, and (**c**) samples prepared with plasma cleaning.

4. Conclusions

The K-means unsupervised ML algorithm was employed to address the challenges associated with optically obscure scratches on nickel-plated AM samples. In this context, the samples were prepared using the L9 orthogonal array TDOE methodology. Due to the nature of Ni-electroless plating, some of the samples exhibited a shiny appearance, making them difficult to analyze accurately when using a digital light microscope. This was primarily due to issues such as light saturation and reflection. To overcome this, SEM was utilized to generate grayscale images and corresponding 3D height maps of the scratched Ni-plating surfaces. Subsequently, the K-means ML clustering approach was applied to visually detect the scratch areas within the SEM images. Through this approach, it was observed that the TDOE methodology resulted in distinct levels of Ni-plating hardness for each of the nine samples. Furthermore, as the scratch force increased, the scratch widths exhibited a non-linear increase, thus highlighting the complex relationship between applied force and scratch width. Our image analysis capabilities highlighted that mid-P nickel coating produced harder coating when compared to low- and high-P content-based nickel coatings. This study also showed that the chempolishing treatment on BJAM produces a higher roughness that impacts the uniformity and quality of nickel coatings. Our research suggests that surface preparation must be chosen with great care to target the specific attributes of electroless nickel coatings, and microscopic high-resolution SEM images should be considered for an adequate understanding of the morphologies that evolve due to interacting parameters. A scratch width analysis with a Taguchi design of experiment should be focused on specific properties. The CP2, PC2, and OC2 samples, where the medium-phosphorous solution was used, appeared to yield harder coatings. Our ML-enabled scratch width analysis was able to capture the differences in various factors leading to the differences in scratch widths and deviations. The difference in standard deviations at each load for each sample category was reflective of the difference in the surface microstructure after different processing techniques and electroless nickel coatings were applied. The K-means clustering approach utilized in this work was able to capture the variation in load. The demonstrated methodology of combining SEM imaging, K-means clustering, and scratch width quantification offered a practical solution in surface analysis when faced with obstacles such as optically obscure scratches. In future work, different clustering and ML approaches may be applied to analyze scratch widths.

Author Contributions: The manuscript was written by B.N.M.; A.C.G. conducted the experiments, performed the K-means analysis, and collected the data. W.D. reviewed and gave feedback about the analysis and on paper refinement. K.L.K. provided equipment support and analysis evaluation. A.E. provided stainless-steel and bronze composite samples that were produced via the binder-jetted method, as well as contributed to the results analysis. Lastly, P.T. supervised, provided guidance, and oversaw the work. All authors have read and agreed to the published version of the manuscript.

Funding: We acknowledge funding support for this course from the National Science Foundation-CREST Award (Contract # HRD-1914751), the Department of Energy/National Nuclear Security Agency (DE-FOA-0003945), and the NASA MUREP grant (80NSSC19M0196). This manuscript has been authored, in part, by UT-Battelle, LLC, under contract DE-AC05-00OR22725 with the US Department of Energy (DOE). The publisher acknowledges the US government's license to provide public access under the DOE Public Access Plan (http://energy.gov/downloads/doe-public-access-plan, accessed on 17 September 2023).

Institutional Review Board Statement: Not applicable.

Informed Consent Statement: Not applicable.

Data Availability Statement: Data will be made available upon reasonable request.

Acknowledgments: We appreciate Lucas Rice's input about the testing protocol. This research was prepared under the MECH 500 Research Methods and Technical Communication course taught by Pawan Tyagi and teaching assistant Wondwosen Demisse in the Fall of 2022.

Conflicts of Interest: The authors declare no conflict of interest. The funders had no role in the design of the study, in the collection, analyses, or interpretation of data, in the writing of the manuscript, or in the decision to publish the result.

References

1. Lundervold, A.S.; Lundervold, A. An overview of deep learning in medical imaging focusing on MRI. *Z. Med. Phys.* **2019**, *29*, 102–127. [CrossRef] [PubMed]
2. Mahesh, B. Machine learning algorithms—A review. *Int. J. Sci. Res.* **2020**, *9*, 381–386.
3. Jiang, J. A survey of machine learning in additive manufacturing technologies. *Int. J. Comput. Integr. Manuf.* **2023**, *36*, 1258–1280. [CrossRef]
4. Parsazadeh, M.; Sharma, S.; Dahotre, N. Towards the next generation of machine learning models in additive manufacturing: A review of process dependent material evolution. *Prog. Mater. Sci.* **2023**, *135*, 101102. [CrossRef]
5. Xames, D.; Torsha, F.K.; Sarwar, F. A systematic literature review on recent trends of machine learning applications in additive manufacturing. *J. Intell. Manuf.* **2022**, *34*, 2529–2555. [CrossRef]
6. Mondal, B.; Mukherjee, T.; DebRoy, T. Crack free metal printing using physics informed machine learning. *Acta Mater.* **2022**, *226*, 117612. [CrossRef]
7. Pattnaik, P.; Sharma, A.; Choudhary, M.; Singh, V.; Agarwal, P.; Kukshal, V. Role of machine learning in the field of Fiber reinforced polymer composites: A preliminary discussion. *Mater. Today Proc.* **2021**, *44*, 4703–4708. [CrossRef]
8. Cearley, D.; Burke, B.; Searle, S.; Walker, M.J. Top 10 strategic technology trends for 2018. *Top* **2016**, *10*, 1–246.
9. Demisse, W.; Xu, J.; Rice, L.; Tyagi, P. Review of internal and external surface finishing technologies for additively manufactured metallic alloys components and new frontiers. *Prog. Addit. Manuf.* **2023**, 1–21. [CrossRef]
10. Ilogebe, A.B.; Waters, C.K.; Elliot, A.M.; Shackleford, C. Morphology of binder-jet additive manufactured structural amorphous metal matrix composites. *Int. J. Eng. Sci.* **2019**, *8*, 15–24.
11. Onler, R.; Koca, A.S.; Kirim, B.; Soylemez, E. Multi-objective optimization of binder jet additive manufacturing of Co-Cr-Mo using machine learning. *Int. J. Adv. Manuf. Technol.* **2022**, *119*, 1091–1108. [CrossRef]
12. Zhu, Y.; Wu, Z.; Hartley, W.D.; Sietins, J.M.; Williams, C.B.; Yu, H.Z. Unraveling pore evolution in post-processing of binder jetting materials: X-ray computed tomography, computer vision, and machine learning. *Addit. Manuf.* **2020**, *34*, 101183. [CrossRef]
13. Roscoe, S.; Cousins, P.D.; Handfield, R. Transitioning additive manufacturing from rapid prototyping to high-volume production: A case study of complex final products. *J. Prod. Innov. Manag.* **2023**, *40*, 554–576. [CrossRef]
14. Almaraz, A.; Estrada, D.; Rajabi-Kouchi, F.; Burgoyne, H.; Mansoor, N.; Koehne, J. *Additive Manufacturing for the Rapid Prototyping of Economical Biosensors*; Boise State University: Boise, ID, USA, 2023.
15. Nys, N.; König, M.; Neugebauer, P.; Jones, M.J.; Gruber-Woelfler, H. Additive Manufacturing as a Rapid Prototyping and Fabrication Tool for Laboratory Crystallizers—A Proof-of-Concept Study. *Org. Process. Res. Dev.* **2023**, *27*, 1455–1462. [CrossRef]
16. Venturi, F.; Taylor, R. Additive Manufacturing in the Context of Repeatability and Reliability. *J. Mater. Eng. Perform.* **2023**, *32*, 6589–6609. [CrossRef]
17. Tyagi, P.; Goulet, T.; Riso, C.; Stephenson, R.; Chuenprateep, N.; Schlitzer, J.; Benton, C.; Garcia-Moreno, F. Reducing the roughness of internal surface of an additive manufacturing produced 316 steel component by chempolishing and electropolishing. *Addit. Manuf.* **2019**, *25*, 32–38. [CrossRef]
18. Townsend, A.; Senin, N.; Blunt, L.; Leach, R.K.; Taylor, J.S. Surface texture metrology for metal additive manufacturing: A review. *Precis. Eng.* **2016**, *46*, 34–47. [CrossRef]
19. Hebert, R.J. Viewpoint: Metallurgical aspects of powder bed metal additive manufacturing. *J. Mater. Sci.* **2016**, *51*, 1165–1175. [CrossRef]
20. Dillard, J.; Grizzle, A.; Demisse, W.; Rice, L.; Klein, K.; Tyagi, P. Alternating chempolishing and electropolishing for interior and exterior surface finishing of additively manufactured (AM) metal components. *Int. J. Adv. Manuf. Technol.* **2022**, *121*, 8159–8170. [CrossRef]
21. Kato, K. Wear in relation to friction—A review. *Wear* **2000**, *241*, 151–157. [CrossRef]
22. Balaraju, J.; Narayanan, T.S.; Seshadri, S. Electroless Ni–P composite coatings. *J. Appl. Electrochem.* **2003**, *33*, 807–816. [CrossRef]
23. Loto, C. *Electroless Nickel Plating—A Review*; Springer: Berlin/Heidelberg, Germany, 2016.
24. Stallard, J.; Poulat, S.; Teer, D. The study of the adhesion of a TiN coating on steel and titanium alloy substrates using a multi-mode scratch tester. *Tribol. Int.* **2006**, *39*, 159–166. [CrossRef]
25. Chernyavsky, D.; Kononenko, D.Y.; Han, J.H.; Kim, H.J.; Brink, J.v.D.; Kosiba, K. Machine learning for additive manufacturing: Predicting materials characteristics and their uncertainty. *Mater. Des.* **2023**, *227*, 111699. [CrossRef]
26. Jain, A.K. Data clustering: 50 years beyond K-means. *Pattern Recognit. Lett.* **2010**, *31*, 651–666. [CrossRef]
27. Greene, D.; Cunningham, P.; Mayer, R. Unsupervised learning and clustering. In *Machine Learning Techniques for Multimedia: Case Studies on Organization and Retrieval*; Springer Science & Business Media: Berlin/Heidelberg, Germany, 2008; pp. 51–90.
28. Ding, C.; He, X. K-means clustering via principal component analysis. In Proceedings of the Twenty-First International Conference on Machine Learning, Banff, AB, Canada, 4–8 July 2004.
29. Kodinariya, T.M.; Makwana, P.R. Review on determining number of Cluster in K-Means Clustering. *Int. J.* **2013**, *1*, 90–95.
30. Sinaga, K.P.; Yang, M.-S. Unsupervised K-Means Clustering Algorithm. *IEEE Access* **2020**, *8*, 80716–80727. [CrossRef]

31. Cohn, R.; Holm, E. Unsupervised Machine Learning Via Transfer Learning and k-Means Clustering to Classify Materials Image Data. *Integr. Mater. Manuf. Innov.* **2021**, *10*, 231–244. [CrossRef]
32. Brent, D.; Saunders, T.A.; Moreno, F.G.; Tyagi, P. Taguchi Design of Experiment for the Optimization of Electrochemical Polishing of Metal Additive Manufacturing Components. In *ASME International Mechanical Engineering Congress and Exposition*; American Society of Mechanical Engineers: New York, NY, USA, 2016. [CrossRef]

Article

Mitigation of Gas Porosity in Additive Manufacturing Using Experimental Data Analysis and Mechanistic Modeling

Satyaki Sinha and Tuhin Mukherjee *

Department of Mechanical Engineering, Iowa State University, Ames, IA 50011, USA; satty51@iastate.edu
* Correspondence: tuhinm@iastate.edu

Abstract: Shielding gas, metal vapors, and gases trapped inside powders during atomization can result in gas porosity, which is known to degrade the fatigue strength and tensile properties of components made by laser powder bed fusion additive manufacturing. Post-processing and trial-and-error adjustment of processing conditions to reduce porosity are time-consuming and expensive. Here, we combined mechanistic modeling and experimental data analysis and proposed an easy-to-use, verifiable, dimensionless gas porosity index to mitigate pore formation. The results from the mechanistic model were rigorously tested against independent experimental data. It was found that the index can accurately predict the occurrence of porosity for commonly used alloys, including stainless steel 316, Ti-6Al-4V, Inconel 718, and AlSi10Mg, with an accuracy of 92%. In addition, experimental data showed that the amount of pores increased at a higher value of the index. Among the four alloys, AlSi10Mg was found to be the most susceptible to gas porosity, for which the value of the gas porosity index can be 5 to 10 times higher than those for the other alloys. Based on the results, a gas porosity map was constructed that can be used in practice for selecting appropriate sets of process variables to mitigate gas porosity without the need for empirical testing.

Keywords: laser powder bed fusion; 3D printing; convective flow; buoyancy; Stokes law; gas porosity index; stainless steel 316; Ti-6Al-4V; Inconel 718; AlSi10Mg

Citation: Sinha, S.; Mukherjee, T. Mitigation of Gas Porosity in Additive Manufacturing Using Experimental Data Analysis and Mechanistic Modeling. *Materials* **2024**, *17*, 1569. https://doi.org/10.3390/ma17071569

Academic Editor: Francesco Iacoviello

Received: 5 March 2024
Revised: 20 March 2024
Accepted: 26 March 2024
Published: 29 March 2024

1. Introduction

Laser powder bed fusion (LPBF) additive manufacturing is capable of printing 3D parts of a wide variety of steels and alloys of nickel, titanium, and aluminum for the aerospace, healthcare, automotive, and energy industries [1–5]. A laser beam selectively scans closely packed layers of powders and creates a molten pool, which after solidification, forms the part. Gas bubbles can originate inside the molten pool from shielding gas, metal vapors, and gases trapped inside powders during atomization [1]. If these gas bubbles are unable to escape from the molten pool before solidification, they can result in gas porosity inside the printed parts [6,7]. Gas porosities can significantly degrade the tensile [8–13] and fatigue properties [14–17] of parts. For example, the presence of porosity can increase the plastic strain during the tensile test to the point where additional plastic deformation is restricted to a smaller cross-sectional area or a region that was not work-hardened [11]. Because of this, tensile properties, such as ultimate tensile strength, are significantly reduced. Porosity is also a determining factor in the case of fatigue performance, as it can serve as a site for the initiation of fatigue cracks [14–17]. Therefore, mitigation of gas porosity is needed to improve the mechanical properties, reliability, and serviceability of metallic parts made by LPBF.

Several attempts have been made to mitigate gas porosity in LPBF parts (Table 1) using experimental techniques [18–25], mechanistic modeling [26–34], and machine learning [35–41]. However, experimental trial-and-error to adjust many process variables for reducing gas porosity is expensive and time-consuming. In addition, this trial-and-error approach does not always guarantee achieving an optimized set of process variables. Post-processing techniques [19,21,23], such as hot isostatic pressing, can reduce porosity but significantly add cost. Numerical models [26–34] have been developed to predict the formation of gas porosity by capturing the underlying physics. However, these models are computationally intensive and often difficult to use in real time. Machine learning models [35–41] can be used in real-time; however, they are often unable to capture the important physical factors causing porosity and need a large volume of high-quality data for a reliable prediction. Thus, the existing approaches (Table 1) using experimental, modeling, and machine learning methods are inadequate to reduce gas porosity in LPBF. Therefore, what is needed and currently unavailable is an integrated theoretical and experimental framework that can identify all important physical factors causing gas porosity and combine them in a comprehensible manner to predict and control the pore formation in LPBF of diverse metallic materials. This article aims to address that need.

Table 1. Several existing approaches to reduce gas porosity in LPBF [18–41].

Approach	Alloy	Example	Ref.
Experimental approach	SS 316	Porosities were eliminated by changing energy density guided by X-ray CT, Optical Microscopy, and Archimedes method.	[16]
	SS 316	Porosities were successfully removed by wire electrical discharge polishing-based post-processing.	[17]
	SS 316	Adjustment of the process variables such as point distance, exposure time, and layer thickness during experiments lowered porosity.	[18]
	Ti6Al4V	Laser post-processing decreased gas pores which was confirmed by micro-CT examination.	[19]
	Ti6Al4V	Gas pores were eliminated by post-process hot isostatic pressing at different temperatures and pressures.	[20]
	AlSi10Mg	A low temperature (350 °C) hot isostatic pressing minimized gas porosity in the manufactured parts.	[21]
	Inconel 718	To reduce the gas porosity, different heat-treatment procedures including aging treatments were used.	[22]
	Inconel 718	Gas porosities were prevented by implementing a high-energy-intensity laser beam that resulted in a larger molten pool.	[23]
Mechanistic modeling approach	SS 316	Gas pores were eliminated by identifying appropriate conditions through thermodynamic calculations and genetic algorithms.	[24]
	SS 316	To reduce gas porosities laser power was varied guided by mechanistic modeling.	[25]
	SS 316	Porosities were successfully removed by varying the volumetric energy density assisted by a numerical model.	[26]
	SS 316	A calculation scheme was introduced that included normalized enthalpy and powder absorptivity measurements to decrease porosity.	[27]
	Ti6Al4V	Process maps were introduced that utilized normalized energy density to reduce porosity.	[28]
	Ti6Al4V	A dimensional analysis helped to reduce porosity and explained variability in defect behavior.	[29]
	AlSi10Mg	A molecular dynamics analysis showed that decreasing the hydrogen content and maximizing the cooling/heating times reduced porosity.	[30]
	AlSi10Mg	Thermal history and graph theory helped to identify appropriate conditions to reduce porosity.	[31]
	AlSi10Mg	Dynamics and mechanisms of pore motion were simulated using a multi-physics model to identify conditions for pore reduction.	[6]
	Inconel 718	A modeling technique was used that coupled laser powder interaction to examine spatter interaction and decreased porosity.	[32]

Table 1. *Cont.*

Approach	Alloy	Example	Ref.
Machine learning approach	SS 316	A novel approach using thermography and deep learning was used to anticipate and reduce local porosity.	[33]
	SS 316	Improved Regression along with Convolutional Neural Networks were used to reduce porosity.	[34]
	Ti6Al4V	A deep learning architecture was employed using heat signals to predict and minimize porosity.	[35]
	Ti6Al4V	A deep learning technique was used for porosity reduction and monitoring that used Convolutional Neural Networks.	[36]
	AlSi10Mg	Effective reduction efforts were aided by porosity-type classification through machine learning.	[37]
	AlSi10Mg	Porosity was reduced through the use of Convolutional Neural Networks that were trained using the molten pool data.	[38]
	Inconel 718	Defect detection in SEM pictures was automated by deep learning, which promoted stochastic development and lowered porosity.	[39]

Here, mechanistic modeling and experimental data analysis were combined to predict and control gas porosity during LPBF of stainless steel 316, Ti-6Al-4V, Inconel 718, and AlSi10Mg that are commonly used in the aerospace, automotive, healthcare, and energy industries. First, the important physical factors that impact the formation of gas porosity were calculated for a broad range of processing conditions. The computed results were rigorously tested against independent experiments. The results were aimed at providing a detailed, comprehendible scientific insight into the pore formation in LPBF. We used the modeling and experimental data to derive a verifiable, user-friendly, dimensionless gas porosity index to predict both the occurrence and amount of gas pores in LPBF parts. Finally, we constructed a process map to help engineers select appropriate processing conditions to mitigate gas porosity. The process map can also provide an in-depth scientific understanding of the effects of LPBF processing conditions on porosity formation. Although the results reported here are for the LPBF process, the methodology can be extended to mitigate gas porosity in laser and electron beam-directed energy deposition as well as in wire arc additive manufacturing processes.

2. Methodology

In this work, a combined approach (Figure 1) of using mechanistic modeling and analysis of experimental data on the occurrence of gas porosity in LPBF parts of various common alloys was used. First, a mechanistic model [42,43] of the LPBF process was tested, calibrated using experimental results, and used to compute temperature fields and molten pool geometry. Then, experimental data [44–60] on porosity during LPBF of stainless steel 316, Ti-6Al-4V, Inconel 718, and AlSi10Mg were collected from the literature. The results from the well-tested model were used to derive and calculate a gas porosity index corresponding to all experimental cases. The mechanistic model, calculation of the gas porosity index, and collection and analysis of experimental data are explained in detail below.

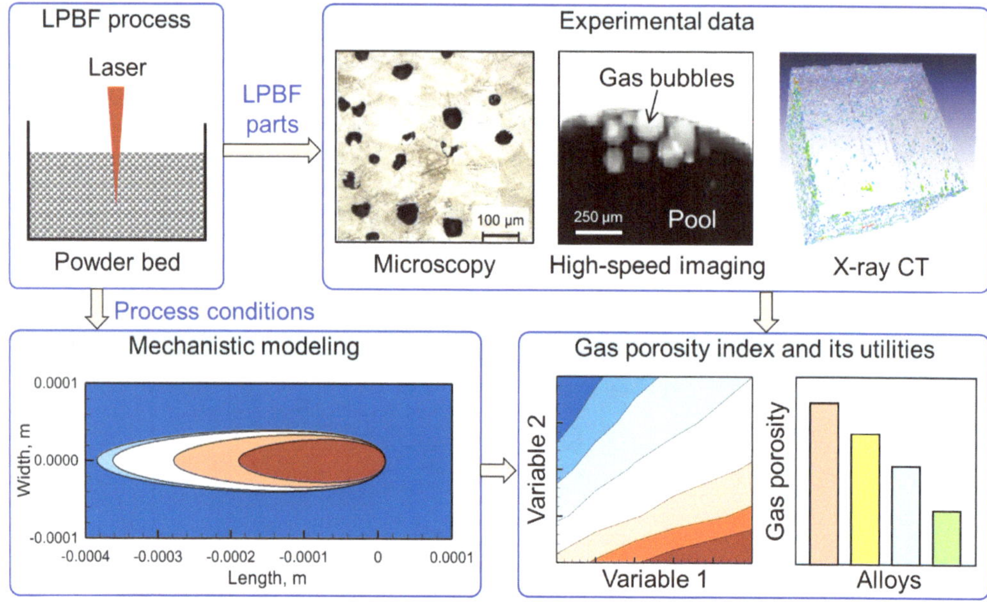

Figure 1. A schematic representation of the methodology used in this work. A combined approach of using mechanistic modeling and analysis of experimental data was implemented to derive and use a gas porosity index for predicting and controlling gas porosity during LPBF of common alloys. The pictures inside the "Experimental data" box are adapted from [7,61,62]. Figures taken from open-access articles [7,61] are under the terms and conditions of the Creative Commons Attribution (CC BY) license. The figure taken from [62] is under the permission obtained from Elsevier.

2.1. Assumptions

The following simplifying assumptions were made for the mechanistic model to make the calculations of temperature and molten pool geometry tractable:

(1)   The laser beam moves at a constant speed in the same direction on a straight path relative to the substrate. The model assumed a quasi-steady state of heat conduction [1,43] where the coordinate along the scanning direction was transformed [1] to capture the effect of scanning speed.

(2)   The laser beam energy was assumed to be focused at a point on the upper surface of the deposit and was applied at a uniform rate [1,43]. It was assumed that the width of the substrate is significantly larger than the track width, and the substrate is much thicker than the depth of the molten pool [1].

(3)   The effects of the convective flow of molten metal, mainly driven by the spatial gradient of surface tension and buoyancy [1,63] on temperature fields were neglected. Heat losses from the surface through convection and radiation [1] were disregarded.

(4)   Thermophysical properties of alloys were considered to be temperature-independent.

(5)   The values of the laser absorptivity were assumed to be constant for a given alloy, even though it is expected to be somewhat influenced by other factors such as processing parameters, the presence of oxide and other surface impurities, surface roughness, and gas composition above the molten pool [64].

(6)   Only conduction mode [2,3] LPBF was considered. Therefore, gas bubbles from unstable keyholes and resulting keyhole porosities [65,66] are not within the scope of this work.

(7)   Effects of gas dissolution [1] in the liquid metal controlled by activity and partial pressure of gas were ignored by assuming the nucleation of gas bubbles on the

solidifying interface. The bubble size was estimated by a pressure balance that ignored the effects of the coalescence of bubbles.

The results from the mechanistic model were compared with a multi-physics, 3D, transient heat transfer and fluid flow model of LPBF to prove that the assumptions are valid. The comparison results are reported in the Supplementary File.

2.2. Calculations of Temperature and Molten Pool Geometry

A mechanistic model of LPBF was used to compute the temperature distributions and the molten pool dimensions using process variables such as laser power, scanning speed, and preheat temperature, as well as alloy properties such as density, thermal conductivity, and specific heat as inputs. The thermophysical properties [2,3,67] of the alloys used in the model are reported in the Supplementary File. Under the same processing conditions, the molten pool shape and size for different alloys can significantly vary depending on these properties [42]. The temperature (T) at any location of the part can be expressed as [43,68]:

$$T = T_0 + \frac{Q}{2\pi\, k_{eff}\, \xi} \exp\left[-\frac{V(\xi + x)}{2\alpha_{eff}}\right] \tag{1}$$

where T_0 indicates the initial or preheat temperature, Q is the laser power absorbed, k_{eff} is the effective thermal conductivity of the powder bed, V is the scanning speed, α_{eff} is the effective thermal diffusivity of the powder bed, ξ is the distance from the laser beam axis, and x represents the coordinate along the scanning direction. The effective powder bed thermophysical properties depend on the properties of both the metal powders and shielding gas [2]. The shielding gas trapped between the closely packed powders and the powder bed's packing efficiency determine the effective thermo-physical properties of the packed powder bed [69]. The traditional solution of heat conduction equation was based on the properties of solid materials [70]. However, in this work, we modified the solution by considering the effective powder bed properties (Equation (1)). The effective thermal diffusivity (α_{eff}) of the powder bed in Equation (1) is represented as [3]:

$$\alpha_{eff} = \frac{k_{eff}}{Cp_{eff}\, \rho_{eff}} \tag{2}$$

where k_{eff} is the effective powder bed thermal conductivity (Equation (1)). Cp_{eff} and ρ_{eff} are the effective specific heat and density of the powder bed, respectively. The effective properties are represented as [3]:

$$k_{eff} = k_s\eta + k_g(1 - \eta) \tag{3}$$

$$Cp_{eff} = \frac{((\rho_s\eta Cp_s) + (\rho_g(1 - \eta)Cp_g))}{\rho_s\eta + \rho_g(1 - \eta)} \tag{4}$$

$$\rho_{eff} = \rho_s\eta + \rho_g(1 - \eta) \tag{5}$$

where η is the powder bed packing efficiency. In Equations (3)–(5), the suffix 's' and 'g' represent the property values for the solid alloy and shielding gas, respectively. The thermophysical properties [2,3,67] of solid alloys and shielding gas [3] and packing efficiency are reported in the Supplementary File.

From the computed temperature field (Equation (1)), the dimensions of the molten pool were extracted by tracking the solidus isotherms of alloys. Calculated temperature and molten pool dimensions were used to derive and compute a gas porosity index for predicting and controlling gas porosity in LPBF, as discussed below.

2.3. Gas Porosity Index and Its Calculations

High-speed imaging [7] of both fusion welding and additive manufacturing processes has revealed that the gas bubbles formed inside the molten pool can result in porosity. For example, Figure 2 shows the presence of gas bubbles inside the molten pool. Gas porosities occur if the gas bubbles are unable to escape out of the molten pool before solidification. Therefore, the time needed for a gas bubble to rise and escape out of the molten pool and the solidification time of the pool are two key time factors affecting the formation of gas porosity. Here, the gas porosity index (τ) was defined as the ratio of the time to rise of the gas bubble (T_R) to the time to solidify (T_S) of the molten pool as:

$$\tau = T_R / T_S \tag{6}$$

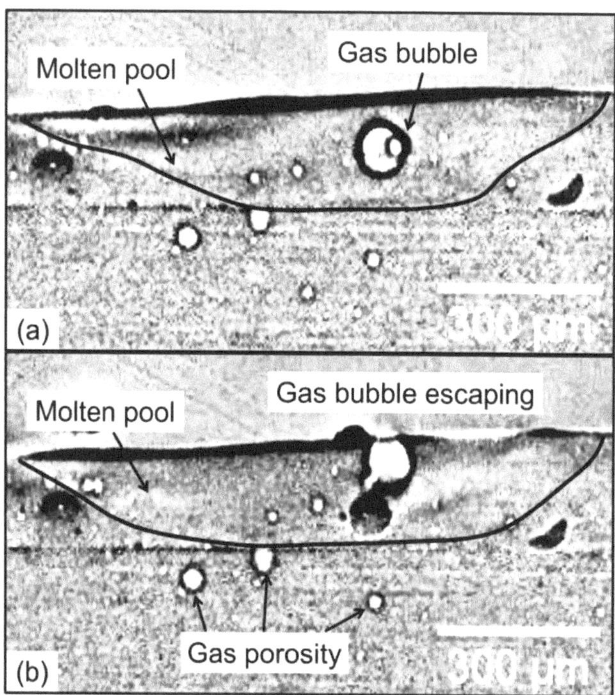

Figure 2. In situ high-speed synchrotron X-ray images showing the formation and dynamics of gas bubbles during directed energy deposition of a nickel-based superalloy. Side views of the molten pool are shown. (**a**) Gas bubbles inside the molten pool. (**b**) Gas bubbles are escaping from the molten pool. The figure is adapted from [7]. Figures are taken from an open-access article [7] under the terms and conditions of the Creative Commons Attribution (CC BY) license.

A high value of τ indicates that a gas bubble takes a long time to escape from the molten pool before solidification resulting in a high susceptibility to gas porosity. The index also captures the effects of process variables and alloy properties on porosity [71–73]. The two aforementioned times were calculated based on the results from the mechanistic model as explained below. A sample calculation of the gas porosity index is provided in the Supplementary File.

2.3.1. Calculation of Time to Solidify

Time to solidify (T_S) indicates the time required for the solidification of the liquid metal pool and can be represented as:

$$T_S = \frac{L}{V} \tag{7}$$

where L is the pool length estimated using the mechanistic model and V is the scanning speed. The time to solidify decreases at a higher scanning speed. This is because an increase in the scanning speed reduces the pool size, and the pool takes less time to solidify.

2.3.2. Calculation of Time to Rise

Time to rise (T_R) indicates the time needed for a gas bubble to rise and escape out of the molten pool and can be written as:

$$T_R = \frac{D}{u_e} \tag{8}$$

where the depth of the pool (D) was computed using the mechanistic model. u_e is the escape velocity of the gas bubble. The calculation assumes that the nucleation of gas bubbles occurs on the solidifying interface near the bottom of the molten pool. It is well-known in the casting and fusion welding literature [74] that the velocity of a gas bubble inside a liquid depends on the size of the bubble. Therefore, we first calculated the size of the bubble by performing a pressure balance inside the liquid metal and used that to estimate the escape velocity as discussed below.

Calculation of Gas Bubble Size by Pressure Balance

The pressure inside a stable gas bubble (P_i) is equal to the sum of the surface tension pressure (P_s) and the liquid pressure (P_l) as [75]:

$$P_i = P_s + P_l \tag{9}$$

Surface tension pressure (P_s) is given by [75]:

$$P_s = \frac{2\sigma}{r} \tag{10}$$

where σ is the surface tension of the molten alloy and r is the radius of a spherical gas bubble. The liquid pressure (P_l) is represented as a summation of the atmospheric pressure (P_a) and the pressure due to the height of the liquid ($\rho g D$) as:

$$P_l = P_a + \rho g D \tag{11}$$

where ρ, g, and D are the density of the liquid metal, the acceleration due to gravity, and pool depth. For a tiny pool in LPBF, $\rho g D$ is negligible. Therefore,

$$P_l = P_a \tag{12}$$

By using Equation (9) and the ideal gas law, the value of the radius of a spherical gas bubble (r) can be calculated as:

$$\frac{4}{3}\pi r^3 \left[P_a + \frac{2\sigma}{r} \right] = RT \tag{13}$$

where R is the universal gas constant and T is the solidus temperature of an alloy.

Escape Velocity of Gas Bubbles

Since the density of gas is much lower than that of the molten liquid, gas bubbles tend to rise inside the liquid pool due to the buoyancy. For small Reynolds numbers, the rising velocity of gas bubbles can be represented by the Stokes law [74]. Here the flow of the molten metal is assumed to be laminar with a low Reynolds number. Therefore, the approximate rising velocity of spherical bubbles called the Stokes velocity (u_s) is given by [76]:

$$u_s = \frac{2}{9} \frac{r^2 \Delta \rho g}{\mu} \tag{14}$$

where r is the radius of the gas bubble calculated using Equation (13), $\Delta \rho$ is the difference in density between the gas and the molten liquid, g is the acceleration due to gravity, and μ is the viscosity of the liquid. The possibility of bubbles escaping from the fusion zone is increased by the low liquid viscosity and large bubble radius [76].

It is well known [1] that the convective flow of liquid metal is mainly driven by the spatial gradient of surface tension (Marangoni effect) and buoyancy. High-speed imaging has shown that the Marangoni effect on the gas bubble dynamics is important only near the pool surface [77]. However, a gas bubble often nucleates near the bottom surface, where buoyancy may play a more crucial role [78]. Because of the greater dominance of the buoyancy force, we compute the convective velocity [79] as:

$$u_c = \sqrt{g \beta \Delta T D} \tag{15}$$

where g is the acceleration due to gravity, β is the coefficient of volumetric expansion, ΔT is the temperature gradient, and D is the depth of the pool computed using the mechanistic model. This convective velocity of liquid metal accelerates the escape velocity of the gas bubbles. Therefore, the escape velocity of gas bubbles (u_e) is the summation of two aforementioned velocities [76] as:

$$u_e = u_S + u_c \tag{16}$$

The value of the escape velocity is used in Equation (8) to estimate the time to rise of the gas bubble.

2.4. Data Collection and Analysis

A total of 93 sets of data on gas porosity formation for four alloys at various processing conditions were collected from the literature [44–60]. Among the 93 sets of data, 60 cases had gas pores and 33 cases were without experimentally detected gas pores. The mechanistic modeling was performed for all 93 cases to calculate the gas porosity index. Using this data, the gas porosity index was tested for the four alloys and the range of variables provided in Table 2. The Supplementary File contains the values of all variables corresponding to the 93 experimental cases for which calculations were performed.

Table 2. Range of process parameters and thermophysical properties of alloys.

Parameters	Range
Laser power (W)	30–331
Scanning speed (mm/s)	50–3400
Pool length (micron)	149–2298
Pool width (micron)	64–664
Time to rise (ms)	5.15–69.75
Time to solidify (ms)	0.18–29.34
Thermal conductivity (W/m-K)	28.1–113.0
Specific heat (J/Kg-K)	409.6–2894.2
Viscosity (Kg/m-s)	0.0013–0.007
Surface tension (N/m)	0.82–1.82

3. Results and Discussion

3.1. Comparison of Molten Pool Geometry of Four Alloys

Both the time needed for a gas bubble to rise and escape out of the molten pool and the solidification time of the pool that determines the gas porosity (Section 2.3) are significantly affected by molten pool dimensions. Since different alloys exhibit a wide variety of molten pool geometries [64], it is important to compare them under the same processing conditions. Figure 3a–d shows the computed temperature distribution on the deposit top surface for the four common alloys. The region surrounded by the isotherm of solidus temperature represents the molten pool. The fusion zone is enclosed by the liquidus isotherm of an alloy. The sky-blue area between the liquidus and solidus isotherms in each figure represents the mushy zone. The laser beam scans from the left to the right direction. Therefore, the molten pool is elongated in the opposite direction of the scanning direction.

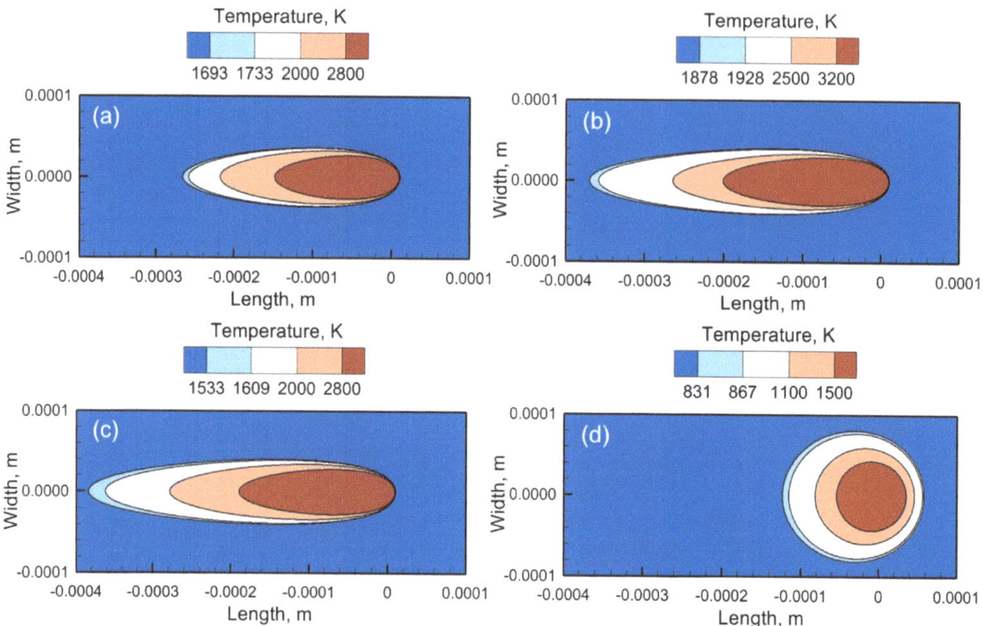

Figure 3. Computed temperature fields and molten pool geometries on the deposit top surface during LPBF of (**a**) Stainless steel 316, (**b**) Ti-6Al-4V, (**c**) Inconel 718, and (**d**) AlSi10Mg using 300 W laser power and 1250 mm/s scanning speed. The results correspond to the location of the laser beam axis at (0,0). The scanning direction is from left to right (along the positive length axis) in all figures. In each figure, the values of the isotherms can be read from the corresponding contour legends.

The different shapes of the molten pool are due to the variation in the thermophysical properties of the four alloys. The molten pools for SS 316, Ti-6Al-4V, and Inconel 718 exhibit a tear-dropped and elongated shape due to rapid scanning. The molten pool for AlSi10Mg is elliptical (Figure 3d) because the heat distribution is nearly uniform in all directions, attributed to its high thermal diffusivity. Consequently, it has a large width and a short length. This elliptical pool has the largest volume compared to the molten pools of the other three alloys. In addition, the temperature inside the AlSi10Mg molten pool is the lowest among the four alloys due to its very high thermal diffusivity. Ti-6Al-4V shows a larger liquid pool (Figure 3b) than SS 316 (Figure 3a) because of its lower density. In addition, a larger difference between the liquidus temperature and solidus temperature of Inconel 718 results in a very elongated mushy zone and molten pool (Figure 3c). Although Ti-6Al-4V and Inconel 718 exhibit similar pool sizes, the temperature inside the molten pool of Ti-6Al-4V is higher due to its lower density than Inconel 718.

Figure 4 compares the calculated and experimentally measured [45] track width of stainless steel 316 deposits made by LPBF at different laser powers and scanning speeds. The proximity of the data points to the 45° line indicates that the computed results agree well with the experiments. The track width for each condition was measured five times along the track, and an average value was reported. The error bars represent the standard deviations. The RMSE value of the prediction is 35.2 microns, which is very similar to the average of the error bars present in the experimental data (27.7 microns). It indicates that the error in prediction is heavily influenced by the uncertainties in the measurement. In addition, several simplifying assumptions in the calculations (Section 2.1) have also contributed to the error. The reasonably good match between the computed and experimental results gives us the confidence to use the mechanistic model to compute the gas porosity index for different alloys and processing conditions.

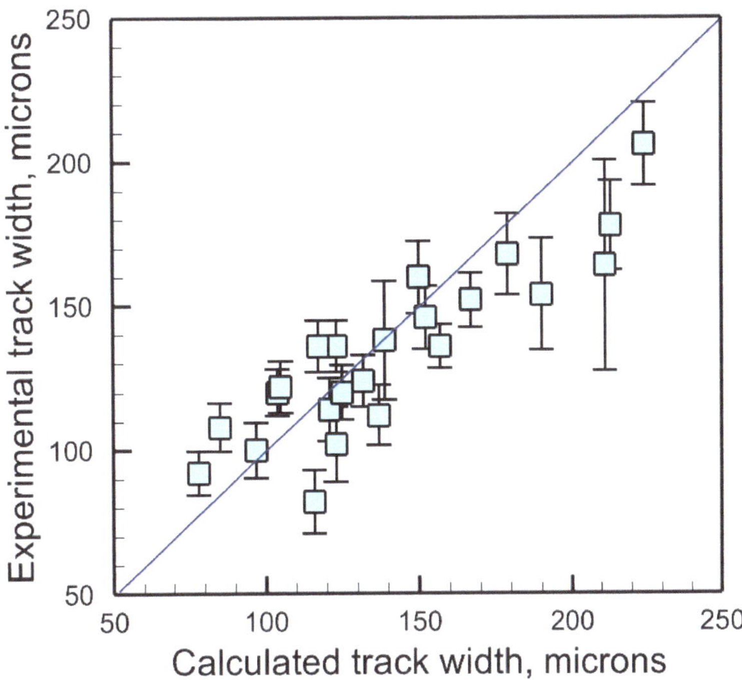

Figure 4. Comparison between the calculated and the experimentally measured [45] track width of stainless steel 316 deposits made by LPBF at different laser powers and scanning speeds. The RMSE value for the experimental width and calculated width is 35.2 microns. The average of the error bars present is calculated as 27.7 microns.

3.2. Prediction of Porosity Using Gas Porosity Index

A gas porosity index (τ) is an easy-to-use, verifiable, and dimensionless indicator (see Section 2.3) that can predict gas porosity defects. There are two main utilities of the gas porosity index. First, it can predict if porosity will form or not under a given set of processing conditions for a particular alloy. Second, if porosity forms, the gas porosity index can provide an approximate quantitative idea of its amount. These two utilities are discussed below.

Figure 5 analyzes the values of the gas porosity index for the 93 experimental cases (see Section 2.4) for four alloys. The values of the gas porosity index are provided in the Supplementary File. We noticed that the gas porosity index values of the four alloys vary widely, primarily due to the differences in their thermophysical properties and pool dimensions (Figure 3). For an easy comparison, we put all values of the gas porosity index on a consistent scale by standardizing the values. For standardization, the difference between each value and the minimum value is divided by the range of the index value for each alloy. The figure shows that the gas porosity index can accurately delineate the cases with pores from the cases where no pores were observed experimentally with an accuracy of 92%. The threshold value delineating the two cases is an essential point of reference for figuring out if the manufactured parts have porosity or not. Three optical micrographs that correlate to particular data points in the figure demonstrate the usefulness of the gas porosity index in accurately predicting the occurrence of gas porosity. The results show that the proposed methodology is consistent with the independent experiments conducted at various processing conditions using different LPBF machines and materials. Using the index, it is possible to identify the appropriate combination of the process variables to minimize porosity.

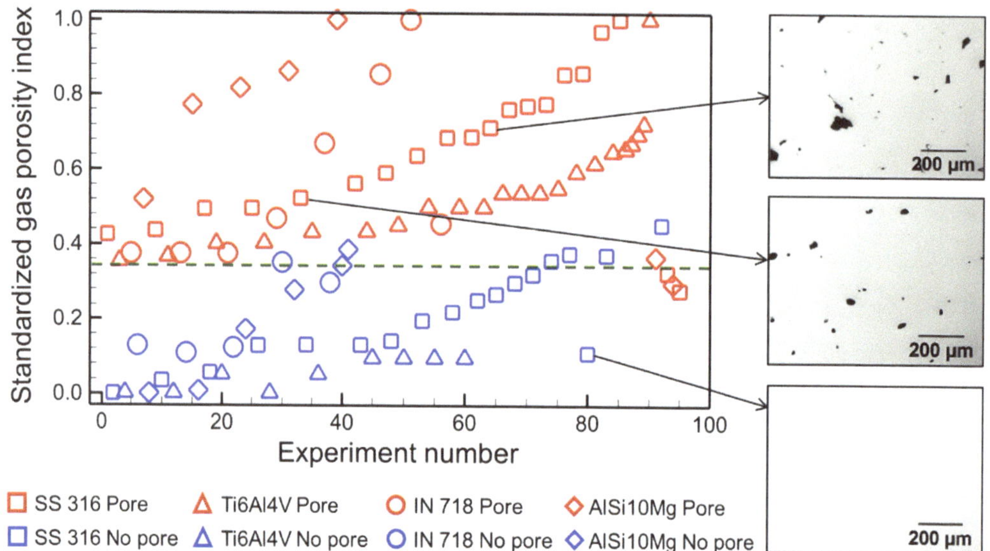

Figure 5. Gas porosity index to predict porosity in LPBF. The figure displays the index values for the 93 experimental cases. The index values are standardized to plot the data for all alloys under the same scale. The threshold value which delineates the pore and no pore cases is shown by a horizontal dashed line. Three optical micrographs [45] with the presence and absence of pores for LPBF of stainless steel 316 are shown corresponding to three experimental data points. The micrographs are taken from an open-access article [45] under the terms and conditions of the Creative Commons Attribution (CC BY) license.

If porosity is expected to form, the gas porosity index can provide an approximate quantitative idea of its amount. Figure 6 shows that the part density decreases as the amount of porosity increases at a higher value of the gas porosity index. In the figure, the percentage of density is plotted instead of porosity because the part density is a more intuitive parameter and easy to measure during experiments. The top right optical micrograph corresponds to a part having a density of 99.5%, which indicates that the part has low porosity, consistent with its gas porosity index value of about 23.09. In contrast, the other optical micrograph shows a density of about 93.39% and a higher gas porosity index of 30.26.

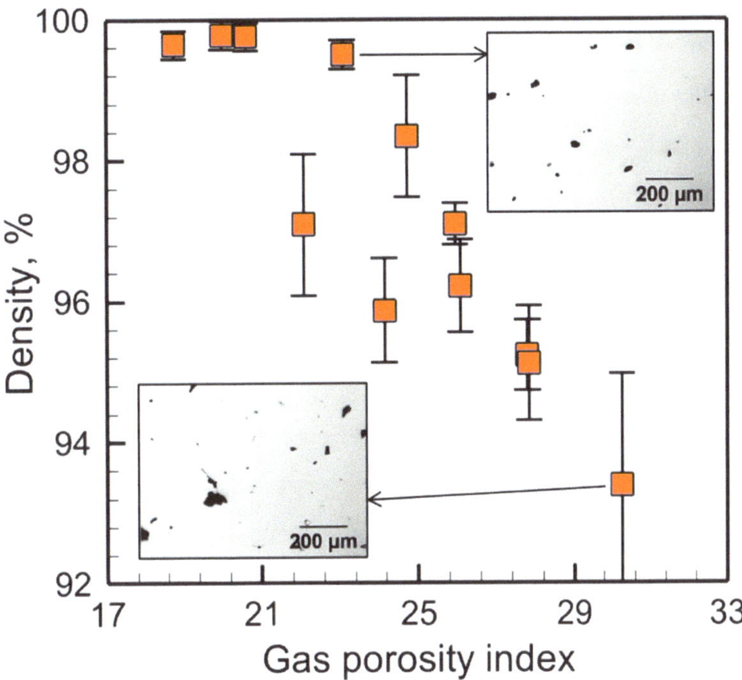

Figure 6. Variation in the volume percentage of density in stainless steel 316 parts made by LPBF with the computed gas porosity index. The reported density values [45] are the average of five measurements and the error bars represent the standard deviation of it. Two optical micrographs [45] with different amounts of pores for LPBF of stainless steel 316 are shown corresponding to two data points of gas porosity index. The micrographs are taken from an open-access article [45] under the terms and conditions of the Creative Commons Attribution (CC BY) license.

This result has important implications for manufacturing processes that depend on attaining accurate part densities. In order to reduce the likelihood of porosity in the finished product, the result emphasizes the significance of predicting the gas porosity index. Manufacturers can optimize their process conditions to obtain desired part densities and improve overall product quality and performance without any time-consuming and expensive trial-and-error.

3.3. Relative Susceptibility of Alloys to Gas Porosity

The gas porosity index provides a quantitative scale for estimating and comparing the relative vulnerabilities of different alloys to gas porosity. Figure 7a compares four commonly used alloys based on their relative susceptibility to gas porosity under a given set of processing conditions. A long, elongated, tear-drop-shaped molten pool for Inconel 718 (Figure 3) allows the gas bubbles a long time to escape. Thus, Inconel 718 is less vulnerable to gas porosity. In contrast, gas bubbles need a long time to escape from a deep, hemispherical pool of AlSi10Mg. Thus, AlSi10Mg is the most susceptible to gas porosity among the four alloys. The high vulnerability of AlSi10Mg to gas porosity has been experimentally observed by many researchers [80,81]. For example, the inset in Figure 7a shows an optical micrograph [82] of an AlSi10Mg part made by LPBF. The part contains a very high amount of porosity that may lead to part rejection. For a different set of experiments [83] on the LPBF of AlSi10Mg, the gas porosity index values were calculated and reported in Figure 7b. A reduction in the ultimate tensile strength was found for the parts with high porosity, as indicated by a large value of the gas porosity index.

The insets show the optical micrographs [83] of the samples containing different amounts of porosity. This result shows that the gas porosity index can also be beneficial to help engineers improve tensile properties by minimizing gas porosity. In addition, the index can be used to construct process maps for shop floor usage to reduce porosity in LPBF parts, as discussed below.

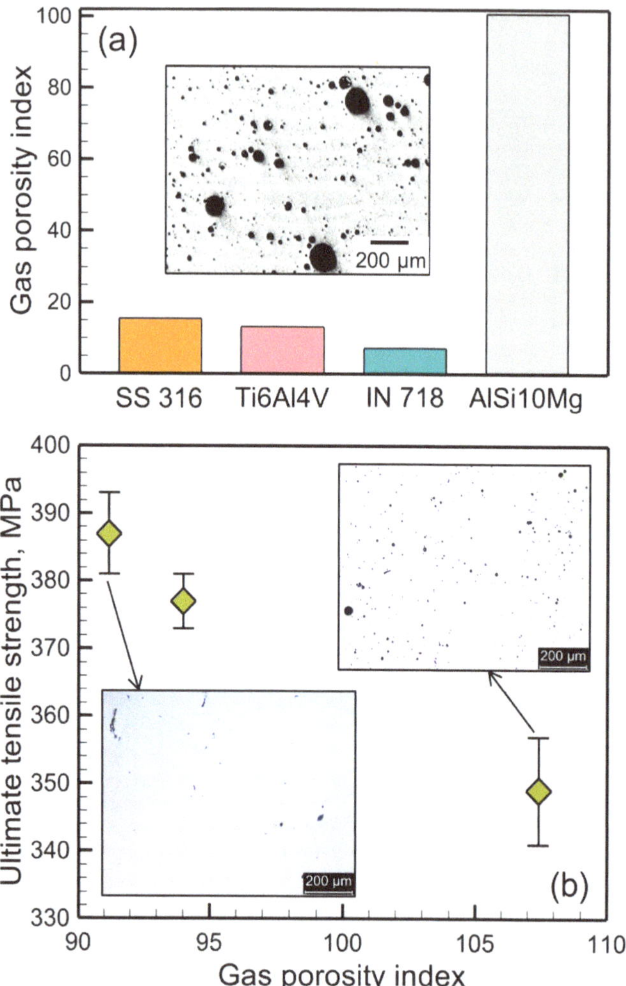

Figure 7. (**a**) Relative susceptibilities of four commonly used alloys to gas porosity evaluated by the computed values of the gas porosity index during LPBF using 300W laser power and 1250 mm/s scanning speed. The inset shows an optical micrograph [82] of an AlSi10Mg part made by LPBF containing a significant amount of gas pores. The micrograph is taken from [82] with permission from Elsevier. (**b**) A reduction in the ultimate tensile strength of AlSi10Mg parts made by LPBF due to the presence of gas porosity. The insets show the optical micrographs [83] of the samples containing different amounts of porosity. The plot is made based on the experimental data reported in [83] where heat input was varied to produce parts with different amounts of porosity. Corresponding values of the gas porosity index were calculated. The micrographs are taken from a thesis [83] available in the public domain.

3.4. Gas Porosity Map

The calculated values of the gas porosity index for different processing conditions and alloys can be utilized to construct gas porosity maps. These maps can indicate the optimum process windows for mitigating gas porosity. Figure 8a shows a gas porosity map, where the contour values represent the magnitude of the gas porosity index during LPBF of stainless steel 316. It is evident that high laser power and slow scanning are beneficial for reducing porosity. In contrast, rapid scanning can lead to fast solidification, resulting in the entrapment of gas bubbles and porosity.

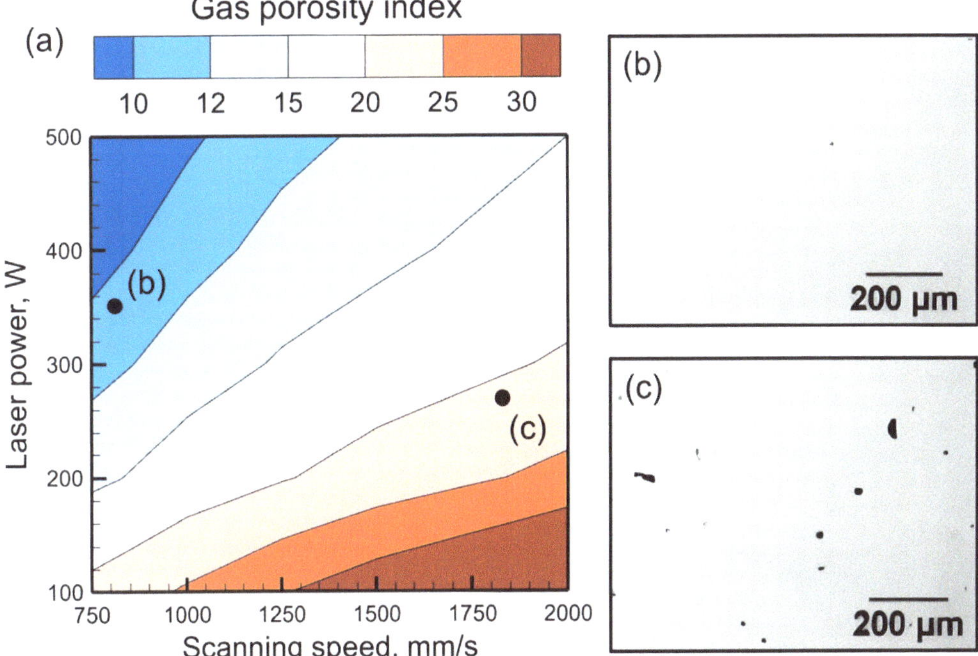

Figure 8. (**a**) A gas porosity map showing the variations in gas porosity index with laser power and scanning speed during LPBF of stainless steel 316. (**b,c**) Two micrographs [45] indicate the presence and absence of porosity for the corresponding conditions in (**a**). The micrographs are taken from an open-access article [45] under the terms and conditions of the Creative Commons Attribution (CC BY) license.

Two micrographs (Figure 8b,c) have been provided to test the map against experimental results [45]. If a laser power of 350 W and a scanning speed of 800 mm/s (Figure 8b) are used, then a dense part is obtained with almost no porosity with a gas porosity index of 17.05. In contrast, a laser power of 275 W and a scanning speed of 1800 mm/s result in a part (Figure 8c) that contains porosity and a gas porosity index of 20.61. This validates the gas porosity map for stainless steel 316 within the range of the process parameters considered in this work. Such maps, when rigorously verified against independent experimental results for various alloys and a wide range of processing conditions, can be made available for real-time prediction of pore formation on the shop floor.

4. Summary and Conclusions

In this work, a combination of mechanistic modeling and experimental data analysis was implemented to derive an easy-to-use, verifiable, dimensionless gas porosity index. The index captured the effects of both process parameters and important alloy properties and included several dominant physical factors causing gas porosity. The results were tested for commonly used alloys: stainless steel 316, Ti-6Al-4V, Inconel 718, and AlSi10Mg. Below are the important findings:

(1) The integrated theoretical and experimental framework identified all important physical factors causing gas porosity. The dimensionless gas porosity index, which is the ratio of the time taken by a gas bubble to escape the molten pool to the time required to solidify the molten pool, delineated the experimental cases with pores from the cases where porosities were not observed with an accuracy of 92%. Higher values of the index indicate that the gas bubbles need a longer time to escape from the molten pool, which increases their susceptibility to gas porosity. Experimental data proved that the number of pores increased at a higher value of the gas porosity index.

(2) Gas bubbles need more time to escape from a larger molten pool. Among the four alloys studied in this work, AlSi10Mg has the largest molten pool under the same processing conditions because of its lowest density. Thus, AlSi10Mg is the most vulnerable to gas porosity among the four alloys. The values of the gas porosity index for AlSi10Mg are 5 to 10 times higher than those for the other alloys. The high susceptibility of AlSi10Mg to gas porosity has also been experimentally observed and reported in the literature. In contrast, an elongated molten pool allows more time for the gas bubbles to escape before they solidify. Inconel 718 exhibits the longest molten pool among the four alloys because it has a large difference between the liquidus and solidus temperatures. Thus, among the four alloys, Inconel 718 is the least susceptible to gas porosity.

(3) The gas porosity process map constructed here showed that for a particular alloy, less heat input at low laser power and fast scanning resulted in a small pool that solidified rapidly, prevented the gas bubbles from escaping, and made the part prone to gas porosity. These process maps, when rigorously tested against experiments, can be made available for shop-floor usage by selecting appropriate processing conditions to reduce porosity without the need for experimental trials.

Supplementary Materials: The following supporting information can be downloaded at: https://www.mdpi.com/article/10.3390/ma17071569/s1. Refs. [2,3,44–59,67,84] are cited in the Supplementary Materials.

Author Contributions: Conceptualization, T.M.; methodology, S.S.; software, S.S. and T.M.; validation, S.S.; formal analysis, S.S.; investigation, S.S. and T.M.; resources, T.M.; data curation, S.S.; writing—original draft preparation, S.S.; writing—review and editing, T.M.; visualization, S.S.; supervision, T.M.; project administration, T.M. All authors have read and agreed to the published version of the manuscript.

Funding: This research received no external funding.

Institutional Review Board Statement: Not applicable.

Informed Consent Statement: Not applicable.

Data Availability Statement: Data used in this research are available in the article and Supplementary File.

Conflicts of Interest: The authors declare no conflicts of interests.

References

1. Mukherjee, T.; DebRoy, T. *Theory and Practice of Additive Manufacturing*; John Wiley & Sons: Hoboken, NJ, USA, 2023.
2. DebRoy, T.; Wei, H.L.; Zuback, J.S.; Mukherjee, T.; Elmer, J.W.; Milewski, J.O.; Zhang, W. Additive manufacturing of metallic components–process, structure and properties. *Prog. Mater. Sci.* **2018**, *92*, 112–224. [CrossRef]
3. Wei, H.L.; Mukherjee, T.; Zhang, W.; Zuback, J.S.; Knapp, G.L.; De, A.; DebRoy, T. Mechanistic models for additive manufacturing of metallic components. *Prog. Mater. Sci.* **2021**, *116*, 100703. [CrossRef]
4. Bhavar, V.; Kattire, P.; Patil, V.; Khot, S.; Gujar, K.; Singh, R. A review on powder bed fusion technology of metal additive manufacturing. In *Additive Manufacturing Handbook*; CRC Press: Boca Raton, FL, USA, 2017; pp. 251–253.
5. Priyadarshi, A.; Shahrani, S.B.; Choma, T.; Zrodowski, L.; Qin, L.; Leung, C.L.A.; Clark, S.J.; Fezzaa, K.; Mi, J.; Lee, P.D.; et al. New insights into the mechanism of ultrasonic atomization for the production of metal powders in additive manufacturing. *Addit. Manuf.* **2024**, *83*, 104033. [CrossRef]

6. Hojjatzadeh, S.M.H.; Parab, N.D.; Yan, W.; Guo, Q.; Xiong, L.; Zhao, C.; Qu, M.; Escano, L.I.; Xiao, X.; Fezzaa, K.; et al. Pore elimination mechanisms during 3D printing of metals. *Nat. Commun.* **2019**, *10*, 3088. [CrossRef] [PubMed]

7. Zhang, K.; Chen, Y.; Marussi, S.; Fan, X.; Fitzpatrick, M.; Bhagavath, S.; Majkut, M.; Lukic, B.; Jakata, K.; Rack, A.; et al. Pore evolution mechanisms during directed energy deposition additive manufacturing. *Nat. Commun.* **2024**, *15*, 1715. [CrossRef] [PubMed]

8. Sanaei, N.; Fatemi, A. Defects in additive manufactured metals and their effect on fatigue performance: A state-of-the-art review. *Prog. Mater. Sci.* **2021**, *117*, 100724. [CrossRef]

9. Attar, H.; Löber, L.; Funk, A.; Calin, M.; Zhang, L.C.; Prashanth, K.G.; Eckert, J. Mechanical behavior of porous commercially pure Ti and Ti–TiB composite materials manufactured by selective laser melting. *Mater. Sci. Eng. A* **2015**, *625*, 350–356. [CrossRef]

10. Olakanmi, E.O.; Cochrane, R.F.; Dalgarno, K.W. A review on selective laser sintering/melting (SLS/SLM) of aluminium alloy powders: Processing, microstructure, and properties. *Prog. Mater. Sci.* **2015**, *74*, 401–477. [CrossRef]

11. Brooks, C.R.; Choudhury, A. *Failure Analysis of Engineering Materials*; McGraw-Hill Education: New York, NY, USA, 2002.

12. Siddique, S.; Imran, M.; Wycisk, E.; Emmelmann, C.; Walther, F. Influence of process-induced microstructure and imperfections on mechanical properties of AlSi12 processed by selective laser melting. *J. Mater. Process. Technol.* **2015**, *221*, 205–213. [CrossRef]

13. Zhang, H.; Zhu, H.; Qi, T.; Hu, Z.; Zeng, X. Selective laser melting of high strength Al–Cu–Mg alloys: Processing, microstructure and mechanical properties. *Mater. Sci. Eng. A* **2016**, *656*, 47–54. [CrossRef]

14. Murchio, S.; Du Plessis, A.; Luchin, V.; Maniglio, D.; Benedetti, M. Influence of mean stress and building orientation on the fatigue properties of sub-unital thin-strut miniaturized Ti6Al4V specimens additively manufactured via Laser-Powder Bed Fusion. *Int. J. Fatigue* **2024**, *180*, 108102. [CrossRef]

15. Berto, F.; Du Plessis, A. (Eds.) *Fatigue in Additive Manufactured Metals*; Elsevier: Amsterdam, The Netherlands, 2023.

16. Pineau, A.; McDowell, D.L.; Busso, E.P.; Antolovich, S.D. Failure of metals II: Fatigue. *Acta Mater.* **2016**, *107*, 484–507. [CrossRef]

17. Tammas-Williams, S.; Withers, P.J.; Todd, I.; Prangnell, P.B. The influence of porosity on fatigue crack initiation in additively manufactured titanium components. *Sci. Rep.* **2017**, *7*, 7308. [CrossRef] [PubMed]

18. Zheng, Z.; Peng, L.; Wang, D. Defect analysis of 316 L stainless steel prepared by LPBF additive manufacturing processes. *Coatings* **2021**, *11*, 1562. [CrossRef]

19. Boban, J.; Ahmed, A. Improving the surface integrity and mechanical properties of additive manufactured stainless steel components by wire electrical discharge polishing. *J. Mater. Process. Technol.* **2021**, *291*, 117013. [CrossRef]

20. AlFaify, A.; Hughes, J.; Ridgway, K. Controlling the porosity of 316L stainless steel parts manufactured via the powder bed fusion process. *Rapid Prototyp. J.* **2019**, *25*, 162–175. [CrossRef]

21. Shen, B.; Li, H.; Liu, S.; Zou, J.; Shen, S.; Wang, Y.; Qi, H. Influence of laser post-processing on pore evolution of Ti–6Al–4V alloy by laser powder bed fusion. *J. Alloys Compd.* **2020**, *818*, 152845. [CrossRef]

22. Rawn, P. Reducing Porosity in LPBF Ti-6Al-4V Alloy by Parameter Optimization and Low Temperature Hot Isostatic Pressing Cycle. Ph.D. Dissertation, Marquette University, Milwaukee, WI, USA, 2023.

23. Santos Macías, J.G.; Zhao, L.; Tingaud, D.; Bacroix, B.; Pyka, G.; van der Rest, C.; Simar, A. Hot isostatic pressing of laser powder bed fusion AlSi10Mg: Parameter identification and mechanical properties. *J. Mater. Sci.* **2022**, *57*, 9726–9740. [CrossRef]

24. Jiang, R.; Mostafaei, A.; Pauza, J.; Kantzos, C.; Rollett, A.D. Varied heat treatments and properties of laser powder bed printed Inconel 718. *Mater. Sci. Eng. A* **2019**, *755*, 170–180. [CrossRef]

25. El Hassanin, A.; Silvestri, A.T.; Napolitano, F.; Scherillo, F.; Caraviello, A.; Borrelli, D.; Astarita, A. Laser-powder bed fusion of pre-mixed Inconel718-Cu powders: An experimental study. *J. Manuf. Process.* **2021**, *71*, 329–344. [CrossRef]

26. Sabzi, H.E.; Maeng, S.; Liang, X.; Simonelli, M.; Aboulkhair, N.T.; Rivera-Díaz-del-Castillo, P.E. Controlling crack formation and porosity in laser powder bed fusion: Alloy design and process optimisation. *Addit. Manuf.* **2020**, *34*, 101360. [CrossRef]

27. Choo, H.; Sham, K.L.; Bohling, J.; Ngo, A.; Xiao, X.; Ren, Y.; Garlea, E. Effect of laser power on defect, texture, and microstructure of a laser powder bed fusion processed 316L stainless steel. *Mater. Des.* **2019**, *164*, 107534. [CrossRef]

28. Bertoli, U.S.; Wolfer, A.J.; Matthews, M.J.; Delplanque, J.P.R.; Schoenung, J.M. On the limitations of volumetric energy density as a design parameter for selective laser melting. *Mater. Des.* **2017**, *113*, 331–340. [CrossRef]

29. Ghasemi-Tabasi, H.; Jhabvala, J.; Boillat, E.; Ivas, T.; Drissi-Daoudi, R.; Logé, R.E. An effective rule for translating optimal selective laser melting processing parameters from one material to another. *Addit. Manuf.* **2020**, *36*, 101496. [CrossRef]

30. Thomas, M.; Baxter, G.J.; Todd, I. Normalised model-based processing diagrams for additive layer manufacture of engineering alloys. *Acta Mater.* **2016**, *108*, 26–35. [CrossRef]

31. Van Elsen, M.; Al-Bender, F.; Kruth, J.P. Application of dimensional analysis to selective laser melting. *Rapid Prototyp. J.* **2008**, *14*, 15–22. [CrossRef]

32. Chen, Y.; Qi, H.; Li, H.; Shen, S.; Yang, Y.; Song, C. Molecular dynamics simulations of the formation and evolution of hydrogen pores during laser powder bed fusion manufacturing. *MRS Commun.* **2021**, *11*, 590–595. [CrossRef]

33. Yavari, R.; Smoqi, Z.; Riensche, A.; Bevans, B.; Kobir, H.; Mendoza, H.; Rao, P. Part-scale thermal simulation of laser powder bed fusion using graph theory: Effect of thermal history on porosity, microstructure evolution, and recoater crash. *Mater. Des.* **2021**, *204*, 109685. [CrossRef]

34. Liljestrand, F. Porosity Closure during Hot Isostatic Pressing of Additively Manufactured Ni-Based Superalloy IN718 Produced by LPBF and EBM. Master's Dissertation, Chalmers University of Technology, Gothernberg, Sweden, 2019. Available online: https://hdl.handle.net/20.500.12380/300734 (accessed on 5 February 2024).

35. Oster, S.; Breese, P.P.; Ulbricht, A.; Mohr, G.; Altenburg, S.J. A deep learning framework for defect prediction based on thermographic in-situ monitoring in laser powder bed fusion. *J. Intell. Manuf.* **2023**, *35*, 1687–1706. [CrossRef]
36. Alamri, N.M.H.; Packianather, M.; Bigot, S. Predicting the porosity in selective laser melting parts using hybrid regression convolutional neural network. *Appl. Sci.* **2022**, *12*, 12571. [CrossRef]
37. Mao, Y.; Lin, H.; Yu, C.X.; Frye, R.; Beckett, D.; Anderson, K.; Agrawal, A. A deep learning framework for layer-wise porosity prediction in metal powder bed fusion using thermal signatures. *J. Intell. Manuf.* **2023**, *34*, 315–329. [CrossRef]
38. Zhang, B.; Liu, S.; Shin, Y.C. In-Process monitoring of porosity during laser additive manufacturing process. *Addit. Manuf.* **2019**, *28*, 497–505. [CrossRef]
39. Massey, C.E.; Moore, D.G.; Saldana, C.J. Porosity Determination and Classification of Laser Powder Bed Fusion AlSi10Mg Dogbones Using Machine Learning. In *Challenges in Mechanics of Time Dependent Materials, Mechanics of Biological Systems and Materials & Micro-and Nanomechanics, Volume 2: Proceedings of the 2021 Annual Conference & Exposition on Experimental and Applied Mechanics*; Springer International Publishing: Berlin/Heidelberg, Germany, 2022; pp. 53–56.
40. Klein, J.; Jaretzki, M.; Schwarzenberger, M.; Ihlenfeldt, S.; Drossel, W.G. Automated porosity assessment of parts produced by Laser Powder Bed Fusion using Convolutional Neural Networks. *Procedia CIRP* **2021**, *104*, 1434–1439. [CrossRef]
41. Mohsin, S.I.; Farhang, B.; Wang, P.; Yang, Y.; Shayesteh, N.; Badurdeen, F. Deep Learning Based Automatic Porosity Detection of Laser Powder Bed Fusion Additive Manufacturing. In Proceedings of the International Conference on Flexible Automation and Intelligent Manufacturing, Porto, Portugal, 18–22 June 2023; Springer Nature Switzerland: Cham, Switzerland, 2023; pp. 328–335.
42. Beuth, J.; Fox, J.; Gockel, J.; Montgomery, C.; Yang, R.; Qiao, H.; Klingbeil, N. Process mapping for qualification across multiple direct metal additive manufacturing processes. In *2013 International Solid Freeform Fabrication Symposium*; University of Texas at Austin: Austin, TX, USA, 2013.
43. Rosenthal, D. Mathematical theory of heat distribution during welding and cutting. *Weld. J.* **1941**, *20*, 220s–234s.
44. Reijonen, J.; Revuelta, A.; Riipinen, T.; Ruusuvuori, K.; Puukko, P. On the effect of shielding gas flow on porosity and melt pool geometry in laser powder bed fusion additive manufacturing. *Addit. Manuf.* **2020**, *32*, 101030. [CrossRef]
45. Diaz Vallejo, N.; Lucas, C.; Ayers, N.; Graydon, K.; Hyer, H.; Sohn, Y. Process optimization and microstructure analysis to understand laser powder bed fusion of 316l stainless steel. *Metals* **2021**, *11*, 832. [CrossRef]
46. Riener, K.; Oswald, S.; Winkler, M.; Leichtfried, G.J. Influence of storage conditions and reconditioning of AlSi10Mg powder on the quality of parts produced by laser powder bed fusion (LPBF). *Addit. Manuf.* **2021**, *39*, 101896. [CrossRef]
47. Kempen, K.; Thijs, L.; Van Humbeeck, J.; Kruth, J.P. Processing AlSi10Mg by selective laser melting: Parameter optimisation and material characterisation. *Mater. Sci. Technol.* **2015**, *31*, 917–923. [CrossRef]
48. Masiagutova, E.; Cabanettes, F.; Sova, A.; Cici, M.; Bidron, G.; Bertrand, P. Side surface topography generation during laser powder bed fusion of $AlSi_{10}Mg$. *Addit. Manuf.* **2021**, *47*, 102230. [CrossRef]
49. Engelhardt, A.; Kahl, M.; Richter, J.; Krooß, P.; Kroll, A.; Niendorf, T. Investigation of processing windows in additive manufacturing of AlSi10Mg for faster production utilizing data-driven modeling. *Addit. Manuf.* **2022**, *55*, 102858. [CrossRef]
50. Khademzadeh, S.; Gennari, C.; Zanovello, A.; Franceschi, M.; Campagnolo, A.; Brunelli, K. Development of micro laser powder bed fusion for additive manufacturing of Inconel 718. *Materials* **2022**, *15*, 5231. [CrossRef] [PubMed]
51. Tran, H.C.; Lo, Y.L.; Le, T.N.; Lau, A.K.T.; Lin, H.Y. Multi-scale simulation approach for identifying optimal parameters for fabrication ofhigh-density Inconel 718 parts using selective laser melting. *Rapid Prototyp. J.* **2022**, *28*, 109–125. [CrossRef]
52. Vastola, G.; Pei, Q.X.; Zhang, Y.W. Predictive model for porosity in powder-bed fusion additive manufacturing at high beam energy regime. *Addit. Manuf.* **2018**, *22*, 817–822. [CrossRef]
53. Narra, S.P.; Rollett, A.D.; Ngo, A.; Scannapieco, D.; Shahabi, M.; Reddy, T.; Lewandowski, J.J. Process qualification of laser powder bed fusion based on processing-defect structure-fatigue properties in Ti-6Al-4V. *J. Mater. Process. Technol.* **2023**, *311*, 117775. [CrossRef]
54. Gordon, J.V.; Narra, S.P.; Cunningham, R.W.; Liu, H.; Chen, H.; Suter, R.M.; Rollett, A.D. Defect structure process maps for laser powder bed fusion additive manufacturing. *Addit. Manuf.* **2020**, *36*, 101552. [CrossRef]
55. Cunningham, R.; Narra, S.P.; Montgomery, C.; Beuth, J.; Rollett, A.D. Synchrotron-based X-ray microtomography characterization of the effect of processing variables on porosity formation in laser power-bed additive manufacturing of Ti-6Al-4V. *JOM* **2017**, *69*, 479–484. [CrossRef]
56. Wu, Z.; Asherloo, M.; Jiang, R.; Delpazir, M.H.; Sivakumar, N.; Paliwal, M.; Mostafaei, A. Study of printability and porosity formation in laser powder bed fusion built hydride-dehydride (HDH) Ti-6Al-4V. *Addit. Manuf.* **2021**, *47*, 102323. [CrossRef]
57. Jaber, H.; Tünde, K. Development of Selective Laser Melting of Ti6Al4V Alloy for Tissue Engineering. *Bánki Közlemények* **2020**, *3*, 19–23.
58. Sanaei, N.; Fatemi, A. Analysis of the effect of internal defects on fatigue performance of additive manufactured metals. *Mater. Sci. Eng. A* **2020**, *785*, 139385. [CrossRef]
59. Emminghaus, N.; Paul, J.; Hoff, C.; Hermsdorf, J.; Kaierle, S. Development of an empirical process model for adjusted porosity in laser-based powder bed fusion of Ti-6Al-4V. *Int. J. Adv. Manuf. Technol.* **2022**, *118*, 1239–1254. [CrossRef]
60. Simchi, A.; Pohl, H. Effects of laser sintering processing parameters on the microstructure and densification of iron powder. *Mater. Sci. Eng. A* **2003**, *359*, 119–128. [CrossRef]
61. Wang, D.; Han, H.; Sa, B.; Li, K.; Yan, J.; Zhang, J.; Liu, J.; He, Z.; Wang, N.; Yan, M. A review and a statistical analysis of porosity in metals additively manufactured by laser powder bed fusion. *Opto-Electron Adv.* **2022**, *5*, 210058–210061. [CrossRef]

62. Du Plessis, A. Effects of process parameters on porosity in laser powder bed fusion revealed by X-ray tomography. *Addit. Manuf.* **2019**, *30*, 100871. [CrossRef]
63. Mundra, K.; DebRoy, T. Toward understanding alloying element vaporization during laser beam welding of stainless steel. *Weld. J.* **1993**, *72*, 1.
64. Tang, M.; Pistorius, P.C.; Beuth, J. Geometric model to predict porosity of part produced in powder bed system. *Mater. Sci. Technol. Proc. (MS&T)* **2015**, *2015*, 129–136.
65. Ren, Z.; Gao, L.; Clark, S.J.; Fezzaa, K.; Shevchenko, P.; Choi, A.; Everhart, W.; Rollett, A.D.; Chen, L.; Sun, T. Machine learning–aided real-time detection of keyhole pore generation in laser powder bed fusion. *Science* **2023**, *379*, 89–94. [CrossRef] [PubMed]
66. Gan, Z.; Kafka, O.L.; Parab, N.; Zhao, C.; Fang, L.; Heinonen, O.; Sun, T.; Liu, W.K. Universal scaling laws of keyhole stability and porosity in 3D printing of metals. *Nat. Commun.* **2021**, *12*, 2379. [CrossRef] [PubMed]
67. Mills, K.C. *Recommended Values of Thermophysical Properties for Selected Commercial Alloys*; Woodhead Publishing: Thorston, UK, 2002.
68. Tang, M.; Pistorius, P.C.; Narra, S.; Beuth, J.L. Rapid solidification: Selective laser melting of AlSi$_{10}$Mg. *JOM* **2016**, *68*, 960–966. [CrossRef]
69. Rombouts, M.; Froyen, L.; Gusarov, A.V.; Bentefour, E.H.; Glorieux, C. Photopyroelectric measurement of thermal conductivity of metallic powders. *J. Appl. Phys.* **2005**, *97*, 024905. [CrossRef]
70. Tang, M.; Pistorius, P.C.; Beuth, J.L. Prediction of lack-of-fusion porosity for powder bed fusion. *Addit. Manuf.* **2017**, *14*, 39–48. [CrossRef]
71. Nath, P.; Mahadevan, S. Probabilistic predictive control of porosity in laser powder bed fusion. *J. Intell. Manuf.* **2021**, *34*, 1085–1103. [CrossRef]
72. Vukkum, V.B.; Gupta, R.K. Review on corrosion performance of laser powder-bed fusion printed 316L stainless steel: Effect of processing parameters, manufacturing defects, post-processing, feedstock, and microstructure. *Mater. Des.* **2022**, *221*, 110874. [CrossRef]
73. Kasperovich, G.; Haubrich, J.; Gussone, J.; Requena, G. Correlation between porosity and processing parameters in TiAl6V4 produced by selective laser melting. *Mater. Des.* **2016**, *105*, 160–170. [CrossRef]
74. Zhou, J.; Tsai, H. Porosity Formation and Prevention in Pulsed Laser Welding. *ASME J. Heat Transfer.* **2007**, *129*, 1014–1024. [CrossRef]
75. Zhao, H.; DebRoy, T. Pore formation during laser beam welding of die-cast magnesium alloy AM60B-mechanism and remedy. *Weld. J.* **2001**, *80*, 204–210.
76. Mukherjee, T.; Elmer, J.W.; Wei, H.L.; Lienert, T.J.; Zhang, W.; Kou, S.; DebRoy, T. Control of grain structure, phases, and defects in additive manufacturing of high-performance metallic components. *Prog. Mater. Sci.* **2023**, *138*, 101153. [CrossRef]
77. Iantaffi, C.; Leung, C.L.A.; Chen, Y.; Guan, S.; Atwood, R.C.; Lertthanasarn, J.; Pham, M.S.; Meisnar, M.; Rohr, T.; Lee, P.D. Oxidation induced mechanisms during directed energy deposition additive manufactured titanium alloy builds. *Addit. Manuf. Lett.* **2021**, *1*, 100022. [CrossRef]
78. Brooks, J.A.; Baskes, M.I.; David, S.A. Advances in welding science and technology. *Met. Park OH ASM Int.* **1986**, *198*, 93.
79. DebRoy, T.; David, S.A. Physical processes in fusion welding. *Rev. Mod. Phys.* **1995**, *67*, 85. [CrossRef]
80. Hastie, J.C.; Kartal, M.E.; Carter, L.N.; Attallah, M.M.; Mulvihill, D.M. Classifying shape of internal pores within AlSi10Mg alloy manufactured by laser powder bed fusion using 3D X-ray micro computed tomography: Influence of processing parameters and heat treatment. *Mater. Charact.* **2020**, *163*, 110225. [CrossRef]
81. Finfrock, C.B.; Exil, A.; Carroll, J.D.; Deibler, L. Effect of hot isostatic pressing and powder feedstock on porosity, microstructure, and mechanical properties of selective laser melted AlSi10Mg. *Metallogr. Microstruct. Anal.* **2018**, *7*, 443–456. [CrossRef]
82. Weingarten, C.; Buchbinder, D.; Pirch, N.; Meiners, W.; Wissenbach, K.; Poprawe, R. Formation and reduction of hydrogen porosity during selective laser melting of AlSi10Mg. *J. Mater. Process. Technol.* **2015**, *221*, 112–120. [CrossRef]
83. Stugelmayer, E.J. Characterization of Process-Induced Defects in Laser Powder Bed Fusion Processed AlSi10Mg Alloy. Ph.D. Dissertation, Montana Tech of The University of Montana, Butte, MT, USA, 2018.
84. Mukherjee, T.; Wei, H.L.; De, A.; DebRoy, T. Heat and fluid flow in additive manufacturing—Part I: Modeling of powder bed fusion. *Comput. Mater. Sci.* **2018**, *150*, 304–313. [CrossRef]

 *materials*

Article

A Random Forest Classifier for Anomaly Detection in Laser-Powder Bed Fusion Using Optical Monitoring

Imran Ali Khan [1,*], Hannes Birkhofer [1], Dominik Kunz [2], Drzewietzki Lukas [3] and Vasily Ploshikhin [1]

[1] Airbus Endowed Chair for Integrative Simulation and Engineering of Materials and Processes (ISEMP), University of Bremen, Am Fallturm 1, 28359 Bremen, Germany; birkhofer@isemp.de (H.B.); ploshikhin@isemp.de (V.P.)

[2] Electro Optical Systems GmbH, Robert-Stirling Ring 1, 82152 Krailling, Germany; dominik.kunz@eos.info

[3] Leibherr-Aerospace Lindenberg GmbH, Pfänderstraße 50-52, 881161 Lindenberg, Germany; lukas.drzewietzki@liebherr.com

[*] Correspondence: khan@isemp.de; Tel.: +49-(0)-421-218-62350

Abstract: Metal additive manufacturing (AM) is a disruptive production technology, widely adopted in innovative industries that revolutionizes design and manufacturing. The interest in quality control of AM systems has grown substantially over the last decade, driven by AM's appeal for intricate, high-value, and low-volume production components. Geometry-dependent process conditions in AM yield unique challenges, especially regarding quality assurance. This study contributes to the development of machine learning models to enhance in-process monitoring and control technology, which is a critical step in cost reduction in metal AM. As the part is built layer upon layer, the features of each layer have an influence on the quality of the final part. Layer-wise in-process sensing can be used to retrieve condition-related features and help detect defects caused by improper process conditions. In this work, layer-wise monitoring using optical tomography (OT) imaging was employed as a data source, and a machine-learning (ML) technique was utilized to detect anomalies that can lead to defects. The major defects analyzed in this experiment were gas pores and lack of fusion defects. The Random Forest Classifier ML algorithm is employed to segment anomalies from optical images, which are then validated by correlating them with defects from computerized tomography (CT) data. Further, 3D mapping of defects from CT data onto the OT dataset is carried out using the affine transformation technique. The developed anomaly detection model's performance is evaluated using several metrics such as confusion matrix, dice coefficient, accuracy, precision, recall, and intersection-over-union (IOU). The k-fold cross-validation technique was utilized to ensure robustness and generalization of the model's performance. The best detection accuracy of the developed anomaly detection model is 99.98%. Around 79.40% of defects from CT data correlated with the anomalies detected from the OT data.

Keywords: machine learning; random forest; quality inspection; laser powder bed fusion; process monitoring; optical tomography; computerized tomography; gas pores; lack of fusion

Citation: Khan, I.A.; Birkhofer, H.; Kunz, D.; Lukas, D.; Ploshikhin, V. A Random Forest Classifier for Anomaly Detection in Laser-Powder Bed Fusion Using Optical Monitoring. *Materials* **2023**, *16*, 6470. https://doi.org/10.3390/ma16196470

Academic Editors: Tuhin Mukherjee and Qianru Wu

Received: 22 August 2023
Revised: 21 September 2023
Accepted: 23 September 2023
Published: 29 September 2023

1. Introduction

In the late 1980s, additive manufacturing technology emerged as a manufacturing tool for application prototypes [1]. Since then, the AM industry has experienced remarkable growth due to its layer-by-layer manufacturing process, which allows for the production of products with complex shapes and various materials [2]. Hence, it plays an important role in many fields, such as aerospace, manufacturing, and automotive. The AM market's expected annual growth in the next five years is projected to surpass 20%, as stated in an industrial insight report from Wohlers's associates in 2020 [3]. Despite significant benefits, quality issues affect the advancement of additive manufacturing technology [4]. One of the key technological challenges to overcome in AM is limited process predictability and repeatability [1].

Metal additive manufacturing techniques using laser powder bed fusion (L-PBF) nowadays provide the highest repeatability and dimensional precision for part production and have thus been extensively investigated in both industry and academia. To manufacture a component, L-PBF methods typically employ the following steps: (1) A layer of metal powder of a specific thickness is placed over the machine's build plate; (2) a laser beam selectively melts the required region within the powder layer; (3) the build plate slides down, and a fresh layer of powder is put onto the build plate. Layer by layer, this procedure is repeated until the part production is complete. The present approach in AM quality assurance is to analyze the component after it is created using computed tomography, which is extremely costly and time-consuming [5]. According to Seifi et al. [6], statistical qualification of AM components based on destructive materials testing may be unacceptably expensive and take over a decade to complete, which is unfeasible, given the tiny batch sizes and time necessary for manufacturing. If defects could be detected in situ, quality assurance costs in metal AM could be reduced significantly.

Porosity is one of the most important defects to avoid, especially for components that require high tensile strength and fatigue resistance. Porosity in L-PBF components can be caused by inadequate melting (i.e., lack of fusion), pre-existing gas holes in metallic powders from the gas-atomizing manufacturing process, and trapping of gas pores during AM processing [7]. Lack of fusion defects in the laser powder bed fusion process refers to irregular and elongated-shaped anomalies that can vary in size from 50 µm to several millimeters. On the other hand, gas pores in L-PBF are spherical in shape and typically range in size from 5 µm to 20 µm [8]. Process anomalies within a layer, which might yield defects such as pores and lack of fusion defects, are closely related to the occurrence of local temperature changes [9]. Optical monitoring data in the form of intensity recordings can reveal these process anomalies which possibly precede defect genesis. Current monitoring systems however produce huge amounts of data that are typically processed only after completion of the printing process.

The introduction of in-situ process monitoring allows for the tracing of defects throughout the process. Process monitoring may be classified into three categories in principle. The first is melt pool monitoring, which monitors the melt pool and its surroundings. The molten pool's size and temperature characteristics provide information on the process's stability and the occurrence of local flaws. The second category examines the entire layer in order to discover defects in various sections of each layer. After scanning, the temperature distribution and surface are observed. The geometric development of the build from slice to slice is considered as the third category [10].

Each of the aforementioned methods generates vast quantities of image data, and the time needed to analyze such large datasets is substantial. Consequently, conducting in-situ data analysis for monitoring purposes in additive manufacturing is currently impractical due to extended processing times. However, a specific branch of artificial intelligence (AI) called machine learning offers a potential solution by enabling rapid and dependable analysis of image data [11]. Process monitoring with the application of ML especially convolutional neural networks (CNN) and random forest classifiers has been utilized successfully for defect detection during the AM process. Baumgarti et al. [2] used in-situ layer-wise images captured by a thermographic camera during the L-PBF process to detect defects using convolutional neural networks. Delamination and uncritical splatters were detected with an accuracy of 96.08%. Grad CAM heat maps were plotted to identify defects. Kwon et al. [12] illustrated the use of CNN for laser power prediction utilizing in-situ layer-wise meltpool images acquired by a high-speed camera during the L-PBF process. The developed CNN model can predict laser power values, which can be utilized to identify problematic positions in AM products without requiring destructive inspections.

ML has grown in popularity in recent years because of its exceptional performance in data tasks such as classification, regression, and clustering [13]. Machine learning is described as computer programming that uses sample data and prior knowledge to maximize a performance criterion [14]. Aside from the traditional application of making predictions through data fitting, the scientific community is exploring new and innovative approaches to integrate ML methods into additive manufacturing. Precise identification, analysis, and prediction of defects hold immense promise in expediting the production of metal AM structures that are both solidly constructed and devoid of defects [15]. Mohr et al. [1] used thermography and optical tomography images for in-situ defect detection during the L-PBF process. A layer-wise OT image is captured using an off-axis CMOS camera, which is similar to the monitoring system utilized in this paper (Section 2.2). CT scans were used to assess the outcomes of OT and thermographic imaging. Only significant defects, such as the lack of fusion void clusters, performed well when compared with the CT data. But for pore detection which is one of the major part defects in additive manufacturing, only 0.7% OT pores and Micro-CT pores overlapped, but 71.4% of thermography anomalies and Micro-CT pores overlapped. For high-quality predictions, ML models require huge training data sets. Due to the high experimental costs, there are restrictions in generating sufficient OT data. As a result, it is ideal to employ a machine learning technique capable of developing an anomaly detection model with a small amount of training data [16]. As a result, there is a need to improve the resolution of the OT system or employ new pore detection approaches using ML techniques for better correlation with micro-CT pores, which is one of the main goals of this research.

The main challenges of developing high-quality machine learning algorithms are Limited data for training, high computational costs, and the lack of generalization to new materials and geometries. The utilization of L-PBF encompasses a wide range of materials and intricate geometries. Nevertheless, the development of machine learning algorithms that can generalize effectively across diverse materials and geometries has a significant challenge. This difficulty arises from the distinct behaviors and characteristics exhibited by each material and geometry, necessitating substantial data and model adaptation. The issue at hand is addressed through the utilization of a traditional machine learning method, specifically the random forest classifier. This choice is made due to its ability to overcome the challenge without necessitating a large volume of training data, unlike more widely used machine learning techniques such as convolutional neural networks [17].

The significance of advancements in data processing algorithms in the field of AM monitoring becomes evident when considering their potential broad impact and applicability. Integrating these algorithms into various monitoring and control systems can enhance process repeatability. This integration can also lead to a reduction in post-processing and non-destructive testing, resulting in cost-effective quality assurance. Conventional quality control methods in L-PBF often involve time-consuming post-processing inspections. However, the utilization of machine learning algorithms can automate the defect detection process by analyzing real-time sensor data and identifying patterns associated with defects [18]. This enables faster and more efficient defect detection, facilitating prompt corrective actions and minimizing the need for extensive post-processing inspections. Ultimately, machine learning offers the ability to swiftly analyze and process in-situ data in L-PBF, thereby enabling accelerated defect detection, real-time monitoring, process optimization, and adaptive control. These advantages collectively contribute to improved efficiency, reduced post-processing requirements, and enhanced overall quality in the L-PBF process [2]. This study aims to contribute to process repeatability and quality assurance through the development of a machine learning algorithm for rapid and reliable anomaly detection leading to defects from monitoring data.

Process invariance and optical noise in the generated OT images make it difficult to identify anomalies. When the amount of data is low for image segmentation random forest technique can be used which is a conventional ML approach. Yaokun Wu and Siddharth Misra [19] demonstrated that RF models outperform neural network approaches in terms of noise tolerance. P. Rajendran et al. [20] also used a random forest classifier to segment brain tumors from MR brain images with an accuracy of 98.37%. In this paper, the focus is on the application of machine learning using optical monitoring data to identify anomalies and validate the detected anomalies using defects obtained using the µCT technique.

2. Materials and Methods

2.1. Material Data

The experiment is conducted on an EOS M 290 laser powder bed fusion machine (L-PBF). A cylindrical metal specimen is built with a diameter of 10 mm and 15.30 mm in length. It consists of 255 layers with a layer thickness of 60 µm. The powder material used is EOS Titanium Ti64, which has a chemical composition corresponding to ASTM F1472 [21] and ASTM F2924 [22]. For additional details regarding the physical, chemical, and thermal properties of the powder, please refer to EOS GmbH [23]. The volume rate and part density of the powder material is 5 mm^3/s and $\approx$4.41 g/cm^3. The operating gas is argon which has a flow rate of approximately 0.6 mbar. The optics of the OT system are designated so that the camera's field of view corresponds to the size of the platform.

2.2. In-Situ Monitoring by Optical Tomography

In-situ monitoring of the L-PBF process is carried out using the optical tomography technique. The OT system used in this study was developed by Electro Optical Systems, (EOS GmbH, Krailling, Munich, Germany) and is known as the EOSTATE-Exposure OT system. During laser powder bed fusion processes powder is melted and three types of emissions are emitted back from the surface such as plasma radiation, thermal radiation, and laser reflection [24]. All radiation with a specified bandwidth is captured by the OT system. Figure 1 shows the schematic overview of the EOSTATE-Exposure OT system. It uses high-resolution CMOS (Complementary metal oxide semiconductor) cameras developed by Exceltas PCO GmbH, which capture signals in the visible and near-infrared spectral range using a band pass filter at 900 nm. The OT system records radiation signals that are proportional to the radiation intensity emitted from the area of the specimen imaged onto the respective pixel element, and are integrated over the entire layer exposure. The basic working principle of this OT system is detailed in [25]. The camera and optics specifications of the OT system are illustrated in Table 1.

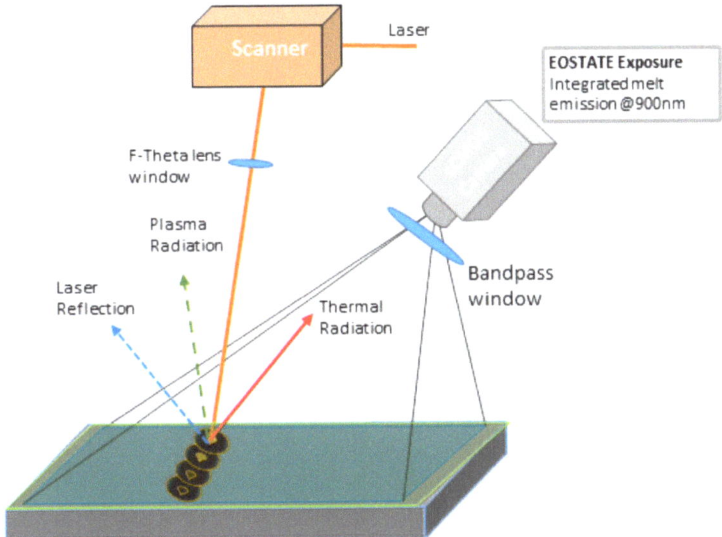

Figure 1. A schematic overview of EOSTATE-Exposure OT system (EOS GmbH).

Table 1. Technical specifications of the EOSTATE-Exposure OT system.

Specifications	Values
Spectral range	887.5 nm–912.5 nm
Camera resolution	2560 × 2160 pixels
Objective lens	8 mm
Frame rate	10 fps
Spatial resolution	125 µm/Pixel
Data interface	USB 3.1

EOSTATE-Exposure OT generates two types of images: integral gray value images formed by combining a sequence of images (approximately 100) during the platform's exposure per single layer, and maximum gray value images formed by taking the maximum intensity value of each pixel during the entire layer exposition. Figure 2 shows the integral optical tomography Figure 2a and maximum optical tomography Figure 2b for the 100th layer of the specimen under normal process conditions. The intensity values in Figure 2 are digital values (DV) which are induced by a combination of overlapping scanning strategies, energy increase, and change in temperatures at the building platform. These intensity values range dynamically for different layers and go up to a value of 40,000 DV. Thus to generate a machine learning model it should be normalized to a scalable range [0–255]. Figure 3 shows the normalized integral OT Figure 3a and maximum OT Figure 3b for the 100th layer of the specimen under normal process conditions. These images capture process variances and possible effects of defects. It also helps in analyzing the homogeneity and stability behavior of the build process. Integral OT images are considered for developing an ML model, as featured in these images are more discrete compared to maximum OT images.

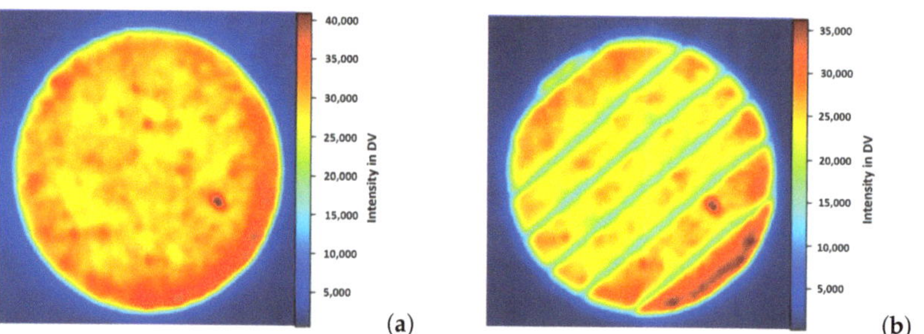

Figure 2. EOSTATE Exposure optical tomography images for the 100th layer under normal process conditions: (**a**) Integral OT image (**b**) Maximum OT image.

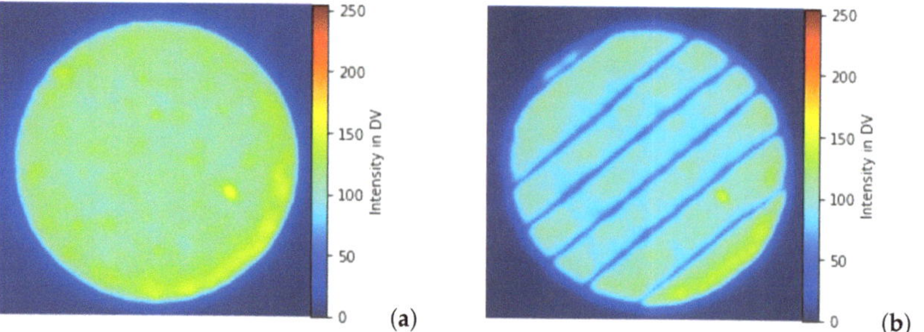

Figure 3. Normalized OT images for 100th layer under normal process conditions: (**a**) Normalized integral OT image (**b**) Normalized maximum OT image.

2.3. *Generation of Artificial Defects by Reducing Laser Power*

In real-world scenarios, it can be challenging to obtain a sufficient amount of data that contains a wide range of naturally occurring defects. Through the deliberate introduction of defects, a meticulously controlled dataset can be produced, encompassing a diverse range of defect types, sizes, and distributions. This allows for more comprehensive training of the machine learning model. This dataset is instrumental in training a machine learning model capable of accurately identifying both artificially induced defects and naturally occurring defects that may exhibit similar characteristics. Consequently, the model's capacity to generalize and effectively detect various types of defects in real-world scenarios is enhanced. To summarize, the deliberate introduction of artificial defects in the laser powder bed fusion process proves to be a beneficial strategy for training and assessing machine learning models designed for defect detection [1]. This practice allows for the creation of controlled datasets, enhances the model's ability to generalize, facilitates accurate performance evaluation, and enables targeted experiments that contribute to a deeper understanding of defect detection in additive manufacturing processes.

In-situ monitoring defects are induced in a cylindrical metal part by applying different processing parameter values at a specific layer height and specific regions called regions of interest (ROI) during the building process. Except for these specific regions, the entire built job is printed with standard process parameters, utilizing a laser wavelength ranging from 1050–1090 µm and a laser beam diameter of 100 µm. At ROI laser power is reduced to 80 watts for four layers, and the laser scan speed and hatch distance are the same as in the standard parameters. Figure 4 shows a normalized integral OT image at the 101st layer height showing ROIs with potential defects due to the reduction of laser power at those specific regions highlighted inside a black box [ROI]. These regions with changes in intensity can be interpreted as anomalies that are caused by changes in temperature and energy density [26]. The entire printed cylinder consists of a total of 8 sections with ROIs in four layers each. Figure 5 shows the isometric view of post-processed CT specimen along with defects caused due to reduced laser power.

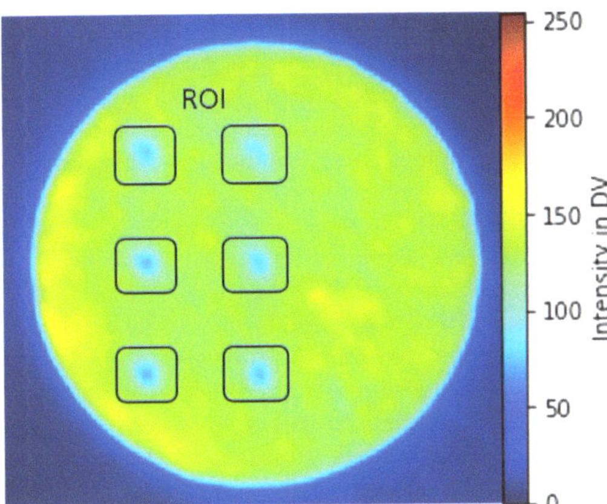

Figure 4. Normalized Integral OT image for layer 101 with induced artifacts.

The anomalies after detection have to be investigated for potential defects. After completion of the L-PBF process, the built-in cylinders were post-processed and examined using the micro-computerized tomography technique. The majority of defects are gas pores and lack of fusion, ranging from 30 to 540 µm in diameter. An algorithm is developed to correlate anomalies from OT data with defects from CT data to prove the potential of the optical monitoring system in identifying defects during L-PBF processes.

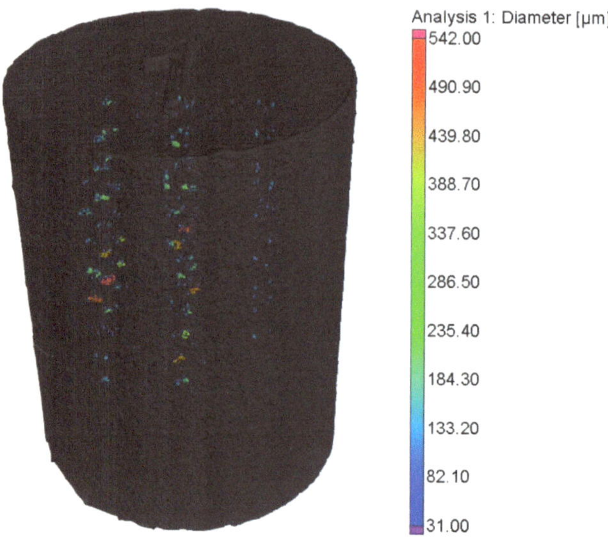

Figure 5. Isometric view of CT specimen along with defects.

2.4. Proposed Model: Random Forest Classifier

Random Forests is an effective machine learning methodology for classification and regression, and it may also be used for image segmentation when training data is limited [27]. RF classifiers have been successfully used for various biomedical image segmentation purposes and this approach can be utilized for defect detection in the additive manufacturing process. Gas pores and lack of fusion, the most critical defects in AM, are considered in this study.

The pre-processing outcomes of optical tomography images revealed that the intensity values of artificially generated defects when normalized, fell within the range of 140 to 170 DV. Interestingly, this range closely aligns with that of non-defect regions, which poses a challenge for image segmentation using histogram-based segmentation, watershed segmentation, or any other direct image segmentation technique. The Random Forest Algorithm is based on the theory of ensemble learning. Ensemble learning is a broad Machine Learning meta-approach that aims to enhance predictive performance by mixing predictions from many models. In layman's terms, it entails fitting many model types to the same data and then using another model to find the optimum approach to combine the predictions. As a result, the Random Forest Algorithm aggregates predictions from decision trees and chooses the best prediction among those trees [28]. Random Forest is defined as a classifier that comprises some decision trees on various subsets of a given dataset and takes the average to enhance that dataset's prediction accuracy. Instead of depending on a single decision tree, the algorithm considers the prediction out of each tree and anticipates the ultimate approach that relies on the majority vote of predictions [28].

A Random forest segmentation (RF_Segm) model for pore detection was developed using OT images and ground truth labels. 100 OT images were considered for training the segmentation model. Ground truth labels were generated for these 100 OT images using the Apeer Annotate platform. The segmentation approach consists of two steps: feature extraction and classification of the derived feature vectors for each pixel in the OT image dataset. The Random Forest classifier was trained to associate certain attributes with each pixel in the OT image dataset. The segmentation workflow includes the following sequential steps:

2.4.1. Image Preprocessing

Preprocessing is an important step prior to feature extraction. It includes RGB to grayscale image conversion and image normalization. Due to the dynamic range of intensity distribution of optical tomography images, it is critical to normalize to minimize non-uniform lighting issues. The normalized images are shown in Figure 3. The normalization method determines the mean and variance of an image, reducing the disparity in illumination. Normalization $f(x,y)$ is formulated as in Equation (1) [29]

$$g(x,y) = \frac{f(x,y) - M_f(x,y)}{\sigma_f(x,y)} \tag{1}$$

where $f(x,y)$ is original image, $M_f(x,y)$ is the estimation of mean of original image and $\sigma_f(x,y)$ is the estimation of the standard deviation.

2.4.2. Feature Extraction

Feature extraction is the process of establishing a set of necessary features, or image characteristics, that form the core element and, when expressed in an efficient or comprehensible manner, provide the necessary information for analysis and segmentation [30]. A total of 42 feature extractors were generated for training an RF_Segm model. General edge detection operators like Sobel, Prewitt, Roberts, and Canny are used as feature extractors. Other than that Gabor filters, Gaussian blur, median filters, and pixel intensity values of the OT images are used to extract features for generating the segmentation model.

Gabor Filter

One of the most well-known feature extraction methods is the Gabor filter. It is made up of wavelet coefficients for various scales and orientations, which makes these features resistant to rotation, translation, distortion, and scaling [31]. In this study, 32 Gabor filters with different orientations and scales were created with a kernel size of 9 × 9. Gabor is a convolutional filter representing a combination of Gaussian and sinusoidal terms. The Gaussian component provides the weights and the sine component provides the directionality. It has excellent localization properties in both the spatial and frequency domains. In the spatial domain, it is a Gaussian-modulated sinusoid, and in the frequency domain, it is a shifted Gaussian. It is represented in Equation (2) [31]:

$$g(x,y,\sigma,\theta,\lambda,\gamma,\phi) = exp\left[\frac{-x'^2 + y'^2\gamma^2}{2\sigma^2}\right] exp\left[i\left[\frac{2\pi x'}{\lambda} + \phi\right]\right] \tag{2}$$

$$x' = x\cos\theta + y\sin\theta \tag{3}$$

$$y' = -x\sin\theta + y\cos\theta \tag{4}$$

In Equations (3) and (4) x and y are image coordinates and other parameters which can be varied to generate different Gabor filters are σ, θ, λ, γ, and ϕ. σ is the standard deviation of the Gaussian envelope. θ is the orientation of the filter. γ describes aspect ratio, $\gamma = 1$ for circular shape, $\gamma < 1$ for elliptical shape. ϕ is the phase offset.

Gaussian Blur

The Gaussian blur feature is obtained by blurring an image using a Gaussian kernel and convolving the image. It functions as a non-uniform low-pass filter, preserving low spatial frequency while reducing image noise and insignificant details. A Gaussian function [32] is formulated as in Equation (5).

$$G(x) = \frac{1}{\sqrt{2\pi\sigma^2}}e^{-\frac{x^2 + y^2}{2\sigma^2}} \tag{5}$$

where x and y are the image coordinates and σ is the standard deviation of the Gaussian distribution. A Gaussian kernel with a standard deviation of 3 and 7 is used to generate feature extractors.

Edge Detection Algorithms

Sobel, Prewitt, and Scharr are first-order derivative techniques of edge detection that can be used for feature extraction from OT images. The Sobel operator enhances the edges of an image by performing a 2-D spatial-gradient operation on it. The operator is made up of a pair of 3×3 convolution kernels that are applied individually to an image to create approximate gradients for each pixel for identifying edges in vertical and horizontal directions. The Prewitt operator finds edges when pixel intensities abruptly fluctuate. It recognizes edges in both the horizontal and vertical axes. Scharr is a filtering method that uses the first derivatives to locate and emphasize gradient edges [33].

Median Filter and Pixel Intensity

A median filter was applied to minimize the amount of noise in the stack of two-dimensional OT images. It is a non-linear digital filter used to smooth images which keeps the edges intact. In addition to all of these image filters, the pixel intensity value of each pixel from the OT image is employed as a feature for segmentation.

2.4.3. Training Random Forest Classifier

Compilation of all feature vectors from the extractors for the selected pixels to create the training and testing data set. By including randomness in training samples and combining the output of various randomized trees into a single classifier, the Random Forest addresses the overfitting and generalization problems. The training samples are down-sampled to improve random tree dependency and reduce training time.

The random forest classifier is trained to identify anomalies from the optical tomography images. Random forest is a pixel-wise segmentation technique where feature extractors are applied on each and every pixel from the OT dataset. A total of 100 images were used for training the model. 100 ground truth labels were manually created by segmenting the anomalies using Apeer Annotate (A free open-source platform from ZEISS) platform. Figure 6 shows the OT images and corresponding ground truth labels for a few images used in the training. The blue-colored regions in the OT images are labeled as anomalies which will be evaluated later with the pores from the CT data. It is evident that the objective is to detect particular blue-colored regions within the image, posing challenges when employing direct image segmentation techniques. The OT images with no anomalies should predict empty gray-scale images. The resolution of a single OT image and mask label is 217×217 pixels. So a single OT image consists of 47,089 pixels. As previously described, there are 40 feature extractors, resulting in a total of ($47{,}089 \times 40$) feature values generated for a single image.

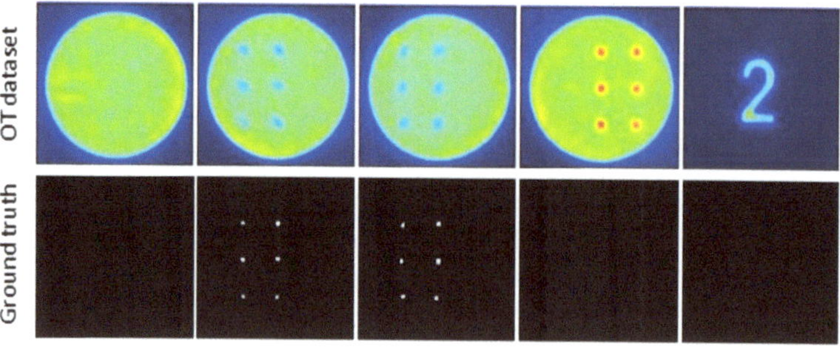

Figure 6. A few OT images and corresponding ground truth labels used in training.

2.4.4. Performance Evaluation on Test Dataset

On the test dataset, the performance of the trained classifier is evaluated. 90% images from the dataset are used for training and the remaining 10% are used for testing. This split is achieved by train_test_split function from sklearn (open source Python library) [34]. The training dataset consists of a total of 47,089 × 90 which is 4,238,010 pixels for training (4,238,010 pixels of OT images and 4,238,010 pixels of mask labels). Similarly, the testing dataset consists of 47,089 × 10 which is 470,890 pixels for testing and validating the developed RF_Segm model. The models are developed for 10, 50, 100, and 1000 estimators to evaluate the prediction accuracy. Training and testing were carried out on the CPU with Intel(R) Xeon(R) CPU E5-1620 v4 @ 3.50 GHz processor with 32 GB RAM.

To ensure that the developed RF_Segm models are not overly dependent on a specific split of the data. Additionally, cross-validation experiments were conducted on one of the RF_Segm models (100 estimators) to ensure the robustness and generalization of the proposed random forest model. The popular K-fold cross-validation technique was used to evaluate the performance of the model in a more rigorous manner. In this approach, the dataset is divided into K equal-sized folds. The model is trained and tested K times, with each fold acting as a test set once and the remaining K-1 folds serving as training. This procedure ensures that the model's performance is evaluated across many data subsets [35].

2.5. Evaluation Using Computed Tomography

Micro-computed tomography is a technique for creating three-dimensional (3D) representations of objects by acquiring multiple X-ray images along an axis of rotation and applying algorithms to reconstruct a 3D model [36]. An industrial 3D micro CT scanner was used to inspect the specimen. Micro CT allows for a comprehensive, non-destructive assessment of the porosity embedded inside AM specimens. The principle of computed tomography is explained in [37]. CT scanning allows for the detection of internal defects in AM parts, including voids, porosity, and cracks. This is achieved by the use of X-rays, which are able to penetrate the part and create a 3D image of the internal structure [38]. The images produced by CT scanning can be used to identify any defects or anomalies in the part. The CT scanner used in this research is located in IABG, Ottobrunn, Germany. It was equipped with a 225 KV micro focus X-ray source and a focal spot size of less than 5 μm and a voxel size of 5 μm with pixel flat panel detector DXR-500L. It has a scanning voltage of 160 kv. The total number of projections is 1440 with a total of 3 frames per projection. The free version of MyVGL which is developed by volume graphics was used to extract layer-wise CT images from the specimens. Additionally, the Register_CT algorithm was created in Matlab programming, employing affine transformation methods to map the coordinates of the extracted CT images onto the optical monitoring images.

2.6. Performance Metrics

The performances of the RF_Segm model with a varying number of estimators were evaluated using various performance measures, such as confusion matrix, accuracy, dice coefficient, precision, recall, and intersection-over-union. The confusion matrix, often referred to as the error matrix, is represented as a matrix that characterizes how well a machine learning model performs when evaluated on a test dataset as shown in Figure 7.

Where TP denotes true positives and is the number of pixels correctly segmented as pores, TN denotes true negatives and is the number of pixels correctly segmented as background, and FP denotes false positives and is the number of pixels incorrectly segmented as pores. FN stands for false negatives and represents the number of pixels that were missed. Accuracy is defined as the proportion of correct estimations to total appraisals. It is concerned with the data set's quality and defects [39], which is defined as follows:

$$Accuracy = \frac{TN + TP}{TP + FP + TN + FN} \tag{6}$$

The dice coefficient calculates the overlapping pixels between the predicted segmentation pixels and the ground truth pixels as follows [40]:

$$Dice\ Coeff = \frac{2 \times TP}{2 \times TP + FP + FN} \tag{7}$$

Precision, also known as sensitivity, is defined as the fraction of pore pixels identified as true-positive pixels in relation to all pixels in an OT image classified by the RF_Segm model, which is defined as follows [40]:

$$Precision = \frac{TP}{TP + FP} \tag{8}$$

The recall is calculated as the proportion of true positive pixels classified by the RF_Segm model vs. pixels labeled by manual labeling, and it is expressed as follows [40]:

$$Recall = \frac{TP}{TP + FN} \tag{9}$$

Intersection over Union (IoU), also known as the Jaccard Index, is defined as the area of intersection between the predicted segmentation map A and the ground truth map B, divided by the area of union between the two maps, and ranges between 0 and 1 [40].

$$IoU = J(A, B) = \frac{|A \cap B|}{|A \cup B|} \tag{10}$$

		Predicted Class	
		Positive	Negative
Actual Class	Positive	TP	FN
	Negative	FP	TN

Figure 7. Understanding the confusion matrix [41].

3. Results and Discussion

3.1. Anomaly Detection Using Random Forest Classifier

In this section, the outcomes of employing a random forest classifier for detecting artificially generated anomalies in the L-PBF process are outlined. First, a summary of the performance metrics and prediction time analysis obtained from the random forest segmentation models is presented. This is followed by an elaborate analysis and interpretation of the results. The findings underscore the proficiency of the random forest classifier in anomaly detection during the L-PBF process, elucidating the pivotal factors that impact its performance.

Performance metrics are utilized to assess the effectiveness of RF_Segm models. Four models are developed with 10, 50, 100, and 1000 number of estimators. The performance of developed anomaly detection models on the test data is evaluated using metrics such as Dice coefficient, precision, recall, accuracy, and intersection over union. These metrics are presented in Table 2, which is detailed in Section 2.6. The RF_Segm model with 1000 estimators demonstrated superior metric values. The highest achieved accuracy Equation (6) was 99.98%, indicating the overall accuracy of the model's predictions. A remarkable IoU score of 71.08 indicated the degree of overlap between the predicted segmentation and the ground truth segmentation. A Dice coefficient of 0.8309 was attained, reflecting the similarity between the predicted and ground truth OT image segmentations. For the RF_Segm model with 1000 estimators, a precision of 0.7705 and a recall of 0.9018 were achieved. These values indicate a minimal number of false negatives compared to false positives, ensuring comprehensive anomaly detection.

Table 2. Performance metrics for RF_Segm model with different number of estimators.

Number of Estimators	Dice Coeff	Precision	Recall	Accuracy [%]	IOU Score
10	0.7068	0.6489	0.7760	96.96	54.66
50	0.7952	0.7334	0.8660	97.96	65.87
100	0.8200	0.7604	0.8899	99.67	69.50
1000	0.8309	0.7705	0.9018	99.98	71.08

The dataset was divided for training (90% dataset) and testing (10% dataset) (Section 2.4.4). To ensure the generalization of the model performance on data splitting, one of the developed anomaly models that is RF_Segm model with 100 estimators was considered for the cross-validation experiment. The data was divided into 10 folds (K = 10), splitting the data into the same shape of 90% training, and 10% testing as used in generating all the anomaly detection models. This approach guarantees that each data point appears in the test set exactly once, reducing the influence of the initial split on model evaluation. Figure 8 shows the plot of metric classification accuracy against each fold. This graph offers a visual depiction of the model's performance variability across various folds. It illustrates that the model consistently achieves accuracies within the range of 99.77% to 99.79% across all ten folds, affirming its suitability for different data splits.

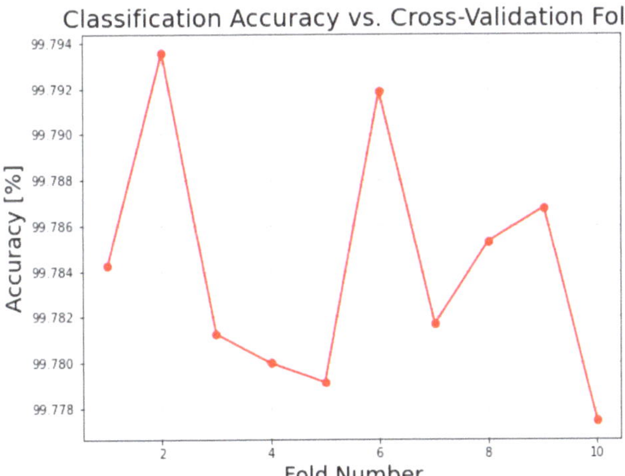

Figure 8. Cross-validation classification accuracy is indicated by the red line across folds for the RF_Segm model with 100 estimators

The confusion matrix (Section 2.6) was calculated for each fold, offering a comprehensive view of the model's performance in terms of true positives, true negatives, false positives, and false negatives. Specifically, the confusion matrix was computed for the test dataset for each fold, encompassing a total of 470,890 pixels. This approach provides a more accurate evaluation of the model's performance on the test dataset.

Subsequently, an average confusion matrix was generated by computing the pixel-wise average (mean) of the individual confusion matrices obtained from all 10 folds of the cross-validation. This average representation consolidates the results and offers a comprehensive view of the model's overall performance.

Figure 9 visually presents the matrix representation of the average confusion matrix. In this matrix, '0' denotes the number of pixels that do not have any anomalies, whereas

'1' reflects the number of pixels with anomalies. This illustration provides valuable insights into the model's performance, allowing us to comprehend its consistency in making accurate and inaccurate predictions, and aids in identifying patterns of errors.

The analysis of the average confusion matrix reveals a predominance of non-anomalous pixels, with a count of ≈467,043 as true positives, accurately identified by the model. Additionally, ≈2829 pixels are correctly recognized as true negatives. On the other hand, there are ≈363 pixels falsely predicted as anomalies (false positives) and ≈653 pixels that are actual anomalies but incorrectly predicted as non-anomalies (false negatives). These numbers highlight the model's strengths and areas for improvement, providing essential metrics to evaluate its performance.

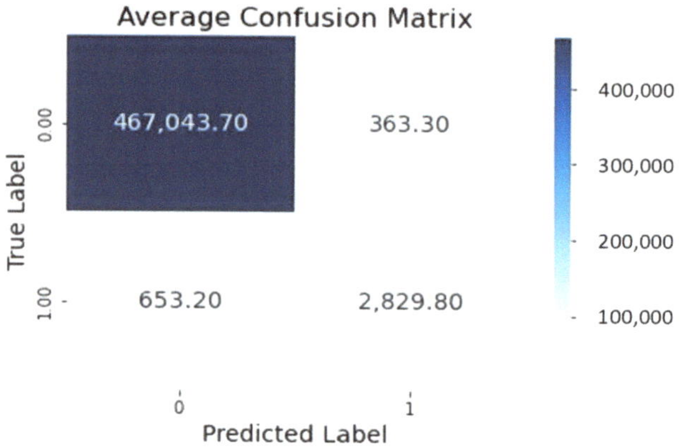

Figure 9. Average confusion matrix: Consistency and performance overview of the RF_Segm model with 100 estimators.

Further analysis of developed models should consider the detection time to strike a balance between precision and detection speed. The time required for training (90% dataset), testing (10% dataset), and prediction time for a single image of the random forest models with different numbers of estimators are tabulated in Table 3. It can be seen that as the number of estimators gets bigger, so does the time required for training, testing, and anomaly prediction time. The prediction time for an anomaly detection model is of significant importance in the L-PBF process. Low prediction time signifies the timely identification of anomalies, improves process efficiency, minimizes costs, enables real-time monitoring, optimizes resource allocation, and facilitates scalability. These combined factors result in improved productivity, decreased defects, and enhanced quality control within L-PBF manufacturing. A prediction time of 40 ms was achieved for the model with 1000 estimators. This detection time goes better with the performance metrics of the RF_Segm model with 1000 estimators when compared to other developed models. Further, if the number of estimators is increased, a point of diminishing returns is reached. At this point, a marginal improvement in performance becomes smaller and does not justify the additional computational resources and time required for training and predicting anomalies.

Table 3. Time required for Training and Testing the RF_Segm model.

No. of Estimators	Training (sec)	Testing (sec)	Prediction Time/Image (sec)
10	100	01	0.011
50	501	02	0.012
100	1115	04	0.014
1000	10,564	35	0.04

In all the generated ML models, a total of 40 feature extractors were utilized in constructing the RF_Segm models. The importance of each feature and the selection of optimal features for model training were deemed crucial. This process, known as Feature Selection in machine learning, involves the removal of less relevant features, thereby simplifying the model, reducing overfitting, and improving computational efficiency [42]. Feature selection based on feature importance contributes to enhancing the model's performance and interpretability. The feature importance diagram, as depicted in Figure 10, illustrates the relative importance of each feature for different estimators. This diagram offers valuable insights into the significance of individual features in the segmentation of anomalies in OT images. Notably, the original pixel values of OT images, Gaussian filter, Median filter, and Gabor24 feature extractors exhibit the highest importance values, indicating a strong relationship with the segmentation label. Overall, the feature importance diagram in the random forest segmentation model provides valuable insights for feature selection, understanding data relationships, model interpretation, error detection, and guidance for future data collection endeavors.

Figure 11 shows the anomaly prediction from OT images for different RF_Segm models developed with different numbers of estimators. In Figure 11, Images A and B are the OT images with artificially induced anomalies and image C is the OT image under normal process conditions. It can be seen that the RF_Segm model with 1000 estimators gives better prediction with respect to models with a lesser number of estimators.

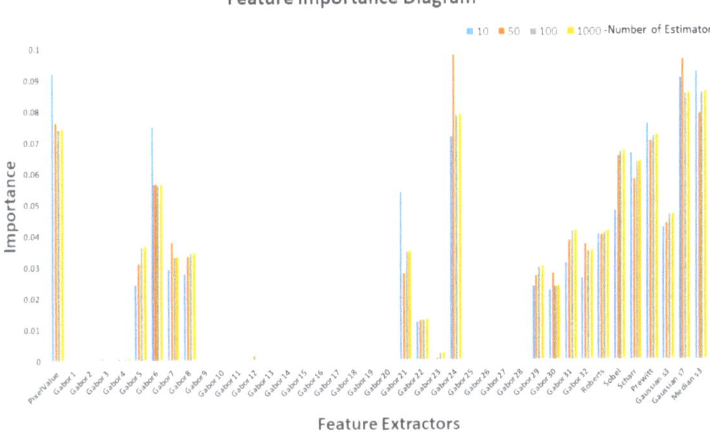

Figure 10. Feature importance diagram for a different number of estimators.

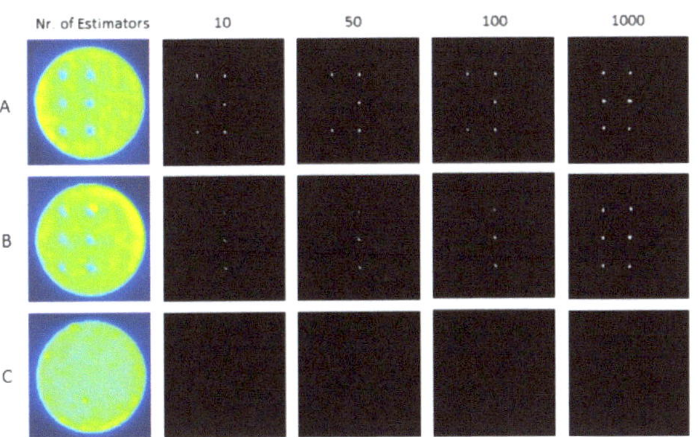

Figure 11. Anomaly prediction in sample optical tomography images (**A–C**) utilizing RF_Segm models with diverse estimator counts.

3.2. Correlation of OT Anomalies with CT Defects

The anomalies detected by the RF_Segm model were evaluated with defects detected in the CT data. Examination of the CT scans revealed the presence of gas pores and a lack of fusion defects in the printed cylinder. These defects exhibited sizes ranging from 5 to 300 µm. Based on their sizes and shapes, the defects were categorized accordingly. Defects with a spherical shape and sizes below 20 µm were designated as gas pores, while irregularly shaped defects measuring between 20 µm and 300 µm were classified as lack of fusion defects [15].

The correlation of OT anomalies and CT defects is carried out using image registration techniques. This technique reduces spatial ambiguity and enables data comparison. Data from different imaging modalities that had varying acquisition setups and spatial resolutions, particularly in the build direction, were overlaid using image registration. It was crucial to emphasize that, in comparison to CT data, which were gathered after the production process, optical monitoring data were acquired during the printing process. Because of this OT data does not account for shrinkage or other deformations that occur after the completion of the building process [1].

An essential part of the registration procedure was the transformation selection. It determines how a certain image is deformed to match the shape of another image dataset. The affine transformation was used to map CT data image coordinates to the OT data image coordinates system. Affine transformation is a type of geometric transformation that combines translation, rotation, similarity, and shear mapping. The developed image registration algorithm called Register_CT as explained in Section 2.5 was used to register the CT image dataset onto the OT monitoring dataset.

Following the successful registration of CT defects onto the OT images, an algorithm was devised for 3D reconstruction using the mapped dataset. The surface rendering module in Matlab was employed to generate a visual representation of the printed cylinder. Figure 12 displays the 3D rendered topography, illustrating the mapping of CT defects onto the OT anomalies. In the figure, blue regions indicate OT anomalies, red regions depict CT defects, and gray color represents the outer geometry of the printed cylinder. Remarkably, approximately 79.4% of CT defects overlapped with the OT anomalies, indicating a strong correlation between the two datasets.

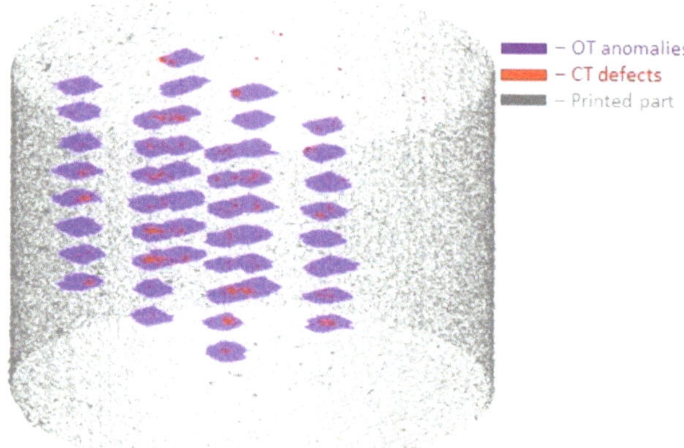

Figure 12. 3-Dimensional rendered surface with overlap of CT defects with detected anomalies.

4. Conclusions

In conclusion, the implemented conventional ML algorithm indicates outstanding abilities in detecting process anomalies within the specified range of intensity values. The experiment was carried out using an EOS M 290 L-PBF machine using EOS Titanium Ti64 as material. Random forest segmentation models were created for a variety of estimators, including 10, 50, 100, and 1000. The RF_Segm model with 1000 estimators obtained an astounding 99.98% accuracy while keeping a fast prediction time of 40 ms. The reported instabilities were analyzed using defects identified using the CT approach to test the algorithm's robustness. 79.4% of the defects identified in the CT data correlated with the anomalies reported by the optical monitoring system, which is promising. This finding emphasizes the proposed random forest segmentation model's potential for quality inspection during L-PBF procedures, outperforming current CT correlation standards in in-situ anomaly identification.

Furthermore, the developed model showcases remarkable efficiency in terms of computational costs, which stands as a significant advantage in utilizing the random forest classifier for anomaly detection model development. Despite the limited training data, consisting of only 100 OT images with corresponding ground truth labels, a segmentation model with an accuracy of 99.98% was successfully created. The model's training process also offers a notable advantage in terms of time requirements. Merely approximately 3 h of computational training was necessary to construct the RF_Segm model with 1000 estimators. This aspect enhances its efficiency and ensures optimal utilization of resources. It is worth highlighting that the model effectively identifies artificially induced defects with reduced laser power parameters and establishes a correlation with defects detected in the CT data.

This paper presents the successful detection of anomalies utilizing the RF_Segm model in the context of in-situ anomaly detection in L-PBF. The anomalies detected in this study were subsequently evaluated and identified as gas pores and lack of fusion defects using the CT technique. These two types of defects are known to significantly impact the fatigue life of printed parts in the L-PBF process. By effectively identifying and characterizing these critical defects, the RF_Segm model contributes to quality assurance and reliability improvement in additive manufacturing processes. The findings of this study highlight the potential of the developed model in enhancing the overall structural integrity and performance of L-PBF-produced components.

In summary, the developed random forest segmentation model, integrated with the optical monitoring system, exhibits exceptional accuracy, swift prediction time, and strong correlation with CT data. Its potential for quality inspection during L-PBF processes demonstrates its efficacy in detecting anomalies and ensuring manufacturing integrity. Further research and validation on larger datasets are warranted to fully exploit the model's capabilities and advance anomaly detection in L-PBF processes.

5. Concluding Remarks

The research has demonstrated the effectiveness of machine learning algorithms in the realm of anomaly detection, particularly in the context of EOS Titanium Ti64 produced by the EOS M 290 L-PBF machine. The developed ML algorithm has showcased remarkable performance, achieving an accuracy rate of 99.98% in identifying anomalies within specified intensity ranges. Notably, it outperforms conventional CT standards, underscoring its potential for enhancing quality assurance processes in the additive manufacturing industry.

Part defects such as gas pores and lack of fusion defects were successfully identified through CT data analysis. which was further correlated with detected anomalies which gave a remarkable correlation accuracy of 79.4%. This underscores the promising capability of optical monitoring systems in enhancing the quality assurance procedures for laser powder bed fusion processes.

Looking ahead, our focus is on the future prospects of integrating machine learning with optical monitoring techniques to further enhance quality assurance in L-PBF processes. Envisioning the utilization of CNN models for faster anomaly detection, harnessing a comprehensive dataset of over 2000 real-time images. Our ongoing efforts will be directed toward improving model robustness and enhancing detection accuracy, paving the way for more reliable and efficient quality control in additive manufacturing.

Author Contributions: Conceptualization, I.A.K., H.B., D.K. and D.L.; methodology, I.A.K.; software, I.A.K.; validation, I.A.K., H.B., D.K. and D.L.; formal analysis, I.A.K. and V.P.; investigation, I.A.K.; resources, D.K. and D.L.; data curation, I.A.K. and H.B.; Writing—original draft preparation, I.A.K.; Writing—review and editing, I.A.K., H.B. and V.P.; visualization, I.A.K.; supervision, H.B. and V.P.; project administration, D.K. and H.B.; funding acquisition, D.K. and V.P. All authors have read and agreed to the published version of the manuscript.

Funding: The authors of this academic research paper would like to express their gratitude for the financial support provided by the German Federal Ministry for Economic Affairs and Energy through the program "Luftfahrtforschungsprogramm LuFo V-3" (PAULA—Prozesse fur die additive Fertigung und Luftfahrtanwendungen, funding code 20W1707E).

Data Availability Statement: The data presented in this study are available on request from the corresponding author. The data are not publicly available due to restrictions from the project partners.

Acknowledgments: The authors would extend our appreciation to EOS GmbH, INDUSTRIEANLA-GEN -BETRIEBSGESELLSCHAFT (IABG) GmbH and Liebherr-Aerospace Lindenberg GmbH for conducting the experiments and their valuable assistance throughout the experimental process. Their contributions have been instrumental in the successful execution of this study.

Conflicts of Interest: The authors declare no conflict of interest.

Abbreviations

The following abbreviations are used in this manuscript:

AM	Additive Manufacturing
OT	Optical Tomography
ML	Machine Learning
μCT	Micro-Computerized Tomography
EOS	Electro-Optical Systems
IABG	INDUSTRIE ANLAGEN-BETRIEBS GESELLSCHAFT
IOU	Intersection-Over-Union
L-PBF	Laser Powder Bed Fusion
CNN	Convolutional Neural Network
CMOS	Complementary metal-oxide-semiconductor
DV	Digital Values
ROI	Regions Of Interest
RF_Segm	Random Forest Segmentation
RGB	Red Green Blue
3D	Three Dimensional
KV	Kilo Volts
TP	True Positives
TN	True Negatives
FP	False Positives
FN	False Negatives
μm	Micrometer
ms	Millisecond

References

1. Mohr, G.; Altenburg, S.J.; Ulbricht, A.; Heinrich, P.; Baum, D.; Maierhofer, C.; Hilgenberg, K. In-Situ Defect Detection in Laser Powder Bed Fusion by Using Thermography and Optical Tomography—Comparison to Computed Tomography. *Metals* **2020**, *10*, 103. [CrossRef]
2. Baumgarti, H.; Tomas, J.; Buettner, R.; Markus Merkel, M. A deep learning-based model for defect detection in laser-powder bed fusion using in-situ thermographic monitoring. *Prog. Addit. Manuf.* **2020**, *5*, 277–285. [CrossRef]
3. Chen, Z.; Han, C.; Gao, M.; Kandukuri, S.Y.; Zhou, K. A review on qualification and certification for metal additive manufacturing. *Virtual Phys. Prototyp.* **2022**, *17*, 382–405. [CrossRef]
4. Abdulhameed, O.; Al-Ahmari, A.; Ameen, W.; Mian, S.H. Additive manufacturing: Challenges, trends, and applications. *Adv. Mech. Eng.* **2019**, *11*, 1687814018822880. [CrossRef]
5. Montazeri, M.; Yavari, R.; Rao, P.; Boulware, P. In-Process Monitoring of Material Cross-Contamination Defects in Laser Powder Bed Fusion. *J. Manuf. Sci. Eng. ASME* **2018**, *140*, 111001. [CrossRef]

6. Seifi, M.; Salem, A.; Beuth, J.; Harrysson, O.; Lewandowski, J.J. Overview of Materials Qualification Needs for Metal Additive Manufacturing. *Miner. Met. Mater. Soc. JOM* **2016**, *68*, 747–764 [CrossRef]

7. Choo, H.; Sham, K.L.; Bohling, J.; Ngo, A.; Xiao, X.; Ren, Y.; Depond, P.J.; Matthews, M.J.; Garlea, E. Effect of laser power on defect, texture, and microstructure of a laser powder bed fusion processed 316L stainless steel. *Mater. Des.* **2019**, *164*, 1264–1275. [CrossRef]

8. Brennan, M.C.; Keist, J.S.; Palmer, T.A. Defects in Metal Additive Manufacturing Processes. *J. Mater. Eng. Perform.* **2021**, *30*, 4808–4818. [CrossRef]

9. Pham, V.T.; Fang, T.H. Understanding porosity and temperature induced variabilities in interface, mechanical characteristics and thermal conductivity of borophene membranes. *Sci. Rep.* **2021**, *11*, 12123. [CrossRef] [PubMed]

10. Yeung, H.; Yang, Z.; Yan, L. A meltpool prediction based scan strategy for powder bed fusion additive manufacturing. *Addit. Manuf.* **2020**, *35*, 2214–8604. [CrossRef]

11. Ravindran, S. Five ways deep learning has transformed image analysis. *Nature* **2022**, *609*, 864–866. [CrossRef]

12. Kwon, O.; Kim, H.G.; Kim, W.; Kim, G.H.; Kim, K. A convolutional neural network for prediction of laser power using melt-pool images in laser powder bed fusion. *IEEE Access* **2020**, *8*, 23255–23263. [CrossRef]

13. Wang, C.; Tana, X.P.; Tora, S.B.; Lim, C.S. Machine learning in additive manufacturing: State-of-the-art and perspectives. *Addit. Manuf.* **2020**, *36*, 2214–8604. [CrossRef]

14. Meng, L.; Park, H.Y.; Jarosinski, W.; Jung, Y.G. Machine learning in additive manufacturing: A Review. *J. Miner. Met. Mater. Soc.* **2020**, *72*, 2363–2377. [CrossRef]

15. Gordon, J.V.; Narra, S.P.; Cunningham, R.W.; Liu, H.; Chen, H.; Suter, R.M.; Beuth, J.L.; Rollett, A.D. Defect structure process maps for laser powder bed fusion additive manufacturing. *Addit. Manuf.* **2020**, *36*, 2214–8604. [CrossRef]

16. Mahapatra, D. Analyzing Training Information from Random Forests for Improved Image Segmentation. *IEEE Trans. Image Process.* **2014**, *23*, 1504–1512. [CrossRef]

17. Aurelia, J.E.; Rustam, Z.; Hartini, S.; Darmawan, N.A. Comparison Between Convolutional Neural Network and Random Forest as Classifier for Cerebral Infarction. In Proceedings of the International Conference on Advanced Intelligent Systems for Sustainable Development, AISC 1417, Marrakech, Morocco, 6–8 July 2020; pp. 930–939.

18. Mahmoud, D.; Magolon, M.; Boer, J.; Elbestawi, M.A.; Mohammadi M.G. Applications of Machine Learning in Process Monitoring and Controls of L-PBF Additive Manufacturing: A Review. *J. Appl. Sci.* **2021**, *11*, 11910. [CrossRef]

19. Wu, Y.; Misra, S. Intelligent Image Segmentation for Organic-Rich Shales Using Random Forest, Wavelet Transform, and Hessian Matrix. *IEEE Geosci. Remote Sens. Lett.* **2020**, *17*, 1144–1147. [CrossRef]

20. Rajendran, P.; Madheswaran, M. Hybrid Medical Image Classification Using Association Rule Mining with Decision Tree Algorithm. *J. Comput.* **2010**, *2*, 2151–9617.

21. ASTM F1472. Available online: https://standards.globalspec.com/std/14343636/ASTM%20F1472 (accessed on 1 November 2020).

22. ASTM International. Available online: https://www.astm.org/standards/f2924 (accessed on 20 October 2021).

23. A Gentle Introduction to k-Fold Cross-Validation. Available online: https://machinelearningmastery.com/k-fold-cross-validation/ (accessed on 3 August 2020).

24. Wang, P.; Nakano, T.; Bai, J. Additive Manufacturing: Materials, Processing, Characterization and Applications. *Crystals* **2022**, *12*, 747. [CrossRef]

25. Zenzinger, G.; Bamberg, J.; Ladewig, A.; Hess, T.; Henkel, B.; Satzger, W. Process Monitoring of Additive Manufacturing by Using Optical Tomography. *AIP Conf. Proc.* **2015**, *1650*, 164–170.

26. Liu, H.; Huang, J.; Li, L.; Cai, W. Volumetric imaging of flame refractive index, density, and temperature using background-oriented Schlieren tomography. *Sci. China Technol. Sci.* **2021**, *64*, 98–110. [CrossRef]

27. Samajpati, B.J.; Degadwala, S.D. Hybrid Approach for Apple Fruit Diseases Detection and Classification Using Random Forest Classifier. In Proceedings of the International Conference on Communication and Signal Processing, Melmaruvathur, India, 6–8 April 2016; pp. 1015–1019.

28. Random Forest: A Complete Guide for Machine Learning. Available online: https://builtin.com/data-science/random-forest-algorithm (accessed on 14 March 2023).

29. Hiew, B.; Teoh, A.B.; Ngo, D.C. Preprocessing of Fingerprint Images Captured with a Digital Camera. In Proceedings of the 2006 9th International Conference on Control, Automation, Robotics and Vision, Singapore, 5–8 December 2006; pp. 1–6.

30. Kumar, K.V.; Jayasankar, T. An identification of crop disease using image segmentation. *Int. J. Pharm. Sci. Res. (IJPSR)* **2019**, *10*, 1054–1064.

31. See, Y.C.; Noor, N.M.; Low, J.L.; Liew, E. Investigation of Face Recognition Using Gabor Filter with Random Forest As Learning Framework. In Proceedings of the TENCON 2017—2017 IEEE Region 10 Conference, Penang, Malaysia, 5–8 November 2017; pp. 1153–1158.

32. Haddad, R.A.; Akansu, A.N. A Class of Fast Gaussian Binomial Filters for Speech and Image Processing. *IEEE Trans. Signal Process.* **1991**, *39*, 723–727. [CrossRef]

33. Image Gradients with OpenCV (Sobel and Scharr). Available online: https://pyimagesearch.com/2021/05/12/image-gradients-with-opencv-sobel-and-scharr/ (accessed on 12 May 2021).

34. Using Train Test Split in Sklearn: A Complete Tutorial. Available online: https://ioflood.com/blog/train-test-split-sklearn/ (accessed on 5 September 2023).

35. 3D Printing with Titanium Alloys. Available online: https://www.eos.info/en/3d-printing-materials/metals/titanium-ti64-ticp (accessed on 21 August 2023).
36. Thompson, A.; Maskery, I.; Leach, R.K. X-ray computed tomography for additive manufacturing: A review. *Meas. Sci. Technol.* **2016**, *27*, 7. [CrossRef]
37. Léonard, F.; Williams, S.T.; Prangnell, P.B.; Todd, I.; Withers, P.J. Assessment by X-ray CT of the effects of geometry and build direction on defects in titanium ALM parts. In Proceedings of the Conference on Industrial Computed Tomography (ICT), Wels, Austria, 19–21 February 2012.
38. Khosravani, M.R.; Reinicke, T. On the Use of X-ray Computed Tomography in Assessment of 3D-Printed Components. *J. Nondestruct. Eval.* **2020**, *39*, 1–17. [CrossRef]
39. Thayumanavan, M.; Ramasamy, A. An efficient approach for brain tumor detection and segmentation in MR brain images using random forest classifier. *Concurr. Eng. Res. Appl.* **2021**, *29*, 266–274. [CrossRef]
40. Minaee, S.; Boykov, Y.; Porikli, F.; Plaza, A.; Kehtarnavaz, N.; Terzopoulos, D. Image Segmentation Using Deep Learning: A Survey. *IEEE Trans. Pattern Anal. Mach. Intell.* **2022**, *44*, 3523–3542. [CrossRef] [PubMed]
41. Markoulidakis, I.; Rallis, I.; Kopsiaftis, G.; Georgoulas, I. Multi-Class Confusion Matrix Reduction method and its application on Net Promoter Score classification problem. In Proceedings of the 14th PErvasive Technologies Related to Assistive Environments Conference (PETRA 21), Corfu, Greece, 29 June–2 July 2021; pp. 412–419. [CrossRef]
42. Miao, J.; Niu, L. A Survey on Feature Selection. *Procedia Comput. Sci.* **2016**, *91*, 919–926. [CrossRef]

Article

Application of Linear Mixed-Effects Model, Principal Component Analysis, and Clustering to Direct Energy Deposition Fabricated Parts Using FEM Simulation Data

Syamak Pazireh , Seyedeh Elnaz Mirazimzadeh and Jill Urbanic *

Department of Mechanical, Automotive and Materials Engineering, University of Windsor, 401 Sunset Ave, Windsor, ON N9B 3P4, Canada; pazireh@uwindsor.ca (S.P.); mirazims@uwindsor.ca (S.E.M.)
* Correspondence: jurbanic@uwindsor.ca

Abstract: The purpose of this study is to investigate the effects of toolpath patterns, geometry types, and layering effects on the mechanical properties of parts manufactured by direct energy deposition (DED) additive manufacturing using data analysis and machine learning methods. A total of twelve case studies were conducted, involving four distinct geometries, each paired with three different toolpath patterns based on finite element method (FEM) simulations. These simulations focused on residual stresses, strains, and maximum principal stresses at various nodes. A comprehensive analysis was performed using a linear mixed-effects (LME) model, principal component analysis (PCA), and self-organizing map (SOM) clustering. The LME model quantified the contributions of geometry, toolpath, and layer number to mechanical properties, while PCA identified key variables with high variance. SOM clustering was used to classify the data, revealing patterns related to stress and strain distributions across different geometries and toolpaths. In conclusion, LME, PCA, and SOM offer valuable insights into the final mechanical properties of DED-fabricated parts.

Keywords: direct energy deposition; machine learning; principal component analysis; self-organizing maps; linear mixed-effects models

Citation: Pazireh, S.; Mirazimzadeh, S.E.; Urbanic, J. Application of Linear Mixed-Effects Model, Principal Component Analysis, and Clustering to Direct Energy Deposition Fabricated Parts Using FEM Simulation Data. *Materials* **2024**, *17*, 5127. https://doi.org/10.3390/ma17205127

Academic Editor: Alexander Yu Churyumov

Received: 27 September 2024
Revised: 15 October 2024
Accepted: 18 October 2024
Published: 21 October 2024

1. Introduction

Direct energy deposition (DED) additive manufacturing (AM) is an advanced manufacturing process used for repairing damaged parts, adding features, and building new parts from 3D model data on a metal additive basis. This process involves joining materials in a stack-by-stack and layer-by-layer manner, with the initial material stock being either in powder or wire form [1]. During the additive process, material is added and energy is applied (laser, beam, arc) to form the melting zone. As the deposition follows the predefined toolpath, the already-built beads solidify as they cool down to lower temperatures, a process that can be tracked and observed using image-filtering tools [2]. The thermal diffusion within the part during the build and cooling-down processes has a significant impact on the intrinsic residual stresses, strains, and distortions of the material [3]. The thermal history and temperature gradients have a microscopic impact on the material's strength through grain formation, which is affected by these temperature gradients [4]. As stresses and strains within fabricated parts vary significantly with temperature gradients, it is essential to pay attention to process parameters [5], geometry [6], and deposition toolpath patterns [7] that affect the temperature history of the built parts.

High fabrication costs, random errors, instability, and significant computational demands in DED necessitate leveraging big data from tests and simulations. Data analysis can help identify the root causes of part quality abnormalities in DED manufacturing, although correlation may be low within post-process data, necessitating in-situ tracing of fusion phenomena [8]. Machine learning has shown promising results in gaining stable fabrication in the laser-directed energy deposition [9]. Convolutional neural networks

(CNNs) have demonstrated potential for developing a predictive model based on a large dataset of real-time image processing to control DED manufacturing [10]. Li et al. collected experimental data from a DED study to predict grain boundary tilt based on thermal gradients, crystal orientation, and the Marangoni effect using an artificial neural network (ANN) structure [11]. The analytical model obtained from a trained ANN provides a valid and fast prediction of grain growth behavior in DED parts.

Machine learning can also recognize patterns, correlations, and dependencies between geometrical and mechanical properties of DED-fabricated parts. A defect classification was proposed by Patil et al. for DED processes [12]. Xu et al. [13] proposed a hybrid approach combining deep learning and mixed-effect modeling, where the random effect accounts for mean temperature variations for real-time defect detection. Unsupervised clustering has also been used in the research [14] to categorize the geometries based on the accuracy of the manufacturing quality.

The toolpath and geometry types have shown an inter-coupled effect on the qualities of DED-fabricated parts [3]. A clustering approach used in [15] investigated stress-distortion feature-based analysis of multiple geometries to determine local point assignments to clusters. This approach provided insight into the physical similarities among edge and internal locations of the observations. However, the aforementioned clustering approach did not account for toolpath effects when generating the FEM-based simulation data.

In the current research, we conducted a statistical analysis across four different geometric shapes, each paired with three distinct toolpath patterns (resulting in twelve separate case studies). The goal was to investigate the impact of toolpath patterns, geometry types, and layering effects on mechanical properties, with a particular focus on residual stresses and strains. Analyses were performed on finite element (FEM) simulation data, which includes element birth and death in a multilayer, multi-track DED physics setting. All node data were extracted. Directional residual stresses, strains, and maximum principal stresses were chosen as the physical features of concern. A correlation analysis was conducted to determine how these features are related. The analysis was applied to the entire dataset as well as to datasets for each geometry and toolpath separately. Then, the linear mixed-effects model was investigated to consider the toolpath pattern, geometry type, and layering effect as fixed variables on the local nodes' data (the observations), with the observations treated as random variables. The linear mixed-effects (LME) model determines the contribution of each fixed variable to the mechanical properties. A principal component analysis (PCA) determines what features have the highest variance and are more informative in DED-processed data. The results are followed by a self-organizing map (SOM) clustering approach to compare the thin and thick longitudinal and transverse wall data at different layers to see how the toolpath and geometry create similar properties within various parts and deposition patterns.

2. Materials and Methods

The case studies, material, and analysis methodology are described in this section.

2.1. Case Studies

Four distinct geometries, each with three separate deposition toolpath patterns, are analyzed in this study. The geometries include a cross-type, a 3-step plate, a 5-step plate, and a rectangle with a hole. For each geometry, three toolpaths are examined: one-way longitudinal, longitudinal zigzag, and one-way transverse. The term "longitudinal" refers to the longitudinal direction of the parts. Figure 1 shows the geometries considered for the research. The schematic of the toolpath on each geometry is depicted with three colored arrows corresponding to the coordinate system. In total, 12 distinct case studies are considered. Hereafter, the x-axis refers to the longitudinal direction, and the y-axis refers to the transverse direction. The APlus add-on from CAD/CAM Mastercam software (Mastercam 2025, V27.0.6876.0) [16] was used to generate the laser metal AM toolpath NC files. All the case studies have the same substrate with a dimension of 90 (mm) $\times$ 60 (mm) $\times$ 7 (mm).

The materials for this study are 316L stainless steel (SS) and A36 steel for the clad and the substrate, respectively.

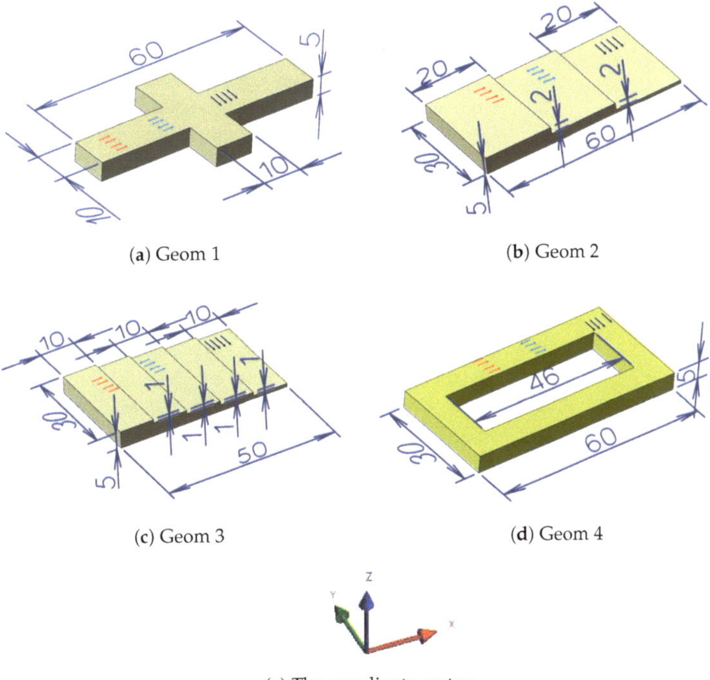

(**a**) Geom 1   (**b**) Geom 2

(**c**) Geom 3   (**d**) Geom 4

(**e**) The coordinate system

Figure 1. The geometries considered for this research include a cross-type, a 3-step plate, a 5-step plate, and a rectangle with a hole, with dimensions in millimeters. The red arrows represent the one-way longitudinal toolpath, the blue arrows represent the longitudinal zigzag toolpath, and the black arrows represent the one-way transverse zigzag for each geometry.

The process parameters of the fabrications are presented in Table 1. In the present study, the term "overlapping" pertains to the region of interlocking between two beads (or two tracks).

Table 1. Process parameters used for the simulations.

Parameter	Data
Laser power	1000 W
Deposition speed	9.1 mm/s
Number of layers	5
Bead width	2 mm
Bead height	1 mm
Bead overlap	23%

The FEM simulation data with bilinear elastoplastic modeling were used as the basis for data analysis in this research. The element birth and death technique, utilizing ANSYS APDL (2023 R2) programming, was employed to activate deposition elements according to the toolpath pattern. A Python script was developed to read the NC files and mesh element data. Then, in the pre-processing step, the order of the activated elements and their corresponding IDs were determined. Once the elements were ordered, the APDL programming added each element individually to mimic the deposition physics. The mesh size for the DED process is determined by the beads' width and depth, adopting a strategy where each element representing the melt zone is activated on an individual basis. The width of each element reflects the bead's non-overlapped area, ensuring that the bead count across the depositions matches the number of deposition tracks extracted from the NC files from APlus software (an Add-Ons of Mastercam). Note that the validity of the FEM model was assessed through a comparison with experimental measurements and a study by the authors in [17]. The results of [17] show that the model accurately captures the stress gradients aligned with the actual physics at the corner sides of the model where heat diffusion is well represented. However, the middle of the model may exhibit an overprediction of residual stress due to an overestimation in the thermal model, yet the stress gradient pattern remains similar to the actual physics. There are limitations in simplifying the model as the material overlap cannot be addressed with the FEM simulations.

The simulations were conducted on the Digital Resource Alliance of Canada cloud accounts, each equipped with 128 GB RAM and 8 CPU cores. Processing and post-processing of all cases took several weeks. An ANSYS post-processing APDL script was developed to extract data for all nodes within the range of simulation sub-steps. The data for all the FEM nodes were stored in a dataset. The current study focused on statistical and machine learning analyses of the data to gain deeper insight into the physics by interpreting the data. These analyses were applied using Python scripts. In the following subsections, the statistical and machine learning methods used in the study are introduced.

2.2. Correlation Analysis

Correlation measures how closely variables or dataset features approximate linear functions. The relationship between two features will always be higher if it is closer to some linear function, so the linear correlation between them will be stronger, and the correlation coefficient will be greater in absolute value.

Considering a dataset with two features: $\vec{x}$ and $\vec{y}$, each with n number of observations, the "Pearson" correlation coefficient is measured by the following:

$$r = \frac{\sum_{i=1}^{n}(x_i - \bar{x})(y_i - \bar{y})}{\sqrt{\sum_{i=1}^{n}(x_i - \bar{x})^2 \sum_{i=1}^{n}(y_i - \bar{y})^2}} \tag{1}$$

where $\bar{x}$ and $\bar{y}$ are the mean values of the features:

$$\bar{x} = \frac{\sum_{i=1}^{n} x_i}{n} \tag{2}$$

$$\bar{y} = \frac{\sum_{i=1}^{n} y_i}{n} \tag{3}$$

The Pearson correlation values, r (Equation (1)), range between -1 and 1, and as the absolute value increases, the correlation between the two features rises.

2.3. Linear Mixed-Effects Model

LMEs are used for regression analyses involving dependent data when data are collected and summarized in groups. The *statsmodels* [18] implementation of LME is primarily group-based, meaning that random effects must be independently realized for responses (observations) in different groups.

Some specific linear mixed-effects models are as follows [18]:

- Random intercepts models: In these models, all responses within a group are additively shifted by a value that is specific to the group.
- Random slopes models: The responses in a group follow a conditional mean trajectory that is linear in the observed covariates, with both slopes (and possibly intercepts) varying by group.
- Variance components models: The levels of one or more categorical covariates are associated with draws from distributions. These random terms additively determine the conditional mean of each observation based on its covariate values.

The coefficients of these models may vary according to the grouping variables used to describe the relationship between the response variable and the independent variables. The mixed-effects model consists of two parts, fixed effects, and random effects. Generally, fixed-effects terms represent linear regression, while random effects represent randomly selected experimental units.

The random coefficients are defined as follows:

$$\mathbf{Y}_{ij} = \beta_0 + \beta_1 \mathbf{Z}_{ij} + \gamma_{0i} + \gamma_{1i} \mathbf{Z}_{ij} + \epsilon_{ij} \tag{4}$$

where $\mathbf{Y}_{ij}$ is the jth measured response (observation) for subject (group) i, and $\mathbf{Z}_{ij}$ is the covariate for this response. The "fixed effects parameters", β_0 and β_1, are shared by all subjects, and the error term ϵ_{ij} is independent of the parameters and distributed with a mean of zero. The "random effects parameters", γ_{0i} and γ_{1i}, also follow a bivariate distribution with a mean of zero.

The *statsmodels MixedLM* treats the entire dataset as a single group to include crossed random effects in a model. The variance components in the model are used to define models with various combinations of crossed and non-crossed random effects. The variable ϵ_{ij} is the normal error with zero means and the values are independent both within and between groups. The detailed procedure can be found in [19].

2.4. Principal Component Analysis

PCA is a machine learning technique for dimensionality reduction that converts a large dataset into a smaller one by transforming the primary features into new features, called principal components, which are combinations of the primary ones. The first few principal components typically retain most of the information from the larger dataset. PCA decomposes a multivariate dataset into a set of successive orthogonal components that explain the maximum amount of variance [20]. Using the singular value decomposition (SVD) of the data to project them to a lower-dimensional space, linear dimensionality reduction is performed. The input data are centered but not scaled for each feature before applying the SVD.

Given a dataset $\mathbf{X}$, with n samples (observations or records) and p features (attributes), the jth principal component can be shown by the following:

$$PCA_j = w_{j1}\mathbf{X}_1 + w_{j2}\mathbf{X}_2 + \cdots + w_{ji}\mathbf{X}_i + \cdots + w_{jp}\mathbf{X}_p \tag{5}$$

where $\mathbf{X}_i$ is the ith feature of the dataset $\mathbf{X}$ and w_{ji} is the weighting (loading) of the jth principal component of the ith feature. As shown in Equation (5), the principal components are a linear combination of the primary features. The loadings can be found using the eigenvectors of matrix Σ, where we have the following:

$$\Sigma = \mathbf{X}^\mathsf{T} \mathbf{X} \tag{6}$$

2.5. Self-Organizing Map

A self-organizing map (SOM) is a neural network unsupervised learning method. This method learns to classify input vectors according to how they are grouped in the input space. Neighboring neurons in the self-organizing map learn to recognize neighboring sections of the input space. Thus, SOM methods learn both the distribution and topology of the input vectors they are trained on. Distances between neurons (clusters or best-matching units (BMUs)) are calculated from their positions with a distance function. In this neural network, all neurons within a certain neighborhood of the winning neuron (BMU) are updated, using the Kohonen rule [21]. The weight vector ($\vec{w}$) of neuron (BMU) ij in the iteration (epoch) t is updated as follows:

$$\vec{w}_{ij}^{t} = \vec{w}_{ij}^{t-1} + \eta^{t-1} f^{t-1} \left(\vec{x} - \vec{w}_{ij}^{t-1} \right) \tag{7}$$

where $\vec{x}$ is the input vector, η is the learning rate, and f is the neighborhood distance function. The details of the algorithm are found in [22].

3. Results and Discussion

3.1. Correlation Analysis

A primary investigation into the interaction between geometry and toolpath on the residual mechanical properties of DED-built parts is based on the correlation between multiple variables, such as stresses and strains. Normal residual stresses, maximum and minimum principal stresses, and total directional strains are selected from the four geometries, each analyzed under three separate toolpath scenarios. The maximum and minimum residual stresses are considered as they play a crucial role in determining the impact of deposition direction on stresses [23]. The correlation heat map of the concatenated data for all the case studies (12 scenarios) is presented in Figure 2. Recall that the x-axis corresponds to the longest length of the shapes and y denotes the transverse direction. The maximum principal stress and the longitudinal residual stress (σ_{xx}) have the highest correlation, 0.73, implying the highest dependency of the optimum toolpath on the longitudinal direction.

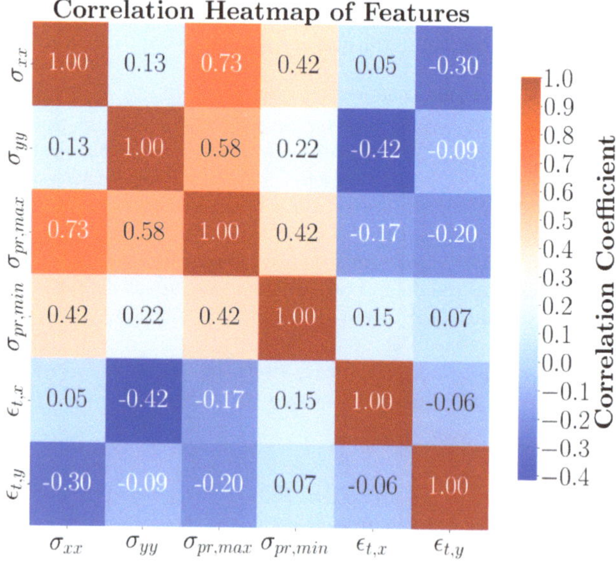

Figure 2. Heat map of the correlation between normal stresses, principal stresses, and total strains for the entire dataset (four geometries, each with three toolpaths).

The results of the entire dataset, including the geometries and the toolpaths, indicate that the longitudinal residual stress has a high correlation with maximum principal residual stress, which significantly impacts the optimum toolpath selection. Figure 3 displays the correlation between features for each geometry individually, incorporating all toolpaths for each geometry. The results show that the step-type geometries (Geom 2 and 3) demonstrate a high correlation between the principal and longitudinal and transverse stresses. It implies that a combination of deposition directions would result in the possible optimum toolpath selection for these geometries. The rectangle part with a big hollow in the middle with thin walls (Geom 4) represents the lowest correlation between the directional stresses and the maximum principal stresses.

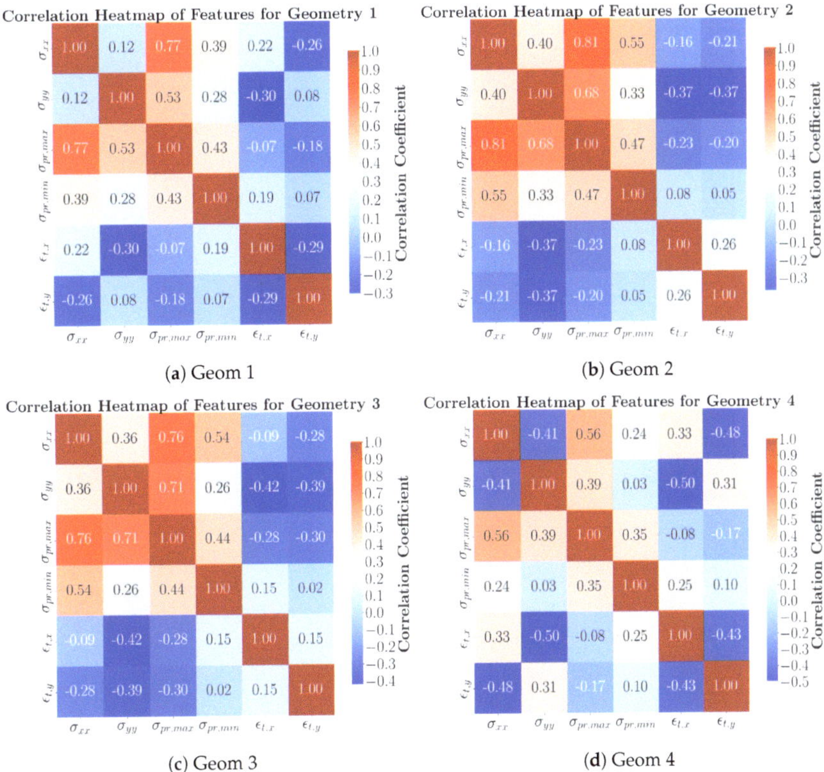

Figure 3. Heat map correlation normal stresses, principal stresses, and total strains for each geometry with three toolpaths.

Similarly, the correlation between features for each toolpath individually, incorporating all geometries, is shown in Figure 4. The toolpath directional stresses are highly correlated with the maximum residual stress. For both the longitudinal one-way and longitudinal zigzag toolpaths, the x-axis residual stresses show a strong correlation with the maximum residual stress. In contrast, the transverse toolpath exhibits a high correlation between the transverse (y-axis) residual stress and the maximum principal stress.

The results reveal that—for thin parts—longitudinal depositions are the determining factor in forming residual stresses, whereas thick parts are affected by both longitudinal and transverse directions. Additionally, the results indicate that the deposition direction generates the highest residual stress in that specific direction.

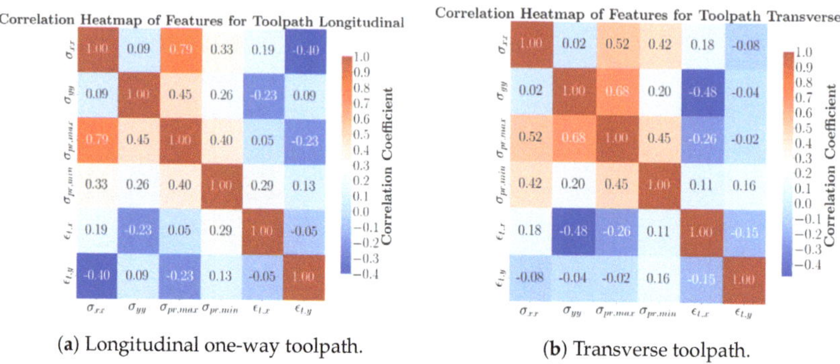

(**a**) Longitudinal one-way toolpath.　　　　(**b**) Transverse toolpath.

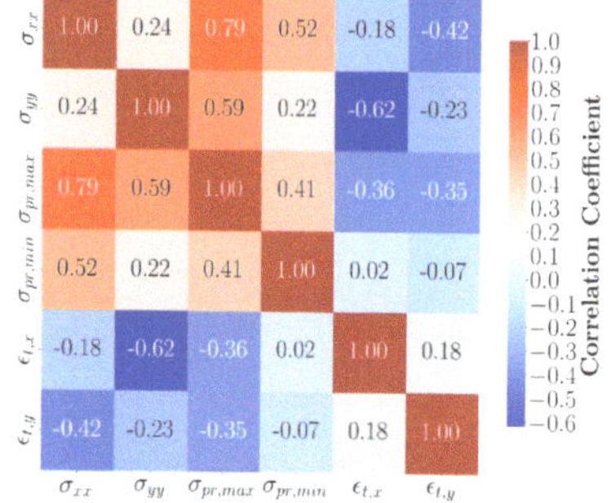

(**c**) Longitudinal zigzag toolpath.

Figure 4. Heat map correlation normal stresses, principal stresses, and total strains for each toolpath with four geometries.

3.2. Linear Mixed-Effects Model

The data reviewed so far indicate that the final mechanical properties of DED-manufactured parts depend on the toolpath, geometry type, and layer. To quantify the contributions of each factor (geometry type, toolpath pattern, and layer number) to the mechanical properties (e.g., residual stresses and strains), a linear mixed-effects model is used. This model considers the fixed effects (geometry, toolpath, and the DED-built layer number) on the random effects (distributed data on the parts). In this context, the random effect refers to the individual local nodes for all the case studies. For this purpose, the *"MixedLM"* module from the *"statsmodels.formula.api"* library [18] of the Python language was used to fit the model. The dependent variables are as follows: σ_{xx}, σ_{yy}, $\sigma_{pr,max}$, $\epsilon_{t,x}$, and $\epsilon_{t,y}$. All the data from the 12 case studies were concatenated into a unique dataset as two additional features of the toolpath pattern, and the geometry types were added to the observations of each data node from the FEM data. The fitting model results for each variable are presented in Tables 2–6. The intercept (reference) case is geometry 1 with the longitudinal toolpath. In comparison with the intercept case, the rest of the case studies are statistically analyzed.

The Coef. parameter in Table 2 for the intercept represents the estimated mean value of σ_{xx} when all other variables are at their reference levels, with the remaining rows showing deviations from the intercept. Geometries 2 to 4 all show decreased levels of σ_{xx}; however, geometry 3 does not display significant change compared to the reference (Geom 1, Longitudinal) as the p-value (0.147) is above 0.05. All coefficients except geometry 3 are statistically significant at the 0.05 level, as indicated by P > |z| values less than 0.05. Only the zigzag pattern appears to have higher stress compared to the reference case. Since the group variance is low (0.003), minimum variation occurs between groups, indicating that most variability in σ_{xx} is explained by fixed effects rather than random effects. The layer, zigzag pattern, and geometry 2 have the highest impacts on σ_{xx}, respectively.

Table 2. Mixed linear model regression results for variable σ_{xx}.

Model:	MixedLM		Dependent Variable:	$Q(\sigma_{xx}(MPa))$		
No. Observations:	36,971		Method:	REML		
No. Groups:	10,206		Scale:	42,248.3905		
Min. group size:	1		Log-Likelihood:	−249,340.2347		
Max. group size:	4		Converged:	Yes		
Mean group size:	3.6					
	Coef.	Std.Err.	z	$P > \lvert z \rvert$	[0.025	0.975]
Intercept ("Geom 1,Longitudinal")	394.452	3.257	121.115	0.000	388.069	400.836
C(Geometry,Treatment)[T.2]	−36.367	3.127	−11.610	0.000	−42.436	−30.297
C(Geometry, Treatment)[T.3]	−4.609	3.220	−1.450	0.147	−10.980	1.642
C(Geometry, Treatment)[T.4]	−9.010	3.113	−2.894	0.004	−15.111	−2.910
C(Toolpath, Treatment)[T.Trans]	−37.963	2.612	−14.537	0.000	−43.082	−32.845
C(Toolpath, Treatment)[T.Zigzag]	12.185	2.643	4.610	0.000	7.005	17.365
Q("Layer")	−38.673	0.658	−58.774	0.000	−39.962	−37.383
Group Var	0.003	0.940				

Table 3. Mixed linear model regression results for variable σ_{yy}.

Model:	MixedLM		Dependent Variable:	$Q(\sigma_{yy}(MPa))$		
No. Observations:	36,971		Method:	REML		
No. Groups:	10,206		Scale:	31,461.3621		
Min. group size:	1		Log-Likelihood:	−244,834.7612		
Max. group size:	4		Converged:	Yes		
Mean group size:	3.6					
	Coef.	Std.Err.	z	$P > \lvert z \rvert$	[0.025	0.975]
Intercept ("Geom 1,Longitudinal")	211.383	2.937	71.963	0.000	205.626	217.141
C(Geometry, Treatment)[T.2]	47.009	2.708	17.359	0.000	41.702	52.317
C(Geometry, Treatment)[T.3]	77.948	2.785	27.990	0.000	72.490	83.406
C(Geometry, Treatment)[T.4]	7.887	2.694	2.927	0.003	2.606	13.167
C(Toolpath, Treatment)[T.Trans]	−18.807	2.747	−7.861	0.000	−23.606	−14.008
C(Toolpath, Treatment)[T.Zigzag]	15.695	2.489	6.306	0.000	10.817	20.573
Q("Layer")	−21.713	0.582	−37.327	0.000	−22.853	−20.573
Group Var	1770.261	0.969				

Regarding variable σ_{yy}, as presented in Table 3, geometries 2 and 3 have the greatest impact, with increased levels of σ_{yy} at 47.0 and 77.9 (MPa), respectively. The effects of the layer and toolpath are lower compared to the impact of geometry. The P-values of all fixed effects are below 0.05, indicating that these effects are significant. The larger group variance (1770.2) compared to the previous model (σ_{xx}) suggests more variability between groups in σ_{yy} compared to σ_{xx}. It suggests that the random effects (local data distribution) are significant on σ_{yy}. A general comparison between the results of Tables 2 and 3 highlights that geometry can have a positive effect on σ_{xx} and a negative impact on σ_{yy}. The layer has a greater influence on σ_{xx}.

The results of Table 4 demonstrate that the geometry has the highest impact on $\sigma_{pr,max}$ compared to the toolpath, while the layer effect is predominant. The analyses of $\epsilon_{t,x}$ and $\epsilon_{t,y}$ in Tables 5 and 6 show positive effects of geometries 2 to 4 on $\epsilon_{t,x}$ and negative effects on $\epsilon_{t,y}$ compared to the intercept (reference), displaying behavior similar to that described for the normal stresses.

Table 4. Mixed linear model regression results for variable $\sigma_{pr,max}$.

Model:	MixedLM		Dependent Variable:	$Q(\sigma_{pr,max}(\text{MPa}))$
No. Observations:	36,971		Method:	REML
No. Groups:	10,206		Scale:	27,322.7092
Min. group size:	1		Log-Likelihood:	−241,291.6528
Max. group size:	4		Converged:	Yes
Mean group size:	3.6			

	Coef.	Std.Err.	z	P > |z|	[0.025	0.975]
Intercept ("Geom 1,Longitudinal")	504.901	2.619	192.757	0.000	499.767	510.034
C(Geometry, Treatment)[T.2]	−33.269	2.515	−13.228	0.000	−38.199	−28.340
C(Geometry, Treatment)[T.3]	9.448	2.589	3.649	0.000	4.373	14.524
C(Geometry, Treatment)[T.4]	6.651	2.503	2.657	0.008	1.744	11.557
C(Toolpath, Treatment)[T.Trans]	−16.531	2.486	−7.870	0.000	−20.648	−12.414
C(Toolpath, Treatment)[T.Zigzag]	13.096	2.128	6.155	0.000	8.926	17.266
Q("Layer")	−34.261	0.529	−64.738	0.000	−35.298	−33.224
Group Var	0.000	0.734				

Table 5. Mixed linear model regression results for variable $\epsilon_{t,x}$.

Model:	MixedLM		Dependent Variable:	$Q(\epsilon_{t,x})$
No. Observations:	36,971		Method:	REML
No. Groups:	10,206		Scale:	0.0001
Min. group size:	1		Log-Likelihood:	118,013.0819
Max. group size:	4		Converged:	Yes
Mean group size:	3.6			

	Coef.	Std.Err.	z	P > |z|	[0.025	0.975]
Intercept ("Geom 1,Longitudinal")	−0.006	0.000	−34.307	0.000	−0.006	−0.005
C(Geometry, Treatment)[T.2]	−0.003	0.000	−23.181	0.000	−0.004	−0.003
C(Geometry, Treatment)[T.3]	−0.003	0.000	−16.873	0.000	−0.003	−0.002
C(Geometry, Treatment)[T.4]	−0.001	0.000	−6.942	0.000	−0.001	−0.001
C(Toolpath, Treatment)[T.Trans]	−0.006	0.000	−41.419	0.000	−0.006	−0.005
C(Toolpath, Treatment)[T.Zigzag]	−0.002	0.000	−15.916	0.000	−0.002	−0.002
Q("Layer")	0.003	0.000	86.830	0.000	0.003	0.003
Group Var	0.000	0.000				

Table 6. Mixed linear model regression results for variable $\epsilon_{t,y}$.

Model:	MixedLM		Dependent Variable:	$Q(\epsilon_{t,y})$
No. Observations:	36,971		Method:	REML
No. Groups:	10,206		Scale:	0.0001
Min. group size:	1		Log-Likelihood:	121,398.0879
Max. group size:	4		Converged:	Yes
Mean group size:	3.6			

	Coef.	Std.Err.	z	P > |z|	[0.025	0.975]
Intercept	−0.017	0.000	−121.048	0.000	−0.018	−0.017
C(Geometry, Treatment)[T.2]	0.003	0.000	23.698	0.000	0.003	0.004
C(Geometry, Treatment)[T.3]	0.005	0.000	33.144	0.000	0.004	0.005
C(Geometry, Treatment)[T.4]	0.003	0.000	21.035	0.000	0.003	0.003
C(Toolpath, Treatment)[T.Trans]	0.006	0.000	50.564	0.000	0.006	0.006
C(Toolpath, Treatment)[T.Zigzag]	−0.003	0.000	−27.517	0.000	−0.003	−0.003
Q("Layer")	0.002	0.000	74.192	0.000	0.002	0.002
Group Var	0.000	0.000				

The similar statistical linear mixed-effects analysis of the toolpath and geometry on the stresses and strains contrasts with the low correlation between the directional stresses and strains mentioned in the previous subsection. To gain a deeper understanding of whether stress or strain can represent the variability in the local data, principal component analysis (PCA) is utilized to identify the most important variables. PCA reduces the data features while introducing new features composed of linear sums of the primary features. The weighting factor of the primary features in the new features reveals which variables have the highest variance. The results of the PCA analysis are described in the next subsection.

3.3. Principal Component Analysis

The PCA class from the Python *scikit-learn* library [24] was used for the PCA machine learning analysis of the dataset of the current study. PCA was implemented to investigate which variables among five—σ_{xx}, σ_{yy}, $\sigma_{pr,max}$, $\epsilon_{t,x}$, and $\epsilon_{t,y}$—show the highest variance and are the most informative. Separate analyses were conducted for each geometry (including all toolpaths) and each toolpath (including all geometries).

The results for the concatenated data of all longitudinal pattern geometries were analyzed to gain an understanding of the "longitudinal scanning" pattern on the variability of the mechanical properties in DED-fabricated parts. Figure 5 shows the scatter plot of the data with two principal component vectors and the bar plot of the explained variance of the data at the middle layer (layer 3). The explained variance of each component indicates how informative the component is. The loading (weight) coefficients of the primary variables for each new principal component variable of the longitudinal toolpath are presented in Table 7. It should be noted that the reason the PCA component vectors are not exactly orthogonal in the 2D plot is that they are projected from a higher-dimensional space (6D in this case) onto a 2D plane.

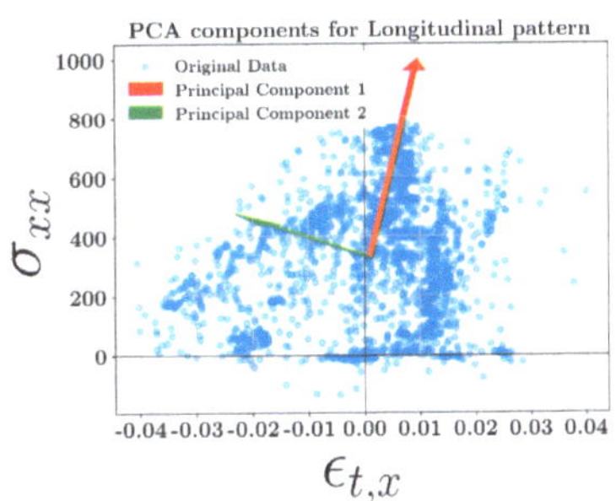
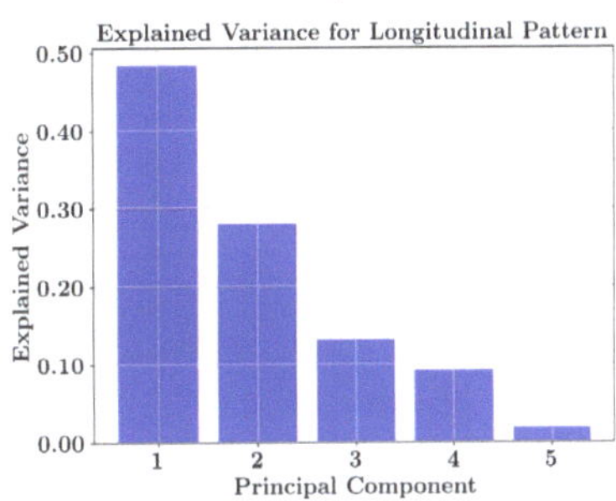

	(a) PCA components vectors	(b) PCA components explained variances

Figure 5. PCA results of the one-way longitudinal pattern at the middle layer; (**a**) scatter data showing two PCA component vectors; (**b**) variance of each new principal component.

Table 7. Loadings (weights) for each principal component—longitudinal toolpath pattern data.

	PCA 1	PCA 2	PCA 3	PCA 4	PCA 5
σ_{xx}	0.81	0.14	0.26	0.23	−0.45
σ_{yy}	−0.09	0.80	−0.51	−0.13	−0.28
$\sigma_{pr,max}$	0.41	0.31	−0.07	0.12	0.85
$\epsilon_{t,x}$	0.20	−0.44	−0.79	0.34	−0.06
$\epsilon_{t,y}$	−0.36	0.21	0.17	0.89	−0.01

The results indicate that the first principal component, which accounts for 48% of the information variance, is highly dependent on σ_{xx}, with a coefficient of 0.81. The second highest variability belongs to σ_{yy}, which has a loading of 0.80 in the second component. ϵ_{xx} and ϵ_{yy} contribute to the principal components with low impact as they mostly affect the third and fourth principal components. Similarly, the variance of $\sigma_{pr,max}$ has a lower contribution than the normal residual stresses. While the maximum principal stress ($\sigma_{pr,max}$)

suggests an optimal approach for toolpath selection, the normal longitudinal stress (σ_{xx}) shows the highest variability at the observed nodes of the parts for the longitudinal pattern. In contrast, the variances for the strains (ϵ_{xx} and ϵ_{yy}) are lower and negligible. Thus, longitudinal stress should be a matter of concern when investigating the longitudinal deposition for a longitudinal geometry.

Figure 6 shows the scatter plot of the data with two principal component vectors and the bar plot of the explained variance of the data at the middle layer (layer 3) for the transverse deposition pattern across all geometries. The loading (weight) coefficients of the primary variables for each new principal component variable of the transverse toolpath are presented in Table 8.

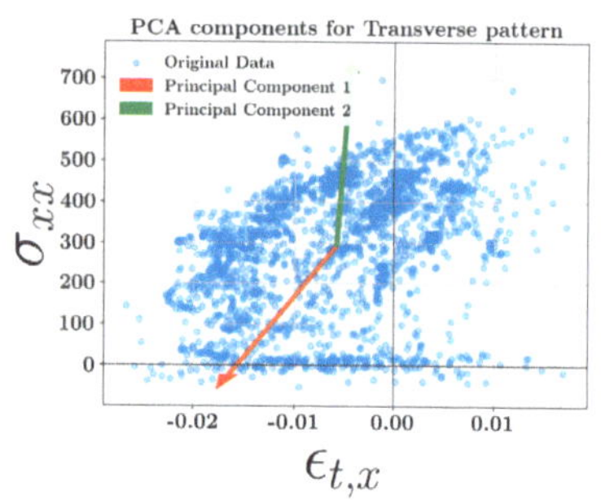

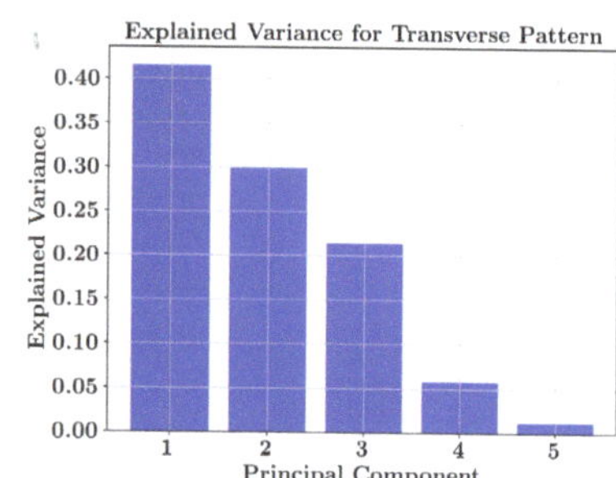

(**a**) PCA components vectors   (**b**) PCA components explained variances

Figure 6. PCA results of the one-way longitudinal pattern at the middle layer; (**a**) scatter data showing two PCA component vectors; (**b**) variance of each new principal component.

Table 8. Loadings (weights) for each principal component—transverse toolpath pattern data.

	PCA 1	PCA 2	PCA 3	PCA 4	PCA 5
σ_{xx}	-0.45	0.61	-0.51	-0.30	0.27
σ_{yy}	0.56	0.56	0.30	0.22	0.49
$\sigma_{pr,max}$	0.16	0.50	-0.13	0.26	-0.80
$\epsilon_{t,x}$	-0.62	0.08	0.28	0.72	0.12
$\epsilon_{t,y}$	0.27	-0.25	-0.74	0.53	0.19

Regarding the data from the transverse deposition for all the geometries, the first principal component has the highest weighting of -0.62 for the longitudinal normal strain ($\epsilon_{t,x}$) followed by the second-highest weighting of 0.56 for the transverse directional residual stress σ_{yy}. Based on this evidence, the geometry shape type, specifically a longitudinal geometry, accounts for a higher directional strain variance than the deposition direction while the directional stress is mostly impacted by the deposition.

Figure 7 shows the scatter plot of the data with two principal component vectors and the bar plot of the explained variance of the data at the middle layer (layer 3) for the longitudinal zigzag deposition pattern of all the geometries. The loading (weight) coefficients of the primary variables for each new principal component variable of the transverse toolpath are presented in Table 9.

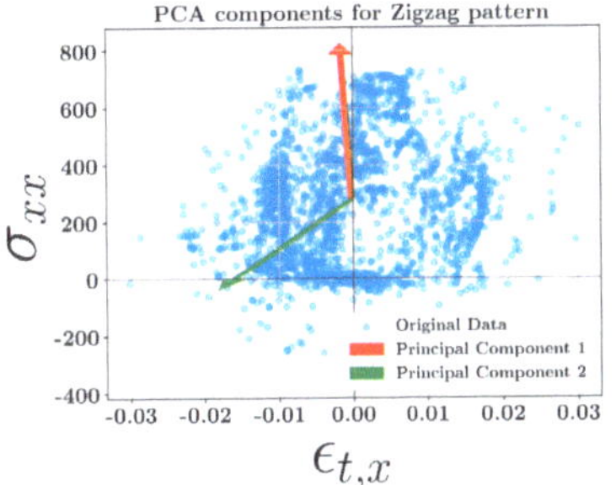

(**a**) PCA components vectors

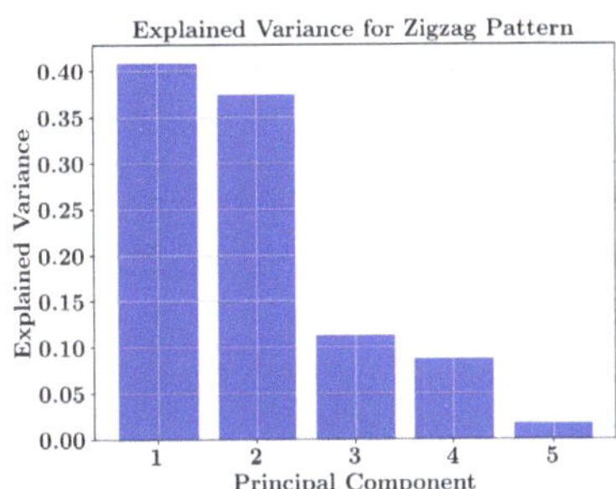

(**b**) PCA components explained variances

Figure 7. PCA results of the one-way longitudinal pattern at the middle layer; (**a**) scatter data showing two PCA component vectors; (**b**) variance of each new principal component.

Table 9. Loadings (weights) for each principal component—Zigzag toolpath pattern data.

	PCA 1	PCA 2	PCA 3	PCA 4	PCA 5
σ_{xx}	0.69	−0.38	−0.16	0.27	−0.53
σ_{yy}	0.27	0.76	0.35	−0.24	−0.42
$\sigma_{pr,max}$	0.63	0.08	0.26	0.06	0.72
$\epsilon_{t,x}$	−0.05	−0.53	0.71	−0.45	−0.12
$\epsilon_{t,y}$	−0.22	0.05	0.52	0.82	−0.07

The PCA results of the longitudinal zigzag show that the directional stress in the longitudinal axis obtains the highest variance and the strains have lower variances compared to the directional stresses. All the toolpath results demonstrate that the second-highest variance occurs in the orthogonal direction of the deposition. This means that longitudinal depositions result in the second highest variances of σ_{yy}, while transverse scanning results in the second highest variance of σ_{xx}, based on the explained variances of the principal components and the individual variable coefficients.

The scatter plots for each geometry, including all toolpaths and the related explained variance bars, are presented in Figures 8–11. The corresponding weighting values are presented in Tables 10–13. The thin geometries (Geoms 1 and 4) show that the highest variance coefficient in the first principal belongs to σ_{xx}, while thick step walls show the highest coefficient pertaining to σ_{yy}. Similar to the toolpath results, the strains contribute to the principal components with low explained variance. It can be concluded that the residual stresses should be the focus due to their high variances at the middle layers of the DED parts, and the strains can be excluded from further multi-dimensional feature studies.

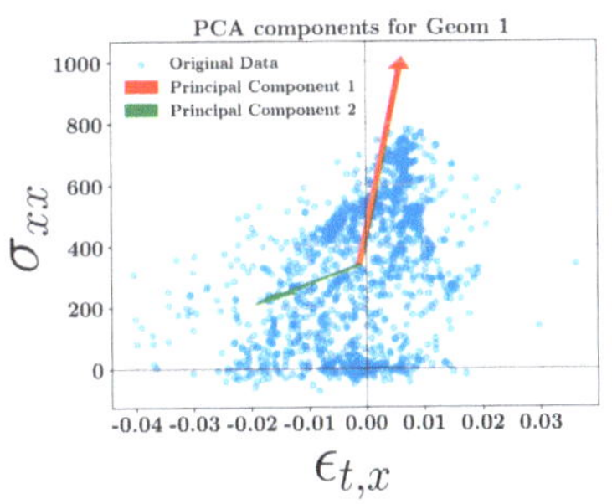

(**a**) PCA components vectors

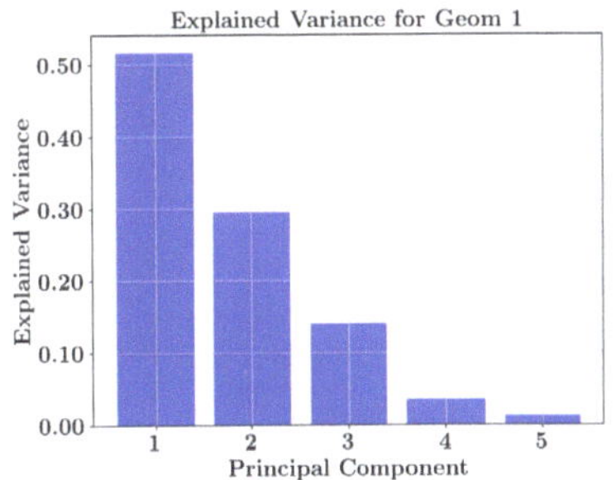

(**b**) PCA components explained variances

Figure 8. PCA component vectors and explained variance bars of geometry 1 at the middle layer.

Table 10. Loadings (weights) for each principal component—Geom 1 data.

	PCA 1	PCA 2	PCA 3	PCA 4	PCA 5
σ_{xx}	0.81	-0.12	-0.35	0.18	-0.41
σ_{yy}	0.21	0.77	0.52	-0.08	-0.28
$\sigma_{pr,max}$	0.46	0.21	-0.05	-0.10	0.86
$\epsilon_{t,x}$	0.20	-0.35	0.26	0.87	-0.11
$\epsilon_{t,y}$	-0.20	0.47	-0.73	-0.44	-0.09

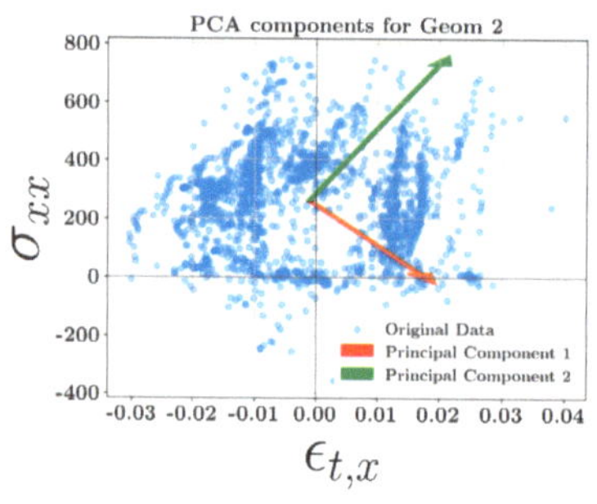

(**a**) PCA components vectors

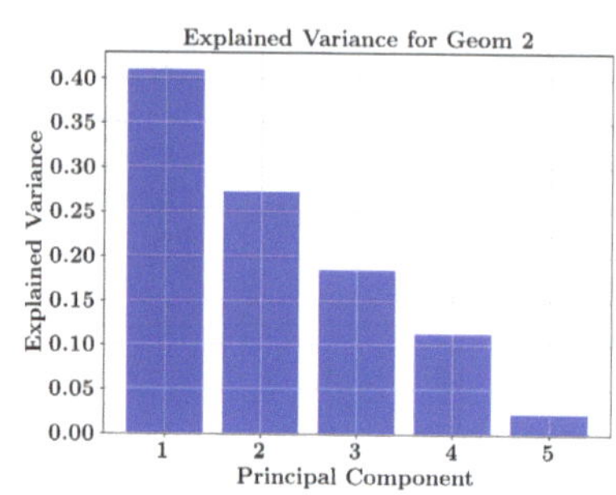

(**b**) PCA components explained variances

Figure 9. PCA component vectors and explained variance bars of geometry 2 at the middle layer.

Table 11. Loadings (weights) for each principal component—Geom 2 data.

	PCA 1	PCA 2	PCA 3	PCA 4	PCA 5
σ_{xx}	-0.33	0.63	0.08	-0.41	-0.57
σ_{yy}	-0.61	-0.29	-0.29	0.55	-0.40
$\sigma_{pr,max}$	-0.55	0.43	-0.01	0.13	0.70
$\epsilon_{t,x}$	0.45	0.52	-0.59	0.41	-0.06
$\epsilon_{t,y}$	0.14	0.25	0.75	0.59	-0.13

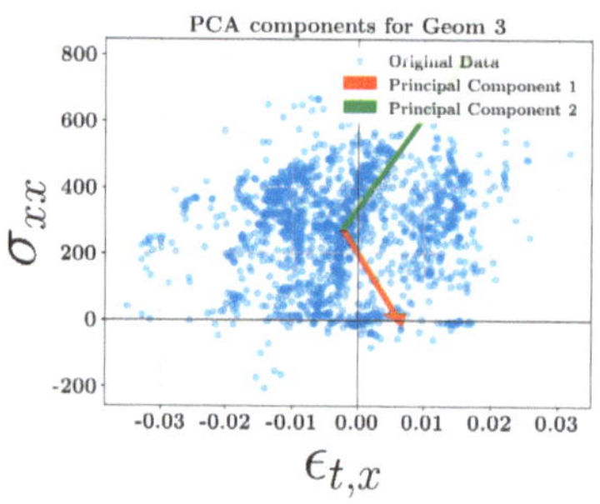

(**a**) PCA components vectors

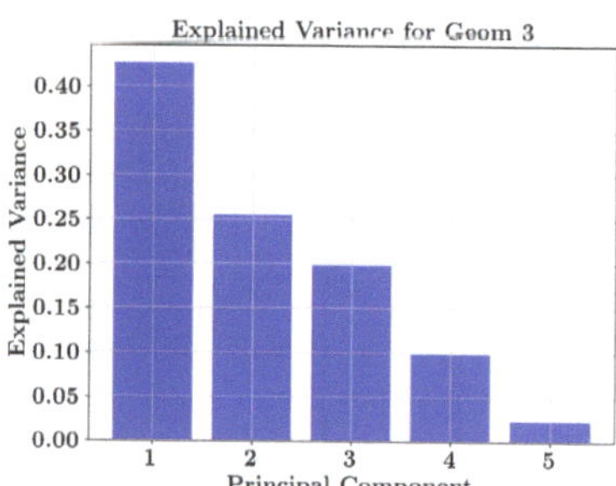

(**b**) PCA components explained variances

Figure 10. PCA component vectors and explained variance bars of geometry 3 at the middle layer.

Table 12. Loadings (weights) for each principal component—Geom 3 data.

	PCA 1	PCA 2	PCA 3	PCA 4	PCA 5
σ_{xx}	-0.36	0.70	0.20	-0.37	0.45
σ_{yy}	-0.64	-0.39	-0.21	0.39	0.50
$\sigma_{pr,max}$	-0.63	0.21	0.06	0.16	-0.73
$\epsilon_{t,x}$	0.25	0.56	-0.36	0.70	0.06
$\epsilon_{t,y}$	0.07	-0.04	0.88	0.45	0.10

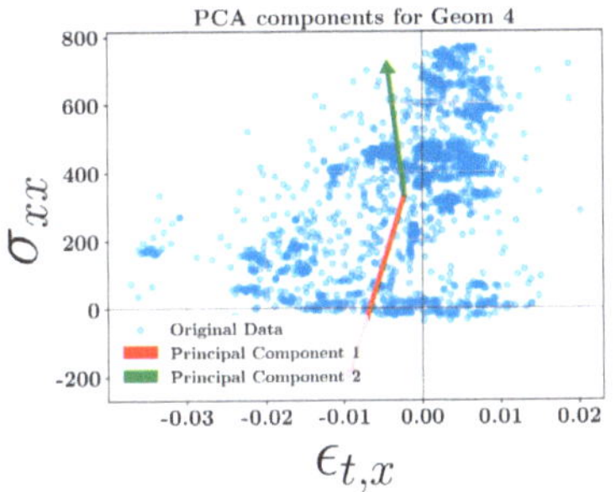

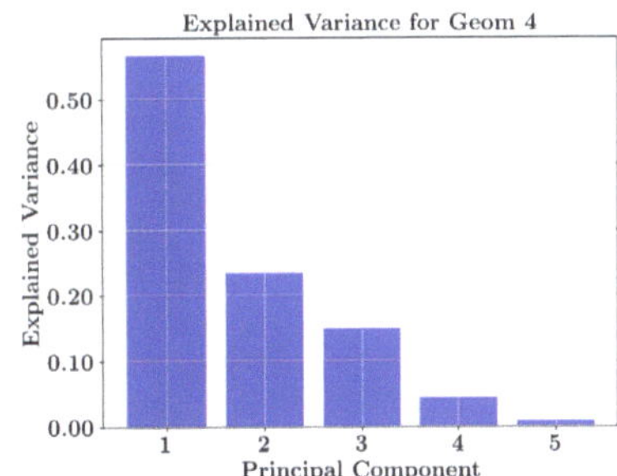

(**a**) PCA components vectors   (**b**) PCA components explained variances

Figure 11. PCA component vectors and explained variance bars of geometry 4 at the middle layer.

Table 13. Loadings (weights) for each principal component—Geom 4 data.

	PCA 1	PCA 2	PCA 3	PCA 4	PCA 5
σ_{xx}	−0.67	0.47	−0.36	0.13	0.44
σ_{yy}	0.55	0.61	0.33	0.02	0.46
$\sigma_{pr,max}$	−0.06	0.63	−0.05	−0.28	−0.72
$\epsilon_{t,x}$	−0.33	−0.10	0.48	0.78	0.21
$\epsilon_{t,y}$	0.36	−0.06	−0.73	−0.55	0.19

3.4. Self-Organizing Map Clustering

The local points dataset from all 12 cases was evaluated using six clusters. The data were trained in Python using the *MiniSom* library [22], with a learning rate of 0.5 and 5000 iterations. The frequency of assigned observations (the FEM local nodes) to each cluster after training the SOM is shown in Figure 12. Clusters 4 and 6 contain the highest frequency of data, while cluster 3 has the fewest records (observations).

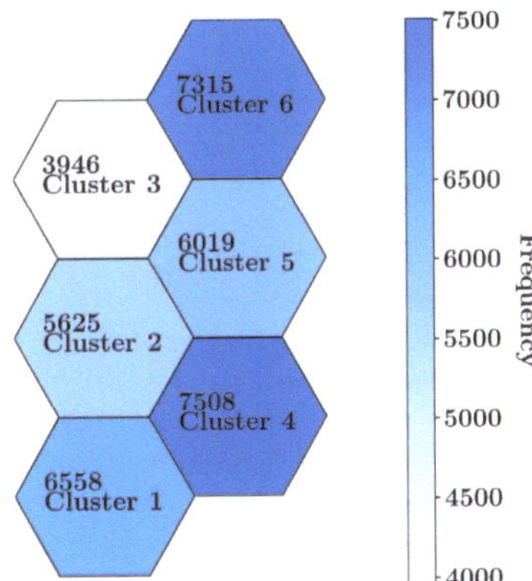

Figure 12. Structure of neurons (clusters) with the number of samples in each cluster for the local dataset.

The weights of each feature in all the clusters (neurons) are presented in Figure 13. The σ_{xx} stress has the highest weight in cluster 4 and $\sigma_{pr,max}$ has the lowest contribution in cluster 3. A visual consideration reveals that the weights of $\sigma_{pr,max}$ are lower compared to the rest of the features. Cluster 3 is mostly affected by $\epsilon_{t,x}$, and cluster 6 is highly impacted by σ_{yy}. Clusters 1 and 4 are predominantly determined by σ_{xx} and $\epsilon_{t,x}$.

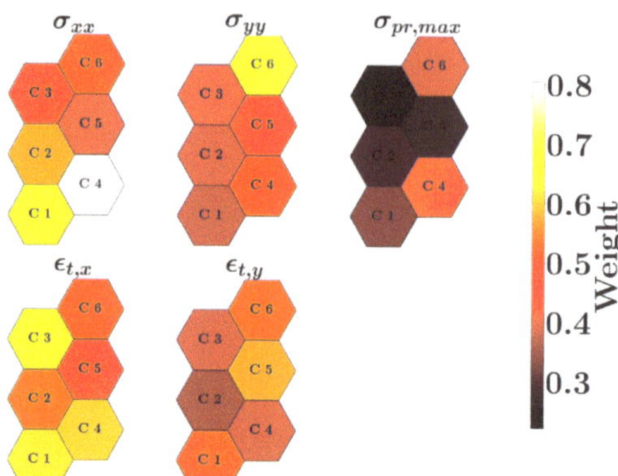

Figure 13. Weight of SOM neural network for each cluster (neuron)—C1 to C6 indicate cluster numbers from 1 to 6.

The features of the center of each cluster of the SOM results along with the defined mechanical status are presented in Table 14. In the contour plots (Figures 14–17), the color represented in each row corresponds to the cluster's color. Cluster 3 (green color) represents the low-stress and high-strain cluster, while Cluster 4 (purple color) represents the high-stress and low-strain cluster.

Table 14. SOM cluster centers—the rows' colors align with the contours plotted in the results.

Cluster	Center σ_{xx}	Center σ_{yy}	Center $\sigma_{pr,max}$	Center $\epsilon_{t,x}$ $(\times 10^3)$	Center $\epsilon_{t,y}$ $(\times 10^3)$	Status
1	385.29	49.29	421.18	5.18	−4.74	High σ_{xx}—Low Strain
2	280.36	66.24	355.53	−2.98	−18.75	High σ_{xx}—High Strain
3	80.19	79.80	245.09	9.89	−15.97	Low Stresses—High Strain
4	591.04	211.21	614.08	3.59	−15.90	High Stresses—High Strain
5	34.97	158.20	325.78	−10.61	1.25	Low σ_{xx}—High Strain
6	186.04	436.02	521.46	−5.26	−3.39	High σ_{yy}—Low Strain

The color of each row demonstrate the cluster color, later shown on the results.

Based on the clustering analysis, the SOM model evaluated the nodes of the FEM results. Figure 14 illustrates the clusters each node is assigned to for geometry 1, encompassing three toolpath patterns at the middle layer (layer 3). It should be noted that the longitudinal one-way and zigzag depositions start at $X = 25$ mm and $Y = 0$ mm, while the transverse one-way deposition starts at $X = 60$ mm and $Y = 20$ mm. The transverse pattern at the third layer (Figure 14b) shows that the intersection zone within the cross-type part lies in a high directional σ_{yy} stress with low strains cluster (cluster 6). The two longitudinal one-way and zigzag patterns (a, c) fall within the all-directional high stresses and high strains cluster (cluster 4) at the intersection. Both clusters at the intersection are characterized by high strains. The corners are assigned to cluster 5, which represents low stresses but high strains.

Figure 15 displays the clusters of each node for geometry 1 at the top surface (layer 5). The longitudinal zigzag pattern has the higher portion of cluster 3 (the green color) that represents the low-stress/low-strain category. The longitudinal one-way pattern exhibits the high transverse stress σ_{yy} cluster assigned to the intersection area while the transverse pattern outcomes a converse result, indicating a high longitudinal stress cluster (cluster 1) at the intersection zone.

A comparison of Figures 14 and 15 ascertains that the intersection might be optimized with a toolpath at the middle layer while worsening at the top layer with the same toolpath.

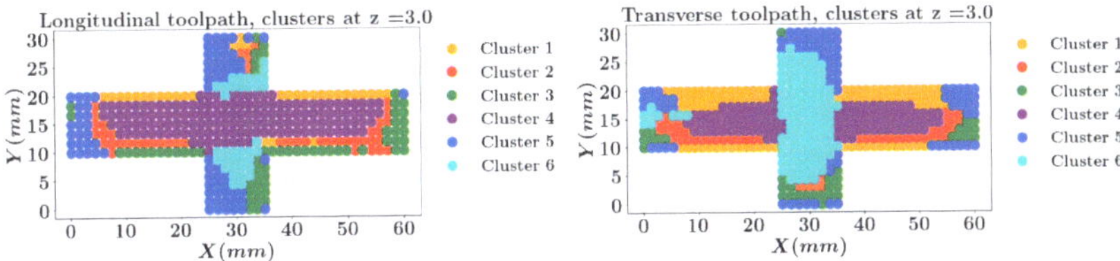

(**a**) Geom 1 longitudinal pattern—layer 3.

(**b**) Geom 1 transverse pattern—layer 3.

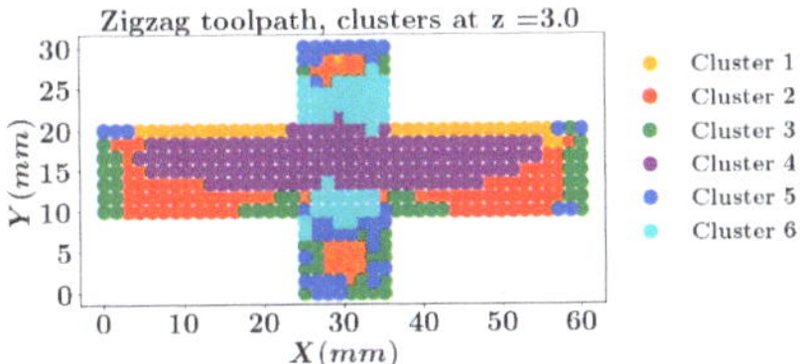

(**c**) Geom 1 zigzag pattern—layer 3.

Figure 14. Local data clustering results for geometry 1 with three different toolpaths at layer 3.

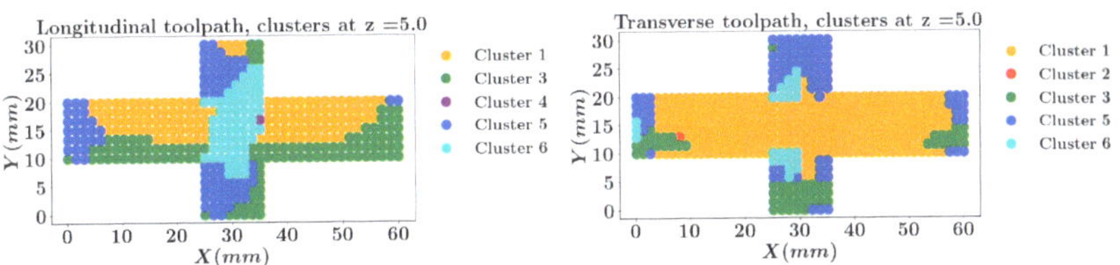

(**a**) Geom 1 longitudinal pattern—layer 5.

(**b**) Geom 1 transverse pattern—layer 5.

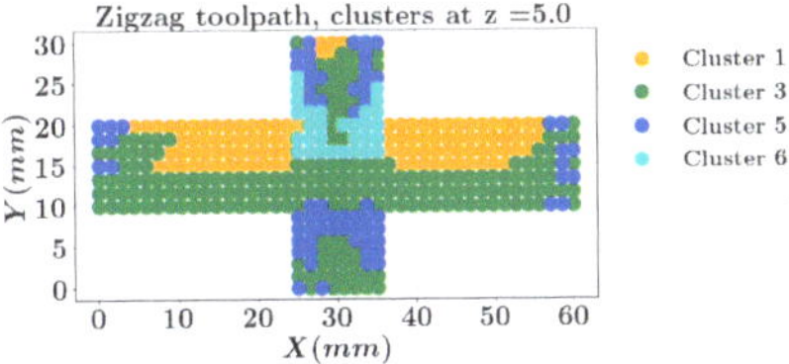

(**c**) Geom 1 zigzag pattern—layer 5.

Figure 15. Local data clustering results for geometry 1 with three different toolpaths at layer 5 (top surface).

The cluster contours for the middle layer and the top layer of geometry 4 are depicted in Figures 16 and 17. Note that the longitudinal one-way and zigzag depositions start at $X = -30$ (mm) and $Y = -15$ (mm) while the transverse one-way deposition starts at $X = 30$ (mm) and $Y = -15$ (mm). Both one-way and zigzag longitudinal patterns at layer 3 (Figure 16a,c) show that the top long wall, where the deposition ends, is predominantly assigned to cluster 4 (the purple color, indicating high stresses and strains), which is less favorable compared to the transverse deposition outcome (Figure 16b). In contrast, the longitudinal patterns feature a low stress/strain cluster (the green color, Figure 17a,c) at the top layer within the first longitudinal wall where the deposition starts, while the transverse deposition shows cluster 1 for the longitudinal wall. However, the transverse deposition concludes in cluster 5 for the short transverse wall.

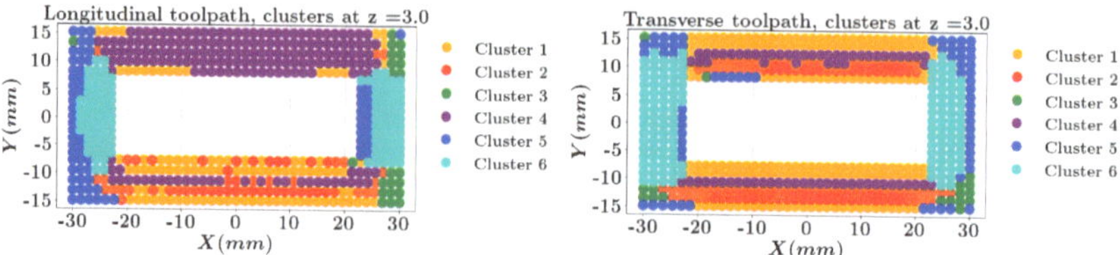

(**a**) Geom 4 longitudinal pattern—layer 3.

(**b**) Geom 4 transverse pattern—layer 3.

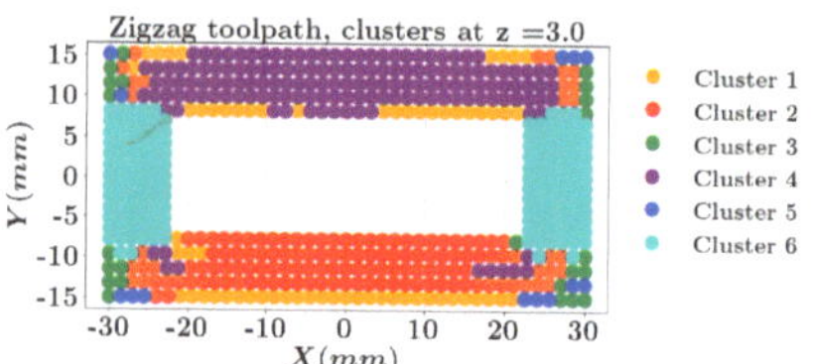

(**c**) Geom 4 zigzag pattern—layer 3.

Figure 16. Local data clustering results for geometry 4 with three different toolpaths at layer 3.

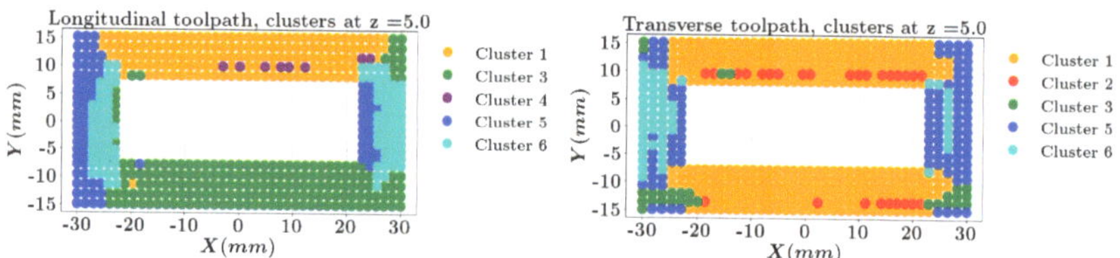

(**a**) Geom 4 longitudinal pattern—layer 5.

(**b**) Geom 4 transverse pattern—layer 5.

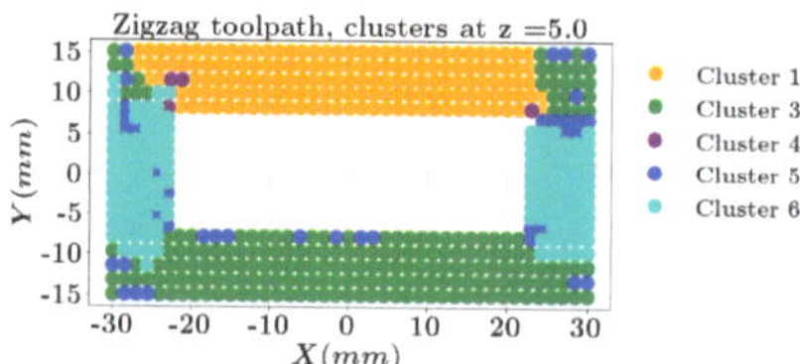

(**c**) Geom 4 zigzag pattern—layer 5.

Figure 17. Local data clustering results for geometry 4 with three different toolpaths at layer 5 (top surface).

4. Conclusions

This research provides a comprehensive analysis of the impact of toolpath patterns, geometry types, and layering effects on the mechanical properties of parts fabricated through DED additive manufacturing using LME, PCA, and SOM clustering. The outcome shows the capability of the methods in quantifying and interpreting the contributions of these factors to residual stresses and strains.

The results indicate that all geometry types, toolpath patterns, and layer numbers significantly impact the distribution of mechanical properties. The toolpath had the highest impact on the longitudinal residual stress, σ_{xx}, while the shape type had the most substantial effect on the transverse residual stress, σ_{yy}. The LME model demonstrated that the layering effect is significant in all cases.

The PCA results further identified that the toolpath pattern has the greatest effect on directional stresses, while the geometry type accounts for the most significant impact on residual strains. It was also shown that residual stresses exhibit higher variance and are more informative than residual strains.

The SOM clustering offers additional insights into the local distribution of stresses and strains, highlighting regions within the parts where specific stress patterns are more likely to occur. The SOM results, based on the mechanical property similarity of the cloud nodes in the parts, suggest that while a toolpath may create a low stress/strain region at one layer, it could deteriorate the mechanical properties at another layer.

Overall, this study underscores the importance of selecting appropriate toolpaths and geometries in DED processes to enhance mechanical properties and reduce defects. The methodologies used demonstrate significant potential in establishing a robust framework for future studies aiming to optimize additive manufacturing processes through data-driven analysis. The conclusions are summarized in Table 15.

Table 15. Summary of key insights and suggestions.

Factor	Effect on Properties	Suggestions
Toolpath Pattern	Highest impact on longitudinal residual stress (σ_{xx})	Use longitudinal toolpaths to minimize longitudinal stresses, especially for thin geometries.
Geometry Type	Highest impact on transverse residual stress (σ_{yy})	Step-type geometries need mixed toolpaths for balancing stress distribution.
Layering Effect	Significant across all mechanical properties	Control layering thickness to minimize internal stress buildup, especially in thick parts.

Author Contributions: All authors contributed to the study's conception and design. Material preparation and data collection were carried out by S.P. and S.E.M. The first draft of the paper was written by S.P. All authors have read and agreed to the published version of the manuscript.

Funding: This research was funded by MITACS Canada (#: IT16938) under an internship position at CAMufacturing Solutions Inc.

Institutional Review Board Statement: Not applicable.

Informed Consent Statement: Not applicable.

Data Availability Statement: The raw data supporting the conclusions of this article will be made available by the authors on request.

Acknowledgments: The authors would like to thank MITACS and CAMufacturing Solutions Inc. for their financial support.

Conflicts of Interest: The authors declare no conflicts of interest.

Abbreviations

The following abbreviations are used in this manuscript:

AM	additive manufacturing
ANN	artificial neural network
CAD	computer-aided design
CNN	convolutional neural network
CPU	central processing unit
DED	direct energy deposition
ELM	linear mixed-effects model
FEM	finite element method
PCA	principal component analysis
RAM	random-access memory
SOM	self-organizing map

References

1. Piscopo, G.; Iuliano, L. Current research and industrial application of laser powder directed energy deposition. *Int. J. Adv. Manuf. Technol.* **2022**, *119*, 6893–6917. [CrossRef]
2. Kriczky, D.A.; Irwin, J.; Reutzel, E.W.; Michaleris, P.; Nassar, A.R.; Craig, J. 3D spatial reconstruction of thermal characteristics in directed energy deposition through optical thermal imaging. *J. Mater. Process. Technol.* **2015**, *221*, 172–186. [CrossRef]
3. Mirazimzadeh, S.E.; Pazireh, S.; Urbanic, J.; Hedrick, B. Investigation of effects of different moving heat source scanning patterns on thermo-mechanical behavior in direct energy deposition manufacturing. *Int. J. Adv. Manuf. Technol.* **2022**, *120*, 4737–4753. [CrossRef]
4. Yao, X.X.; Ge, P.; Li, J.Y.; Wang, Y.F.; Li, T.; Liu, W.W.; Zhang, Z. Controlling the solidification process parameters of direct energy deposition additive manufacturing considering laser and powder properties. *Comput. Mater. Sci.* **2020**, *182*, 109788. [CrossRef]
5. Mathoho, I.; Akinlabi, E.T.; Arthur, N.; Tlotleng, M. Impact of DED process parameters on the metallurgical characteristics of 17-4 PH SS deposited using DED. *CIRP J. Manuf. Sci. Technol.* **2020**, *31*, 450–458. [CrossRef]
6. Keist, J.S.; Palmer, T.A. Role of geometry on properties of additively manufactured Ti-6Al-4V structures fabricated using laser based directed energy deposition. *Mater. Des.* **2016**, *106*, 482–494. [CrossRef]
7. Nazemi, N.; Urbanic, R.J. A numerical investigation for alternative toolpath deposition solutions for surface cladding of stainless steel P420 powder on AISI 1018 steel substrate. *Int. J. Adv. Manuf. Technol.* **2018**, *96*, 4123–4143. [CrossRef]
8. Hartmann, S.; Vykhtar, B.; Möbs, N.; Kelbassa, I.; Mayr, P. IoT-Based Data Mining Framework for Stability Assessment of the Laser-Directed Energy Deposition Process. *Processes* **2024**, *12*, 1180 . [CrossRef]
9. Wang, M.; Kashaev, N. On the maintenance of processing stability and consistency in laser-directed energy deposition via machine learning. *J. Manuf. Syst.* **2024**, *73*, 126–142. [CrossRef]
10. Mochi, V.H.; Núñez, H.H.L.; Ribeiro, K.S.B.; Venter, G.S. Real-time prediction of deposited bead width in L-DED using semi-supervised transfer learning. *Int. J. Adv. Manuf. Technol.* **2023**, *129*, 5643–5654. [CrossRef]
11. Li, J.; Sage, M.; Guan, X.; Brochu, M.; Zhao, Y.F. Machine Learning-Enabled Competitive Grain Growth Behavior Study in Directed Energy Deposition Fabricated Ti6Al4V. *JOM* **2020**, *72*, 458–464. [CrossRef]
12. Patil, D.B.; Nigam, A.; Mohapatra, S.; Nikam, S. A Deep Learning Approach to Classify and Detect Defects in the Components Manufactured by Laser Directed Energy Deposition Process. *Machines* **2023**, *11*, 854. [CrossRef]
13. Xu, R.; Huang, S.; Song, Z.; Gao, Y.; Wu, J. A deep mixed-effects modeling approach for real-time monitoring of metal additive manufacturing process. *IISE Trans.* **2024**, *56*, 945–959. [CrossRef]
14. Khanzadeh, M.; Rao, P.; Jafari-Marandi, R.; Smith, B.K.; Tschopp, M.A.; Bian, L. Quantifying Geometric Accuracy With Unsupervised Machine Learning: Using Self-Organizing Map on Fused Filament Fabrication Additive Manufacturing Parts. *J. Manuf. Sci. Eng* **2017**, *140*, 4038598. [CrossRef]
15. Mirazimzadeh, S.E.; Pazireh, S.; Urbanic, J.; Jianu, O. Unsupervised clustering approach for recognizing residual stress and distortion patterns for different parts for directed energy deposition additive manufacturing. *Int. J. Adv. Manuf. Technol.* **2023**, *125*, 5067–5087. [CrossRef]
16. CNC Software Inc. *Mastercam*; CAD/CAM Software; CNC Software Inc.: Tolland, CT, USA, 2024.
17. Mirazimzadeh, S.E.; Mohajernia, B.; Pazireh, S.; Urbanic, J.; Jianu, O. Investigation of residual stresses of multi-layer multi-track components built by directed energy deposition: Experimental, numerical, and time-series machine-learning studies. *Int. J. Adv. Manuf. Technol.* **2024**, *130*, 329–351. [CrossRef]
18. Python. Statsmodels, Statistical Models in Python, Mixed Linear Models (MixedLM). 2024. Available online: https://www.statsmodels.org/stable/mixed_linear.html#module-statsmodels.regression.mixed_linear_model (accessed on 1 May 2024).
19. Lindstrom, M.; Bates, D. Newton Raphson and EM algorithms for linear mixed effects models for repeated measures data. *J. Am. Stat. Assoc.* **1988**, *83*, 1014–1022.
20. Python. Principal Component Analysis (PCA). 2024. Available online: https://scikit-learn.org/stable/modules/decomposition.html (accessed on 1 May 2024).
21. Kohonen, T. Self-organized formation of topologically correct feature maps. *Biol. Cybern.* **1982**, *43*, 59–69. [CrossRef]
22. Vettigli, G. MiniSom: Minimalistic and NumPy-Based Implementation of the Self Organizing Map. 2018. Available online: https://github.com/JustGlowing/minisom (accessed on 1 May 2024).

23. Tam, K.M.M.; Mueller, C.T. Additive Manufacturing Along Principal Stress Lines. *3D Print. Addit. Manuf.* **2017**, *4*, 63–81. [CrossRef]
24. Pedregosa, F.; Varoquaux, G.; Gramfort, A.; Michel, V.; Thirion, B.; Grisel, O.; Blondel, M.; Prettenhofer, P.; Weiss, R.; Dubourg, V.; et al. Scikit-learn: Machine Learning in Python. *J. Mach. Learn. Res.* **2011**, *12*, 2825–2830.

Article

A Data-Driven Framework for Direct Local Tensile Property Prediction of Laser Powder Bed Fusion Parts

Luke Scime [1,*], Chase Joslin [2], David A. Collins [3], Michael Sprayberry [1], Alka Singh [1], William Halsey [1], Ryan Duncan [2], Zackary Snow [1], Ryan Dehoff [2] and Vincent Paquit [1]

[1] Electrification and Energy Infrastructure Division, Oak Ridge National Laboratory, Oak Ridge, TN 37830, USA; sprayberryma@ornl.gov (M.S.); singhar@ornl.gov (A.S.); halseywh@ornl.gov (W.H.); snowzk@ornl.gov (Z.S.); paquitvc@ornl.gov (V.P.)

[2] Manufacturing Science Division, Oak Ridge National Laboratory, Oak Ridge, TN 37830, USA; joslincb@ornl.gov (C.J.); duncanrk@ornl.gov (R.D.); dehoffrr@ornl.gov (R.D.)

[3] Materials Science and Technology Division, Oak Ridge National Laboratory, Oak Ridge, TN 37830, USA; collinsda@ornl.gov

* Correspondence: scimelr@ornl.gov

Abstract: This article proposes a generalizable, data-driven framework for qualifying laser powder bed fusion additively manufactured parts using part-specific in situ data, including powder bed imaging, machine health sensors, and laser scan paths. To achieve part qualification without relying solely on statistical processes or feedstock control, a sequence of machine learning models was trained on 6299 tensile specimens to locally predict the tensile properties of stainless-steel parts based on fused multi-modal in situ sensor data and a priori information. A cyberphysical infrastructure enabled the robust spatial tracking of individual specimens, and computer vision techniques registered the ground truth tensile measurements to the in situ data. The co-registered 230 GB dataset used in this work has been publicly released and is available as a set of HDF5 files. The extensive training data requirements and wide range of size scales were addressed by combining deep learning, machine learning, and feature engineering algorithms in a relay. The trained models demonstrated a 61% error reduction in ultimate tensile strength predictions relative to estimates made without any in situ information. Lessons learned and potential improvements to the sensors and mechanical testing procedure are discussed.

Keywords: laser powder bed fusion; tensile properties; machine learning; in situ monitoring

Citation: Scime, L.; Joslin, C.; Collins, D.A.; Sprayberry, M.; Singh, A.; Halsey, W.; Duncan, R.; Snow, Z.; Dehoff, R.; Paquit, V. A Data-Driven Framework for Direct Local Tensile Property Prediction of Laser Powder Bed Fusion Parts. *Materials* **2023**, *16*, 7293. https://doi.org/10.3390/ma16237293

Academic Editor: Alessandro Dell'Era

Received: 3 October 2023
Revised: 18 November 2023
Accepted: 21 November 2023
Published: 23 November 2023

1. Introduction

As a new class of manufacturing processes, metal additive manufacturing (AM) [1] holds significant promise for the rapid production of small-to-medium quantities of parts with complex geometries and internal structures [2]. For industries producing safety-critical components, part qualification is an integral step of any manufacturing process [3]. Qualification frameworks for traditionally manufactured components typically fall into one or more of the following paradigms: (1) destructive or nondestructive testing of a representative sample of the larger population of manufactured components or of designated coupons to detect process drift [4] (i.e., statistical quality control), (2) maintenance of a robustly defined, in-control manufacturing process [5] (i.e., process qualification) combined with a set of materials specifications for the feedstock or workpiece [6] (e.g., usage of A- and B-basis allowables), and (3) non-destructive evaluation (NDE) [7,8] of the entire population of manufactured components (i.e., part-specific qualification). Because these traditional qualification approaches are often incompatible with additively manufactured components, this work proposes a data-driven qualification framework that leverages in situ data to directly predict localized static tensile properties.

Laser-powder bed fusion (L-PBF) is a widely adopted metal AM process in which a laser beam is used to melt powder feedstock into a stack of two-dimensional 2D cross-sections of the intended part geometry [1,2]. Many current and anticipated business cases for L-PBF components [2] are not compatible with traditional qualification paradigms. For example, because one advantage of AM is its ability to make small production runs of customized designs [2], many L-PBF designs are not produced in quantities conducive to population-based qualification. Furthermore, as articulated by Seifi et al. [3], L-PBF's relative novelty, geometry-dependent process dynamics, and the nature of localized re-solidification of the feedstock challenges many of the assumptions required to qualify the process and materials in a way that is agnostic to part geometry. Finally, components optimized for L-PBF often have complex internal geometries [2], rough surfaces [9], or are manufactured with high atomic number alloys [10], complicating traditional post-build NDE techniques [8].

Fortunately, the layer-wise nature of L-PBF, along with the incremental re-solidification of discrete volumes within a layer, provide unique opportunities for in situ process monitoring [11] because each sub-volume is directly observable at some point during the manufacturing process. This facet of AM offers a significant advantage over many NDE techniques, which struggle to spatially resolve flaws in three dimensions, particularly for complex geometries. Such in situ data can also be used to construct a *component digital twin* [12] of individual components. As defined by Grieves and Vickers [13], a component digital twin is a virtual copy of its *physical twin*—a specific instance of a manufactured component that can be simulated [14]. This work develops new models to convert in situ data into localized material property predictions that could be used to instantiate a component digital twin.

Critically, properties predicted by these models should be localized (i.e., valid for a sub-volume of a component) so the models can be generalized to arbitrary component geometries. Scime et al. [15] proposed combining deep learning (DL), machine learning (ML) [16], and feature engineering [17] algorithms in an augmented intelligence relay (AIR) such that each algorithm solves a sub-problem within the overall data workflow. Over the last decade, the AM in situ process monitoring community has made significant strides in applying signal processing and computer vision techniques to both temporal [18–22] and spatial [23–27] data for the purposes of anomaly and flaw detection. A growing number of researchers are studying co-registered multi-modal sensor data stacks [28,29] and leveraging DL algorithms to achieve pixel-wise anomaly and flaw segmentation [28,30,31]. Although rarely generalized or presented in the context of a larger qualification framework, visualization of in situ data [32], including its registration to ex situ measurements [33,34], is also relatively common in the literature. In contrast, relatively little work has been reported regarding direct property prediction for localized sub-volumes, which is the focus of this work.

Most tensile property measurement and prediction research reported in the literature [35] for metal AM is similar to that of Lavery et al. [36], which correlates tensile measurements to laser processing parameters, post-build treatments such as hot isostatic pressing (HIP), and porosity content measured post-build. Similarly, Kusano et al. [37] extracted microstructural features from scanning electron microscopy (SEM) images of L-PBF-processed Ti-6Al-4V and performed multiple linear regression to fit tensile property prediction models to these feature vectors. Thematically similar, Hayes et al. [38] used constitutive modeling to predict the yield strength of Ti-6Al-4V processed via directed energy deposition (DED) based on microstructural features measured post-build. A related area of research garnering significant attention is the prediction of fatigue life based on flaw populations measured post-build using either x-ray computed tomography (XCT) of the part or destructive microscopy of witness coupons. In one example of this approach, Romano et al. [39] developed analytical models capable of predicting the fatigue properties of L-PBF-processed AlSi10Mg.

When considering only studies that use in situ sensor data to inform material property prediction models, the authors determined that most existing work has focused on polymer fused deposition modeling (FDM) processes, such as that performed by Zhang et al. [40]. Interestingly, Seifi et al. [41] performed DL analyses of melt pool thermal images to detect flaws in DED-processed Ti-6Al-4V. The resulting data were then used to inform traditional fatigue life models. Importantly, this approach required not only detecting the presence and location of each flaw but also estimating their size. Bisht et al. [42] correlated the relative occurrence of anomalies observed in on-axis photodiode data with the measured plastic elongation of L-PBF-processed Ti-6Al-4V, thus demonstrating a potentially viable approach that jumps directly from in situ data processing to localized property predictions. Finally, among the most similar work is that of Xie et al. [43], which encoded in situ thermal history measurements into engineered features and predicted local tensile properties for thin-walled Inconel 718 DED components using a neural network trained on tensile data from extracted tensile specimens.

This work uses a relay of machine-learned algorithms (Section 2.1) to predict local static tensile properties based on in situ data for L-PBF components. Computer vision techniques were used to process the in situ sensor data and to spatially register it to the ex situ mechanical test results (Section 2.2). Localized property prediction required a bespoke build strategy (Section 2.6) and drove many of the decisions regarding specimen geometry and post-processing methodologies. An extensive cyberphysical infrastructure was implemented to enable robust spatial tracking of thousands of individual tensile specimens (Section 2.7) and facilitate the public availability of the entire 230 GB Peregrine v2023-11 dataset [44] used in this research. Neural networks were designed to first segment anomalies apparent in the in situ sensor data (Section 2.3) and then to predict local tensile properties based on human-engineered feature vectors (Section 2.10). Model validation and testing performance results are presented in Sections 3.1 and 3.2, respectively. Ultimately, the goal of the proposed approach is to support future qualification paradigms that rely more heavily on the standardization of in situ sensor suites and validated algorithms and less heavily on certification of "locked-down" manufacturing processes and material specifications.

2. Materials and Methods

2.1. Experimental Conditions and Data Analysis Framework

Experiments were performed at the Manufacturing Demonstration Facility (MDF) located at Oak Ridge National Laboratory (ORNL) in Oak Ridge, Tennessee. Specimens were printed using stainless-steel (SS) 316L powder on a Concept Laser M2 (General Electric, General Electric Additive, Cincinnati, OH, USA) L-PBF printer. The Concept Laser M2 had two 400 W laser modules with overlapping fields of operation, and a compliant recoater blade was used to spread the powder. Algorithms were developed in Python v3.7 with relevant dependencies, including TensorFlow v1.13.1, OpenCV v3.4.1, and Scikit-image v0.18.1. Computations were benchmarked on a desktop computer equipped with two Quadro RTX 5000 (Nvidia Corporation, Santa Clara, CA, USA) graphical processing cards, two 16-core 2.10 GHz processors, and 256 GB of volatile memory. The AM terminology used in this document complies with ISO/ASTM 52900:2015 [1] where appropriate.

Because in situ data from L-PBF processes contain complex, multi-modal, contextually dependent information [28,45], it is not easily interpreted solely by physics-based models or human-designed heuristics. Therefore, the authors propose that machine-learned models are the most viable approach for translating in situ data to localized property predictions. Decomposing the data workflow into a relay enabled the use of both ML and DL models, even when the ground truth tensile properties were expensive to collect. This is possible because physics-informed decisions could be made at each interface throughout the relay to reduce the complexity of the feature encodings, which must be learned from the experimental data. For example, features encoding the laser scan vector length within a sub-volume were explicitly designed rather than learned based on the a priori knowledge that the scan length might affect solidification conditions [46] and, therefore, local material properties. The use of a relay provided other advantages, including (1) computationally efficient translation of data across spatial size scales, (2) improved interpretability of property predictions, (3) improved model generalizability, and (4) increased opportunities for modularity within the framework.

Figure 1 depicts the implemented AIR, including elements developed in prior work (blue) and downstream elements (magenta), which are beyond the scope of this paper. Starting at N1, natively temporal data (i.e., laser scan order) were spatially mapped as rasterized images for each print layer, while natively spatial data (i.e., visible-light images of the powder bed) entered the AIR directly at N2. At N3, the in situ data streams and design intent information (i.e., part geometry) were co-registered and placed into a common coordinate system. At N4, a subset of the fused sensor data was processed by a DL image segmentation model to produce an anomaly mask for each print layer. Training this DL model (feedback loop between N4 and N3) occurred pixel-wise using approximately 180 million ground-truth classifications acquired via expert annotation [28]. Up to this point in the relay, the spatial resolution of the data was on the order of 100 μm. At N5, the print volume was demarcated into 1 mm *super-voxels*, defined here as a sub-volume of a component for which local material property predictions can be made by the AIR. Associated with each super-voxel are (1) an engineered feature vector encoding the in situ sensor data, (2) anomaly segmentation metrics, (3) part geometry, and (4) proxy representations of the local thermal history. Finally, an ML model was trained (feedback loop between N6 and N5) using 6299 tensile tests to predict the local tensile properties at N6 based on the super-voxel feature vectors. The cyberphysical infrastructure that supported the AIR is referred to as the *digital platform*, and it allowed the in situ and ex situ data to be tracked as *digital threads* [15].

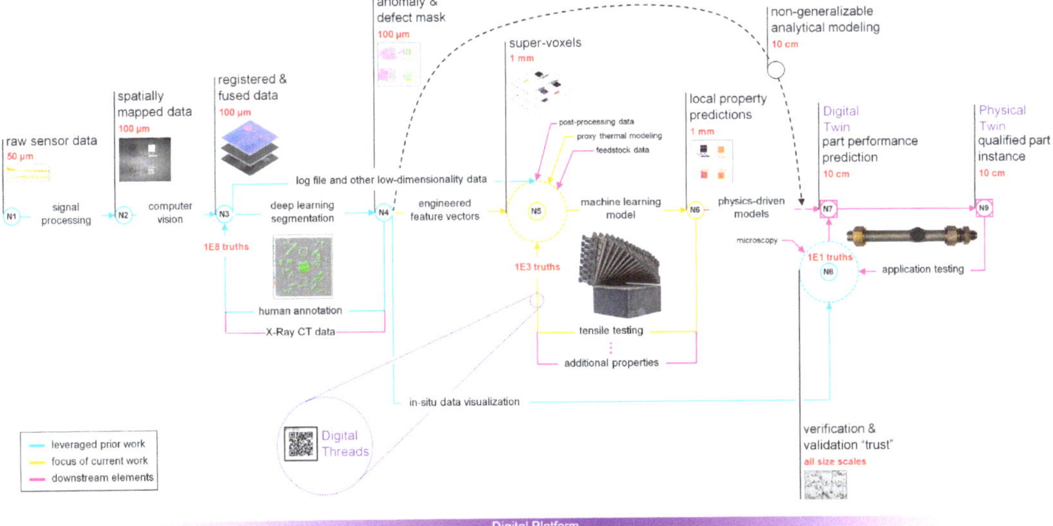

Figure 1. Graphical representation of the implemented AIR. This paper focuses on nodes N5 and N6 (orange), leveraging prior work by the authors shown at nodes N1, N2, N3, N4, and N8 (blue). Approximate spatial resolutions of data at each node are shown in red. Approximations for the number of ground truth values required for each feedback loop are also shown in red. Key elements of the digital platform that supported the implementation of the relay are shown in purple. The black dashed line from N4 to N7 represents a possible alternate pathway for predicting localized properties.

While beyond the scope of this work, downstream elements of the AIR should be considered to understand how the current research fits into the ultimate goal of achieving part-specific qualification. For example, a component digital twin consisting of a canonical finite element model of the component could be instantiated at N7 using the local material properties predicted at N6. This contrasts with traditional finite element meshes, which typically use material properties derived from a statistical analysis of aggregated historical data.

2.2. Multi-Modal Data Collection, Co-Registration, and Fusion

To enable local tensile property predictions, data from multiple in situ and a priori sources were spatially mapped (N2) and co-registered (N3) within a common coordinate system defined by the computer-aided design (CAD) model of each build. The CAD model was first converted into the standard triangle language (STL) file format and sliced using the Magics Image Build Processor (Materialise, Leuven, Belgium) into a set of binary layer images (Figure 2a). Each binary image was globally thresholded to identify the locations of the printed geometries, and a two-pass, 2.5D implementation of a standard 2D connected-components algorithm [47] was performed to uniquely segment each printed part within the 3D build volume. Each component was automatically assigned an identifier based on its position within the 3D build volume, and this information was shared with the digital platform, which generated a globally unique identifier. The universal coordinate system was defined in reference to ISO/ASTM 52900:2015 [1], with the positive x-axis oriented left–right from the perspective of the printer operator, the positive y-axis oriented front–back, and the positive z-axis oriented vertically along the build direction. The powder recoating and shield gas flow directions are parallel to the negative x-axis, as shown in Figure 2c.

The original equipment manufacturer (OEM) quality monitoring (QM) coating [48] camera captured two visible-light 5 mega-pixel (MP) images during each layer: one after melting was complete and one after powder had been spread across the print bed. Because this camera was mounted at an angle relative to the normal vector of the print bed, distorted images were produced. To correct this distortion and transform the image data into the common reference frame, a calibration pattern consisting of a 15×15 grid of cones with a 12 mm center-to-center spacing was printed and imaged (Figure 2b). The planar center of each cone was automatically detected and used to calculate a homographic transformation matrix using the least median of squares (LMedS) [49] method implemented by OpenCV. To increase the contrast between the printed dots and the surrounding powder and to facilitate the automatic detection of their centroids, multiple consecutive post-melt images were background-subtracted, and their difference images were composited together. After Gaussian blurring to further reduce noise, Otsu's method [50] was used to threshold the dot grid image. A connected-components analysis of the binary image enabled outlier dots to be rejected based on size, aspect ratio, areal fill, and center-to-center distance criteria. Finally, a bounding rectangle was fit to the dot grid to identify the four outer corner dots, and each dot was matched to its presumed corresponding dot in the target dot pattern.

The transformed image was then cropped to a physical size of 245×245 mm, which encompassed the printable area of the Concept Laser M2 and produced a calibrated image size of 1842×1842 pixels with a spatial resolution of approximately 130 μm. The resolving power of the camera setup ranged from 220 to 280 μm across the powder bed, as measured using a USAF 1951 camera resolution target (Edmund Optics, Barrington, NJ, USA). The registration error between the CAD geometries and the imaging data was estimated at approximately 250 μm. To compensate for uneven lighting conditions over the print area, a smoothed image of an anomaly-free powder spread was used to generate a lighting correction mask. Figure 2c shows a fused representation of the two visible-light images after the perspective and lighting corrections have been applied, along with the nominal CAD geometry. Unless otherwise specified, all following images of the print area are in the same coordinate system and have the same field of view as introduced in Figure 2c.

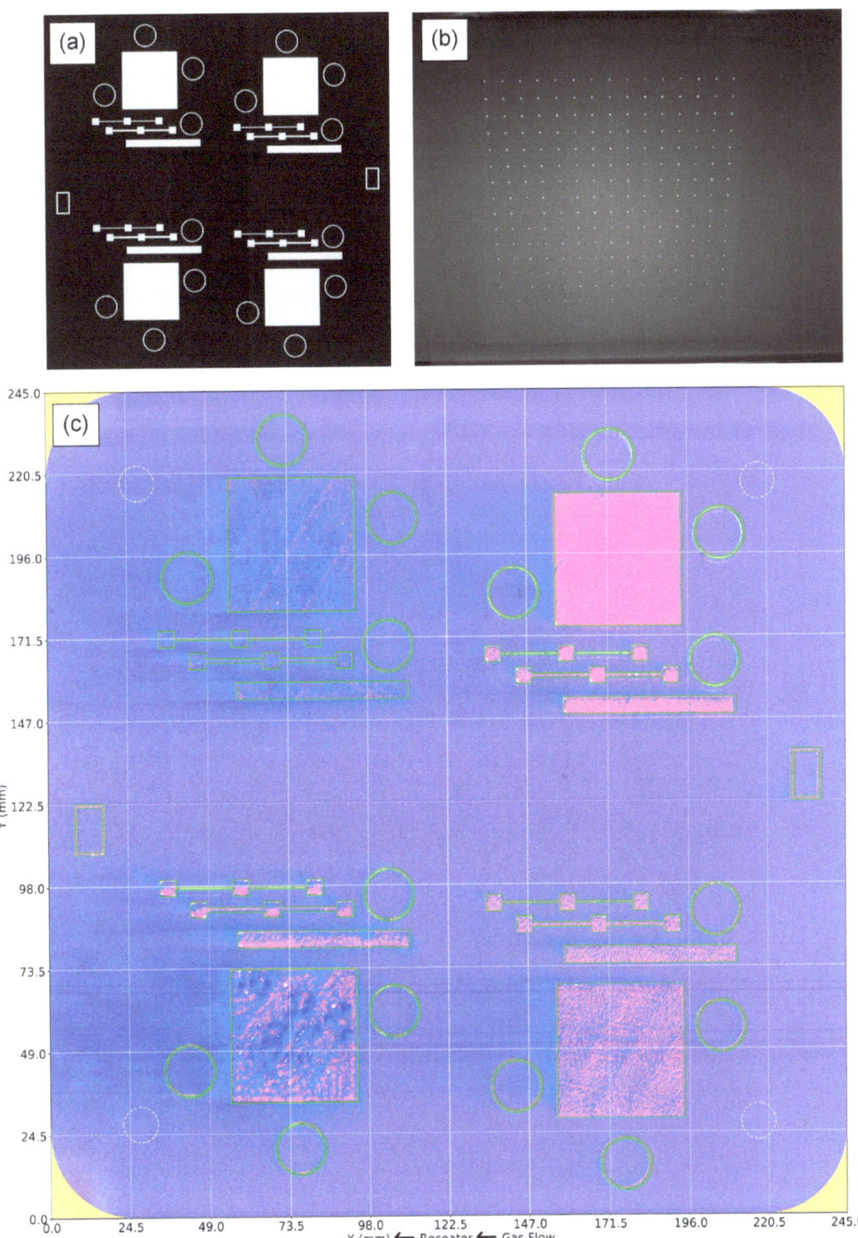

Figure 2. (**a**) A binary image representing the nominal CAD geometry (white) for layer 650 (32.50 mm) of build B1.2. (**b**) A raw, post-melt, QM coating layer image showing the printed calibration cone grid. The raw images are affected by perspective and lens distortions and suffer from uneven lighting across the print area. (**c**) A false-color image fusing the calibrated post-melt and post-spreading visible-light images from layer 650 (32.50 mm) of build B1.2. The nominal CAD geometry is indicated by the green outlines, and typical "keep-out" regions of the print area are marked in beige. Arrows indicate recoating and shield gas flow directions. The dynamic range of the composite image has been modified to accentuate features of interest, such as the horizontal streaks from the recoating mechanism.

Unlike the inherently spatial visible-light images, the OEM log file recorded machine health metrics temporally at sampling rates on the order of 5 Hz. The temporal log file values used included (1) the total layer time measured in seconds, (2) the top and (3) bottom argon flow rates measured in cubic meters per hour, (4) the oxygen percentage within the build chamber, (5) the temperatures of the build plate and (6) bottom argon gas flow measured in degrees Celsius, and (7) the gas ventilator flow rate measured in cubic meters per hour. For the purposes of this work, these low-frequency temporal data were spatially mapped by assigning the average of the values recorded over the duration of a given print layer to that print layer. Laser scan path information was recovered from the OEM QM Meltpool [48] system, which records laser location data at approximately 40 kHz in a technical data management streaming (TDMS) file format (National Instruments, Austin, TX, USA). Spatial mapping and registration of laser scan order data were achieved using the methods described by Halsey et al. [51]. However, in the previously reported implementation, on-axis photodiode data were used to filter out the "skywriting" that occurs at the laser beam turnaround points, as well as the "jump lines" between printed components. To improve the reliability of this artifact removal process for this effort, skywriting detection was instead performed using the laser travel vector to detect the turnaround points, as well as an empirically derived travel duration to identify the surrounding turnaround region. Similarly, jump lines were detected based on an empirically derived laser speed threshold. Figure 3 shows a visualization of the laser scan path within a single layer of a build. The QM Meltpool data used included the laser module and the laser scan path. Data from the QM Meltpool on-axis photodiodes and melt pool cameras were not available for this specific printer.

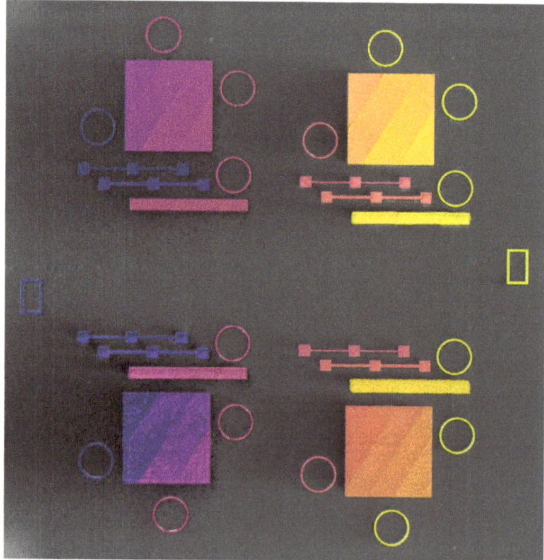

Figure 3. A false-color image fusing the laser scan path information with the calibrated post-melt visible-light image. The color map represents the time since the start of the layer, with darker blue regions melted first and lighter yellow regions melted last. The laser stripe boundaries are clearly visible as diagonal discontinuities in the color map. These data are from layer 650 (32.50 mm) of build B1.2.

2.3. Powder Bed Anomaly Segmentation and Training Methodologies

While the CAD geometries, log file data, and laser scan path information were directly incorporated into the super-voxels at N5, the two visible-light layer images were first processed by a dynamic segmentation convolutional neural network (DSCNN) DL algorithm at N4. The DSCNN converted the high-dimensional multi-modal image data into a lower-dimensional embedding that encoded salient features across multiple size scales in a latent space. That is, the DSCNN performed a semantic segmentation [52] task and produced pixel-wise classifications of various anomalous and nominal L-PBF conditions. Because the DSCNN architecture was previously reported by Scime et al. [28], only the germane differences in its implementation are noted here.

Prior work demonstrated the DSCNN's ability to classify a variety of powder bed anomaly classes across multiple powder bed printer types [28]. Eight classes were identified as potentially relevant to tensile property prediction, examples of which are shown in Figure 4. The *powder* and *printed* classes represent the two nominal L-PBF conditions, respectively, indicating the unfused and fused material. *Recoater streaking* generally occurs when the recoating mechanism is either damaged or dragging debris across the powder bed and is visually characterized as a long streak parallel to the recoating direction [53]. *Edge swelling* appears as small regions of the part, typically corners and edges, elevating above the spread powder layer. Although edge swelling is common in L-PBF, even under nominal processing conditions, certain process parameter changes and local part geometries can increase its occurrence [54,55]. *Debris*, in the context of this work, generally indicates low energy density melt parameters prone to causing lack-of-fusion porosity [56]. *Super-elevation* of large regions of the part above the powder layer can occur either as the result of warping caused by residual thermal stresses [57] or improper powder dosing. *Soot* refers to spatter particles [58] that have landed on the powder bed. For the imaging system used in this work, soot generally manifests as clouds of dark particles. In the context of this work, *excessive melting* indicates high energy density melt parameters prone to causing keyhole porosity [59] and is visually characterized and labeled by a bubbling of the part surface.

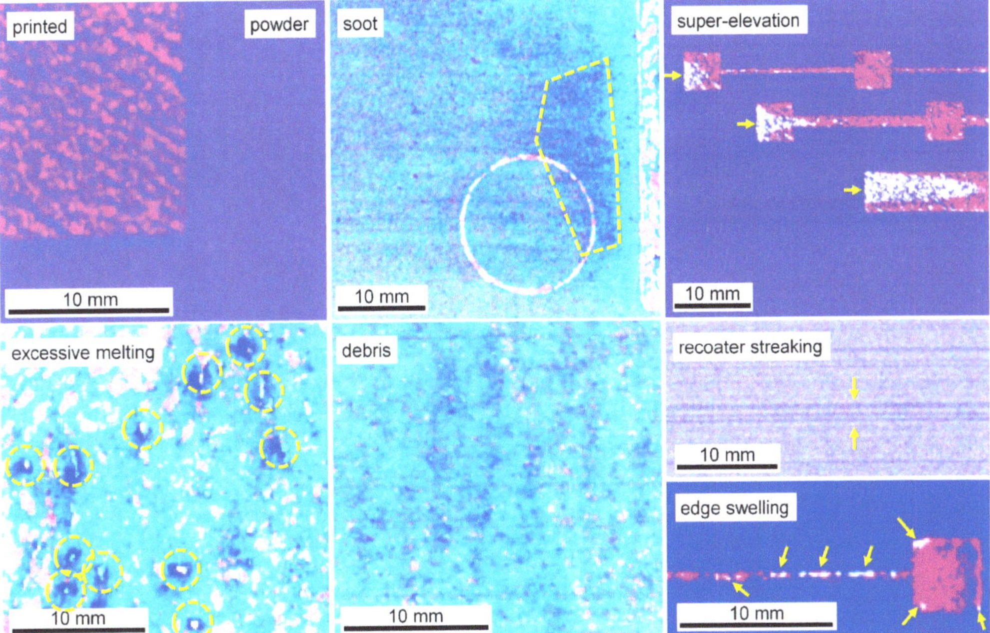

Figure 4. False-color images fusing the post-melt and post-spreading visible-light images of each of the eight powder bed classes that were segmented by the DSCNN. The dynamic range of each image has been modified to accentuate features of interest. Annotations are presented in yellow to highlight the relevant sensor indications. Some classes have well-defined boundaries, such as edge swelling and excessive melting, while others have nebulous boundaries, such as soot and debris, and are delineated as entire regions during annotation.

Ground truth training data for the DSCNN were collected through expert annotation of 180 million pixels across 38 print layers sourced from 23 different Concept Laser M2 builds, all using SS 316L feedstock. For this task, annotators leveraged a purpose-designed graphical interface incorporating image calibration, data fusion, dynamic range scaling, standard drawing tools (e.g., brushes, lassos, flood fills), intensity thresholding options, and ML-assisted annotation capabilities. While it is also possible to train the DSCNN using ground truths collected from ex situ characterization (e.g., flaw detections from XCT), this approach is beyond the scope of this work. The manual annotation procedure is further described in Scime et al. [28], and example annotations and associated data can be downloaded from the Peregrine v2021-03 [60] dataset.

Data augmentation was used to increase the size of the training set without requiring the collection of additional ground truths [61]. The augmentation mechanisms applied during training differed slightly from those reported in Scime et al. [28]; they consisted of (1) global intensity shifts with magnitudes uniformly distributed between −15% and +15% of the dynamic range of each image channel, (2) additive Gaussian noise distributions with variances equal to 0.0001% and 0.001% of the dynamic range of each image channel, and, new in this work, (3) spatial shifts of each image tile by up to 20 pixels in each direction. The spatial shift augmentation technique was included to increase the total amount of training data and to reduce artifacts at the edges of the tiles. The DSCNN was trained with a cross-entropy loss function weighted by the inverse of the class frequencies as specified by Equation (6) in Scime et al. [28]. Additional pixel-wise weights were applied to disincentivize the optimization function from spatially expanding less common classes (e.g., edge swelling) at the expense of more common classes. This re-balancing was achieved by increasing the weight of the ground truths for powder and printed pixels located near an interface with a less common class. This weight adjustment, w, is given by Equations (1)–(3).

$$m = \frac{MIN\left(0, \frac{N_{all}}{N_{int}}\right) - 1}{\phi}, \tag{1}$$

$$b = 1 - m, \tag{2}$$

where ϕ is a saturation distance set at 1% of the image size, N_{all} is the total number of pixels in the image, and N_{int} is the number of pixels closer to an interface than the saturation distance.

$$w(i,j) = m[MAX(1, \phi - [D_1 \cup D_2](i,j))] + b, \tag{3}$$

where $w(i,j)$ is the interfacial weight adjustment at pixel (i,j), and D_1 and D_2 are the distance transforms [62] for the combined powder and printed annotation masks.

Other noteworthy changes from [28] include the preservation of the native bit depth (16 bits) of the visible-light images and the implementation of training early-stop [63] criteria based on detecting a plateau in the epochal validation accuracy. Additionally, two heuristics were applied to the segmentation mask produced by the trained DSCNN. The first heuristic converted excessive melting segmentations further than 750 µm from the CAD geometry to debris. This mitigates observed confusion between these classes and is justifiable because excessive melting definitionally cannot occur beyond a melted part. The second heuristic extends recoater streaking segmentations horizontally across the print area. This is necessary because, although recoater streaking is readily apparent in the unfused powder, it is generally difficult to observe over the top of a printed part. However, the literature suggests that L-PBF recoater streaks often extend into the parts themselves [34], especially when a compliant recoater is used. Specifically, this heuristic is triggered by DSCNN recoater streaking segmentations with horizontal dimensions larger than 5 mm, and only pixels initially classified by the DSCNN as either powder or printed material are overwritten by the heuristic.

2.4. Performance of the Dynamic Segmentation Convolutional Neural Network

The overall performance of the DSCNN architecture is extensively documented in Scime et al. [28]. Table 1 reports the true-positive (TP) and false-positive (FP) validation rates for the specific DSCNN model and training dataset used for the AIR. The significant TP performance improvements (e.g., from 17.5% to 85.8% for soot) relative to those reported by Scime et al. [28] are the result of the increased size of the training database and the modified training procedures described above.

Table 1. The TP and FP validation performance metrics for the specific DSCNN model used. The training and validation dataset splits are 90% and 10%, respectively. FP rates are typically higher for spatially small classes owing to the chosen balance between the class-wise and interfacial boundary loss weighting schemes.

Class	TP (%)	FP (%)
Powder	97.3	0.8
Printed	98.6	5.1
Recoater streaking	87.4	62.8
Edge swelling	95.1	34.2
Debris	95.9	26.2
Super-elevation	98.5	3.0
Soot	85.8	15.5
Excessive melting	94.8	72.2

Figure 5 shows an example of a segmented layer. Note that while heuristics were enabled for Figure 5, they are not included in the validation metrics. The average DSCNN inference time is 1.7 s, and the layer-wise connected-component analysis of the CAD geometry is typically less than 2 s, depending on the geometry. Loading the visible-light images into computer memory and performing the image calibration procedure may be performed in parallel with DSCNN inference.

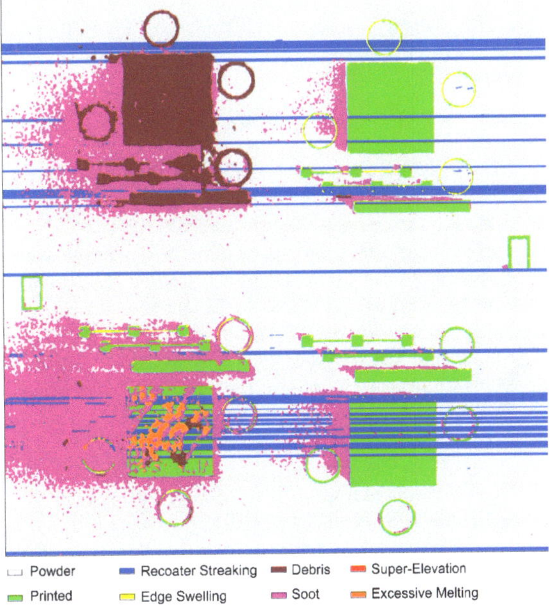

Figure 5. Pixel-wise segmentation results from the trained DSCNN for layer 650 (32.50 mm) from build B1.2. Anomaly classes are indicated by color.

2.5. Specimen Design

Four factors influenced the mechanical specimen design: (1) the material properties of interest, (2) the importance of measuring representative material properties across multiple as-printed part geometries, (3) the compatibility of the physical specimen size with the computational constraints of the models, and (4) adherence to accepted material characterization standards. First, tensile properties, including yield strength (YS), ultimate tensile strength (UTS), uniform elongation (UE), and total elongation (TE), were selected as prediction targets for this work. Whereas other material properties, such as fatigue life, are expected to be comparatively more sensitive to processing anomalies [39], the shorter testing cycles for room temperature static tensile tests enabled the collection of many more ground truths. Additionally, static tensile properties were of interest for nuclear power applications [64], and there is sufficient literature indicating that variations in L-PBF processing conditions (e.g., geometric feature size, solidification conditions, and void-type flaws) could induce variations in these properties [36,65–68]. Next, to maintain the generalizability of the ML models to multiple component geometries and local thermal histories, the tensile specimens were extracted from a set of larger as-printed geometries instead of being printed in their final shape, as has been conducted in prior high-volume tensile testing work by Roach et al. [65]. For clarity, specimens extracted from as-printed *parts* will be referred to as *samples* throughout the remainder of this manuscript. Because the spatial resolution of the ground truth tensile data was controlled by the size of the specimen's gauge section, the standard [69] subsize SS-J3 geometry shown in Figure 6 was selected, as described in Appendix A.

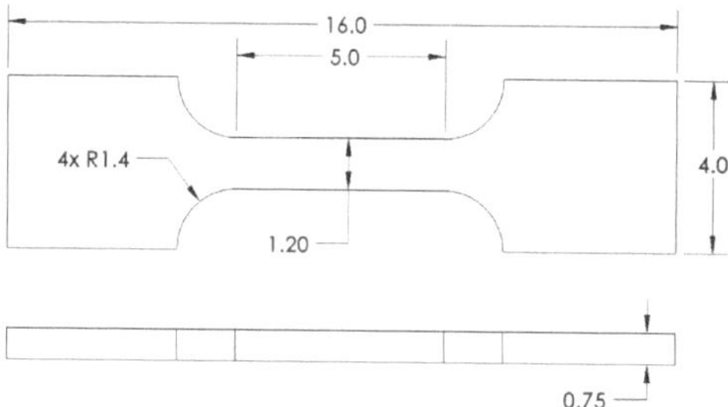

Figure 6. Key nominal dimensions of the SS-J3 tensile specimens. All values are in millimeters.

The SS-J3 samples were extracted from four different printed part geometries designated as SSJ3-A, SSJ3-B, SSJ3-C, and SSJ3-D, as indicated in Figure 7. The CAD models of each geometry were adjusted to achieve as-printed dimensions as close as possible to the reported nominal dimensions. The SS-J3 samples were distributed along the nominal build direction with vertical center-to-center spacings of 19 mm. The SSJ3-A and SSJ3-B geometries incorporated buttresses to increase part stiffness and maintain the dimensional accuracy of the parts during printing, heat treatment, and machining. Note that while the SSJ3-C and SSJ3-D parts produced samples with as-printed and machined surfaces, the SSJ3-A parts produced samples with only as-printed surfaces, and the SSJ3-B parts produced samples with only machined surfaces.

2.6. Build Design and Conditions

For all builds in this work, the nominal print layer thickness was maintained at 50 μm, no preheating was performed, and argon was used as the shield gas. The feedstock was TruForm (Praxair Surface Technologies, Indianapolis, IN, USA) SS 316L powder sourced from a single lot. The manufacturer reported D10, D50, and D90 powder particle diameter values of 20 μm, 31 μm, and 43 μm, respectively. The chemical composition of the virgin powder, as reported by the manufacturer, is provided in Table 2.

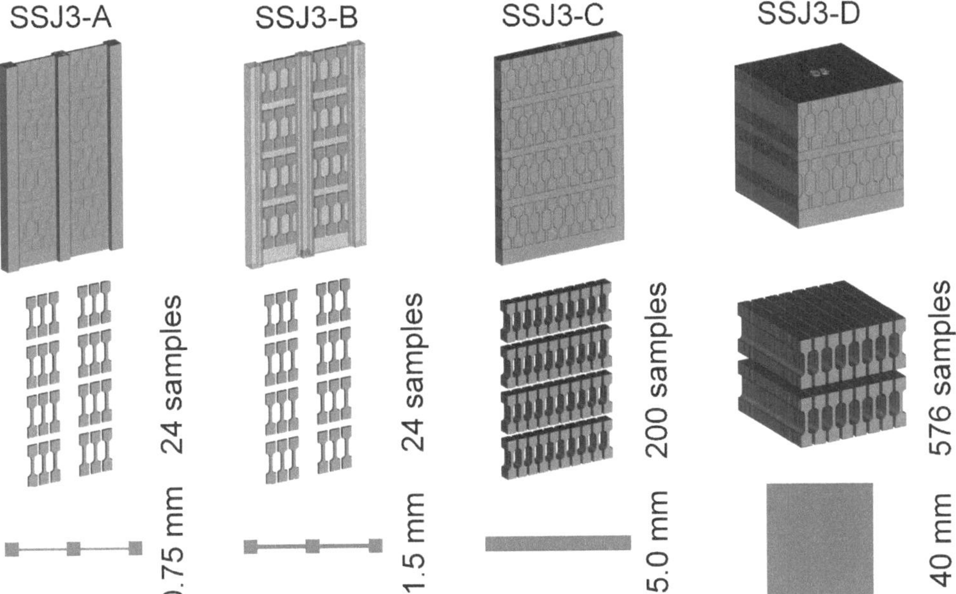

Figure 7. An isometric view of the CAD model for each of the as-printed SSJ3 part geometries (**top row**), the 3D layouts of the SS-J3 samples relative to their corresponding printed part volumes (**middle row**), and a top-down view of the part geometries (**bottom row**). The number of potential samples and the nominal wall thickness for each geometry are given in the middle and bottom rows, respectively.

Table 2. Chemical composition of the SS 316L powder lot as reported by Praxair. Values are given in weight percent.

C	Co	Cr	Cu	Fe	Mn	Mo	N	Ni	O	P	S	Si	Other
<0.005	0.08	17.01	0.00	Bal	1.29	2.48	0.01	12.67	0.03	<0.005	0.005	0.59	<0.1

The bulk regions of the SS-J3 samples were melted using the laser raster process parameter sets defined in Table 3. The *nominal* parameter set was provided as the default for SS 316L by Concept Laser, the *best* parameter was chosen to minimize porosity, the *lack-of-fusion* parameter was known to produce significant lack-of-fusion (*LOF*) porosities, and the *keyhole* parameter was chosen to induce keyholing pores. It was expected that the varying energy densities and solidification cooling rates of these parameter sets would result in substantially different as-printed microstructures and void-type flaw populations, as shown in other work [67]. Table 3 also defines the *soot* parameter, which did not directly melt any SS-J3 samples but was instead used to generate abnormally large quantities of soot near some of the tensile samples, with the goal of introducing spatter-induced porosities as observed by Schwerz et al. [23] and others. The default post-contour parameter set provided by Concept Laser was applied with a laser beam power of 120 W, a laser beam speed of 220 mm/s, and a laser spot size of 70 μm.

A total of five L-PBF builds were performed to generate 6299 SS-J3 tensile specimens used in the feedback loop connecting N6 and N5. Additional builds were performed for camera calibration, algorithm development, DSCNN training, process parameter development, specimen design, heat treatment development, and specimen extraction and tracking procedure development as described in [70]. Design requirements for the five builds discussed in this work include (1) facilitating the extraction of thousands of SS-J3 tensile specimens from trackable locations, (2) capturing the range of process and part variability expected to occur during L-PBF manufacturing, and (3) generating a range of local tensile properties caused by various mechanisms hypothesized to correlate to signatures observable in the available in situ sensor data. The number SS-J3 samples extracted from each build and the variable build conditions which were expected to result in variable tensile properties are summarized in Table 4.

Table 3. Laser raster parameter sets used across the tensile sample builds, as provided by Concept Laser or determined through internal testing.

Parameter Set	Laser Beam Power (W)	Laser Beam Speed (mm/s)	Hatch Spacing (μm)	Nominal Laser Spot Size (μm)	Stripe Width (mm)	Scan Rotation (Degrees/Layer)
Nominal	370	1350	90	130	10	67
Best	380	800	110	125	18	67
LOF	290	1200	150	50	18	67
Keyhole	290	800	70	125	18	67
Soot	290	1200	70	50	18	90

Table 4. Each build was designed to produce hundreds of SS-J3 tensile specimens which capture representative process variation expected during L-PBF manufacturing. The number of samples and the variable process conditions experienced by the samples for each build are summarized below.

Build ID	Number of Extracted Samples	Varied Location within the Build Volume	Varied Local Part Geometry	Varied Laser Module	Varied Laser Raster Process Parameters	Overhang Angle Relative to the z-Axis	Included Spot Generating Parts	Argon Flow Rate Setpoint (m³/h)	Powder Dosing Factor (%)	Used a Damaged Recoater Blade	Contours Enabled
B1.1	503	X	X	X		0°		40	200		X
B1.2	2705	X	X		X	0°		40	200		
B1.3	813	X	X	X	X	30°		40	200		X
B1.4	694	X	X	X		0°	X	25–40	200		X
B1.5	1584	X	X	X		0°		40	5–200	X	X

Figure 8 shows an isometric view of each build, along with the part layout, process parameter assignments, and laser module assignments. At a high level, B1.1 was designed to produce baseline process conditions, B1.2 was designed to capture the effects of variable laser raster processing parameters, B1.3 was designed to represent overhanging geometries, B1.4 was designed to capture the effects of spatter particles and decreased argon gas flow, and B1.5 was designed to capture the effects of recoater blade damage and powder short feeds. Intermediate visualizations (N8) of the in situ data and DSCNN segmentation results for each build are provided in Appendix B for additional context.

2.7. Sample Extraction and Tracking

The printed parts were first heat-treated while attached to their build plates to relieve residual thermal stresses induced during printing [57]. Relieving these stresses reduced the amount of distortion experienced by the parts during sample extraction, improving both the dimensional tolerances of the specimens and the fidelity of the registration between the tensile test results and the in situ data. Full heat treatment details are provided in Appendix C.

Following heat treatment, the build plate and associated parts were bead-blasted to remove oxide scaling and to provide clean touch-off surfaces to define the part origins for the wire electrical discharge machining (EDM) operation. The parts were separated from the build plate using an AQ750LH (Sodick, Yokohama, Japan) wire EDM and were then fixtured individually for SS-J3 sample extraction. The 3D locations of the SS-J3 samples were predefined in a CAD model, and a tool path was generated using the ESPRIT 2021 (Hexagon, Stockholm, Sweden) computer-aided machining (CAM) software. This CAD model was also sliced using the Magics Image Build Processor for registration with the in situ data (Section 2.2). The parts were sectioned into sheets (Figure 9a), with each individual SS-J3 specimen still attached via a single Table The samples were then manually separated from the surrounding material and placed into individual bags labeled with a quick-response (QR) code (Figure 9b). When scanned, each QR code linked the physical sample to its unique identifier within the digital platform and allowed the retrieval of its digital thread (Figure 9c). This cyberphysical infrastructure substantially aided the robust tracking of the in situ and ex situ data associated with thousands of unique samples.

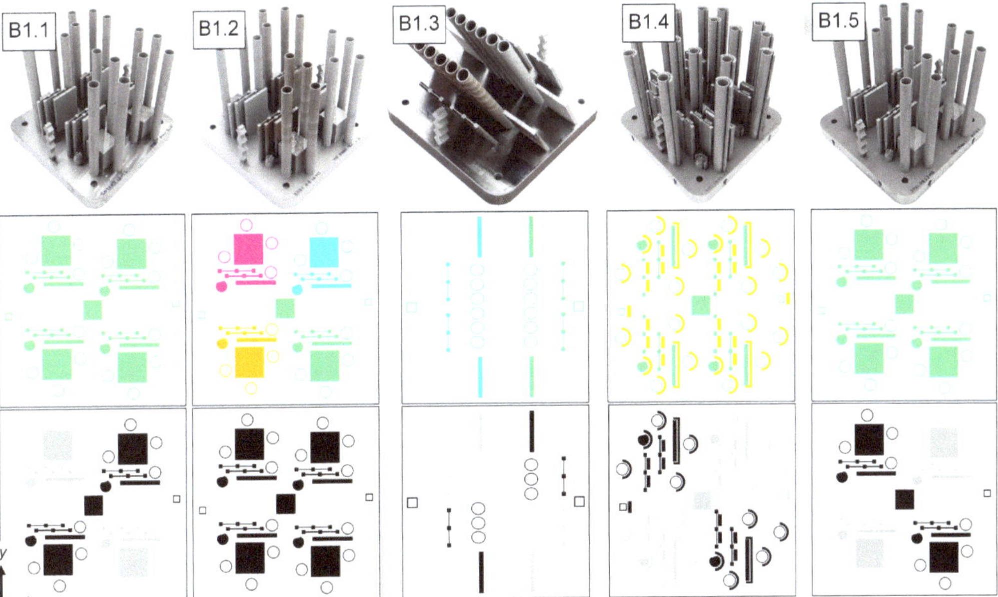

Figure 8. Each column shows the build layout for a given build. The top row contains isometric views of each build after the print was completed. The middle row colors each part by the process parameter set, with nominal, best, LOF, keyhole, and soot indicated by green, blue, purple, orange, and yellow, respectively. The bottom row colors each part by the laser module, with the first laser module indicated by black and the second laser module indicated by gray. Note that the yellow, soot-generating parts in B1.4 are located upstream of the gas flow relative to the SS-J3 samples.

2.8. Tensile Testing Procedure

Tensile tests were performed using the ASTM E8/E8M [71] procedure with some modifications to facilitate the high volume of testing. Testing was performed across two load frames (TestResources, Shakopee, MN, USA) configured with 500 lbf static-rated load cells calibrated as prescribed by ASTM E4 [72]. The width and thickness of each SS-J3 gauge section were measured using calipers, while the length was assumed to be the nominal value of 5 mm because of the difficulty of accurately measuring this dimension. Samples were installed in a shoulder-loading tensile configuration, preloaded to a nominal load of between 10 N and 50 N, and then continuously loaded under displacement control at a rate of 0.5 mm/min (nominally 10% strain/min). The load and crosshead displacement were recorded at a rate of 10 Hz until the SS-J3 specimen either fractured or the measured load fell below 10 N. Representative engineering stress–strain curves from samples printed in B1.2 are shown in Figure 10.

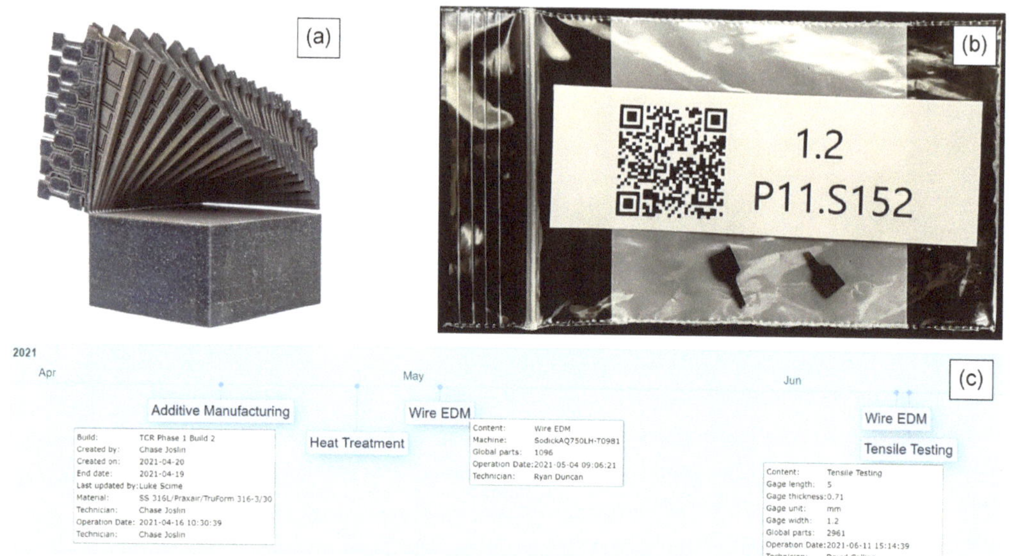

Figure 9. (**a**) Example of SS-J3 samples partially extracted from an SSJ3-D part. Samples remain attached to the sheets in a known configuration until they can be individually labeled. (**b**) Example of a tested SS-J3 sample stored in a bag marked with a unique QR code. The QR code contains information about the printer, build, parent part, and sample location such that it can be uniquely tracked across the digital platform. (**c**) A screenshot of a web-based visual representation of the digital thread for the tensile sample in (**b**). The digital thread provides a record of the operations applied to a sample and its parent part, beginning with printing and ending with tensile testing.

Tensile properties were algorithmically calculated from the raw load-displacement data. First, the curves were adjusted to account for crosshead displacement measured prior to preloading by shifting the data origin to the initial load measurement and then removing all measurements that were zero load or lower. Then, load and crosshead displacements were converted to engineering stress and strain, respectively, using the initial dimensions of the gauge section. The elastic linear region was identified using datapoints with stress values between 5 and 50 MPa, and the YS was calculated using the canonical 0.2% offset procedure. The UTS was defined as the maximum engineering stress measured during testing, whereas UE was the engineering strain measured from yielding until the point of maximum stress. The TE was defined as the total engineering strain measured from loading until failure of the specimen. Any specimens with UTS values lower than 50 MPa were considered failed tests and were rejected; the lowest non-rejected UTS value was 80 MPa.

2.9. Selected Tensile Test Results

Table 5 summarizes the range of tensile property values measured across the five builds, along with reference properties from ASTM A240 [73] and for wrought SS 316L as measured by Byun et al. [64] using similarly sized SS-J2 specimens. Byun et al. [64] also report the effects of a comparable post-build stress-relief heat treatment on the static tensile properties of additively manufactured SS 316L, observing a 46 MPa reduction in YS and no significant differences in UTS, UE, or TE relative to the as-built condition. As desired, the measured tensile properties span a wide range of values across the intentionally varied processing conditions. The minimum YS, UTS, UE, and TE are substantially lower than both the A240 specifications and the wrought properties, whereas the maximum values exceed these baselines. To estimate an appropriate reporting precision for the measured tensile values, the standard deviations were calculated for the non-surface SS-J3 samples extracted from the SSJ3-D part geometry printed with the BEST process parameters in build B1.2. These samples were selected because they were expected to have the lowest true variation in tensile properties, due to uniform thermal conditions and a low flaw density. The measured standard deviations were 16.6 MPa, 15.6 MPa, 1.73%, and 2.92% for YS, UTS, UE, and TE, respectively. Therefore, all tensile values are reported with precisions of 10 MPa, 10 MPa, 1%, and 1%, chosen based on the nearest order of magnitude to the corresponding standard deviation. The complete set of tensile results are available in the Peregrine v2023-11 dataset [44].

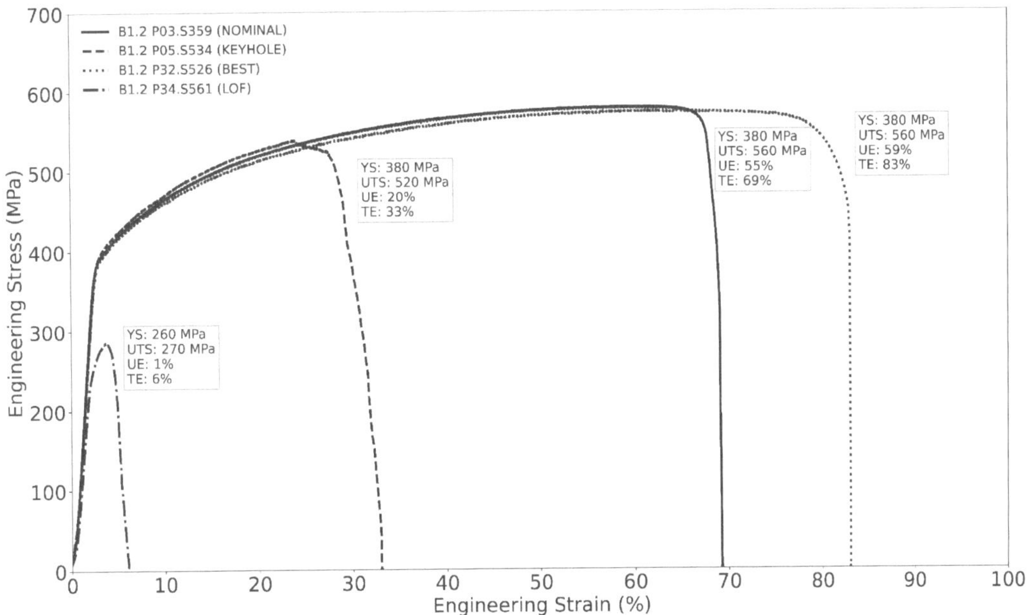

Figure 10. Representative engineering stress–strain curves for four SS-J3 samples extracted from four different SSJ3-D parts printed in B1.2. Each sample was printed with a different parameter set, as indicated in the legend. The calculated values for YS, UTS, UE, and TE are reported next to their corresponding curve. The SS-J3 specimens printed with the best and nominal parameters demonstrate superior static tensile properties compared to the specimens printed with the keyhole and LOF parameters. Strain was approximated by normalizing the crosshead displacement to the nominal gauge length.

Table 5. Literature values compared to the range of measured YS, UTS, UE, and TE values observed across all the samples.

Source	YS (MPa)	UTS (MPa)	UE (%)	TE (%)
ASTM A240 [73]	170	480	40	N/A
Wrought [64]	261	562	66.0	72.8
B1.1–B1.5	70–420	80–610	0–69	4–94

Figure 11 shows the mean tensile properties measured for the B1.2 samples extracted from the four SSJ3-D parts, separated by the laser processing parameter set. As expected, the samples printed with the best parameters have the highest tensile properties, followed by the nominal samples. The samples printed with the LOF parameters have drastically lower tensile properties, particularly for UE and TE, which are 56 and 71 percentage points lower than the best values, respectively. Whereas the magnitudes of the keyhole UE and TE values only experience relative reductions of 54% and 55%, respectively, compared to the corresponding BEST values, their standard deviations demonstrate relative increases of 410% and 240%, respectively. This observation suggests that bulk parts produced using the KEYHOLE parameter set may be expected to have substantially increased local variation in their elongation behavior throughout their printed volume as a result of the stochastic formation of individual keyhole pores [59].

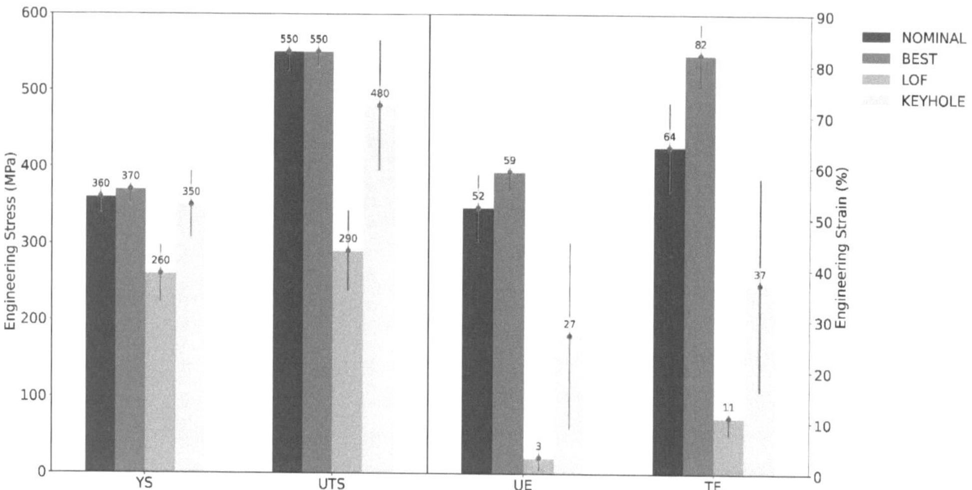

Figure 11. Mean tensile properties measured for all the B1.2 samples extracted from the four SSJ3-D parts, separated by the laser processing parameter set. Each red error bar represents one standard deviation.

Figure 12 shows the mean tensile properties measured for the nominal B1.1 samples separated by the as-printed part geometry (i.e., SSJ3-A, SSJ3-B, and SSJ3-D) from which the samples were extracted. The lowest tensile properties were observed for the SSJ3-A thin wall geometry, for which all the extracted SS-J3 samples had two as-printed surfaces. The highest tensile properties were observed for the SSJ3-B thin wall geometry, for which all the extracted samples had only machined surfaces. Additionally, the SSJ3-B thin wall is expected to have a different thermal history than the SSJ3-D blocks, which may result in beneficial microstructural differences [66]. However, this has not been specifically investigated for these samples.

Figure 13 shows the mean tensile properties measured across selected subsets from builds B1.3, B1.4, and B1.5. Considering B1.3, the best overhang-adjacent surface (OAS) samples extracted from the SSJ3-B parts have substantially lower tensile properties compared to the corresponding best bulk and top surface (BTS) samples. This indicates that printed material immediately adjacent to an overhanging design feature can be expected to have noticeably reduced tensile properties. Conversely, the nominal subset from B1.4 suggests that the increased amount of soot does not substantially impact the static tensile properties relative to the nominal data from B1.2. Similarly, comparing the nominal no-short-feed (NSF) and nominal short-feed (SF) subsets from B1.5 indicates that the intentional powder short feeds (caused by the reduction in dosing factor) had no significant impact on any of the measured tensile properties. Further analysis of this large quantity of tensile data is beyond the scope of this work.

2.10. Construction of the Super-Voxels

To allow an ML model (N6) to learn an accurate transfer function between the in situ data and the tensile properties measured during testing, the in situ sensor data and part geometry information were encoded into engineered feature vectors corresponding to discrete volumes designated as super-voxels (N5). The encoding was designed to contain the information required for a generalizable ML model to make property predictions at the super-voxel size scale. To this end, the printer's build volume was demarcated into a fixed grid of rectangular prisms, with each super-voxel measuring 1.0×1.0 mm in the x-y plane and 3.5 mm along the z-axis. Several factors, including empirical measures of model performance, were considered when determining an appropriate super-voxel size. First, the super-voxels were sized to approximately match the volume of the SS-J3 gauge sections, which were, in turn, sized based on the criteria enumerated in Section 2.5 and Appendix A. Of secondary importance, computational memory restrictions placed a lower bound on the super-voxel volume. Figure 14 shows a portion of a print layer divided into a super-voxel grid, with super-voxels extending 70 layers in the vertical print direction.

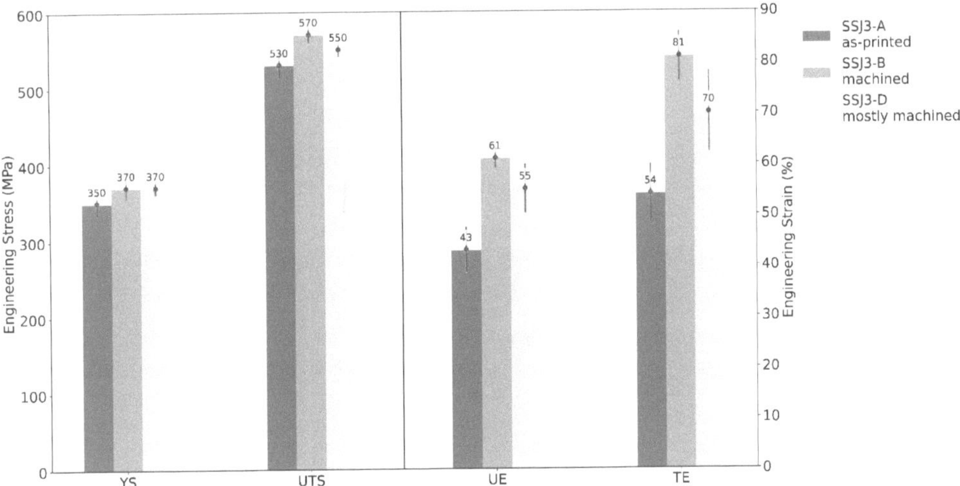

Figure 12. Mean tensile properties measured for all the nominal B1.1 samples separated by the as-printed part geometries. Each red error bar represents one standard deviation.

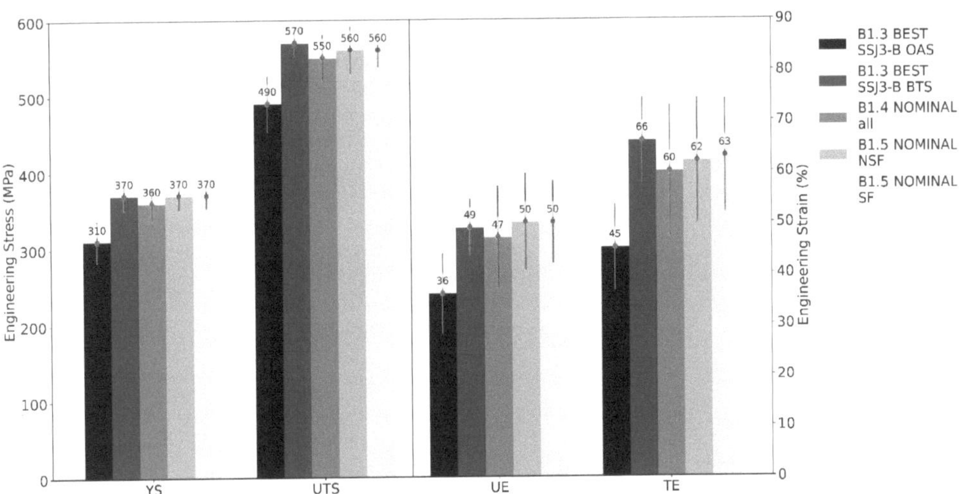

Figure 13. Mean tensile properties measured from several selected subsets of samples from builds B1.3, B1.4, and B1.5. Each red error bar represents one standard deviation.

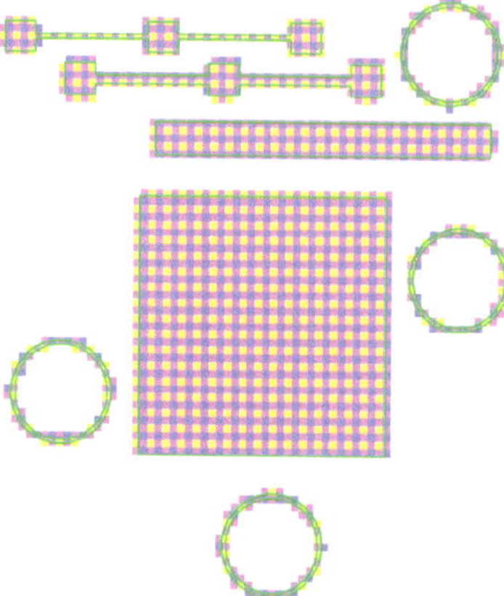

Figure 14. Intersection of the fixed super-voxel grid with a set of CAD geometries (green lines). Only super-voxels that overlap the CAD geometry are shown. Each super-voxel is 1 mm on a side and is colored so that its boundaries with neighboring super-voxels are visible.

Each feature vector was composed of 21 engineered features calculated based on the CAD geometries, pixel-wise DSCNN segmentations of the post-melt and post-spreading visible-light layer images, sensor data recorded in the printer log file, and laser scan path data. Each feature is enumerated, described, and justified in Appendix D. Figure 15 shows six example feature maps for a sub-region of B1.2. The planar resolution of the geometry information, DSCNN results, and laser scan path data was 130 μm (Section 2.2), which is substantially smaller than the super-voxel size. Therefore, these pixel-wise values were first averaged in the *x-y* plane to produce a single value per feature, per super-voxel, per layer. To mitigate computational edge effects, only pixels contained within the CAD geometry were considered in the planar averaging. Similarly, the vertical resolution of these features and the log file data was equal to the 50 μm layer thickness. Therefore, averaging in the vertical direction was accomplished with a sliding window filter with a stride equal to the height of the super-voxels. For super-voxels intersecting the edges of the printed part geometries, the averages are weighted proportionally to the number of pixels contained within the intersection of the super-voxel and the CAD geometry at each layer.

As discussed in more detail in the following section, the training data for the tensile property prediction model consisted of the feature vectors for valid super-voxels intersecting the SS-J3 gauge sections. Super-voxels were considered invalid if any in situ data were missing or if less than 10% of their area overlapped with the gauge section of a given SS-J3 sample. To improve training stability, the raw feature values are zero-center normalized on [−1,1] based on the minimum and maximum values observed across the combined training and validation dataset. Generally, the potential for an ML model to learn meaningful discrimination between material volumes increases as the number of in situ sensor modalities and engineered features increases. However, the total number of features must be balanced with the challenges of performing enough experiments (Section 2.6) to capture the expected variation across each feature axis.

2.11. Architecture and Training of the Property Prediction Model

The voxelized property prediction model (VPPM) implements a perceptron [74] ML algorithm and was used to predict the local tensile properties at N6 based on the engineered feature vectors described in Section 2.10. The ground truth training targets are the zero-center normalized tensile measurements reported in Section 2.9. Perceptrons are shallow neural networks [16] that have been successfully applied to a wide range of regression problem sets. The perceptron is not necessarily the optimal ML model for this stage of the relay, but it serves as a viable proof-of-concept because this work primarily focuses on framework development, data collection, and feature engineering, leaving model optimization for future publications. A separate VPPM was trained to predict each of the four tensile properties: YS, UTS, UE, and TE. The perceptron architecture is reported in Table 6. Along with the training procedure, the perceptron architecture is identical for each of the VPPMs. The fully connected and dropout neural network layers used by the VPPM are described in Krizhevsky et al. [75].

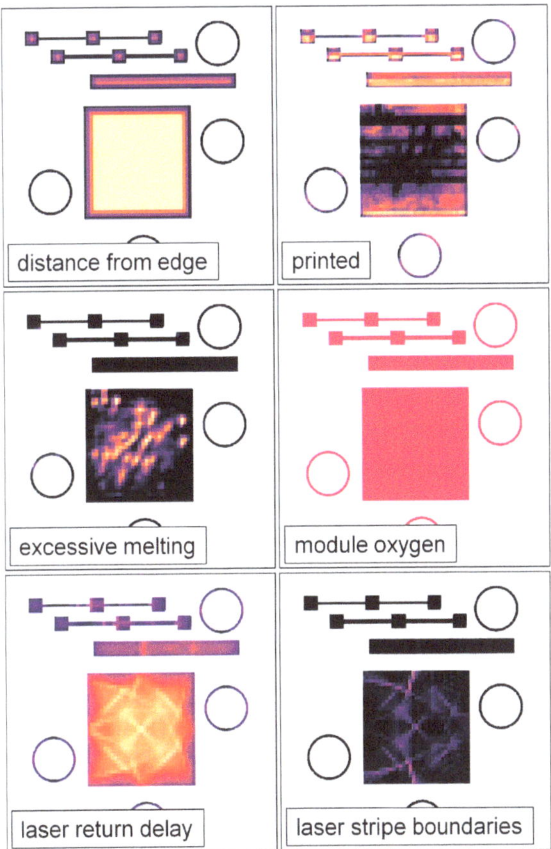

Figure 15. Normalized visualizations of six super-voxel features with brighter and darker regions, respectively, indicating larger and smaller values. These super-voxels encode data from the keyhole parts for layers 639 (31.95 mm) through 709 (35.45 mm) of B1.2. For ease of interpretation, super-voxels not fully contained within the CAD geometry have been cropped. Note that the feature encoding the distance from the part edge saturates in the center of the SSJ3-D part. Also note that the module oxygen value is uniform for all super-voxels fully contained within the same layer range.

Table 6. VPPM perceptron architecture. Columns indicate the type of network layer, the number of input channels (C_i), and the number of output channels (C_o). The variable n_{feats} represents the number of elements in the super-voxel feature vectors and is fixed at 21 for this work.

Layer	C_i	C_o
Fully connected [75]	n_{feats}	128
10% dropout [75]	128	128
Fully connected	128	1

The VPPMs were trained using backpropagation [76] and the Adam optimizer [77] with a loss function that minimized the L_1 (i.e., absolute) error between the predicted values and the ground truths. The kernel weights of both fully connected layers were initialized from zero-centered normal distributions with a standard deviation of 0.1. Each VPPM was trained using a mini-batch size of 1000 super-voxels and a learning rate of 1×10^{-8}, with tests indicating that the model performance was insensitive to these hyperparameters. The first and second exponential decay rates of the Adam optimizer were set at 0.9 and 0.999, respectively, and the epsilon value was fixed at 1×10^{-4}.

The 6299 SS-J3 tensile samples available for training corresponded to 29,680 unique super-voxel feature vectors, with an average of 4.7 super-voxels per SS-J3 gauge section. The mini-batch size was 1000 super-voxels, with 20% of the data used for validation during each repetition of a 5-fold cross-validation procedure [78]. Cross-validation was performed to reduce the sensitivity of the model performance to the randomly determined split between training and validation datasets. For each fold repetition, training proceeded until the validation error plateaued. Bifurcation of the data into the training and validation sets was performed sample-wise so that all super-voxels associated with the same ground truth were assigned to the same set. Note the implication that the same ground truth target, i.e., measured SS-J3 tensile property value, is associated with multiple super-voxels, the exact number of which depends on the intersection of the fixed super-voxel grid with the sample geometry in 3D space. This condition caps the maximum possible model performance and, in extremis, could cause instabilities in the training process because each ground truth cannot be correlated to a unique feature vector. Therefore, minimizing the number of super-voxels contained by each gauge section was a key computational consideration for setting the appropriate super-voxel size. Ultimately, this limitation is dependent upon the localizability of the ex situ testing data and is further considered in the Discussion section. Although VPPM validation accuracy might be improved by calculating each feature vector based on the totality of the SS-J3 gauge volume instead of for super-voxels on a fixed grid, such an approach would not be representative of model test conditions (i.e., predicting local properties for a printed component of arbitrary geometry) because the prediction volumes would be differently sized, and certain edge effects (e.g., varying overlap between the super-voxels and the part geometry) would not be accounted for in the training or validation sets.

At model test time, each VPPM applied its learned weights and biases to each feature vector, predicting a single value for each super-voxel that was then denormalized and converted to real YS, UTS, UE, or TE units. For part qualification, it is valuable to identify predictions that may be extrapolated instead of interpolated to ensure that human decision-makers are involved whenever conditions exceed the scope of the models' training. This area of study is commonly referred to as *out-of-distribution detection* [79] and is a nontrivial determination in high-dimensional space. For this work, a super-voxel was considered out-of-distribution if the value of any one of the individual features was below the minimum or above the maximum values observed for that feature within the training set. Although this bounding heuristic is a necessary check, it is not strictly sufficient. A more sufficient approach might consist of a clustering analysis [80] of the training set paired with an empirically determined maximum acceptable distance between a new feature vector and the observed clusters. However, advanced out-of-distribution detection methods are reserved for future work. Computationally, any predictions for out-of-distribution super-voxels were converted to not-a-number (NaN) values and were reported separately.

3. Results

3.1. Validation Performance

The primary performance metrics used to evaluate the VPPMs, and by extension, the entire property prediction pipeline, were based on the root mean square (RMS) validation errors averaged over five training folds. Because each gauge section contains several super-voxels, multiple unique predictions were made for sub-volumes within each sample. These overlapping predictions were collapsed to a single value by calculating the minimum of these predictions. Taking the minimum value provides a conservative estimate of the local material properties, which best informs the part qualification process. Averaged over the five folds, only 0.05% of super-voxels were considered out of distribution, indicating that the training sets successfully captured the process variability observed in the validation set.

Table 7 reports the average RMS error and standard deviations for each VPPM trained using the full set of 21 features. In relative terms, the RMS errors ranged between 7.1% and 13.2% of their corresponding observed property ranges. It is worthwhile to compare the VPPM errors to the RMS error produced by naïvely predicting the average of the ground truth property values observed across the entire dataset. Importantly, this comparison demonstrates error reductions between 30% and 48% over the naïve approach, which assumes average tensile properties throughout an entire L-PBF part, with the largest improvement reported for the UTS predictions. Next, the total RMS error can be considered as the summation of the intrinsic measurement error (i.e., aleatoric uncertainty) $RMSE_I$ of the tensile test and the model error $RMSE_{M-VPPM}$ as shown in Equation (4), where the intrinsic measurement error cannot be predicted by any model and is estimated by calculating the standard deviation (equivalent to an RMS error in this situation) for the subset of BEST SS-J3 samples extracted from the non-edge regions of the SSJ3-D part produced in B1.2 (see Section 2.9 for justification). This intrinsic error is also a component of the naïve RMS error ($RMSE_{\text{naïve}}$) as shown in Equation (5) and can, therefore, be separated from both the VPPM and the naïve predictions. After separating the intrinsic error, the relative reductions in the errors improve to 57%, 61%, 49%, and 46% for YS, UTS, UE, and TE, respectively. Because this final metric considers both the distribution of the data due to the process variation across the entire sample set and the spread associated with intrinsic variations in the tensile testing procedure, the authors propose the use of this or similar metrics when comparing these VPPM prediction results with those reported for similar works in the future. Although this is beyond the scope of this work, a more rigorous analysis of the summation of the error terms [81] may further improve the utility of such performance metrics.

$$RMSE_{VPPM} = RMSE_I + RMSE_{M-VPPM} \qquad (4)$$

$$RMSE_{\text{naïve}} = RMSE_I + RMSE_{M-\text{naïve}} \qquad (5)$$

Table 7. The mean and standard deviations of the 5-fold validation RMS errors for the full-featured VPPM compared naïve predictions and the estimated intrinsic tensile measurement error. All table entries are equivalent to RMS errors.

	YS (MPa)	UTS (MPa)	UE (%)	TE (%)
Full-featured VPPM	24.7 ± 1.0	38.3 ± 0.9	9.0 ± 0.3	11.9 ± 0.1
Naïve predictions	35.4 ± 1.4	74.2 ± 1.8	16.1 ± 0.4	19.7 ± 0.2
Reduction from naïve	10.7	38.9	7.1	7.8
Measurement error	16.6	15.6	1.7	2.9

To infer the relative importance of the different features, additional VPPMs were trained using three subsets of the available features: the first set was composed of only the CAD (Table A2) and scan path information (Table A5), the second set was composed of only data from the printer log file (Table A4), and the third set was composed of only the DSCNN segmentation results (Table A3). Comparing the results presented in Table 8 to those in Table 7, it is apparent that the VPPMs with access to only the CAD information, scan path information, and printer log file data have predictive powers comparable to the naïve approach (Table 7). However, including these features in the full-featured VPPM results in a minor reduction in the RMS errors relative to the VPPMs trained with the DSCNN segmentations alone. To estimate the sensitivity of the model performance to the size of the training set, an ablation study was performed in which the size of the training set was artificially reduced to 20% of the available data. As shown in Table 8, the RMS errors remain essentially unchanged, suggesting that significantly fewer tensile tests could be performed without negatively affecting the predictive capabilities. Reducing the size of the training set below 20% was not explored in this work and will require careful consideration to ensure that a representative sampling of the expected process variations is maintained. Additional validation metrics may also be required to properly measure any increases in the prediction error for relatively rare events and process conditions.

Table 8. Mean and standard deviations of the 5-fold validation RMS errors for each trained VPPM under each set of ablated training conditions. The first row of Table 7, indicating the final VPPM performance, is duplicated here for ease of reference.

	Fraction	YS (MPa)	UTS (MPa)	UE (%)	TE (%)
Full-featured VPPM	1.0	24.7 ± 1.0	38.3 ± 0.9	9.0 ± 0.3	11.9 ± 0.1
CAD and scan path	1.0	35.4 ± 1.5	76.3 ± 2.1	16.9 ± 0.5	20.1 ± 0.2
Printer log file	1.0	34.4 ± 1.5	73.5 ± 1.9	16.8 ± 0.5	19.8 ± 0.2
DSCNN classifications	1.0	25.7 ± 1.0	40.6 ± 0.8	9.0 ± 0.3	12.2 ± 0.2
Full-featured VPPM	0.2	24.7 ± 1.0	38.4 ± 0.9	9.1 ± 0.4	12.2 ± 0.2

The following results use VPPMs that were trained using the full feature set from the same representative fold iteration. Figure 16 plots curves similar in function to receiver operating characteristics (ROC) curves [82] for each of the four VPPMs. The y-axis reports the percentage of validation samples with RMS errors less than the error threshold, given as a percentage of the observed validation range of the corresponding tensile value, on the x-axis. For example, 81% of the validation samples have UTS RMS errors less than 8.0% of the observed validation range of 470 MPa. As suggested by the metrics above, the UTS VPPM demonstrates the strongest predictive ability, whereas the UE and TE VPPMs demonstrate the weakest performance. Interestingly, although the UE VPPM slightly outperforms the TE VPPM at lower error thresholds, its relatively longer tail at higher error thresholds suggests that a small number of samples have measured UE values that are particularly difficult for the trained VPPM to predict accurately.

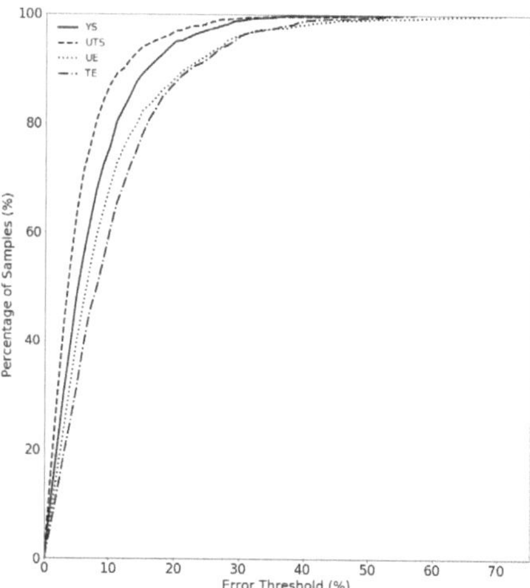

Figure 16. Validation ROC-like curves for each of the four VPPMs, as indicated in the legend. The y-axis reports the percentage of validation samples with RMS errors less than the error threshold given on the x-axis. The RMS errors are reported as a percentage of the observed range of the corresponding tensile value within the validation set.

Figure 17 shows correlation plots for each of the four VPPMs for a selected validation fold. In these plots, the *x*-axis reports the ground truth tensile measurement, and the *y*-axis reports the predicted tensile property value. This representation of the validation accuracy can be considered a 2D histogram, with the colormap representing the number of SS-J3 samples present in each bin. If all the VPPM predictions were correct, then only the bins along the diagonal line (with a slope of unity) would be brightly colored. Note that the ground truth YS and UTS measurements are primarily bimodal, whereas the UE and TE measurements exhibit a substantially more uniform distribution of values spread across the observed range. At the time of writing, the cause of this difference in distribution behavior has not been determined.

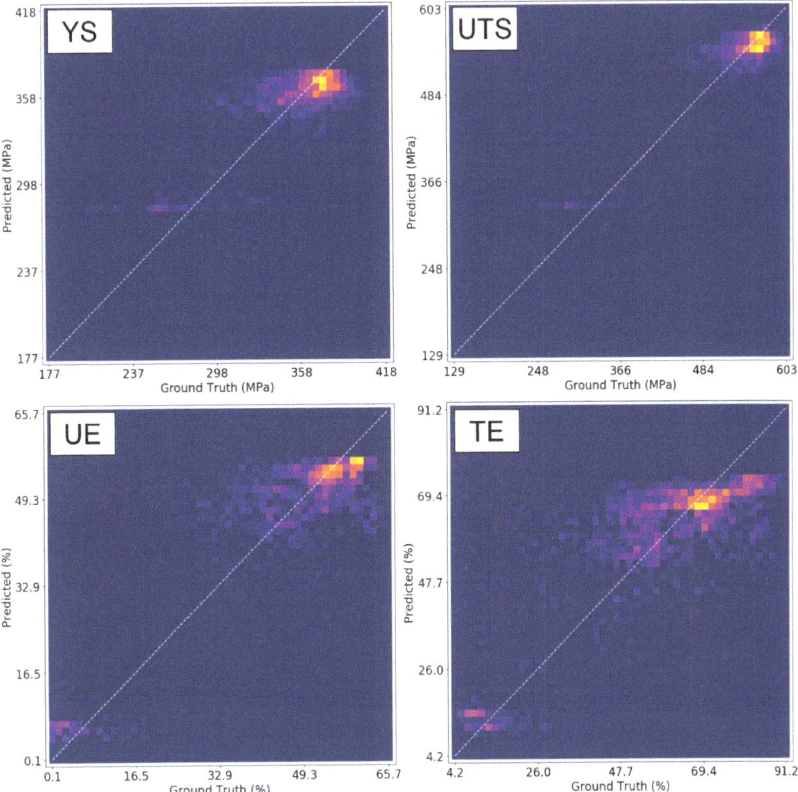

Figure 17. Normalized correlation plots showing VPPM predictions for YS, UTS, UE, and TE vs. the corresponding measured ground truth values for a selected validation fold. The color map indicates the number of SS-J3 samples present in each bin, with more brightly colored bins containing more samples and the darkest bins containing no samples. If each VPPM performed perfectly, then all the datapoints would lie on the diagonal dashed line with a slope of unity.

To focus on the outlying predictions, Figure 18 shows a scatter plot of predicted UTS versus the ground truth values for a selected validation fold. Each datapoint represents a single SS-J3 gauge section, and some are labeled with the automatically generated unique identifier matching the sample's physical QR code and tracked by the digital platform. If all the UTS VPPM predictions were correct, then all the datapoints would lie on the diagonal line with a slope of unity.

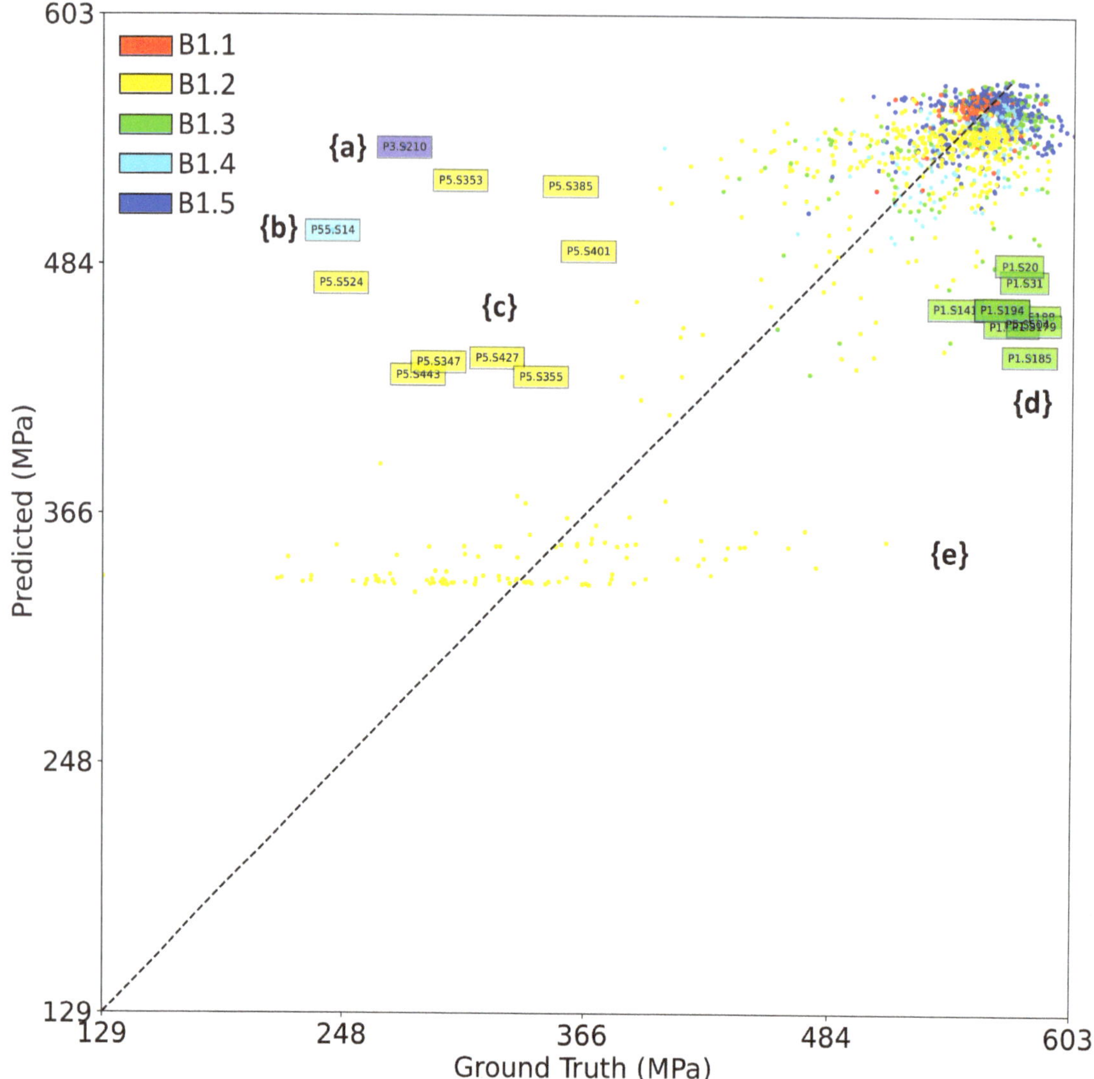

Figure 18. Scatter plot showing the UTS VPPM predictions vs. the corresponding measured ground truth values for a selected validation fold. Each datapoint represents a single SS-J3 sample tracked by a unique identifier using the digital platform. The datapoints are colored by the source build as identified in the legend. If the UTS VPPM performed perfectly, then all the datapoints would lie on the diagonal dashed line with a slope of unity. Selected subsets of the outlying datapoints are marked by letters enclosed by curly brackets and are discussed in the text.

Outlier set {a} represents a single tensile sample, designated as P3.S210, from build B1.5 that is predicted to have a UTS of 540 MPa but has a measured UTS of 280 MPa. Sample P3.S210 was extracted from the bulk region of an SSJ3-D part printed with nominal process parameters. As shown in Figure 19, P3.S210 is surrounded by several other samples with UTS values that are substantially lower than the rest of the printed part. Samples in this group were not tested sequentially, so it is likely that this outlier represents real variation in the printed material (as opposed to a tensile testing artifact) not properly modeled by the VPPM. Future investigation of the fracture surfaces may indicate the root cause of this behavior.

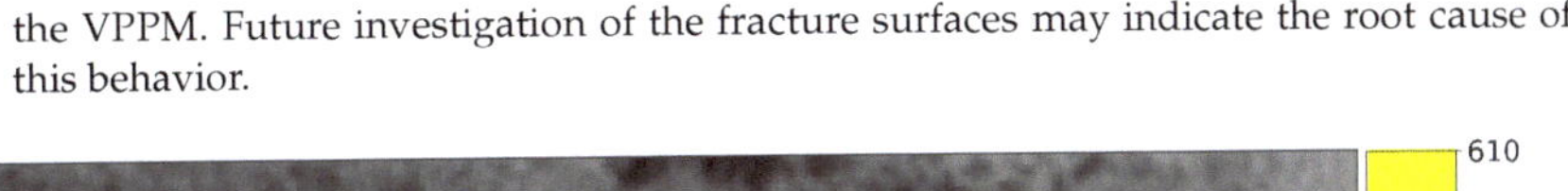

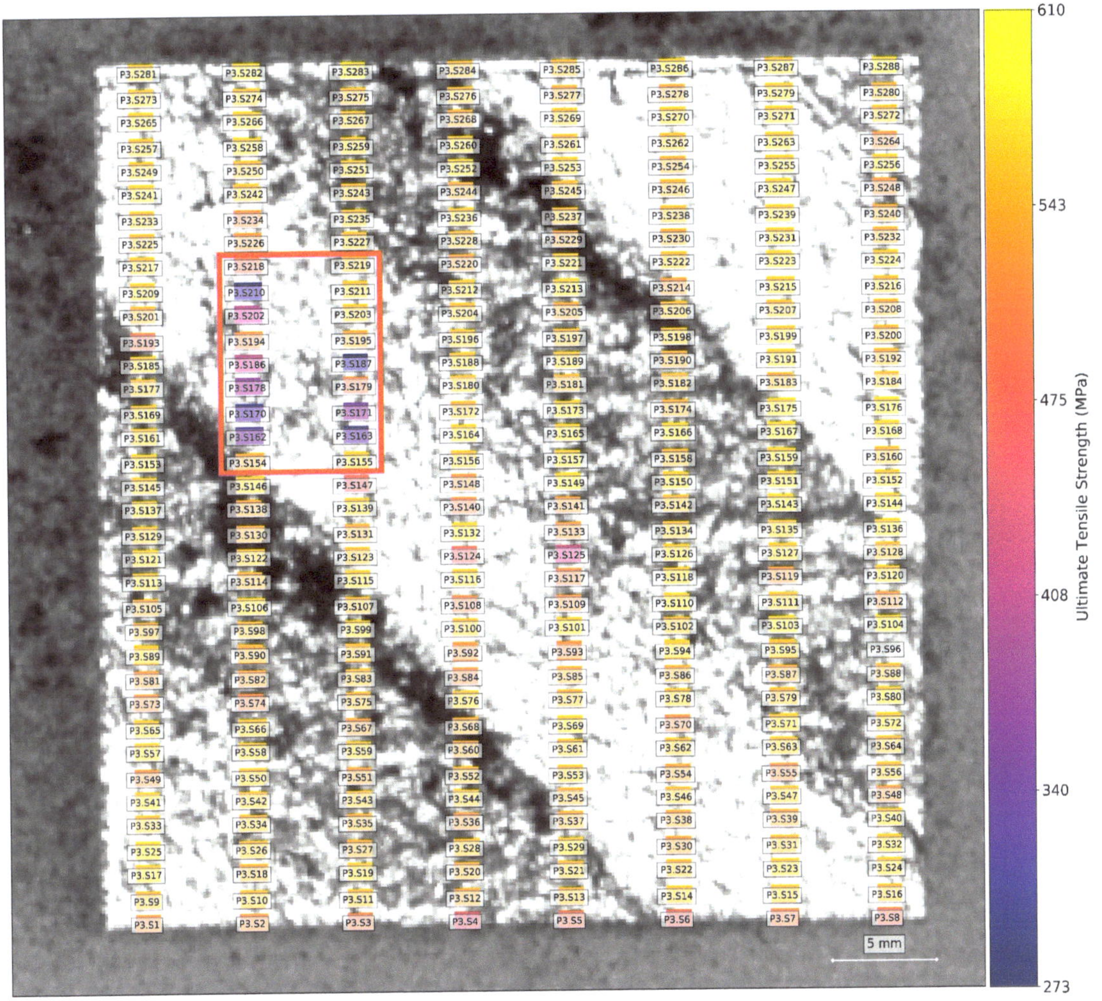

Figure 19. Measured UTS values for 288 uniquely identified SS-J3 tensile specimens extracted from an SSJ3-D part printed in B1.5, spatially registered to the corresponding CAD coordinate system. For reference, test results are overlaid on top of the visible-light post-melt image captured for the layer located in the center of their gauge sections. Within the color space, brighter samples represent a higher measured UTS, and darker samples represent a lower measured UTS. Note the cluster of relatively low UTS measurements highlighted by the red bounding box.

Outlier set {b} also represents a single tensile sample (P55.S14 from a nominal SSJ3-B part printed in B1.4) that is predicted to have a UTS of 500 MPa but has a measured UTS of 240 MPa. In contrast to set {a}, the samples surrounding P55.S14 were all measured to have much higher UTS values, including sample P55.S17, which is located at the same distance from its closest soot generator and is expected to have a similar thermal history. Additional investigation (e.g., microstructural characterization), which is beyond the scope of this paper, would be necessary to determine if this measurement represents a true variation in the material properties or if it is an artifact of the tensile testing procedure.

Outlier set {c} represents a group of samples extracted from an SSJ3-D part (designated P5) produced using keyhole parameters in B1.2, which were measured to have UTS values ranging from approximately 250 to 450 MPa. Although the VPPM successfully predicts some of this variation, it is not fully captured by the model. The authors hypothesize that the relatively large vertical dimension of the SS-J3 gauge sections is detrimental to the predictive performance of the VPPMs for material with significant populations of keyhole porosity because it is difficult to correlate the volumetric tensile measurements with individual in situ sensor indications of keyholing.

Next, outlier set {d} represents a group of samples extracted from an SSJ3-C part (designated P1) printed using best parameters in B1.3, which were measured to have UTS values of approximately 570 MPa but were predicted to have a UTS as low as 440 MPa. A manual review of the sensor data for P1 revealed that significantly more soot was generated in the vicinity of this part than is typical for parts produced using the best parameters or for the other parts printed in B1.3. This soot was correctly segmented by the DSCNN, and the increased soot levels were encoded into the corresponding super-voxel feature vectors. Unlike other cases within this dataset (e.g., the LOF parts), these increased soot levels did not correlate with a lower measured UTS, resulting in the lower-than-correct VPPM predictions.

Finally, outlier set {e} encompasses samples extracted from the LOF parts produced in B1.2. These samples exhibited extremely low UTS values (Figure 11) because of the significant levels of lack-of-fusion porosity present within the parts. Significant variation in UTS is also observed, likely the result of the variable number of pores that intersect with a given SS-J3 gauge section. After a manual review of the sensor data, the authors consider it unlikely that this variation could be more accurately modeled, given the available in situ sensor data. Fortunately, the VPPM still correctly recognizes that these samples have much lower strengths than nominal L-PBF SS 316L material, which would be a sufficient threshold for many part qualification scenarios.

3.2. Testing Performance

Extraction of the CAD geometries, spatial registration of the visible-light camera data, and anomaly segmentation by the DSCNN were performed in real-time on a network-attached server during the printing operation, and they collectively required several seconds of processing time per print layer, as discussed in Section 2.4. The time required to calculate the feature vectors for all the super-voxels within a build depended upon the volume of printed material and ranged from 12 to 19 h for the five builds reported in this work and was performed post-build. Once the feature vectors were generated, the prediction of the local tensile properties using the trained VPPMs was trivial, requiring less than a minute for each build. Given the early-stop criteria, VPPM training required less than five minutes for each fold. Therefore, the most computationally expensive portion of this AIR was the generation of the super-voxel feature vectors. Currently, super-voxel generation is dominated by the time required to retrieve the scan path information from the TDMS files and spatially map the corresponding temporal data into the common spatial coordinate system [51]. The authors expect that additional computational optimizations can be applied to reduce this time burden significantly.

Figure 20 shows the local UTS predictions for a set of layers from B1.2 alongside the UTS values measured from the SS-J3 samples extracted from the same vertical region of that build. Most noticeably, the models correctly predict the low strength of the LOF parts in the back-left quadrant, although the exact values are nearly a uniform average of the observed variation for these parts, as discussed in Section 3.1. Importantly, the property variation measured for the keyhole SSJ3-D part in the front-left quadrant is also captured, with structures appearing qualitatively analogous to the DSCNN segmentations of excessive melting shown in Figure 15. Furthermore, despite the measured UTS differences being relatively small, the models also correctly predicted slightly lower strengths for the thin-walled SSJ3-A parts, as well as other parts with similarly thin cross-sections. Interestingly, within the nominal SSJ3-D part in the back-right quadrant, regions of slightly lower strength were predicted that qualitatively overlap with several of the structures observed in the scan path feature maps (Figure 15); additional investigation would be required to determine the veracity of this predicted shift in UTS distributions.

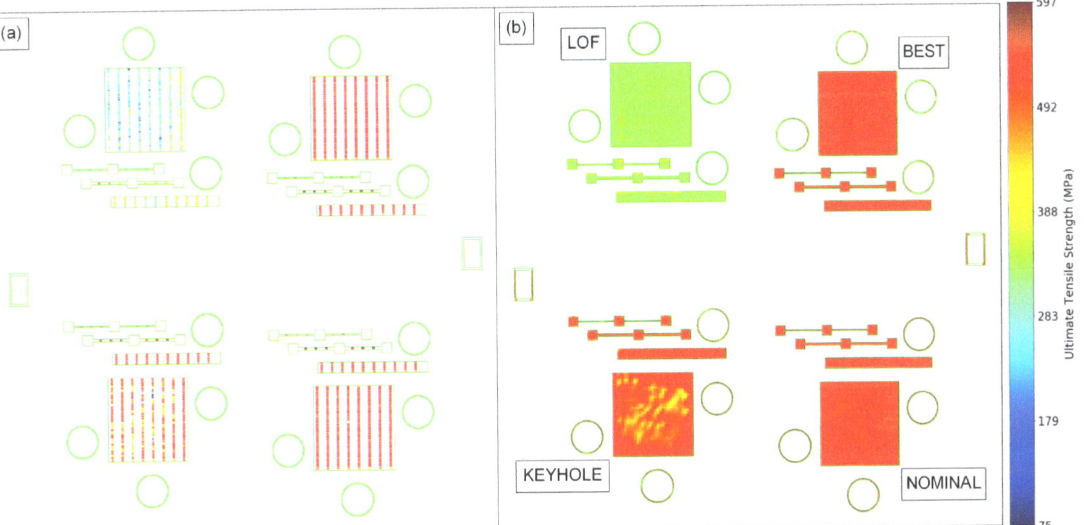

Figure 20. (**a**) Measured UTS values from the SS-J3 samples located between layer 639 (31.95 mm) and layer 709 (35.45 mm) of B1.2. (**b**) VPPM-predicted local UTS values for the corresponding layers as a color map, with white super-voxels indicating regions beyond the part geometries or detected instances of out-of-distribution super-voxels. The nominal CAD geometry is indicated by the green outlines, and both images show the full 245 × 245 mm print area.

As a final example, Figure 21 shows the local UTS predictions within a region of interest for a set of layers from B1.3 alongside the UTS values measured from the corresponding SS-J3 samples. The models correctly predict that the UTS was, on average, substantially reduced in the thin-walled SSJ3-A parts and for super-voxels immediately adjacent to the overhanging surface of the thicker SSJ3-C parts (Figure 13). Of course, with access to entire volumes of localized tensile property predictions, a wealth of additional visualizations and analyses are possible. However, the authors reserve a more detailed exploration of volumetric property predictions for future work.

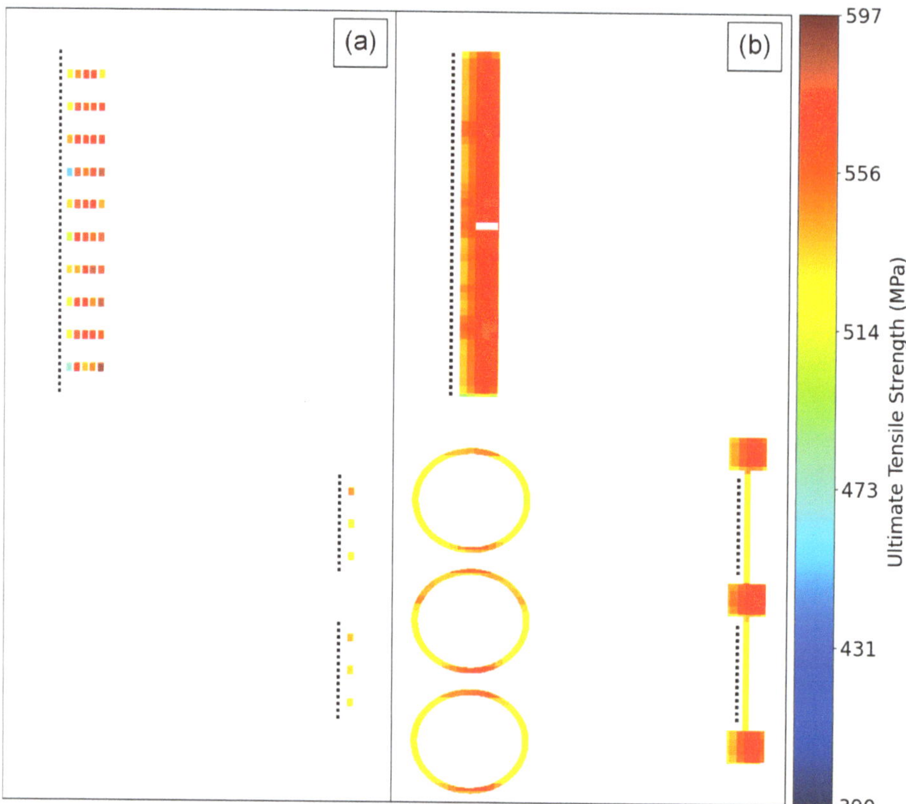

Figure 21. (**a**) Measured UTS values from the SS-J3 samples located within a region of interest and between layer 1136 (56.80 mm) and layer 1206 (60.30 mm) of B1.3. The region of interest lies within the back-right quadrant of the powder bed, cropped so that the right side of the image is 122.5 mm in length. (**b**) VPPM-predicted local UTS values for the corresponding layers as a color map, with white super-voxels indicating regions beyond the part geometries or detected instances of out-of-distribution super-voxels. Overhanging surfaces of interest are indicated by dashed vertical lines.

4. Discussion

In this study, the authors demonstrated several key components of a novel L-PBF part qualification framework. The results of the tensile tests reported in Section 2.9 make apparent the interdependencies between local part geometry, melting parameters, and process conditions, which challenge efforts to create traditional manufacturing design rules for L-PBF AM. Fortunately, as shown in Sections 3.1 and 3.2, the relay of machine-learned models (i.e., the AIR) achieved direct localized prediction of tensile properties for L-PBF printed parts based on data collected in situ during the printing process. The presented framework is designed to be geometry- and process-parameter-agnostic and, therefore, substantially generalizable. Importantly, each step in the proposed property prediction pipeline is modular, allowing the iterative improvement of each component to address specific limitations. During this study, the authors identified four areas that should be addressed before similar AIR-based pipelines are applied to instance-wise part qualification.

First, the resolution of the visible-light camera is currently limited, with a resolving power of only 280 μm and many features that might indicate sub-surface pores or other flaws are not currently observable. Other work has demonstrated the successful use of substantially higher resolution visible-light cameras for this purpose, such as the analysis by Snow et al. [83], which used a 36.3 MP sensor to image a similarly sized powder bed. Similarly, multiple lighting conditions can be used to cast shadows across the powder bed and increase the contrast of topological features associated with various processing anomalies [83]. The incorporation of temporally integrated near-infrared (NIR) camera data, as explored by Schwerz [23], may allow the DSCNN to segment additional anomaly classes, such as individual spatter particles interfering with the melt tracks [84]. Indeed, the performance of the current models is impressive, given that a low-cost, OEM-standard sensor suite was used. The authors expect that significant performance improvements can be achieved with improved sensor modalities.

Second, precise localization of the tensile test results remains a challenge. Even using the subsize SS-J3 specimens, the measured tensile properties apply to a relatively large volume of printed material compared to the size of stochastic flaws such as keyhole porosity. A potential solution is to use digital image correlation (DIC) [85] to further localize the failure location within the specimen's gauge section. Additionally, a subset of the SS-J3 specimens could be rotated during printing so that their long axis is perpendicular to the vertical build direction. This approach could enable better correlation between tensile measurements and certain laser scan path features and vertically isolated flaws such as powder short feeds. Further shrinking the gauge section is also possible but will have to be considered carefully to avoid introducing additional mechanical testing artifacts. Relatedly, intrinsic tensile measurement errors could be reduced by using DIC to observe the true strain (as opposed to using crosshead displacement as a proxy), measuring the initial gauge section dimensions using an optical silhouette or laser profilometer technique, and using a robotic arm to insert the specimens into the tensile load frames repeatably.

Third, based on the results of this work, each of the DL, ML, and feature encoding steps can now be improved. Currently, all the DSCNN annotations are generated manually, but some flaw indications observable in the NIR [23] sensor modality could be annotated automatically using XCT as the ground truth [83]. Furthermore, in this work, local thermal histories are only represented in the super-voxel feature vectors via proxy, but the use of contemporary analytical thermal models could instead directly encode the thermal histories, as shown by Donegan et al. [46] and Stump and Plotkowski [86]. Similarly, the current feature vectors do not explicitly discriminate between super-voxels with machined vs. as-printed surfaces. This limitation was most apparent in the predictions for the SSJ3-A and SSJ3-B part types, both of which had similar wall thicknesses but different surface conditions. Indeed, this situation demonstrates the need for the digital thread to include information beyond just the printing operation itself (i.e., post-processing steps). Finally, the VPPM perceptrons were trained using a canonical L_1 loss function, which may bias the model to learn average responses at the expense of predicting rare events. Applying a loss weighting scheme, such as that described for the DSCNN in Section 2.3, or implementing a different ML model type altogether, may improve the ability to model the effects of rare processing conditions. Notably, the existence of several outlying predictions suggests that the VPPM is not overfitting to specific fluctuations in the data. For example, outlier {a} is representative of several co-located SS-J3 specimens (Figure 19) with unusually low measured UTS values but in situ sensor signatures highly similar those of specimens with much higher UTS values. The VPPM incorrectly predicts the UTS for all nine of these outlying specimens, including those in the training set. If the VPPM were overfitting, it is expected that it would instead correctly predict outliers in its training set, despite the lack of physically significant variation in their corresponding input features. Therefore, the depth of this neural network could potentially be increased to improve performance while still producing a generalizable model.

Lastly, if any data-driven L-PBF qualification framework is to be adopted by an industrial user base, it cannot be prohibitively expensive to implement at scale. Therefore, the generalizability of the proposed framework to different L-PBF printers, feedstock materials, and material properties is a critical consideration. For discussion, the authors have grouped potential model transfer situations into three categories of increasing difficulty. In the first category, the authors hypothesize that a relay of pre-trained models could be directly transferred to other similar L-PBF printers using the same alloy, assuming that the in situ sensor suite is held constant. In the next category, while the proposed data analysis framework could be applied to a different alloy or material property, this approach will certainly require replication of a significant testing campaign and a retraining of the machine-learned models. Fortunately, the ablation results presented in Section 3.1 suggest that the number of mechanical tests could be reduced by at least 80% without significantly degrading the predictive capabilities. However, additional research is necessary to determine the minimum number of specimens required to fully represent the possible process conditions within the training set, and alternate measures of VPPM performance may be needed to ensure that the error rate for predicting rare but safety-critical conditions is accurately quantified. It is also apparent that some material properties (e.g., fatigue life) are substantially more expensive to measure at scale than static tensile properties, so the bypass between N4 and N7 (Figure 1) utilizing analytical modeling may be the more industrially scalable option for such properties. Alternatively, recent federated learning strategies such as those proposed by Mehta and Shao [30] offer a potential solution to this problem whereby expensive specimen characterization could be performed across multiple locations but used to jointly train common models without compromising the intellectual property of any of the participants. In the final category, it must be recognized that the difficulty of observing certain anomalies in situ will vary across alloy systems, just as the localizability of the ex situ measurements will depend on the material property of interest. In these situations, alternate data-driven qualification frameworks should be explored.

5. Conclusions

In this work, the authors implemented a relay of machine-learned models to further digital twin-informed, instance-wise qualification of parts printed using L-PBF processes. The presented modular architecture was designed to be flexible, providing a conceptual framework for future AM part qualification efforts. To support these efforts, the authors printed and tested over six thousand tensile specimens, each of which was uniquely tracked and spatially registered to the in situ sensor data using a cyberphysical digital platform. The co-registered in situ data and tensile test results are approximately 230 GB in size and are available in the Peregrine v2023-11 dataset [44]. Layer-wise visible-light images of the powder bed were segmented using a DSCNN, the results of which were combined with other in situ data and then encoded into human-engineered feature vectors using a combination of image segmentation neural networks and classical signal processing and computer vision algorithms. Localized tensile properties were then predicted at the super-voxel scale by training a perceptron neural network with the ground truth tensile measurements.

The viability of this framework was evaluated with multiple performance metrics, demonstrating significant error reductions relative to traditional, naïve estimates of several tensile properties, with the best performance improvements observed for UTS. An ablation study indicated that the in situ layer-wise visible-light powder bed images contained the majority of the information used by the trained models to predict the tensile properties, while the log file data, laser scan paths, and CAD information enabled only moderate improvements. Several sets of outliers were explored in detail to better understand the generalizability of the learned models and to motivate future work. Finally, tensile properties were predicted for the entire volumes of all five builds, enabling the qualitative assessment of the framework's ability to capture experimentally observed trends, such as the effects of process parameters, wall thickness, and overhanging surfaces.

Three primary areas have been identified for future work. First, the authors will make iterative improvements, such as increasing the visible-light imaging sensor resolution and incorporating a temporally integrated NIR imaging system as an additional in situ sensor modality. Additionally, the use of a robotic tensile testing system and DIC will enable increased sample throughput and will provide more accurate measurements of the specimen gauge sections to reduce the overall intrinsic tensile measurement error. Changes to the specimen design and build layout will also improve the localizability of the ground truth data and will prioritize building smaller sets of samples distributed across a wider range of builds. Next, the authors will seek to quantify the generalizability of the framework by applying it to other alloys, L-PBF printers, sensor modalities, and material properties. Because this set of work will span efforts across multiple national laboratories, it may also explore techniques such as federated learning, which will be critical for the industry to adopt any artificial intelligence-based qualification frameworks relying on extensive training datasets. Finally, software tools and experiments must be developed to demonstrate the ability of the proposed approach to estimate the performance of a given instance of a part within a specific application context. Ultimately, the authors believe that this work demonstrates one of the first viable approaches for direct, localized material property prediction based on in situ sensor data collected during an L-PBF printing process.

Author Contributions: Conceptualization, C.J., L.S., Z.S. and M.S.; methodology, L.S.; software, W.H., L.S., A.S., Z.S. and M.S.; validation, L.S.; formal analysis, L.S. and M.S.; investigation, D.A.C., R.D. (Ryan Duncan) and C.J.; resources, V.P.; data curation, D.A.C., W.H., L.S., A.S. and M.S.; writing—original draft preparation, L.S.; writing—review and editing, D.A.C., W.H., L.S. and Z.S.; visualization, L.S. and Z.S.; supervision, R.D. (Ryan Dehoff), V.P. and L.S.; project administration, L.S.; funding acquisition, R.D. (Ryan Dehoff) and V.P. All authors have read and agreed to the published version of the manuscript.

Funding: This research was sponsored by the Advanced Materials and Manufacturing Technologies (AMMT) and Transformational Challenge Reactor (TCR) programs, which are supported by the US Department of Energy (DOE) Office of Nuclear Energy. Funding support was also provided by DOE's Advanced Materials and Manufacturing Technologies Office (AMMTO).

Institutional Review Board Statement: Not applicable.

Informed Consent Statement: Not applicable.

Data Availability Statement: The in situ powder imaging data, machine health sensor data, laser scan paths, tensile test results, and all associated metadata used in this work can be downloaded from https://doi.ccs.ornl.gov/ui/doi/452 (released on 28 September 2023). Representative training data for the DSCNN can be downloaded from https://doi.ccs.ornl.gov/ui/doi/341 (released on 23 April 2021) and https://doi.ccs.ornl.gov/ui/doi/417 (released on 13 February 2023).

Acknowledgments: The authors would like to thank the staff at the MDF, particularly Keith Carver, Fred List III, and Joseph Simpson, for their assistance with designing the build layouts and experimental conditions. Alex Huning from the Nuclear Energy and Fuel Cycle Division at ORNL provided valuable guidance on the part qualification process for nuclear applications. Kevin Hanson of the Materials Science and Technology Division at ORNL performed the heat treatments. Annabelle Le Coq and Kory Linton from the Nuclear Energy and Fuel Cycle Division and T.S. Byun from the Materials Science and Technology Division at ORNL assisted with the selection of the tensile specimen design. Meimei Li from the Nuclear Science and Engineering Division at Argonne National Laboratory assisted with the development of the heat treatment procedure. Technical reviews of this document were provided by Alex Plotkowski and James Haley from ORNL's Materials Science and Technology Division and ORNL's Electrification and Energy Infrastructure Division, respectively.

Conflicts of Interest: Luke Scime, Vincent Paquit, William Halsey, and Chase Joslin are authors of patent #11,458,542 relating to the DSCNN. Some of the software tools used in this work are licensed by ORNL to industry partners, and several of the authors receive royalties from these licensing agreements. The funders had no role in the design of the study, in the collection, analyses, or interpretation of data, in the writing of the manuscript, or in the decision to publish the results.

Appendix A

Because the resolution of the in situ sensor data is on the order of 100 μm, as are many of the void-type flaws [87] and surface features [65] expected to impact the tensile properties, a relatively small gauge section was expected to facilitate correlation of the ground truth measurements to in situ sensor signatures. Additionally, smaller specimen sizes allowed samples to be extracted from a more diverse set of part geometries, particularly thin-walled structures with thicknesses relevant to the nuclear power industry. However, inherent uncertainties in the spatial registration of the gauge sections to the in situ data, as well as physical challenges associated with performing tensile tests on extremely small specimens, ultimately limit the minimum size of the gauge section. Given these considerations, a standard sub-size specimen was selected for mechanical testing. Existing literature indicates that tensile properties measured using sub-size specimens may have higher variances and may not be fully indicative of bulk properties for L-PBF components, generally providing a conservative estimate of the bulk tensile properties [65]. Therefore, additional corrections may be required when applying the local property predictions at N7 to a full part; this topic is addressed in the Discussion section but is broadly beyond the scope of this work. Finally, it was advantageous for parallel research efforts that the tensile specimens maintain a form factor that is compatible with the standard canisters used for irradiation studies at ORNL's High Flux Isotope Reactor (HFIR). Therefore, an SS-J3 geometry was ultimately selected, with nominal dimensions specified by ORNL's generic metal irradiation specimen standard [69].

Appendix B

Build B1.1, presented in Figure A1, shows the volume percentage of each part that was classified as printed material by the DSCNN. While all parts were printed with the nominal process parameters, the DSCNN classifications varied substantially across the print bed and the part geometries. One objective of this work is to determine the correlation between the sensor variation observed in a nominally processed build and the measured static tensile properties.

Each SSJ3-D part in build B1.2 was printed with a different parameter set. Figure A2 visualizes the layer-wise distributions of the printed, edge swelling, debris, and excessive melting DSCNN classes throughout the volumes of these four SSJ3-D parts. For the histogram of each part, the frequency of a class within a print layer was calculated as a percentage of the cross-sectional area of the CAD geometry at that layer. These results were then grouped such that layers with similar class frequencies were placed in the same bin. Parts with similar in situ sensor (visible-light camera) signatures have overlapping distributions, while parts with different signatures show differentiation between the distributions for one or more DSCNN classes. The four process parameter sets used to manufacture each of these SSJ3-D parts resulted in significant shifts in the anomalies segmented by the DSCNN. As expected, the LOF and keyhole SSJ3-D parts show the highest percentages of the debris and excessive melting classes, respectively. The SSJ3-D part printed with the best parameters has the highest percentage of printed material, whereas the edge swelling distributions are similar for the nominal, best, and keyhole SSJ3-D parts.

Figure A3 shows a post-build picture of several overhanging B1.3 parts with noticeable differences in surface oxidation, a plot of DSCNN soot classifications, and a plot of several shield gas metrics as functions of build height. Note that the seven instances of increased soot segmentation correlate with the seven distinct discoloration bands in the as-built parts. The authors hypothesize that as the oxygen concentration within the build chamber increased, oxidation of the spatter particles increased [88]. The darker spatter particles were easier for the DSCNN to segment as soot because of the increased contrast within the layer-wise images. When the oxygen concentration exceeded a threshold of approximately 0.18%, the printer automatically increased the argon gas flow rate across the top of the build chamber from 90 m^3/h to 93 m^3/h until the oxygen level returned to below 0.04%. At this point, the gas flow returned to its nominal setpoint, and the oxygen level was again allowed to climb, a pattern that resulted in the observed periodic behavior.

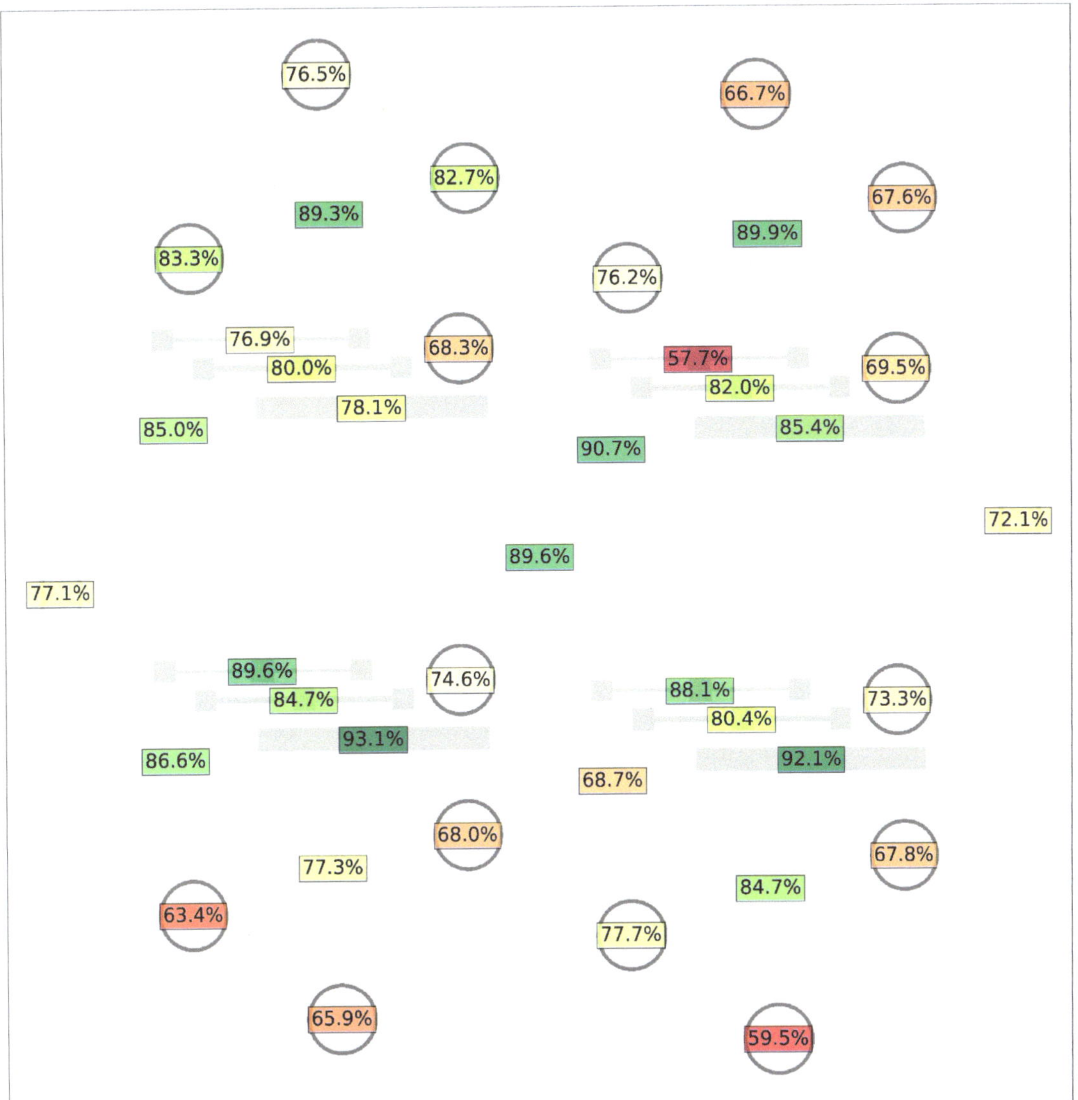

Figure A1. The volume percentage of each part from B1.1 that is classified as printed material by the DSCNN. Parts with lower percentages of printed material are shaded red, and parts with higher percentages are shaded green. Significant variation in this metric is present despite uniform process parameters.

Figure A4 shows a heat map of DSCNN soot segmentations projected through the height of B1.4 onto the *x-y* plane, with brighter regions indicating higher percentages of soot. This visualization conveys the correlation between the right-to-left gas flow and the deposition of spatter particles across the print area. It is also apparent that, as intended, parts printed with the soot parameters (indicated by white asterisks) deposited significant amounts of spatter on top of the SSJ3-A, SSJ3-B, and SSJ3-C parts.

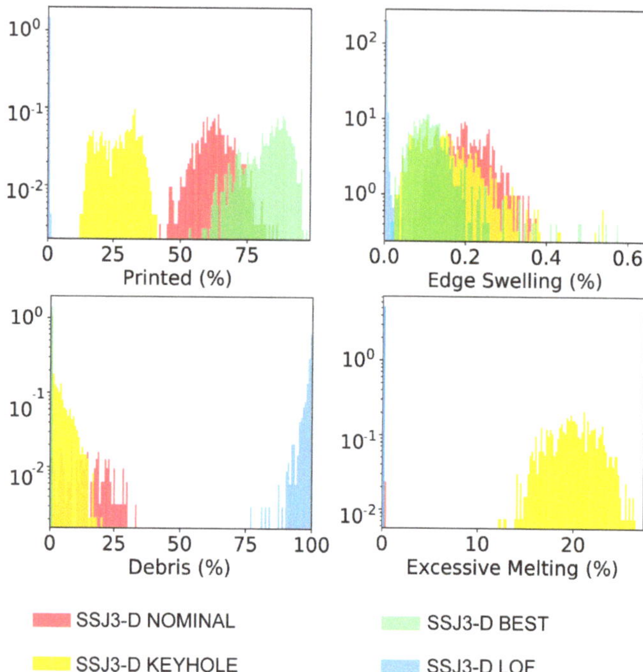

Figure A2. Histograms showing the distribution of four DSCNN classes throughout the volumes of the four SSJ3-D parts printed in B1.2. The y-axes are a log scaling of the bin density. These histograms represent data from layer 25 (1.25 mm) through layer 819 (40.95 mm). The four process parameter sets result in significant shifts in several of the distributions. For reference, one layer of the DSCNN results for this build is shown in Figure 5.

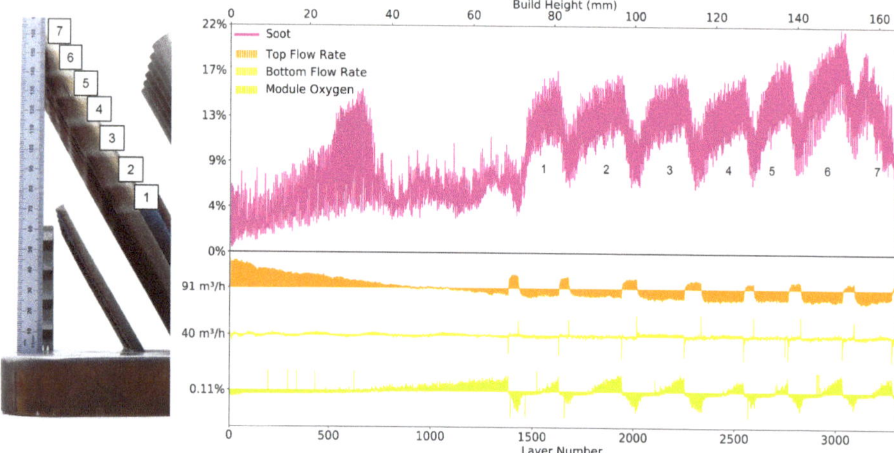

Figure A3. On the left, a picture of B1.3 taken after it was removed from the build chamber. Note the seven distinct numbered regions of increased oxidation on several of the parts. On the right, DSCNN soot segmentations (upper) and gas flow metrics (lower) are plotted as functions of layer height. Soot percentages are normalized by the cross-sectional area of the CAD geometry at each layer. Values from the log file are presented as deviations from their means, which are listed along the y-axis.

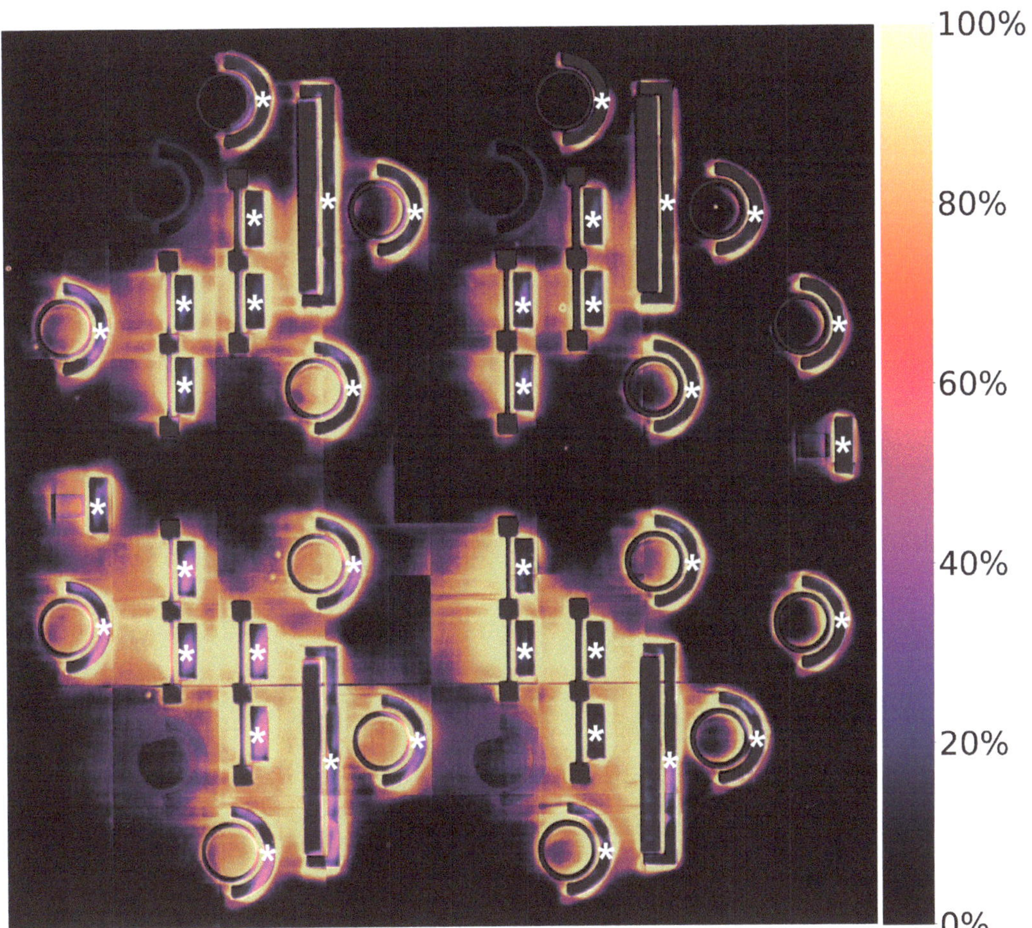

Figure A4. Projection of DSCNN soot segmentations from layer 0 through layer 1599 (79.95 mm) of build B1.4. The brightest regions indicate that soot was identified in 100% of the layers, and the darkest regions indicate that no soot was identified in that location for any of the layers. The shielding gas flows from right to left. White asterisks indicate the parts printed with the SOOT parameters; the remaining parts were printed with nominal parameters, as indicated in Figure 8. Tiling artifacts [28] and minor spotting on the camera's viewport are also apparent in this visualization.

Figure A5 shows a 3D rendering of build B1.5, with DSCNN segmentations of super-elevation highlighted in red. The two artificially induced powder short feeds were detected as multiple consecutive layers of super-elevation by the DSCNN. Considering powder consolidation effects and typical powder packing densities [89], a typical powder layer is expected to be approximately 70 μm thick, whereas these anomalous layers are expected to result in powder layers up to 170 μm thick, depending on the location within the print area.

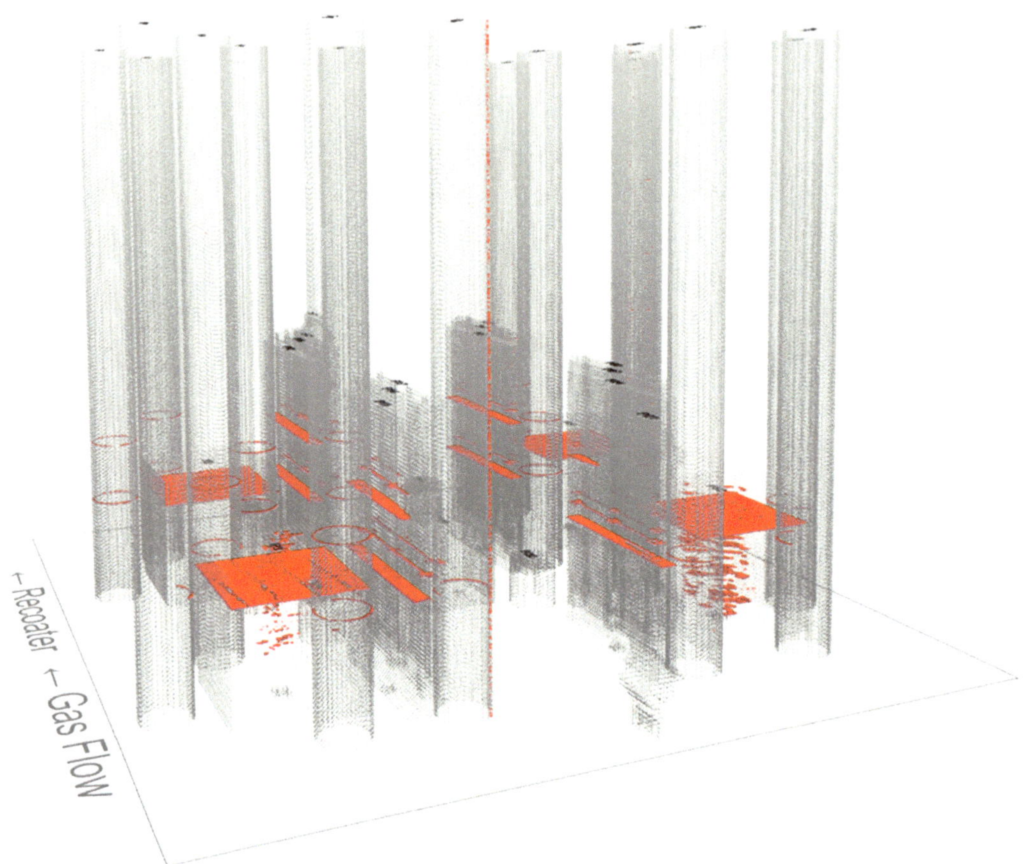

Figure A5. 3D rendering of B1.5, with part CAD geometries shown in gray and DSCNN super-elevation segmentations shown in red. Super-elevation is primarily observed at the two build heights (33.05 and 52.00 mm), where artificial powder short feeds were induced. DSCNN segmentations of super-elevation are highlighted in red.

Appendix C

Because of a focus on nuclear power applications, the maximum soak temperature was specified to retain the excellent as-built creep properties of the L-PBF-processed SS 316L material, as identified by Li et al. [90]. Soak times ranging from 0.5 to 24 h were then tested to minimize the total geometric distortion. The final furnace profile is specified in Table A1. To ensure that the larger SSJ3 parts reached the target soak temperature for the specified duration, a thermocouple was inserted into a 1-inch printed cube located in the middle of each build plate. All heat treatments were performed in ambient air.

Table A1. The post-build heat treatment profile applied to all five builds. The profile was designed to reduce part distortions due to residual thermal stresses while maintaining creep behavior appropriate for nuclear power applications.

Segment	Description
1	Ramp up at 10 °C/min to 650 ± 10 °C
2	Soak at 650 ± 10 °C for 24 ± 0.5 h
3	Furnace cool to 100 ± 20 °C

Appendix D

Table A2. Set of features calculated based on part CAD geometries and the minimum and maximum values observed for the combined training and validation sets.

	Feature	Min	Max	Justification and Description
1	Distance from edge (pixels)	1.115	27.000	Thermal histories [46] and porosity populations [58] for a volume are influenced by its distance from the part edge. Additionally, the surface roughness [9] of the SS-J3 samples is highly dependent upon whether it was extracted from the surface or the bulk of its parent part. To represent the planar distance of a super-voxel from a part edge, a distance transform [62] is applied to the CAD geometry with values allowed to saturate above 3.0 mm. A Gaussian blur with a 1.0 mm kernel and a standard deviation of 0.5 mm is then applied to mitigate computational edge effects, and a planar maximum is then performed within each super-voxel.
2	Distance from overhang (layers)	26.305	71.000	Thermal histories [91,92] for a volume are influenced by its distance above an overhanging surface. This distance is calculated for a vertical column of pixels with values allowed to saturate above 71 layers. A Gaussian blur with a 1.0 mm kernel and a standard deviation of 0.5 mm is then applied to mitigate computational edge effects.
3	Build height (mm)	8.875	72.775	Thermal histories for a volume can be influenced by its vertical distance away from the build plate [93]. This feature is calculated using the nominal print layer thickness and the known layer number.

Table A3. Set of features calculated based on DSCNN classifications of post-melt and post-spreading visible-light image data and the minimum and maximum values observed for the combined training and validation sets.

	Feature	Min	Max	Justification and Description
4	Powder	0.000	0.632	Descriptions of DSCNN classes and their corresponding
5	Printed	0.000	1.000	physical mechanisms are provided in Section 2.3. Each feature is
6	Recoater streaking	0.000	0.984	calculated by applying a Gaussian blur with a 1.0 mm kernel
7	Edge swelling	0.000	0.393	and a standard deviation of 0.5 mm to the binary mask of each
8	Debris	0.000	1.000	class. The resultant values represent the distance-weighted
9	Super-elevation	0.000	0.033	fractions of the surrounding area belonging to each class. The
10	Soot	0.000	0.945	Gaussian kernel also facilitates the encoding of soot, which is
11	Excessive melting	0.000	0.955	most easily segmented on the powder surrounding a part.

Table A4. Set of features calculated from the sensor values recorded in the printer log file and the minimum and maximum values observed for the combined training and validation sets.

	Feature	Min	Max	Justification and Description
12	Layer print time (s)	45.3	155.9	The selected log file variables encode sensor values which may
13	Top gas flow Rate (m^3/h)	62.5	99.8	be correlated to differences in thermal history (features 12, 16,
14	Bottom gas flow rate (m^3/h)	24.9	40.1	and 17) [94], part oxidation (features 13, 14, 15, and 18) [88],
15	Module oxygen (%)	0.000	0.148	and laser attenuation (features 13, 14, and 18) [95,96]. For each
16	Build plate temperature (°C)	27	39	sensor, all the values recorded during a layer are averaged
17	Bottom flow temperature (°C)	41	60	together, weighted by the temporal persistence of each
18	Actual ventilator flow rate (m^3/h)	25.0	40.0	individual sensor reading.

Table A5. Set of features calculated from the laser scan path data and the minimum and maximum values observed for the combined training and validation sets.

	Feature	Min	Max	Justification and Description
19	Laser module	0.000	1.000	Despite OEM calibration, the laser beam diameter and laser power may not behave identically between the two laser modules, potentially resulting in differences in thermal history, as well as melt pool morphology and size [97]. This feature encodes the laser module used to melt a given super-voxel.

Table A5. *Cont.*

	Feature	Min	Max	Justification and Description
20	Laser return delay	0.020	0.750	The amount of time between adjacent laser passes can influence the thermal history and the melt pool morphology and size [46]. To calculate a proxy for this metric, minimum and maximum filters with 1.0 mm kernels are applied to maps of the melt time since the start of each layer, and their pixel-wise difference is then calculated. A saturation value was chosen empirically to prevent saturation (excluding the stripe boundaries) for all the scan strategies used across the five builds.
21	Laser stripe boundaries	0.018	9.940	Thermal histories, porosity populations, and melt pool morphologies and sizes may be different at the interfaces between laser stripes [46]. To encode the locations of stripe boundaries, Sobel filters [47] were applied along both image axes to maps of the melt time since the start of each layer. The results are combined into a single pixel-wise value using the root mean square sum of the two Sobel filter responses.

References

1. *ISO/ASTM 52900-15*; Standard Terminology for Additive Manufacturing. ASTM International: West Conshohocken, PA, USA, 2015. Available online: https://compass.astm.org/download/ISOASTM52900.23551.pdf (accessed on 3 May 2017).
2. Wohlers, T.; Mostow, N.; Campbell, I.; Diegel, O.; Kowen, J.; Fidan, I. *Wohlers Report 2022: 3D Printing and Additive Manufacturing Global State of the Industry*; Wohlers Associates: Fort Collins, CO, USA, 2022.
3. Seifi, M.; Gorelik, M.; Waller, J.; Hrabe, N.; Shamsaei, N.; Daniewicz, S.; Lewandowski, J.J. Progress Towards Metal Additive Manufacturing Standardization to Support Qualification and Certification. *JOM* **2017**, *69*, 439–455. [CrossRef]
4. Vijayaram, T.R.; Sulaiman, S.; Hamouda, A.M.S.; Ahmad, M.H.M. Foundry quality control aspects and prospects to reduce scrap rework and rejection in metal casting manufacturing industries. *J. Mater. Process. Technol.* **2006**, *178*, 39–43. [CrossRef]
5. Xie, M.; Goh, T.N.; Ranjan, P. Some effective control chart procedures for reliability monitoring. *Reliab. Eng. Syst. Saf.* **2002**, *77*, 143–150. [CrossRef]
6. Battelle Memorial Institute. *Metallic Materials Properties Development and Standardization (MMPDS-12)*; Battelle Memorial Institute: Columbus, OH, USA, 2017. Available online: https://app.knovel.com/hotlink/toc/id:kpMMPDSM86/metallic-materials-properties/metallic-materials-properties (accessed on 22 June 2022).
7. Liu, Z.; Forsyth, D.S.; Komorowski, J.P.; Hanasaki, K.; Kirubarajan, T. Survey: State of the Art in NDE Data Fusion Techniques. *IEEE Trans. Instrum. Meas.* **2007**, *56*, 2435–2451. [CrossRef]
8. United States Department of Defense. Nondestructive Evaluation System Reliability Assessment. 2009. Available online: http://everyspec.com/MIL-HDBK/MIL-HDBK-1800-1999/MIL-HDBK-1823A_33187/ (accessed on 25 July 2022).
9. Whip, B.; Sheridan, L.; Gockel, J. The effect of primary processing parameters on surface roughness in laser powder bed additive manufacturing. *Int. J. Adv. Manuf. Technol.* **2019**, *103*, 4411–4422. [CrossRef]
10. Ziabari, A.; Dubey, A.; Venkatakrishnan, S.; Frederick, C.; Bingham, P.; Dehoff, R.; Paquit, V. High Resolution X-Ray CT Reconstruction of Additively Manufactured Metal Parts using Generative Adversarial Network-based Domain Adaptation in AI-CT. *Microsc. Microanal.* **2021**, *27*, 2940–2942. [CrossRef]
11. Grasso, M.; Colosimo, B.M. Process defects and in situ monitoring methods in metal powder bed fusion: A review. *Meas. Sci. Technol.* **2017**, *28*, 044005. [CrossRef]
12. Kritzinger, W.; Karner, M.; Traar, G.; Henjes, J.; Sihn, W. Digital Twin in manufacturing: A categorical literature review and classification. *IFAC-Pap.* **2018**, *51*, 1016–1022. [CrossRef]
13. Grieves, M. Digital Twin: Manufacturing Excellence through Virtual Factory Replication. 2015. Available online: https://www.researchgate.net/publication/275211047_Digital_Twin_Manufacturing_Excellence_through_Virtual_Factory_Replication (accessed on 17 May 2021).
14. Gunasegaram, D.R.; Murphy, A.B.; Matthews, M.J.; DebRoy, T. The case for digital twins in metal additive manufacturing. *J. Phys. Mater.* **2021**, *4*, 040401. [CrossRef]
15. Scime, L.; Singh, A.; Paquit, V. A scalable digital platform for the use of digital twins in additive manufacturing. *Manuf. Lett.* **2021**, *31*, 28–32. [CrossRef]
16. LeCun, Y.; Bengio, Y.; Hinton, G. Deep learning. *Nature* **2015**, *521*, 436–444. [CrossRef] [PubMed]
17. Xiao, J.; Hays, J.; Ehinger, K.A.; Torralba, A. SUN database: Large-scale scene recognition from abbey to zoo. In Proceedings of the 2010 IEEE Computer Society Conference on Computer Vision and Pattern Recognition, San Francisco, CA, USA, 13–18 June 2010; pp. 3485–3492.
18. Forien, J.B.; Calta, N.P.; DePond, P.J.; Guss, G.M.; Roehling, T.T.; Matthews, M.J. Detecting keyhole pore defects and monitoring process signatures during laser powder bed fusion: A correlation between in situ pyrometry and ex situ X-ray radiography. *Addit. Manuf.* **2020**, *35*, 101336. [CrossRef]
19. Grasso, M.; Colosimo, B.M. A statistical learning method for image-based monitoring of the plume signature in laser powder bed fusion. *Robot. Comput. Integr. Manuf.* **2019**, *57*, 103–115. [CrossRef]

20. Khanzadeh, M.; Bian, L.; Shamsaei, N.; Thompson, S.M. Porosity Detection of Laser Based Additive Manufacturing Using Melt Pool Morphology Clustering. In Solid Freeform Fabrication Proceedings, Proceedings of the 27th Annual International Solid Freeform Fabrication Symposium—An Additive Manufacturing Conference, Austin, TX, USA, 8–10 August 2016; University of Texas at Austin: Austin, TX, USA, 2016; pp. 1487–1494.

21. Shevchik, S.A.; Kenel, C.; Leinenbach, C.; Wasmer, K. Acoustic emission for in situ quality monitoring in additive manufacturing using spectral convolutional neural networks. *Addit. Manuf.* **2017**, *21*, 598–604. [CrossRef]

22. Coeck, S.; Bisht, M.; Plas, J.; Verbist, F. Prediction of lack of fusion porosity in selective laser melting based on melt pool monitoring data. *Addit. Manuf.* **2019**, *25*, 347–356. [CrossRef]

23. Schwerz, C.; Raza, A.; Lei, X.; Nyborg, L.; Hryha, E.; Wirdelius, H. In-situ detection of redeposited spatter and its influence on the formation of internal flaws in laser powder bed fusion. *Addit. Manuf.* **2021**, *47*, 102370. [CrossRef]

24. Croset, G.; Martin, G.; Josserond, C.; Lhuissier, P.; Blandin, J.-J.; Dendievel, R. In-situ layerwise monitoring of electron beam powder bed fusion using near-infrared imaging. *Addit. Manuf.* **2021**, *38*, 101767. [CrossRef]

25. Pagani, L.; Grasso, M.; Scott, P.J.; Colosimo, B.M. Automated layerwise detection of geometrical distortions in laser powder bed fusion. *Addit. Manuf.* **2020**, *36*, 101435. [CrossRef]

26. Caltanissetta, F.; Grasso, M.; Petrò, S.; Colosimo, B.M. Characterization of in-situ measurements based on layerwise imaging in laser powder bed fusion. *Addit. Manuf.* **2018**, *24*, 183–199. [CrossRef]

27. Kleszczynski, S.; Jacobsmühlen, J.Z.; Sehrt, J.T.; Witt, G. Error detection in laser beam melting systems by high resolution imaging. In Solid Freeform Fabrication Proceedings, Proceedings of the 23rd Annual International Solid Freeform Fabrication Symposium, Austin, TX, USA, 6–8 August 2012; University of Texas at Austin: Austin, TX, USA, 2012; pp. 975–987. Available online: http://www.scopus.com/inward/record.url?eid=2-s2.0-84898410758&partnerID=40&md5=712ce94d7c84e8ff7c4cd478 2a58ba39 (accessed on 20 June 2019).

28. Scime, L.; Siddel, D.; Baird, S.; Paquit, V. Layer-wise anomaly detection and classification for powder bed additive manufacturing processes: A machine-agnostic algorithm for real-time pixel-wise semantic segmentation. *Addit. Manuf.* **2020**, *36*, 101453. [CrossRef]

29. Imani, F.; Gaikwad, A.; Montazeri, M.; Rao, P.; Yang, H.; Reutzel, E. Process mapping and in-process monitoring of porosity in laser powder bed fusion using layerwise optical imaging. *J. Manuf. Sci. Eng. Trans. ASME.* **2018**, *140*, 4040615. [CrossRef]

30. Mehta, M.; Shao, C. Federated learning-based semantic segmentation for pixel-wise defect detection in additive manufacturing. *J. Manuf. Syst.* **2022**, *64*, 197–210. [CrossRef]

31. Cannizzaro, D.; Varrella, A.G.; Paradiso, S.; Sampieri, R.; Chen, Y.; Macii, A.; Patti, E.; Di Cataldo, S. In-situ defect detection of metal Additive Manufacturing: An integrated framework. *IEEE Trans. Emerg. Top. Comput.* **2021**, *10*, 74–86. [CrossRef]

32. Steed, C.A.; Halsey, W.; Dehoff, R.; Yoder, S.L.; Paquit, V.; Powers, S. Falcon: Visual analysis of large, irregularly sampled, and multivariate time series data in additive manufacturing. *Comput. Graph.* **2017**, *63*, 50–64. [CrossRef]

33. McNeil, J.L.; Sisco, K.; Frederick, C.; Massey, M.; Carver, K.; List, F.; Qiu, C.; Mader, M.; Sundarraj, S.; Babu, S.S. In-Situ Monitoring for Defect Identification in Nickel Alloy Complex Geometries Fabricated by L-PBF Additive Manufacturing. *Metall. Mater. Trans. A.* **2020**, *51*, 6528–6545. [CrossRef]

34. Abdelrahman, M.; Reutzel, E.W.; Nassar, A.R.; Starr, T.L. Flaw detection in powder bed fusion using optical imaging. *Addit. Manuf.* **2017**, *15*, 1–11. [CrossRef]

35. Kouraytem, N.; Li, X.; Tan, W.; Kappes, B.; Spear, A.D. Modeling process–structure–property relationships in metal additive manufacturing: A review on physics-driven versus data-driven approaches. *J. Phys. Mater.* **2021**, *4*, 032002. [CrossRef]

36. Lavery, N.P.; Cherry, J.; Mehmood, S.; Davies, H.; Girling, B.; Sackett, E.; Brown, S.G.R.; Sienz, J. Effects of hot isostatic pressing on the elastic modulus and tensile properties of 316L parts made by powder bed laser fusion. *Mater. Sci. Eng. A.* **2017**, *693*, 186–213. [CrossRef]

37. Kusano, M.; Miyazaki, S.; Watanabe, M.; Kishimoto, S.; Bulgarevich, D.S.; Ono, Y.; Yumoto, A. Tensile properties prediction by multiple linear regression analysis for selective laser melted and post heat-treated Ti-6Al-4V with microstructural quantification. *Mater. Sci. Eng. A.* **2020**, *787*, 139549. [CrossRef]

38. Hayes, B.J.; Martin, B.W.; Welk, B.; Kuhr, S.J.; Ales, T.K.; Brice, D.A.; Ghamarian, I.; Baker, A.H.; Haden, C.V.; Harlow, D.G.; et al. Predicting tensile properties of Ti-6Al-4V produced via directed energy deposition. *Acta Mater.* **2017**, *133*, 120–133. [CrossRef]

39. Romano, S.; Brückner-Foit, A.; Brandão, A.; Gumpinger, J.; Ghidini, T.; Beretta, S. Fatigue properties of AlSi10Mg obtained by additive manufacturing: Defect-based modelling and prediction of fatigue strength. *Eng. Fract. Mech.* **2018**, *187*, 165–189. [CrossRef]

40. Zhang, J.; Wang, P.; Gao, R.X. Deep learning-based tensile strength prediction in fused deposition modeling. *Comput. Ind.* **2019**, *107*, 11–21. [CrossRef]

41. Seifi, S.H.; Yadollahi, A.; Tian, W.; Doude, H.; Hammond, V.H.; Bian, L. In Situ Nondestructive Fatigue-Life Prediction of Additive Manufactured Parts by Establishing a Process–Defect–Property Relationship. *Adv. Intell. Syst.* **2021**, *3*, 2000268. [CrossRef]

42. Bisht, M.; Ray, N.; Verbist, F.; Coeck, S. Correlation of selective laser melting-melt pool events with the tensile properties of Ti-6Al-4V ELI processed by laser powder bed fusion. *Addit. Manuf.* **2018**, *22*, 302–306. [CrossRef]

43. Xie, X.; Bennett, J.; Saha, S.; Lu, Y.; Cao, J.; Liu, W.K.; Gan, Z. Mechanistic data-driven prediction of as-built mechanical properties in metal additive manufacturing. *Npj Comput. Mater.* **2021**, *7*, 86. [CrossRef]

44. Scime, L.; Joslin, C.; Collins, D.; Halsey, W.; Duncan, R.; Paquit, V. *A Co-Registered In-Situ and Ex-Situ Tensile Properties Dataset from a Laser Powder Bed Fusion Additive Manufacturing Process (Peregrine v2023-11)*; Oak Ridge National Laboratory (ORNL): Oak Ridge, TN, USA, 2023; Volume 8, p. 2001425. [CrossRef]

45. Petrich, J.; Snow, Z.; Corbin, D.; Reutzel, E.W. Multi-Modal Sensor Fusion with Machine Learning for Data-Driven Process Monitoring for Additive Manufacturing. *Addit. Manuf.* **2021**, *48*, 102364. [CrossRef]

46. Donegan, S.P.; Schwalbach, E.J.; Groeber, M.A. Zoning additive manufacturing process histories using unsupervised machine learning. *Mater. Charact.* **2020**, *161*, 110123. [CrossRef]

47. Szeliski, R. *Computer Vision*; Springer: London, UK, 2011. [CrossRef]

48. ConceptLaser. Quality Management. 2020. Available online: https://www.ge.com/additive/sites/default/files/2018-02/1708_QM-pm_EN_update_1__lowres_einzel.pdf (accessed on 14 December 2020).

49. Massart, D.L.; Kaufman, L.; Rousseeuw, P.J.; Leroy, A. Least median of squares: A robust method for outlier and model error detection in regression and calibration. *Anal. Chim. Acta.* **1986**, *187*, 171–179. [CrossRef]

50. Otsu, N. A Threshold Selection Method from Gray-Level Histograms. *IEEE Trans. Syst. Man. Cybern.* **1979**, *9*, 62–66. [CrossRef]

51. Halsey, W.; Rose, D.; Scime, L.; Dehoff, R.; Paquit, V. Localized Defect Detection from Spatially Mapped, In-Situ Process Data With Machine Learning. *Front. Mech. Eng.* **2021**, *7*, 767444. [CrossRef]

52. Long, J.; Shelhamer, E.; Darrell, T. Fully convolutional networks for semantic segmentation. In Proceedings of the IEEE Conference on Computer Vision and Pattern Recognition, Boston, MA, USA, 7–12 June 2015; pp. 3431–3440. [CrossRef]

53. Craeghs, T.; Clijsters, S.; Yasa, E.; Kruth, J.-P. Online quality control of selective laser melting. In Solid Freeform Fabrication Proceedings, Proceedings of the 22nd Solid Freeform Fabrication (SFF) Symposium, Austin, TX, USA, 8–10 August 2011; University of Texas at Austin: Austin, TX, USA, 2011; pp. 212–226. Available online: http://utwired.engr.utexas.edu/lff/symposium/proceedingsarchive/pubs/Manuscripts/2011/2011-17-Craeghs.pdf (accessed on 20 June 2019).

54. Jacobsmuhlen, J.Z.; Kleszczynski, S.; Witt, G.; Merhof, D. Detection of elevated regions in surface images from laser beam melting processes. In *IECON 2015—41st Annual Conference of the IEEE Industrial Electronics Society*; IEEE: New York, NY, USA, 2016; pp. 1270–1275. [CrossRef]

55. Sames, W.J. Additive Manufacturing of Inconel 718 Using Electron Beam Melting: Processing, Post-processing, & Mechanical Properties. Ph.D. Thesis, Texas A&M University, College Station, TX, USA, 2015.

56. Tang, M.; Pistorius, P.C.; Beuth, J.L. Prediction of lack-of-fusion porosity for powder bed fusion. *Addit. Manuf.* **2017**, *14*, 39–48. [CrossRef]

57. Dunbar, A.J.; Denlinger, E.R.; Heigel, J.; Michaleris, P.; Guerrier, P.; Martukanitz, R.; Simpson, T.W. Development of experimental method for in situ distortion and temperature measurements during the laser powder bed fusion additive manufacturing process. *Addit. Manuf.* **2016**, *12*, 25–30. [CrossRef]

58. Khairallah, S.A.; Anderson, A.T.; Rubenchik, A.; King, W.E. Laser powder-bed fusion additive manufacturing: Physics of complex melt flow and formation mechanisms of pores, spatter, and denudation zones. *Acta Mater.* **2016**, *108*, 36–45. [CrossRef]

59. Gong, H.; Rafi, K.; Gu, H.; Starr, T.; Stucker, B. Analysis of defect generation in Ti–6Al–4V parts made using powder bed fusion additive manufacturing processes. *Addit. Manuf.* **2014**, *1*, 87–98. [CrossRef]

60. Scime, L.; Paquit, V.; Joslin, C.; Richardson, D.; Goldsby, D.; Lowe, L. *Layer-Wise Imaging Dataset from Powder Bed Additive Manufacturing Processes for Machine Learning Applications (Peregrine v2021-03)*; Oak Ridge National Laboratory: Oak Ridge, TN, USA, 2021. [CrossRef]

61. Ronneberger, O.; Fischer, P.; Brox, T. U-Net: Convolutional Networks for Biomedical Image Segmentation. In Proceedings of the Medical Image Computing and Computer-Assisted Intervention—MICCAI 2015: 18th International Conference, Munich, Germany, 5–9 October 2015; pp. 1–8. [CrossRef]

62. Fisher, R.; Perkins, S.; Walker, A.; Wolfart, E. Distance Transform. *Univ. Edinb. Sch. Inform.* 2003. Available online: https://homepages.inf.ed.ac.uk/rbf/HIPR2/distance.htm (accessed on 10 May 2019).

63. Prechelt, L.; When, E.S.-B. Early Stopping—But When? In *Neural Networks: Tricks of the Trade*; Springer: Berlin/Heidelberg, Germany, 1998; pp. 55–69. [CrossRef]

64. Byun, T.S.; Garrison, B.E.; McAlister, M.R.; Chen, X.; Gussev, M.N.; Lach, T.G.; Le Coq, A.; Linton, K.; Joslin, C.B.; Carver, J.K.; et al. Mechanical behavior of additively manufactured and wrought 316L stainless steels before and after neutron irradiation. *J. Nucl. Mater.* **2021**, *548*, 152849. [CrossRef]

65. Roach, A.M.; White, B.C.; Garland, A.; Jared, B.H.; Carroll, J.D.; Boyce, B.L. Size-dependent stochastic tensile properties in additively manufactured 316L stainless steel. *Addit. Manuf.* **2020**, *32*, 101090. [CrossRef]

66. Godec, M.; Zaefferer, S.; Podgornik, B.; Šinko, M.; Tchernychova, E. Quantitative multiscale correlative microstructure analysis of additive manufacturing of stainless steel 316L processed by selective laser melting. *Mater. Charact.* **2020**, *160*, 110074. [CrossRef]

67. Sofinowski, K.A.; Raman, S.; Wang, X.; Gaskey, B.; Seita, M. Layer-wise engineering of grain orientation (LEGO) in laser powder bed fusion of stainless steel 316L. *Addit. Manuf.* **2021**, *38*, 101809. [CrossRef]

68. Dryepondt, S.; Nandwana, P.; Fernandez-Zelaia, P.; List, F. Microstructure and high temperature tensile properties of 316L fabricated by laser powder-bed fusion. *Addit. Manuf.* **2021**, *37*, 101723. [CrossRef]

69. Howard, R.H. *S16-18-FUSSAM01: Generic Metal Irradiation Specimens*; Oak Ridge National Laboratory: Oak Ridge, TN, USA, 2016.

70. Scime, L.; Sprayberry, M.; Collins, D.; Singh, A.; Joslin, C.; Duncan, R.; Simpson, J.; List, F.; Carver, K.; Huning, A.; et al. *Report on Diagnostic and Predictive Capabilities of the TCR Digital Platform*; Oak Ridge National Laboratory: Oak Ridge, TN, USA, 2021. Available online: https://www.osti.gov/biblio/1831630-report-diagnostic-predictive-capabilities-tcr-digital-platform (accessed on 17 May 2021).
71. *ASTM E8/E8M*; Standard Test Methods for Tension Testing of Metallic Materials. ASTM International: West Conshohocken, PA, USA, 2016.
72. *ASTM E4*; Standard Practices for Force Verification of Testing Machines. ASTM International: West Conshohocken, PA, USA, 2009.
73. *ASTM A240/A240M-20a*; Standard Specification for Chromium and Chromium-Nickel Stainless Steel Plate, Sheet, and Strip for Pressure Vessels and for General Applications. ASTM International: West Conshohocken, PA, USA, 2020.
74. Rosenblatt, F. *The Perceptron—A Perceiving and Recognizing Automation*; Cornell Aeronautical Laboratory: Buffalo, NY, USA, 1957; Available online: https://blogs.umass.edu/brain-wars/files/2016/03/rosenblatt-1957.pdf (accessed on 17 May 2021).
75. Krizhevsky, A.; Sutskever, I.; Hinton, G.E. ImageNet Classification with Deep Convolutional Neural Networks. *Adv. Neural Inf. Process. Syst.* **2012**, *25*, 1–9. [CrossRef]
76. Deshpande, A. A Beginner's Guide to Understanding Convolutional Neural Networks. 2016. Available online: https://adeshpande3.github.io/adeshpande3.github.io/A-Beginner\T1\textquoterights-Guide-To-Understanding-Convolutional-Neural-Networks/ (accessed on 7 February 2018).
77. Kingma, D.P.; Ba, J. Adam: A Method for Stochastic Optimization. *ICLR* **2014**, *1631*, 58–62.
78. Kuhn, M.; Johnson, K. *Applied Predictive Modeling*, 5th ed.; Springer: New York, NY, USA, 2013. [CrossRef]
79. Ren, J.; Liu, P.J.; Fertig, E.; Snoek, J.; Poplin, R.; DePristo, M.A.; Dillon, J.V.; Lakshminarayanan, B. Likelihood Ratios for Out-of-Distribution Detection. In Proceedings of the 33rd International Conference on Neural Information Processing Systems, Vancouver BC Canada, 8–14 December 2019; pp. 14707–14718.
80. Likas, A.; Vlassis, N.; Verbeek, J.J. The global k-means clustering algorithm. *Pattern Recognit.* **2003**, *36*, 451–461. [CrossRef]
81. Lane, B.; Whitenton, E.; Madhavan, V.; Donmez, A. Uncertainty of temperature measurements by infrared thermography for metal cutting applications. *Metrologia* **2013**, *50*, 637–653. [CrossRef]
82. Fawcett, T. An introduction to ROC analysis. *Pattern Recognit. Lett.* **2005**, *27*, 861–874. [CrossRef]
83. Snow, Z.; Diehl, B.; Reutzel, E.W.; Nassar, A. Toward in-situ flaw detection in laser powder bed fusion additive manufacturing through layerwise imagery and machine learning. *J. Manuf. Syst.* **2021**, *59*, 12–26. [CrossRef]
84. Nassar, A.R.; Gundermann, M.A.; Reutzel, E.W.; Guerrier, P.; Krane, M.H.; Weldon, M.J. Formation processes for large ejecta and interactions with melt pool formation in powder bed fusion additive manufacturing. *Sci. Rep.* **2019**, *9*, 5038. [CrossRef]
85. International Digital Image Correlation Society; Jones, E.M.C.; Iadicola, M.A. (Eds.) *A Good Practices Guide for Digital Image Correlation*; International Digital Image Correlation Society: Wilson, CT, USA, 2018. [CrossRef]
86. Stump, B.; Plotkowski, A. An adaptive integration scheme for heat conduction in additive manufacturing. *Appl. Math. Model.* **2019**, *75*, 787–805. [CrossRef]
87. Voisin, T.; Calta, N.P.; Khairallah, S.A.; Forien, J.-B.; Balogh, L.; Cunningham, R.W.; Rollett, A.D.; Wang, Y.M. Defects-dictated tensile properties of selective laser melted Ti-6Al-4V. *Mater. Des.* **2018**, *158*, 113–126. [CrossRef]
88. Spears, T.G.; Gold, S.A. In-process sensing in selective laser melting (SLM) additive manufacturing. *Integr. Mater. Manuf. Innov.* **2016**, *5*, 16–40. [CrossRef]
89. Jacob, G.; Donmez, A.; Slotwinski, J.; Moylan, S. Measurement of powder bed density in powder bed fusion additive manufacturing processes. *Meas. Sci. Technol.* **2016**, *27*, 115601. [CrossRef]
90. Li, M.; Zhang, X.; Wei-Ying, C.; Heidet, F. *Location-Dependent Mechanical Property Evaluation on Additively Manufactured Materials*; Argonne National Lab: Argonne, IL, USA, 2021.
91. Montgomery, C. *The Effect of Alloys, Powder, and Overhanging Geometries in Laser Powder Bed Additive Manufacturong*; Carnegie Mellon University: Pittsburgh, PA, USA, 2017. [CrossRef]
92. Wei, L.C.; Ehrlich, L.E.; Powell-Palm, M.J.; Montgomery, C.; Beuth, J.; Malen, J.A. Thermal conductivity of metal powders for powder bed additive manufacturing. *Addit. Manuf.* **2018**, *21*, 201–208. [CrossRef]
93. Fisher, B. *Part Temperature Effects in Powder Bed Fusiosn Additive Manufacturing of Ti-6Al-4V*; Carnegie Mellon University: Pittsburgh, PA, USA, 2018. [CrossRef]
94. Krauss, H.; Zeugner, T.; Zaeh, M.F. Layerwise monitoring of the Selective Laser Melting process by thermography. *Phys. Procedia* **2014**, *56*, 64–71. [CrossRef]
95. Ladewig, A.; Schlick, G.; Fisser, M.; Schulze, V.; Glatzel, U. Influence of the shielding gas flow on the removal of process by-products in the selective laser melting process. *Addit. Manuf.* **2016**, *10*, 1–9. [CrossRef]
96. Bidare, P.; Bitharas, I.; Ward, R.M.; Attallah, M.M.; Moore, A.J. Fluid and particle dynamics in laser powder bed fusion. *Acta Mater.* **2018**, *142*, 107–120. [CrossRef]
97. Francis, Z. *The Effects of Laser and Electron Beam Spot Size in Additive Manufacturing Processes*; Carnegie Mellon University: Pittsburgh, PA, USA, 2017. [CrossRef]

MDPI AG
Grosspeteranlage 5
4052 Basel
Switzerland
Tel.: +41 61 683 77 34

Materials Editorial Office
E-mail: materials@mdpi.com
www.mdpi.com/journal/materials